MOLECULAR PHYSICS AND QUANTUM CHEMISTRY HANDBOOK

Molecular Physics and Quantum Chemistry Handbook

Editor:

Maria Emilova Velinova, Ph.D.

ArclerPress

www.arclerpress.com

Molecular Physics and Quantum Chemistry Handbook

Editor: Maria Emilova Velinova, Ph.D.

708 3rd Avenue, 6th Floor
New York
NY 10017
United States of America
www.arclerpress.com

ISBN: 978-1-68094-428-0

Library of Congress Control Number: 2016951122

About the Editor

Maria Emilova Velinova, Ph.D.

Maria Velinova is Ph.D. holder in Quantum chemistry at the University of Sofia since April 2012. Her major research experience is in the field of Computational Chemistry, especially in statistical mechanics methods applied to different sorts of biomolecules. Member of the Laboratory of Quantum and Computational Chemistry at the University of Sofia.

Contents

List of Contributors

Venera Khoromskaia
Max-Planck-Institute for Dynamics of Complex Systems, Magdeburg, Germany
Max-Planck Institute for Mathematics in the Sciences, Inselstr. 22-26, Leipzig, Germany

Boris N. Khoromskij
Max-Planck Institute for Mathematics in the Sciences, Inselstr. 22-26, Leipzig, Germany

Cleydson B. R. Santos
Laboratory of Modeling and Computational Chemistry, Federal University of Amapá, Macapá, Brazil
Postgraduate Program in Biotechnology and Biodiversity-Network BIONORTE, Campus Universitário Marco Zero, Macapá, Brazil
Laboratory of Drug Research, School of Pharmaceutical Sciences, Federal University of Amapá, Macapá, Brazil

Cleison C. Lobato
Laboratory of Modeling and Computational Chemistry, Federal University of Amapá, Macapá, Brazil

Francinaldo S. Braga
Laboratory of Modeling and Computational Chemistry, Federal University of Amapá, Macapá, Brazil

Sílvia S. S. Morais
Laboratory of Statistical and Computational Modeling, University of the State of Amapá, Macapá, Brazil

Cesar F. Santos
Laboratory of Modeling and Computational Chemistry, Federal University of Amapá, Macapá, Brazil

Caio P. Fernandes
Laboratory of Modeling and Computational Chemistry, Federal University of Amapá, Macapá, Brazil
Laboratory of Drug Research, School of Pharmaceutical Sciences, Federal University of Amapá, Macapá, Brazil

Davi S. B. Brasil
Laboratory of Modeling and Computational Chemistry, Federal University of Amapá, Macapá, Brazil
Institute of Technology, Federal University of Pará, Belém, Brazil

Lorane I. S. Hage-Melim
Laboratory of Modeling and Computational Chemistry, Federal University of Amapá, Macapá, Brazil

Williams J. C. Macêdo
Laboratory of Modeling and Computational Chemistry, Federal University of Amapá, Macapá, Brazil
Postgraduate Program in Biotechnology and Biodiversity-Network BIONORTE, Campus Universitário Marco Zero, Macapá, Brazil

José C. T. Carvalho
Laboratory of Modeling and Computational Chemistry, Federal University of Amapá, Macapá, Brazil
Postgraduate Program in Biotechnology and Biodiversity-Network BIONORTE, Campus Universitário Marco Zero, Macapá, Brazil
Laboratory of Drug Research, School of Pharmaceutical Sciences, Federal University of Amapá, Macapá, Brazil

Ibon Alkorta
Instituto de Química Médica, C.S.I.C., Juan de la Cierva, 3, E-28006 Madrid

José Elguero
Instituto de Química Médica, C.S.I.C., Juan de la Cierva, 3, E-28006 Madrid

Peter Maxwell
Manchester Institute of Biotechnology (MIB), 131 Princess Street, Manchester M1 7DN, UK
School of Chemistry, University of Manchester, Oxford Road, Manchester M13 9PL, UK

Ángel Martín Pendás
Departamento de Quimica Fisica y Analitica, Universidad de Oviedo, E-33006 Oviedo, Spain

Paul L. A. Popeliera
School of Chemistry, University of Manchester, Oxford Road, Manchester M13 9PL, UK

Xuan L. Liu
Department of Materials Science and Engineering, The Pennsylvania State University, University Park, Pennsylvania, 16802, United States of America

Thomas Gheno
Department of Mechanical Engineering and Materials Science, University of Pittsburgh, Pittsburgh, Pennsylvania, 15261, United States of America

Bonnie B. Lindahl
Department of Materials Science and Engineering, KTH Royal Institute of Technology, SE-100 44, Stockholm, Sweden

Greta Lindwall
Department of Materials Science and Engineering, The Pennsylvania State University, University Park, Pennsylvania, 16802, United States of America

Brian Gleeson
Department of Mechanical Engineering and Materials Science, University of Pittsburgh, Pittsburgh, Pennsylvania, 15261, United States of America

ZiKui Liu
Department of Materials Science and Engineering, The Pennsylvania State University, University Park, Pennsylvania, 16802, United States of America

Mekala R
Department of Physics, Sri Sarada College for Women, Salem, Tamil Nadu, India

Mathammal R
Department of Physics, Sri Sarada College for Women, Salem, Tamil Nadu, India

Sangeetha M
Department of Physics, Sri Sarada College for Women, Salem, Tamil Nadu, India

Nguyen Xuan Truonga
aInstitut für Optik und Atomare Physik, Technische Universität Berlin, Hardenbergstraße 36, D-10623 Berlin, Germany

Marco Savoca
Institut für Optik und Atomare Physik, Technische Universität Berlin, Hardenbergstraße 36, D-10623 Berlin, Germany

Dan J. Harding
Institut für Physikalische Chemie, Georg-August-Universität Göttingen, Tammannstraße 6, D-37077 Göttingen, Germany
Department of Dynamics at Surfaces, Max-Planck-Institut für Biophysikalische Chemie, Am Faßberg 11, D-37077 Göttingen, Germany

André Fielicke
Institut für Optik und Atomare Physik, Technische Universität Berlin, Hardenbergstraße 36, D-10623 Berlin, Germany

Otto Dopfer
Institut für Optik und Atomare Physik, Technische Universität Berlin, Hardenbergstraße 36, D-10623 Berlin, Germany

E. Hywel Evans
School of Geography, Earth, and Environmental Sciences, University of Plymouth, Drake Circus, Plymouth, UK PL4 8AA

Jorge Pisonero
University of Oviedo, Faculty of Science, Department of Physics, c/ Calvo Sotelo s/n, 33006 Oviedo, Spain

Clare M. M. Smith
St. Ambrose High School, Blair Road, Coatbridge, Lanarkshire, UK ML5 2EW

Rex N. Taylor
Ocean and Earth Science, University of Southampton, NOC, Southampton, UK SO14 3ZH

Rawad Halabi
Faculty of Science, Beirut Arab University, Beirut, Lebanon

Mahmoud Korek
Faculty of Science, Beirut Arab University, Beirut, Lebanon

Eric Paquet
Vaccine Program, National Research Council, 1200 Montreal Road, Ottawa, ON, Canada K1A 0R6
School of Electrical Engineering and Computer Science, University of Ottawa, 800 King Edward Road, Ottawa, ON, Canada K1N 6N5

Herna L. Viktor
School of Electrical Engineering and Computer Science, University of Ottawa, 800 King Edward Road, Ottawa, ON, Canada K1N 6N5

Sachin Patodia
Centre for Converging Technologies, University of Rajasthan, Jaipur 302004, Rajasthan, India

Ashima Bagaria
Centre for Converging Technologies, University of Rajasthan, Jaipur 302004, Rajasthan, India

Deepak Chopra
Department of Chemistry, Indian Institute of Science Education and Research, Bhopal 462066, India

Rui Li
School of Highway, Chang'an University, Xi'an 710064, China

Hui Du
School of Transportation Engineering, Southeast University, Nanjing 210096, China

Zepeng Fan
School of Highway, Chang'an University, Xi'an 710064, China

Jianzhong Pei
School of Highway, Chang'an University, Xi'an 710064, China

Antreas C. Kalli
Department of Biochemistry, University of Oxford, Oxford, United Kingdom

Mark S. P. Sansom
Department of Biochemistry, University of Oxford, Oxford, United Kingdom

Reinhart A. F. Reithmeier
Department of Biochemistry, University of Toronto, Toronto, Canada

Daniel Wüstner
Department of Biochemistry and Molecular Biology, University of Southern Denmark, Odense M DK-5230, Denmark

Heinz Sklenar
Theoretical Biophysics Group, Max Delbrück Center for Molecular Medicine, Robert-Rössle-Str. 10, Berlin D-13125, Germany

Zhe Zhang
Physik-Department T38, Technische Universität München, James-Franck-Str. 1, 84748 Garching, Germany

Christina E. M. Schindler
Physik-Department T38, Technische Universität München, James-Franck-Str. 1, 84748 Garching, Germany

Oliver F. Lange
Biomolecular NMR and Munich Center for Integrated Protein Science, Department Chemie, Technische Universität München, Lichtenbergstr. 4, 85748 Garching, Germany

Martin Zacharias
Physik-Department T38, Technische Universität München, James-Franck-Str. 1, 84748 Garching, Germany

Jens Nielsen
Research Software Development Team, Research IT Services, University College London, Torrington Place, London WC1E 6BT, United Kingdom

Mayeul d'Avezac
Research Software Development Team, Research IT Services, University College London, Torrington Place, London WC1E 6BT, United Kingdom

James Hetherington
Research Software Development Team, Research IT Services, University College London, Torrington Place, London WC1E 6BT, United Kingdom

Michail Stamatakis
Department of Chemical Engineering, University College London, Torrington Place, London WC1E 7JE, United Kingdom

Preface

This book aims to report recent advances and applications in the field of the molecular physics and quantum chemistry. The molecular physics and quantum chemistry study how to predict and clarify the structure, the properties and the dynamics of atoms and molecules. Selections of theoretical and experimental techniques are included beside their application to systems at different level of complexity and nature.

The fundamental equation in the quantum chemistry is the Schrödinger equation that is unbearable to be solved exactly for real systems. For this reason, the development of proper models for the description of the effects of electron correlation is key component of the theory. The methods in the quantum chemistry can be divided into those that aim at solving the Schrödinger equation by assuming a certain mathematical form for a wave function (i.e. Hartree-Fock methods) and those that do not engage a wave function explicitly (i.e. DFT, Density Functional Theory). These methods and their applications in different fields are discussed in the first section of the book with a broader discussion given to DFT methods - the most prominent approaches of the modern quantum chemistry. The second part of the book looks at experimental aspects of the molecular physics including various types of spectroscopy. This latter studies the interaction between electromagnetic radiation in all its forms and matter. The interaction can induce electronic excitations, molecular vibrations or nuclear spin orientations. The calculations of the spectrum of the molecules by theoretical methods are also discussed. The remaining content of this book focuses on methods that follow the laws of classical mechanics as the Molecular Dynamics (MD) method and Monte Carlo (MC) simulations. The MD method is a numerical technique of statistical mechanics for integration of the equations of motion for a many-particles system. On the other hand, the objective of a Monte Carlo simulation is to generate an ensemble of representative configurations under specific thermodynamics conditions for a complex macromolecular system applying random perturbations to the system. Both of the methods produce trajectories which can be used to evaluate various structural, transport, and thermodynamic properties of the system.The exposition in the book is made from first principles to classical methods to support a better understanding of the potentialities, restrictions and applications of the methods used in molecular physics and quantum chemistry. Crossing the boundaries between several computational and experimental techniques, this book aims to be of interest to a broad auditory, including experimental and theoretical physicists, chemists and biologist.

Editor
Maria Emilova Velinova, Ph.D.

CHAPTER 1

TENSOR NUMERICAL METHODS IN QUANTUM CHEMISTRY: FROM HARTREE–FOCK TO EXCITATION ENERGIES

Venera Khoromskaia[a,b] and Boris N. Khoromskij[b]

[a]Max-Planck-Institute for Dynamics of Complex Systems, Magdeburg, Germany

[b]Max-Planck Institute for Mathematics in the Sciences, Inselstr. 22-26, Leipzig, Germany

We resume the recent successes of the grid-based tensor numerical methods and discuss their prospects in real-space electronic structure calculations. These methods, based on the low-rank representation of the multidimensional functions and integral operators, first appeared as an accurate tensor calculus for the 3D Hartree potential using 1D complexity operations, and have evolved to entirely grid-based tensor-structured 3D Hartree–Fock eigenvalue solver. It benefits from tensor calculation of the core Hamiltonian and two-electron integrals (TEI) in O(n log n) complexity using the rank-structured approximation of basis functions, electron densities and convolution integral operators all represented on 3D $n \times n \times n$ Cartesian grids. The algorithm for calculating TEI tensor in a form of the Cholesky decomposition is based on multiple factorizations using algebraic 1D "density fitting" scheme, which yield an almost irreducible number of product basis functions involved in the 3D convolution integrals, depending on a threshold $\varepsilon > 0$. The basis functions are not restricted to separable Gaussians, since the analytical integration is substituted by high-precision tensor-structured numerical quadratures. The tensor approaches to post-Hartree–Fock calculations for the MP2 energy correction and for the Bethe–Salpeter excitation energies, based on using low-rank factorizations and the reduced basis method, were recently introduced. Another direction is towards the tensor-based Hartree–Fock numerical scheme for finite lattices, where one of the numerical challenges is the summation of electrostatic potentials of a large number of nuclei. The 3D grid-based tensor method for calculation of a potential sum on a $L \times L \times L$ lattice manifests the linear in L computational work, O(L), instead of the usual $O(L^3 \log L)$ scaling by the Ewald-type approaches.

Citation: Khoromskaia, V, Khoromskij, BN. Tensor numerical methods in quantum chemistry: from Hartree–Fock to excitation energies. Phys. Chem. Chem. Phys., 2015 17, 31491-31509. DOI: 10.1039/C5CP01215E.

INTRODUCTION

The problems of numerical modeling the many-particle interactions in large molecular systems, lattice structured metallic clusters and crystals, proteins and nanomaterials are the most challenging tasks in modern computational physics and chemistry. The traditional approaches for these multidimensional problems are restricted to the concepts which have well recognized limitations for computations with higher accuracy and for larger non-periodic molecular systems, as well as for efficient calculation of excited states. Recent advances in numerical analysis for multidimensional problems and significant achievements in high performance computing suggest new creative approaches to these problems.

The Hartree–Fock (HF) equation governed by the 3D integral-differential operator[29,67] is one of the basic models for ab initio calculation of the ground state energy of molecular systems. It is a strongly nonlinear eigenvalue problem (EVP) in a sense, that one should solve the equation in a self-consistent way when the integral part of the governing operator depends on the solution itself. Multiple strong singularities in the electron density of a molecule due to nuclear cusps impose strong requirements on the accuracy of calculations.

Commonly used numerical methods for solution of the Hartree–Fock equation are based on the analytical computation of the arising two-electron integrals (convolution type integrals in R^3) in the problem adapted naturally separable Gaussian-type bases,[1,17] by using erf-function expansions. This rigorous approach resulted in a number of efficient implementations which are widely used in computational quantum chemistry. The success of the analytical integration methods stems from the big amount of precomputed information based on the physical insight including the construction of problem adapted atomic orbitals basis sets and elaborate nonlinear optimization for calculation of density fitting basis. The known limitations of this approach appear due to a strong dependence of the numerical efficiency on the size and quality of the chosen Gaussian basis sets, that might be crucial for larger molecular clusters and heavier atoms.

The intention to replace or assist the analytical calculations for the Hartree–Fock problem by a data-sparse grid-based numerical schemes has a long history. First success was the grid-based numerical method for the diatomic molecules in ref. 66, though this approach was not feasible to compact (3D) molecules. The wavelet multiresolution schemes[13] capable for the accurate representation of nuclear cusps, have been applied to electronic structure simulations,[21,28,63] and recently this approach was further advanced due to achievements in high performance supercomputing.[8,18,20,63] However, due to extensive computational resources, the entirely wavelet-based or sparse-grid approaches[24,74] are limited so far only to rather small atomic systems with few electrons.[2] Tensor hyper-contraction decompositions in density fitting schemes have been analyzed in ref. 32 and 59.

The newly developed tensor-structured numerical methods, both the name and the concept, appeared during the work on the grid-based tensor approach to the solution of the 3D Hartree–Fock problem.[38,39,42] The central point is the representation of d-variate functions and integral/differential operators on large $n^{\otimes d}$ grids and their approximation in the low-rank tensor formats, which allows numerical calculations in O(dn) complexity instead of $O(n^d)$ by conventional methods.

In this paper we summarize the main benefits of the tensor numerical methods in electronic structure calculations, and discuss further prospects of this approach in several directions, such as

- Algebraic directional density fitting and tensor factorization of the two-electron integrals in Hartree–Fock calculations;
- Tensor decompositions in the MP2 energy correction, and in calculation of excitation energies based on the Bethe–Salpeter equation;
- Fast tensor summation of electrostatic potentials on large 3D lattices for efficient calculation of the Fock matrix in the case of lattice-structured systems;
- Fast method for calculation of the interaction energy of a large number of Coulombic potentials on a 3D lattice.

Notice that basic rank-structured tensor formats such as the canonical (PARAFAC/CANDECOMP) and Tucker tensor decompositions have been since long used in the computer science for the quantitative analysis of correlations in the experimental multidimensional data arrays in data processing and chemometrics, see ref. 12, 50 and 64 and references therein. In 2006 the exceptional properties of the Tucker decomposition for the discretized multidimensional functions have been revealed in ref. 36 and 38, where it was proven that for a class of function-related tensors the approximation error of the Tucker decomposition decays exponentially with respect to the Tucker rank. This gives the opportunity to represent the discretized multidimensional functions and integral (convolution) operators in an algebraically separable form and thus reduce the numerical treatment of the multidimensional transforms to 1D operations. It was shown that the number of vectors in such a representation depends only logarithmically on the size of the d-dimensional grid, $O(n^d)$, used for discretization of multivariate functions.

The above results have led to the idea to calculate the Hartree potential and the Coulomb and exchange operators by numerical quadratures[39] using Gaussian-type basis functions discretized on 3D Cartesian grids and the fast tensor-product convolution.[35] In this way, the efficient low-rank canonical tensor representation to the Newton kernel was an essential contribution.[7] To reduce the initial rank of the electron density, that is quadratically proportional to the number of GTO basis functions, the canonical-to-Tucker tensor transform was invented,[38,39] which made computations for large tensor grids and extended molecules tractable (even in Matlab).

The initial version of tensor-structured algorithms for solving Hartree–Fock equation employed the 3D grid-based calculation of the Coulomb and exchange integral operators "on-the-fly", thus avoiding precomputation and storage of the TEI tensor.[40,42] In particular, it was shown that tensor calculus allows to reduce the 3D convolution integrals to combinations of 1D convolutions, and 1D Hadamard and scalar products. Besides, these results promoted spreading of the tensor-structured methods in the community of numerical analysis,[22,27,37] and further development of the tree-tensor formats like tensor-train[57] and hierarchical Tucker representations.[26]

Further development of tensor methods in electronic structure calculations was due to the fast algorithm for the grid-based computation of the TEI tensor[44] in $O(N_b^3)$ storage in the number of basis functions N_b. The fourth order TEI tensor is calculated in a form of the Cholesky factorization by using the algebraic 1D "density fitting" scheme, which applies to the product basis functions. Imposing the low-rank tensor representation of the product basis functions and the Newton convolving kernel all discretized on $n \times n \times n$ Cartesian grid, the 3D integral transforms are calculated in O(n log n) complexity. Given the factorized TEI, the update of the Column and exchange parts in the Fock matrix reduces to the cheap algebraic operations. Other steps are tensor calculation of the core Hamiltonian and the efficient MP2 energy correction scheme,[45] which all together gave rise to the black-box Hartree–Fock solver.[46]

Due to the grid representation of basis functions basis sets are now not restricted to Gaussian-type orbitals and allowed to be any well-separable function defined on a grid. The tensor-based Hartree–Fock solver is competitive in computational time and accuracy with the standard packages based on analytical calculation of integrals. High accuracy is attained owing to easy calculations on large 3D grids up to $n^3 = 10^{18}$, so that the resolution with mesh size h of the order of atomic radii, 10^{-4} Å, is possible.

Motivated by tensor decompositions in the MP2 scheme, the new approach for calculation of the excited states in the framework of the Bethe–Salpeter equation was recently introduced,[6]that employs the reduced basis method in combination with low-rank tensor approximations.

Further developments of the tensor methods in multi-particle simulations are focused on the large lattice structures in a box and nearly periodic systems, for which the tensor approach may reduce computational costs dramatically.[48] One of the challenging problems in the numerical treatment of the crystalline-type molecular clusters is summation of a large number of electrostatic potentials distributed on a non-periodic (finite) lattice. The novel grid-based method for summation of the long-range potentials in the canonical and Tucker formats[47,49] works on L × L × L lattices with the computational cost O(L) instead of $O(L^3 \log L)$ by the traditional Ewald-type methods.[19] The required precision is guaranteed by employing large 3D Cartesian grids for representation of potentials. The method remains efficient for lattices with non-rectangular geometries and in the presence of multiple defects.[48,49]

In particular, we present fast tensor method for the summation of the long-range interaction potentials on a 3D lattice by using the assembled vectors of their canonical and Tucker tensor representations. The important application of this approach to calculation of interaction energy of the Coulombic potentials on a lattice with sub-linear cost in the lattice size is described in detail.

The rest of the paper is organized as follows. In Section 2, we overview the basic tensor formats and show why the orthogonal Tucker tensor decomposition, originating from computer science and data processing, became useful for the treatment of the multidimensional functions and operators in numerical analysis. In particular, Sections 2.3 and 2.4 discuss the matrix product states (tensor train) formats and the novel quantics tensor approximation method, respectively, while Section 2.5 addresses the basic tensor-structured multilinear algebra operations. Section 3 describes tensor calculus of the multidimensional convolution transform on examples of the Hartree potential (Section 3.1) and the TEI tensor (Section 3.2), and recalls the main building blocks in the tensor-based black-box Hartree–Fock solver. Section 4 shows benefits of the tensor approach in MP2 calculations and in calculation of the lowest part of the excitation energies in the framework of the Bethe–Salpeter equation. Section 5 describes the benefits of tensor methods in applications to lattice-type molecular systems. Appendix discusses some computational details on the Canonical-to-Tucker tensor transform, the Galerkin discretization scheme for the nonlinear Hartree–Fock equation, and the basics of the low-rank canonical tensor representation for the Newton kernel.

RANK-STRUCTURED TENSOR REPRESENTATIONS OF DISCRETIZED FUNCTIONS AND OPERATORS

In this section we discuss shortly why the rank-structured tensors, which were traditionally used in experimental data processing, appeared to be useful for the separable representation of multivariate functions and operators represented on tensor product grids. There is a numerous literature on application of the canonical and Tucker tensor decompositions in chemometrics and data processing, see ref. 12, 31, 50, 64 and 68 and references therein. In the recent decade the class of matrix product states type tensor formats became popular in the simulations of quantum spin systems and quantum molecular dynamics. We refer to ref. 37 and 62 for recent literature surveys on commonly used tensor formats.

Canonical and Tucker Tensor Formats

We consider a tensor of order d, as a real multidimensional array $\mathbf{A} = [a_{i1,\ldots,id}] \in R^{n_1\times\ldots\times n_d}$ numbered by a d-tuple index set,† with multi-index notation $\mathbf{i} = (i_1,\ldots,i_d)$, $il \in \{1,\ldots,nl\}$, $l = 1,\ldots,d$. It is an element of the linear vector space $R^{n_1\times\cdots\times n_d}$ equipped with the Euclidean scalar product,

$$\langle \mathbf{A}, \mathbf{B} \rangle = \sum_{i_1=1}^{n_1} \cdots \sum_{i_d=1}^{n_d} a_{i_1,\ldots,i_d} b_{i_1,\ldots,i_d}.$$

Euclidean vectors and matrices are the special case of dth order tensors. For a general tensor, the required storage scales exponentially in the dimension, $n_1 n_2 \ldots n_d$, (the so-called "curse of dimensionality"). To get rid of exponential scaling in the dimension, one can apply the rank-structured separable representations of multidimensional tensors. The simplest separable element is given by rank-1 canonical tensor,

$U = \mathbf{u}^{(1)} \otimes \cdots \otimes \mathbf{u}^{(d)} \in R^{n_1\times\cdots\times n_d}$,

with entries

$$u_{i_1,\ldots,i_d} = u_{i_1}^{(1)} \ldots u_{i_d}^{(d)},$$

requiring only $n_1 + \cdots + n_d$ numbers to store it.

A tensor in the R-term canonical format is defined by

$$\mathbf{U} = \sum_{k=1}^{R} c_k \mathbf{u}_k^{(1)} \otimes \ldots \otimes \mathbf{u}_k^{(d)}, \quad c_k \in \mathbb{R}, \tag{1}$$

where $\mathbf{u}_k^{(\ell)} \in \mathbb{R}^{n_\ell}$ are normalized vectors, and R is called the canonical rank of a tensor. It is convenient to introduce the so-called side matrices $\mathbf{U}^{(\ell)} = \left[\mathbf{u}_1^{(\ell)} \ldots \mathbf{u}_R^{(\ell)}\right] \in \mathbb{R}^{n_\ell \times R}$, l = 1, 2, 3, obtained by concatenation of the canonical vectors $\mathbf{u}_k^{(\ell)}$, k = 1,…,R. Now the storage cost is bounded by dRn. For $d \geq 3$ computation of the canonical rank of a tensor **U**, i.e. the minimal number R in representation (1) and the respective decomposition, is an N–P hard problem. In the case d = 2 the representation (1) is merely a rank-R matrix.

We say that a tensor **A** is represented in the rank-**r** orthogonal Tucker format with the rank parameter $\mathbf{r} = (r_1,\ldots, r_d)$, if

$$\mathbf{A} = \sum_{\nu_1=1}^{r_1} \cdots \sum_{\nu_d=1}^{r_d} \beta_{\nu_1,\ldots,\nu_d} \mathbf{v}_{\nu_1}^{(1)} \otimes \ldots \otimes \mathbf{v}_{\nu_d}^{(d)}, \quad \ell = 1, \ldots, d, \tag{2}$$

where $\left\{\mathbf{v}_{\nu_\ell}^{(\ell)} \in \mathbb{R}^{n_\ell}\right\}$ represents the set of orthonormal vectors, and $\beta = [\beta_{v1,\ldots,vd}] \in R^{r_1 \times \cdots \times r_d}$ is the Tucker core tensor. The storage cost for the Tucker tensor is bounded by drn + r^d, r = max r l. Often the equivalent notation for the Tucker tensor format is used,

$$\mathbf{A} = \beta \times_1 V^{(1)} \times_2 V^{(2)} \times \cdots \times_d V^{(d)}, \tag{3}$$

where × l denotes the contraction along the mode l and orthogonal matrices $V^{(l)}$ incorporate the set of vectors $\left\{\mathbf{v}_{\nu_\ell}^{(\ell)} \in \mathbb{R}^{n_\ell}\right\}$. In the case d = 2, the orthogonal Tucker decomposition is equivalent to the singular value decomposition (SVD) of a rectangular matrix. Fig. 1 and 2 visualizes the canonical and Tucker tensors in the case d = 3. Canonical tensor decomposition leads to an ill-posed problem and, up to our best knowledge, there are no robust algebraic algorithms for the canonical approximation of an arbitrary tensor. For some classes of analytic functions, the explicit low-rank approximation can be constructed in analytic form by using sinc-quadrature approximation to the Laplace transform. The robust methods for Tucker decomposition are based on the orthogonal projections using the higher order singular value decomposition[16] (HOSVD) that is the generalization of the singular value decomposition (SVD) of matrices.

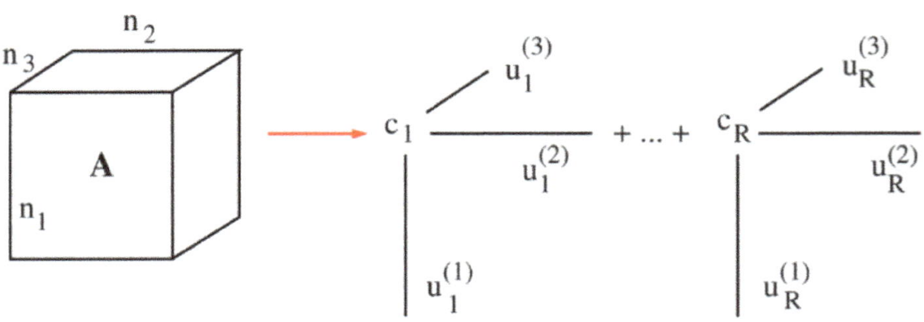

Figure 1: The canonical tensor decomposition for d = 3.

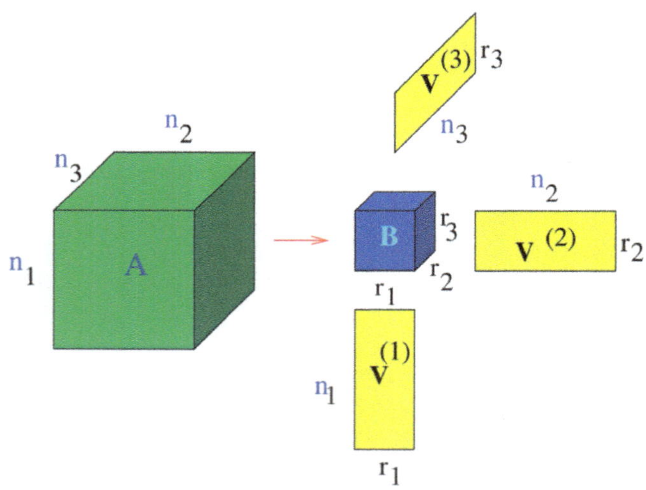

Figure 2: Visualizing the Tucker tensor for d = 3.

The rank-structured decomposition (approximation) of multidimensional tensors provides means for the separation of variables in the discretized representation of multivariate functions and operators, and thus the possibility to substitute multidimensional algebraic transforms by univariate operations. Notice that in the computer science community these possibilities were restricted to moderate-dimensional tensors of small mode size (with rather large rank parameters) obtained from the experimental data sets.

Tucker Decomposition for Function Related Tensors. Canonical-to-Tucker Approximation

It was proven and shown in numerical simulations,[36,38] that the rank of the (fixed accuracy) Tucker approximation to some function related tensors depends only logarithmically on the size of the discretization grid. In particular, it was shown that for tensors resulting from the grid discretization of physically relevant multidimensional functions one can find algebraically a reduced subspace in each mode of the 3D tensor, which lead to a separable approximation of the function by a tensor product of a small number of vectors.

For a given continuous function f: Ω → R, $\Omega := \prod_{\ell=1}^{d} [a_\ell, b_\ell] \subset \mathbb{R}^3$, we introduced the function related 3rd order tensor, obtained by Galerkin discretization in a volume box using n × n × n3D Cartesian grid. In particular, the low-rank tensor approximations were calculated for functions of the Slater type f(x) = exp (−α||x||), Newton kernel $f(x) = \frac{1}{\|x\|}$, Yukawa potential $f(x) = \frac{e^{-\alpha\|x\|}}{\|x\|}$, and the Helmholtz potential $f(x) = \frac{\cos \alpha\|x\|}{\|x\|}$, x ∈ R^3.

Numerical tests demonstrated that the error of the rank-**r** (with **r** = (r,r,r)) Tucker approximation applied to these third order tensors decays exponentially with respect to the Tucker rank r. Moreover, for fixed approximation error, the Tucker rank r scales as O(log n) for above mentioned functions.[35]

Fig. 3 (bottom) shows the exponential convergence of the Tucker tensor approximation in the Tucker rank r of the Slater function in the relative Frobenius norm, $E_{FN} = \frac{\|A - A_r\|}{\|A\|}$. Fig. 3 and 4 demonstrate that the exponential decay of approximation error is nearly the same for both a single potential and for a sum of the same potentials distributed in nodes of a cubic lattice. This property of the Tucker decomposition will be further gainfully applied to the fast lattice summation of interaction potentials.

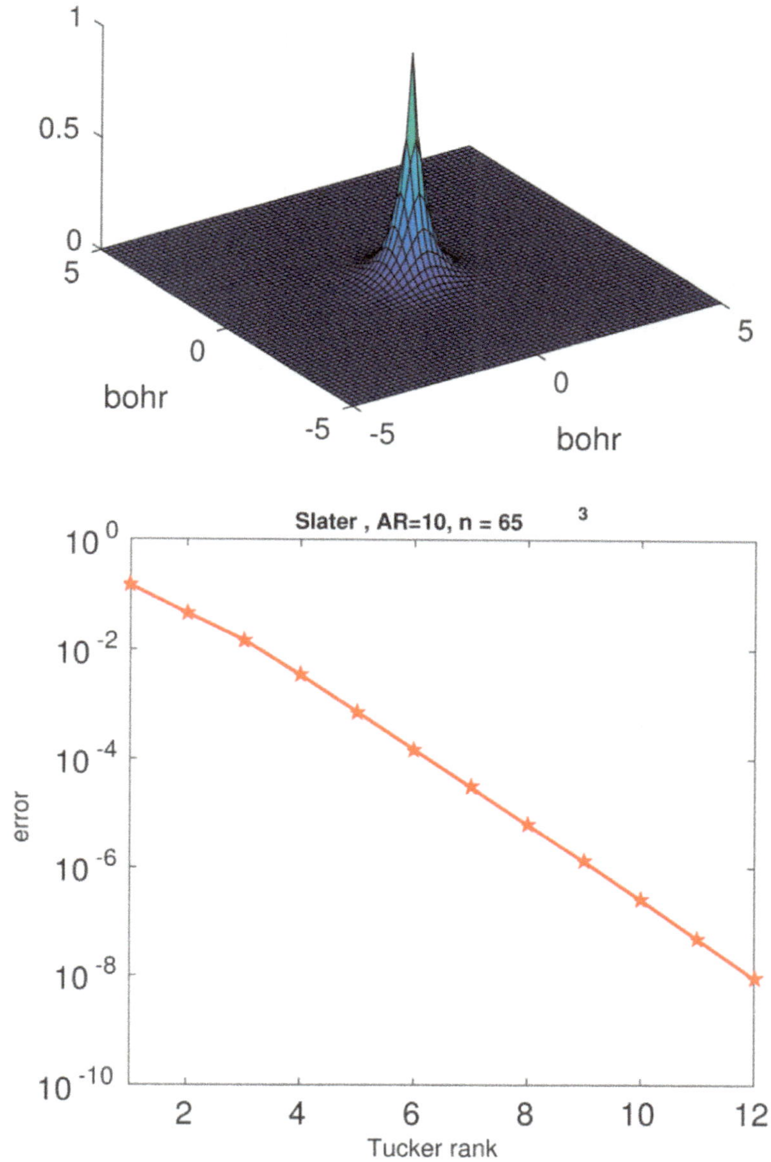

Figure 3: The Tucker approximation error in the Frobenius norm vs. the Tucker rank r for a single Slater function.

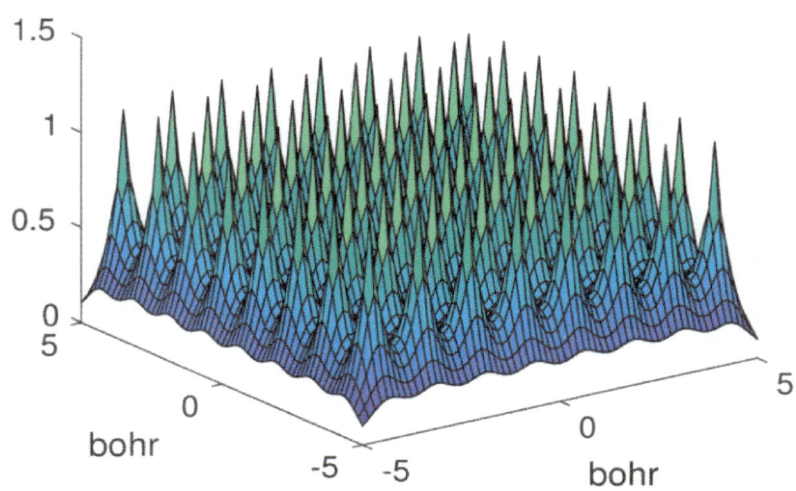

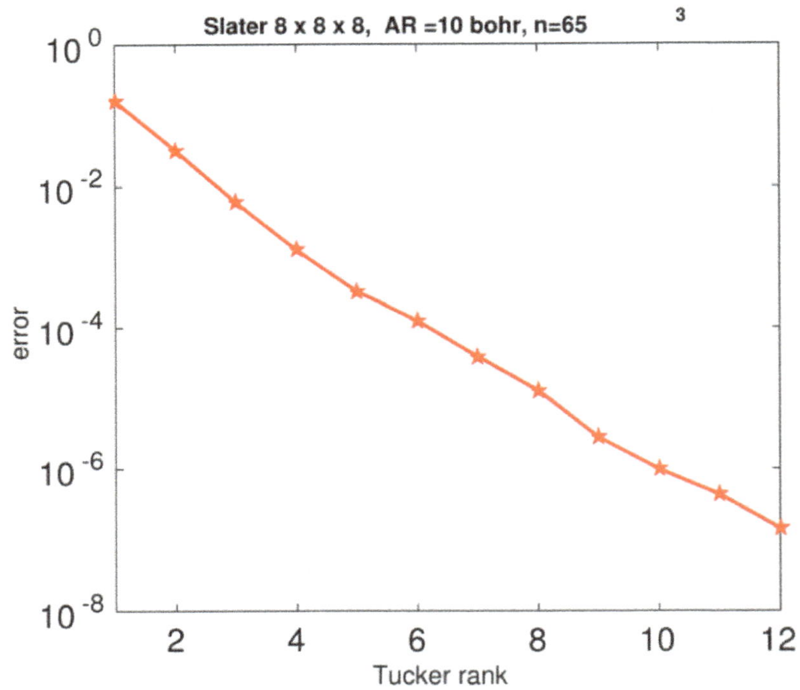

Figure 4: The Tucker approximation error in the Frobenius norm vs. the Tucker rank r for a sum of 512 Slater potentials over the cubic 8 8 8 lattice (right).

Motivated by these observations, we invented the canonical-to-Tucker (C2T) decomposition for function related tensors,[38] based on the reduced higher order singular value decomposition (RHOSVD) (Theorem 2.5, ref. 39). The canonical-to-Tucker algorithm (see Appendix) combined with the Tucker-to-canonical transform serves for reducing the ranks of canonical tensors with large R. Fig. 5 in its ALS part shows the algorithmic step for l = 1, which is repeated for every mode l = 1, 2, 3 (see Appendix). The computational work for the multigrid tensor decomposition C2T algorithm introduced in ref. 39 exhibits linear complexity scaling with respect to all input parameters, O(Rnr).

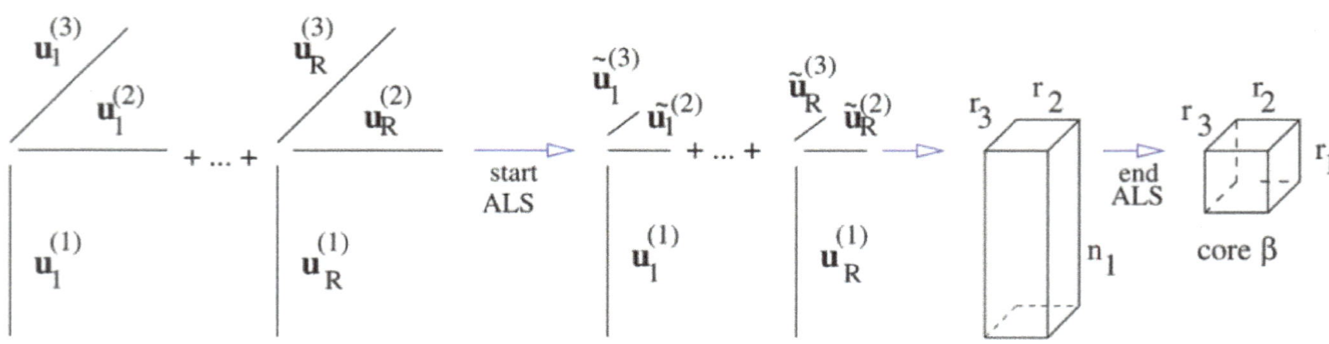

Figure 5: Part of the canonical-to-Tucker decomposition algorithm for mode l = 1.

The multigrid Tucker decomposition algorithm applied to full format tensors[39] leads to the complexity scaling $O(n^3)$, whereas its standard version (commonly used HOSVD algorithm in computer science) scales as $O(n^4)$, becoming intractable even for moderate sizes of tensors.

We conclude by notice that the optimized Tucker and canonical tensor approximations can be computed by the alternating least square (ALS) iteration with initial guess obtained by HOSVD/RHOSVD approximations.

Matrix Product States/Tensor Train Format

The product-type representation of dth order tensors, which is called in the physical literature as the matrix product states (MPS) decomposition (or more generally, tensor network states models), was introduced and successfully applied in DMRG quantum computations,[62,69,73] and, independently, in quantum molecular dynamics as the multilayer (ML) MCTDH methods.[54,71]MPS type representations reduce the storage complexity to $O(dr^2N)$, where r is the maximal rank parameter.

In the recent years the various versions of the MPS-type tensor approximations were discussed and further investigated in mathematical literature including the hierarchical dimension splitting,[36] the tensor train (TT),[14,57] the quantics-TT (QTT),[34] as well as the hierarchical Tucker representation,[26] which belongs to the class of tensor network states model.

The TT format is the particular form of MPS type factorization in the case of open boundary conditions. For a given rank parameter $\mathbf{r} = (r_0,\ldots,r_d)$, the rank-**r** TT format contains all elements $\mathbf{A} = [a_{i1,\ldots,id}] \in R^{n_1 \times \cdots \times n_d}$, which can be represented as the contracted products of 3-tensors, that in the index notation takes a form,

$$a_{i_1,\ldots,i_d} = \sum_{\alpha_1=1}^{r_1} \cdots \sum_{\alpha_d=1}^{r_d} \mathbf{a}^{(1)}_{\alpha_1}(i_1)\mathbf{a}^{(2)}_{\alpha_1,\alpha_2}(i_2)\ldots\mathbf{a}^{(d)}_{\alpha_{d-1}}(i_d)$$

$\equiv A^{(1)}(i_1)A^{(2)}(i_2)\ldots A^{(d)}(i_d)$,

specified by the set of column vectors, $\mathbf{a}^{(\ell)}_{\alpha_\ell,\alpha_{\ell+1}} \in \mathbb{R}^{n_\ell}$, (l = 1,…,d), or equivalently by the vector-valued $r_l \times r_{l+1}$ matrices, $A^{(\ell)} = \left[\mathbf{a}^{(\ell)}_{\alpha_\ell,\alpha_{\ell+1}}\right]$, (i.e., 3-tensors), cf.(2). The latter representation is written in the matrix product form, explaining the notion MPS, where $A^{(l)}(i_l)$ is $r_{l-1} \times r_l$ matrix.

Fig. 6 illustrates the TT representation of a 5th-order tensor, where each particular entry is factorized as a product of five matrices, $a_{i1,i2,\ldots,i5} = A^{(1)}(i_1)A^{(2)}(i_2)\ldots A^{(5)}(i_5)$, where, for example, $A^{(2)}(i_2) \in R^{r_1 \times r_2}$.

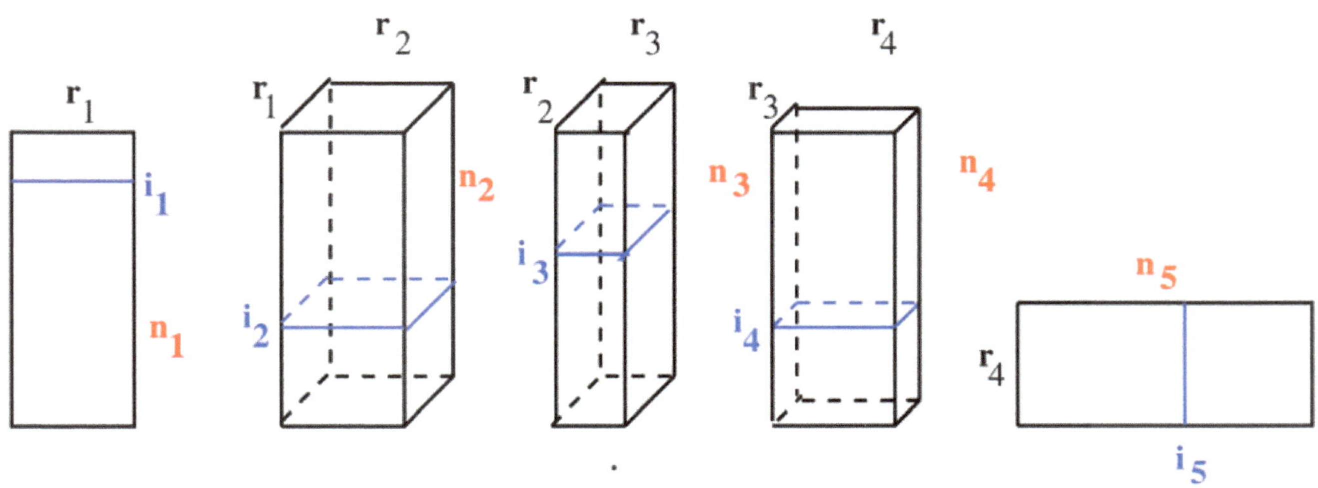

Figure 6: Visualizing 5th-order MPS/TT tensor.

The rank-structured tensor formats like canonical, Tucker and MPS/TT-type decompositions also apply to matrices. For example, the d-dimensional FFT matrix over $N^{\otimes d}$ grid can be implemented on the rank-k tensor with the linear-logarithmic cost $O(dkN \log_2 N)$, due to the rank-1 factorized representation

$$F^{(d)}_N = (F^{(1)}_N \otimes I\ldots\otimes I)(I \otimes F^{(2)}_N\ldots\otimes I)\ldots(I \otimes I\ldots\otimes F^{(d)}_N) \equiv F^{(1)}_N \otimes\ldots\otimes F^{(d)}_N,$$

where $F^{(\ell)}_N \in \mathbb{R}^{N\times N}$ represents the univariate FFT matrix along mode l.

Quantics Tensor Approximation of Functional Vectors

In the case of large mode size, the asymptotic storage cost for a class of function related N–d tensors can be reduced to O(d log N) by using quantics-TT (QTT) tensor approximation method.[34] The QTT-type approximation of an N-vector with $N = 2^L$, $L \in N$, is defined as the tensor decomposition (approximation) in the canonical, TT or more general formats applied to a tensor obtained by the dyadic folding (reshaping) of the target vector to an L-dimensional $2 \times\cdots\times 2$ data array (tensor) that is thought as an element of the L-dimensional quantized tensor space.

In the vector case, i.e. for d = 1, a vector $\mathbf{x} = [x_i] \in R^N$, with $N = 2^L$, is reshaped to its quantics (quantized) image in $\otimes_{j=1}^{L}\mathbb{R}^2$ by dyadic folding,

$$\mathcal{F}_{2,L} : \mathbf{x} \to \mathbf{Y} = [y_{\mathbf{j}}] \in \otimes_{j=1}^{L}\mathbb{R}^2, \quad \mathbf{j} = (j_1,\ldots,j_L),\ j_\nu \in \{1,2\},$$

where for fixed i, we have $y_{\mathbf{j}} := x_i$, with $j_\nu = j_\nu(i) = C_{\nu-1}$ ($\nu = 1,\ldots,L$) being defined via binary coding, i.e. the coefficients $C_{\nu-1} \in \{0, 1\}$ are found from the binary representation of $i - 1$,

$$i - 1 = C_0 + C_1 2^1 + \cdots + C_{L-1}2^{L-1} \equiv \sum_{\nu=1}^{L}(j_\nu - 1)2^{\nu-1}$$

Next figure visualizes the QTT approximation process.

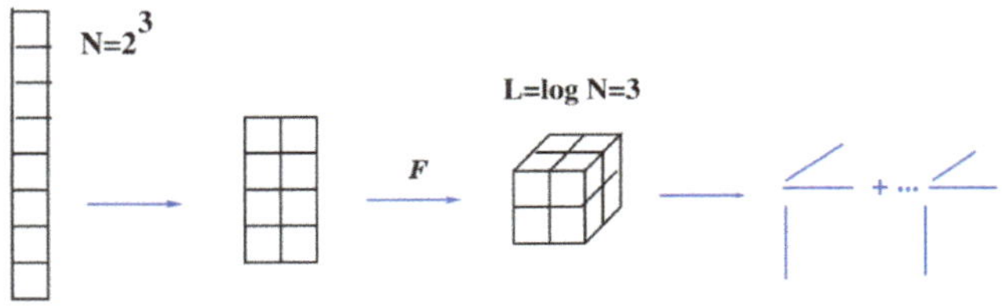

Suppose that the quantics image for an N-vector, i.e. an element of L-dimensional quantized tensor space $\otimes_{j=1}^{L} \mathbb{R}^2$ with $L = \log_2 N$, can be represented (approximated) by the low-rank canonical or TT tensor of order L, thus introducing the QTT approximation of an N-vector. Given rank parameters $\{r_k\}$ ($k = 1,\ldots,L$), the QTT approximation of an N-vector requires a number of representation parameters estimated by

$$2r^2 \log_2 N \ll N, \quad \text{where} \quad r_k \le r, \quad k = 1,\ldots, L,$$

providing log-volume complexity scaling in the size of initial vector, N. For $d > 1$ the construction is similar.[34]

The power of QTT approximation method is explained by the theoretical substantiation of the QTT approximation properties discovered in ref. 34 and establishing the perfect rank-r decomposition for the wide class of function-related tensors obtained by sampling continuous functions over uniform (or properly refined) grid:

- $r = 1$ for complex exponents,
- $r = 2$ for trigonometric functions and plain-waves,
- $r \le m + 1$ for polynomials of degree m,
- r is a small constant for wavelet functions, Gaussians, etc. all independently on the vector size N.

Notice that the notion quantics (or quantized) tensor approximation (with a shorthand QTT) originally introduced in 2009, see ref. 34, is reminiscent of the entity "quantum of information", that mimics the minimal possible mode size ($n = 2$) of the quantized image.

Concerning the matrix case, it was first found in ref. 56 by numerical tests that in some cases the dyadic reshaping of an $N \times N$ matrix with $N = 2^L$ may lead to a small TT-rank of the resultant matrix rearranged to the tensor form. The efficient low-rank QTT representation for a class of discrete multidimensional elliptic operators (matrices) and their inverse was proven in ref. 33. Moreover, based on the QTT approximation, the important algebraic matrix operations like FFT, convolution and wavelet transforms can be implemented by superfast algorithms in $O(\log_2 N)$-complexity, see survey paper[37] and references therein.

Rank-Structured Tensor Operations in 1D Complexity

The rank-structured tensor representation provides 1D complexity of multilinear operations with multidimensional tensors. Rank-structured tensor representation provides fast multi-linear algebra with linear complexity scaling in the dimension d.

For given canonical tensors **A** and **B** as in (1), with ranks R_a and R_b, respectively, their Euclidean scalar product can be computed by univariate operations

$$\langle \mathbf{A}, \mathbf{B} \rangle = \sum_{i=1}^{R_a} \sum_{j=1}^{R_b} c_i c_j \prod_{\ell=1}^{d} \left\langle \mathbf{a}_i^{(\ell)}, \mathbf{b}_j^{(\ell)} \right\rangle, \tag{4}$$

at the expense $O(dnR_aR_b)$.

The Hadamard (entrywise) product of tensors **A**, **B** is defined by $\mathbf{Y} = [y_{i1,\ldots,id}] := \mathbf{A} \odot \mathbf{B}$, where $y_{i1,\ldots,id} = a_{i1,\ldots,id} b_{i1,\ldots,id}$. For canonical tensors **A** and **B** given in form (1), the Hadamard product is calculated in $O(dnR_aR_b)$ operations by 1D entrywise products of vectors,

$$\mathbf{A} \odot \mathbf{B} = \sum_{i=1}^{R_a} \sum_{j=1}^{R_b} c_i c_j \left(\mathbf{a}_i^{(1)} \odot \mathbf{b}_j^{(1)}\right) \otimes \cdots \otimes \left(\mathbf{a}_i^{(d)} \odot \mathbf{b}_j^{(d)}\right). \tag{5}$$

Summation of two tensors in the canonical format $\mathbf{C} = \mathbf{A} + \mathbf{B}$ is performed by a simple concatenation of their factor matrices, $A^{(\ell)} = \left[\mathbf{a}_1^{(\ell)}, \ldots, \mathbf{a}_{R_a}^{(\ell)}\right]$ and $B^{(\ell)} = \left[\mathbf{b}_1^{(\ell)}, \ldots, \mathbf{b}_{R_b}^{(\ell)}\right]$,

$$C^{(\ell)} = \left[\mathbf{a}_1^{(\ell)}, \ldots, \mathbf{a}_{R_a}^{(\ell)}, \mathbf{b}_1^{(\ell)}, \ldots, \mathbf{b}_{R_b}^{(\ell)}\right] \in \mathbb{R}^{n_\ell \times (R_a + R_b)}. \tag{6}$$

The rank of the resulting canonical tensor increases up to $R_c = R_a + R_b$.

In electronic structure calculations, the 3D convolution transform with the Newton kernel, $\frac{1}{\|x-y\|}$, is the most computationally expensive operation. The tensor method to compute convolution over large $n \times n \times n$ Cartesian grids in $O(n \log n)$ complexity was introduced in ref. 35. Given canonical tensors **A**, **B**, their convolution product is represented by the sum of tensor products of 1D convolutions,

$$\mathbf{A} \times \mathbf{B} = \sum_{i=1}^{R_a} \sum_{j=1}^{R_b} c_i c_j \left(\mathbf{a}_i^{(1)} * \mathbf{b}_j^{(1)}\right) \otimes \left(\mathbf{a}_i^{(2)} * \mathbf{b}_j^{(2)}\right) \otimes \left(\mathbf{a}_i^{(3)} * \mathbf{b}_j^{(3)}\right) \tag{7}$$

where $\mathbf{a}_k^{(\ell)} * \mathbf{b}_m^{(\ell)}$ denotes the univariate convolution product of n-vectors. The cost of tensor convolution in both storage and time is estimated by $O(R_aR_b n \log n)$. The resulting algorithm considerably outperforms the conventional 3D FFT-based approaches of complexity $O(n^3 \log n)$, see numerics in ref. 39.

The sequences of rank-structured operations on matrices and vectors normally lead to the increase of tensor ranks, usually being multiplied or added after each operation. The necessary rank reduction in the Tucker and MPS type formats can be implemented by stable algorithms based on the higher order SVD. In the physical community the HOSVD-type algorithms are known since longer as the Schmidt decomposition.[62,69,70] In the case of canonical tensors the rank reduction can be performed by the RHOSVD algorithm based on the canonical-to-Tucker and then Tucker-to-canonical transforms described and analyzed in ref. 38, 39 and 49 (see Section 2.2 and Appendix), which demonstrated the stable behavior for most of examples in the Hartree–Fock calculations we considered so far. The stability conditions for the RHOSVD have been discussed in ref. 39 and 49.

TENSOR CALCULUS FOR THE HARTREE–FOCK EQUATION

Calculation of Multi-Dimensional Integrals

Tensor-structured calculation of the multidimensional convolution integral operators with the Newton kernel have been introduced in ref. 39, 40 and 42, where on the examples of the Hartree and exchange operators in the Hartree–Fock equation, it was shown that calculation of the 3D and 6D convolution integrals can be reduced to a combination of 1D Hadamard products, 1D convolutions and 1D scalar products.

The molecule is embedded in a certain fixed computational box $\Omega = [-b,b]^3 \in R^3$, as in Fig. 7. For a given discretization parameter $n \in N$, we use the equidistant $n \times n \times n$ tensor grid $\omega_{3,n} = \{x_i\}$, $\mathbf{i} \in J := \{1,\ldots,n\}^3$, with the mesh-size $h = 2b/(n + 1)$. In calculations of integral terms, the Gaussian basis functions $g_k(x)$, $x \in R^3$, are approximated by sampling their values at the centers of discretization intervals, as in Fig. 8, using one-dimensional piecewise constant basis functions $g_k(x) \approx \overline{g}_k(x) = \prod_{\ell=1}^{3} \overline{g}_k^{(\ell)}(x_\ell)$, $l = 1, 2, 3$, yielding their rank-1 tensor representation,

$$\mathbf{G}_k = \mathbf{g}^{(1)}_k \otimes \mathbf{g}^{(2)}_k \otimes \mathbf{g}^{(3)}_k \in R^{n \times n \times n}, \quad k = 1,\ldots,N_b. \qquad (8)$$

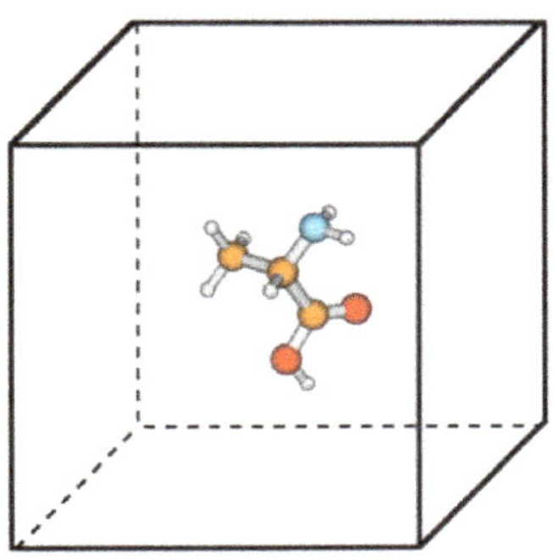

Figure 7: Glycine amino acid in a computational box.

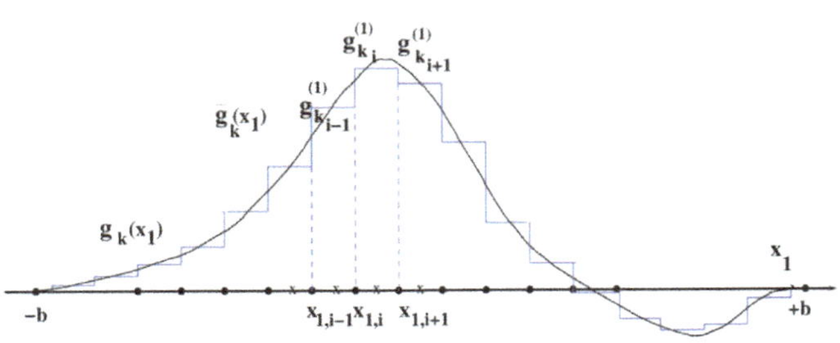

Figure 8: Approximation of the Gaussian-type basis function by a piecewise constant function.

Let us consider the tensor calculation of the Hartree potential

$$V_{\mathrm{H}}(x) := \int_{\mathbb{R}^3} \frac{\rho(y)}{\| x - y \|} \mathrm{d}y,$$

and the corresponding Coulomb matrix,

$$J_{km} := \int_{\mathbb{R}^3} g_k(x) g_m(x) V_{\mathrm{H}}(x) \mathrm{d}x, \quad k, m = 1, \ldots N_{\mathrm{b}} \quad x \in \mathbb{R}^3,$$

where the electron density, $\rho(x) = 2 \sum_{a=1}^{N_{\mathrm{orb}}} (\varphi_a)^2$, is represented in terms of molecular orbitals $\varphi_a(x) = \sum_{k=1}^{N_{\mathrm{b}}} c_{a,k} g_k(x)$. Given the discrete tensor representation of basis functions (8), the electron density is approximated using 1D Hadamard products of rank-1 tensors (instead of product of Gaussians),

$$\rho \approx \Theta = 2 \sum_{a=1}^{N_{\mathrm{orb}}} \sum_{k=1}^{N_{\mathrm{b}}} \sum_{m=1}^{N_{\mathrm{b}}} c_{a,m} c_{a,k} \left(\mathbf{g}_k^{(1)} \odot \mathbf{g}_m^{(1)} \right) \otimes \left(\mathbf{g}_k^{(2)} \odot \mathbf{g}_m^{(2)} \right) \otimes \left(\mathbf{g}_k^{(3)} \odot \mathbf{g}_m^{(3)} \right) \in \mathbb{R}^{n \times n \times n}.$$

Further, the representation of the Newton kernel $\frac{1}{\| x - y \|}$ by a canonical rank-R_N tensor[7] is used (see Appendix for details),

$$\mathbf{P}_R = \sum_{q=1}^{R_N} \mathbf{p}_q^{(1)} \otimes \mathbf{p}_q^{(2)} \otimes \mathbf{p}_q^{(3)} \in \mathbb{R}^{n \times n \times n}. \qquad (9)$$

Since large ranks make tensor operations inefficient, the multigrid canonical-to-Tucker and Tucker-to-canonical algorithms should be applied to reduce the initial rank of $\Theta \mapsto \Theta'$ by several orders of magnitude, from $N_b^2/2$ to $R_\rho \ll N_b^2/2$. Then the 3D tensor representation of the Hartree potential is calculated by using the 3D tensor product convolution, which is a sum of tensor products of 1D convolutions,

$$V_{\mathrm{H}} \approx \mathbf{V}_{\mathrm{H}} = \Theta' * \mathbf{P}_R = \sum_{m=1}^{R_\rho} \sum_{q=1}^{R_N} c_m \left(\mathbf{u}_m^{(1)} * \mathbf{p}_q^{(1)} \right) \otimes \left(\mathbf{u}_m^{(2)} * \mathbf{p}_q^{(2)} \right) \otimes \left(\mathbf{u}_m^{(3)} * \mathbf{p}_q^{(3)} \right).$$

The Coulomb matrix entries J_{km} are obtained by 1D scalar products of $\mathbf{V}_H$ with the Galerkin basis consisting of rank-1 tensors,

$$J_{km} \approx \langle \mathbf{G}_k \odot \mathbf{G}_m, \mathbf{V}_H \rangle, \quad k, m = 1,\ldots N_b.$$

The cost of 3D tensor product convolution is $O(n \log n)$ instead of $O(n^3 \log n)$ for the standard benchmark 3D convolution using the 3D FFT. Table 1 shows CPU times (s) for the Matlab computation of V_H for H_2O molecule[39] on a SUN station using 8 Opteron Dual-Core/2600 processors (times for 3D FFT for $n \geq 1024$ are obtained by extrapolation). C2T shows the time for the canonical-to-Tucker rank reduction. In a similar way, the algorithm for 3D grid-based tensor-structured calculation of 6D integrals in the exchange potential operator $K_{km} = \sum_{a=1}^{N_{\mathrm{orb}}} K_{km,a}$ was introduced inref. 40, with

$$K_{km,a} := \int_{\mathbb{R}^3} \int_{\mathbb{R}^3} g_k(x) \frac{\varphi_a(x)\varphi_a(y)}{|x - y|} g_m(y) \mathrm{d}x\mathrm{d}y, \quad k, m = 1, \ldots N_{\mathrm{b}}$$

the contribution from the a-th orbital are approximated by tensor anzats,

$$K_{km,a} := \left\langle \mathbf{G}_k \odot \left[\sum_{\mu=1}^{N_{\mathrm{b}}} c_{\mu a} \mathbf{G}_\mu \right], \left[\mathbf{G}_m \odot \sum_{\nu=1}^{N_{\mathrm{b}}} c_{\nu a} \mathbf{G}_\nu \right] * \mathbf{P}_R \right\rangle$$

Table 1: Times (s) for the 3D tensor product convolution vs. 3D FFT convolution

n^3	1024^3	2048^3	4096^3	8192^3	16384^3
FFT_3	~6000	—	—	—	~2 years
$C \times C$	8.8	20.0	61.0	157.5	299.2
C2T	6.9	10.9	20.0	37.9	86.0

Here, the tensor product convolution is first calculated for each ath orbital, and then scalar products in canonical format yield the contributions to entries of the exchange Galerkin matrix from the a-th orbital.

These algorithms were employed in the first tensor-structured solver using 3D grid-based evaluation of the Coulomb and exchange matrices in 1D complexity at every step of SCF EVP iteration.[41,42] A sequence of dyadically refined 3D Cartesian grids was used for reducing time in first iterations, with an ε convergence criterion for switching to larger grids. This is a nonstandard computational scheme avoiding calculation of the two-electron integrals. The accuracy for small molecules like H_2O and CH_4 was of the order of 10^{-4} Hartree. Though time performance of this solver was not compatible with the standard Hartree–Fock packages it was the first proof of concept for the tensor numerical methods.

3D Grid-Based Calculation of the Two-Electron Integrals

The basic tensor-structured Hartree–Fock solver employs the factorized 3D grid-based calculation of the two-electron integrals tensor, $\mathbf{B} = [b_{\mu\nu\kappa\lambda}]$, in the form

$$b_{\mu\nu\kappa\lambda} = \int_{\mathbb{R}^3}\int_{\mathbb{R}^3} \frac{g_\mu(x)g_\nu(x)g_\kappa(y)g_\lambda(y)}{\| x-y \|} dx\,dy \tag{10}$$

$= \langle \mathbf{G}_\mu \odot \mathbf{G}_\nu, \mathbf{P}_R * (\mathbf{G}_\kappa \odot \mathbf{G}_\lambda)\rangle_{n^{\otimes 3}}$,

by using univariate tensor operations. Introduce the side matrices $G^{(l)}$ representing on the grid the full set of canonical vectors composing the products of the Gaussian basis functions, $\{\mathbf{G}_\mu \odot \mathbf{G}_\nu\}$,

$$G^{(\ell)} = \left[\mathbf{g}_\mu^{(\ell)} \odot \mathbf{g}_\nu^{(\ell)}\right]_{1\le\mu,\nu\le N_b} \in \mathbb{R}^{n\times N_b^2} \quad \ell = 1,2,3,$$

where in most of our Hartree–Fock calculations grids of size $n^3 = 32 \times 10^3$ or $n^3 = 64 \times 10^3$ have been used. It was found that the large matrices $G^{(l)}$ of size $n \times N_b^2$ (e.g. $N_b^2 = 40\,000$ for Alanine amino acid) can be approximated with high accuracy by low rank matrices with the rank parameter bounded by $Rl \le N_b$.[44,45] The corresponding low-rank factorizations ("1D density fitting") in the form (for l = 1, 2, 3)

$$G^{(\ell)} \cong U^{(\ell)} V^{(\ell)T}, \quad U^{(\ell)} \in \mathbb{R}^{n\times R_\ell}, \quad V^{(\ell)} \in \mathbb{R}^{N_b^2\times R_\ell}, \tag{11}$$

is computed by the truncated Cholesky decomposition of the symmetric, positive definite $G^{(l)}G^{(l)T}$ (see ref. 44 and 45 for more details).

Based on factorization (11), the number of convolutions in (10) is reduced dramatically from N_b^2 to N_b at most (say from 40 000 to 200). In fact, using canonical factors from the rank-R canonical tensor $\mathbf{P}_R$ representing the Newton kernel (9) (see (31) in Appendix) we, first, precompute the set of "convolution" matrices for every space variable l, l = 1, 2, 3,

$$M_k^{(\ell)} = U^{(\ell)^T}\left(\mathbf{p}_k^{(\ell)} *_n U^{(\ell)}\right) \in \mathbb{R}^{R_\ell\times R_\ell}, \quad k = 1,\ldots,R, \tag{12}$$

which includes the convolution products with $Rl \le N_b$ column vectors of the matrix $U^{(\ell)} \in \mathbb{R}^{n\times R_\ell}$ instead of $N_b^2/2$ convolutions in the initial formulation.

Given matrices $M_k^{(\ell)}$, then the resulting 4-th order TEI tensor is represented in a matrix form

$$B = \mathrm{mat}(\mathbf{B}) \cong B_\varepsilon := \sum_{k=1}^{R} \odot_{\ell=1}^3 V^{(\ell)} M_k^{(\ell)} V^{(\ell)^T} \in \mathbb{R}^{N_b^2\times N_b^2}.$$

The above nonstandard factorization of the TEI matrix B allows to reduce dramatically the computational cost of the standard Cholesky factorization schemes[3,5,30] applied to the reduced-rank symmetric positive definite matrix B. In fact, the low-rank Cholesky decomposition of B is calculated in the following sequence. First, the diagonal elements of B are calculated as

$$B(i,i) = \sum_{k=1}^{R} \odot_{\ell=1}^3 V^{(\ell)}(i,:) M_k^{(\ell)} V^{(\ell)}(:,i)^T.$$

Then the selected columns of the matrix B, required for the rank-truncated Cholesky factorization scheme, are computed by the following fast tensor operations

$$B(:,j^*) = \sum_{k=1}^{R} \odot_{\ell=1}^3 V^{(\ell)} M_k^{(\ell)} V^{(\ell)}(:,j^*)^T,$$

leading to representation of the matrix B in the form of rank-R_B approximate decomposition[44,45]

$$B := \left[b_{\mu\nu,\kappa\lambda}\right] \approx LL^T, \quad \text{with } L \in \mathbb{R}^{N_b^2\times R_B},\ R_B \sim N_b.$$

This algorithm is much faster than the direct Cholesky decomposition of the matrix B with on-the-fly computation of the required column vectors.

Table 2 represents times (s) for 3D grid-based calculation of the directional density fitting and the TEI tensor (electron repulsion integrals) for H_2O molecule in a box $[-20,20]^3$ bohr^3, performed in Matlab on a 2-Intel Xeon Hexa-Core/2677. Time for convolution integrals in (12) scales almost linearly in the 1D grid-size n as expected by theory.

Table 2: Times (s) for 3D grid-based calculation of the directional density fitting and the TEI for H_2O molecule

n^3	$16\,384^3$	$32\,768^3$	$65\,536^3$	$131\,072^3$
Mesh (bohr)	0.0024	0.0012	6×10^{-4}	3×10^{-4}
Had. prod.	1.6	3.4	12	19
Density fit.	0.5	1.0	2.0	4.0
3D conv. time	69	151	698	496
Chol. time	2.2	2.2	2.2	2.2

Note that the algorithm for fast tensor computation of TEI is described in detail for the case of rank-1 GTO basis functions in form (8). However, this approach can be also applied to basis sets with low separation rank, say, to Slater-type basis functions or some kind of agglomerated basis. The advantage of use the "more compressed" basis is the essential reduction in the number of basis functions, N_b, that enters to the size of TEI tensor in the fourth power. The payoff for such a reduction in the tensor size is the additional costs for calculation of the particular entries of TEI tensor in (10), since in that case the number of terms in tensor representation of the scalar products in (10) increases quadratically in the separation rank of basis functions. The resulting effect from implementation of the reduced basis sets (with larger separation ranks) depends on subsequent (post) Hartree–Fock calculations and it will be studied in detail in the forthcoming paper.

Core Hamiltonian

The Galerkin representation of the 3D Laplace operator in the nonlocal Gaussian basis $\{g_k(x)\}_{1\leq k\leq N_b}$, $x \in R^3$, leads to the fully populated matrix $A_g = [a_{km}] \in R^{N_b \times N_b}$. Tensor calculation of the matrix entries a_{km} for the discrete Laplacian A_g in the separable Gaussian basis is reduced to 1D matrix operations[46] involving the FEM Laplacian Δ_3, defined on $n \times n \times n$ grid,

$$\Delta_3 = \Delta_1^{(1)} \otimes I^{(2)} \otimes I^{(3)} + I^{(1)} \otimes \Delta_1^{(2)} \otimes I^{(3)} + I^{(1)} \otimes I^{(2)} \otimes \Delta_1^{(3)},$$

where $\Delta_1 = \frac{1}{h}\text{tridiag}\{-1, 2, -1\}$. Specifically, we have

$a_{km} = \langle \Delta_3 \mathbf{G}_k, \mathbf{G}_m \rangle$,

where $\mathbf{G}_k$ is the tensor representation of Gaussian basis functions using the piecewise linear finite elements (see Fig. 9).

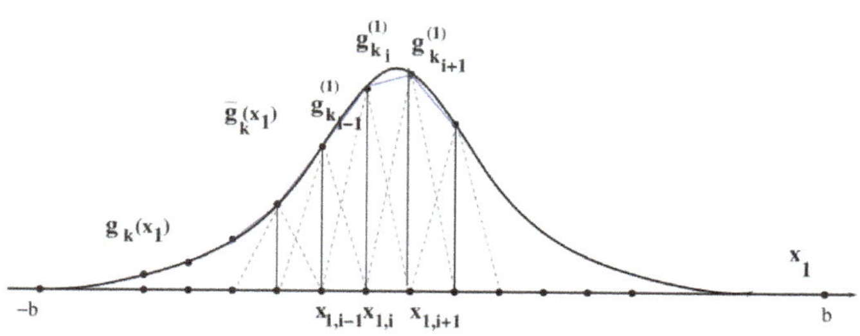

Figure 9: Discretization of a Gaussian by piecewise linear finite elements.

In the case of large $n \times n \times n$ grids, this calculation can be implemented with logarithmic cost in n by using the low-rank QTT representation of the large matrix Δ_3, see ref. 33, 43 and 46.

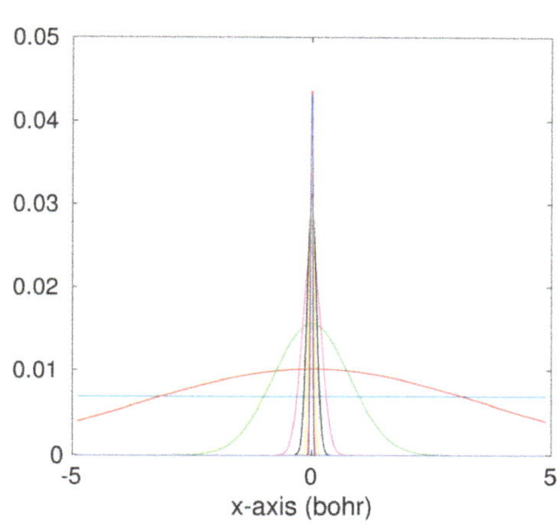

Figure 10: Several vectors of the canonical representation of the Newton kernel along one of variables.

For tensor calculation of the nuclear potential operator

$$V_c(x) = -\sum_{\alpha=1}^{M} \frac{Z_\alpha}{\| x - a_\alpha \|}, \quad Z_\alpha > 0, \quad x, a_\alpha \in \mathbb{R}^3,$$

we apply the rank-1 windowing operator, $W_\alpha = W^{(1)}_\alpha \otimes W^{(2)}_\alpha \otimes W^{(3)}_\alpha$, for shifting the reference Newton kernel $\tilde{\mathbf{P}}_R \in R^{2n\times 2n\times 2n}$ according to the coordinates of nuclei in a molecule (see Section 5 and Appendix). Then the resulting nuclear potential, $\mathbf{P}_c \in {}^{n\times n\times n}$, is obtained as a direct tensor sum of shifted potentials,[46]

$$\mathbf{P}_c = \sum_{\alpha=1}^{M} Z_\alpha \mathcal{W}_\alpha \widetilde{\mathbf{P}}_R$$

$$= \sum_{\alpha=1}^{M} Z_\alpha \sum_{q=1}^{R} \mathcal{W}_\alpha^{(1)} \widetilde{\mathbf{p}}_q^{(1)} \otimes \mathcal{W}_\alpha^{(2)} \widetilde{\mathbf{p}}_q^{(2)} \otimes \mathcal{W}_\alpha^{(3)} \widetilde{\mathbf{p}}_q^{(3)}.$$

This leads to the following representation of the Galerkin matrix, $V_c = [\bar{v}_{km}]$, by tensor operations

$$\bar{v}_{km} = \int_{\mathbb{R}^3} V_c(x) \bar{g}_k(x) \bar{g}_m(x) dx \approx \langle \mathbf{G}_k \odot \mathbf{G}_m, \mathbf{P}_c \rangle, \quad 1 \leq k, m \leq N_b.$$

Fig. 10, shows several vectors of the canonical representation of the Coulomb kernel along one of variables. Fig. 11, represents the cross-section of the resulting nuclear potential $\mathbf{P}_c$ for C_2H_5OH molecule.

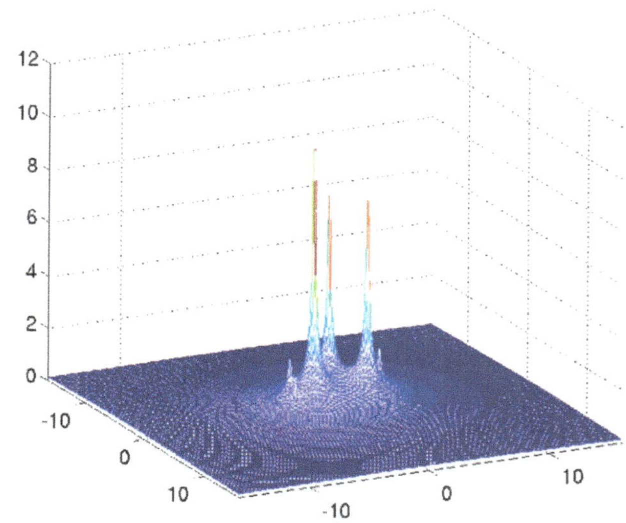

Figure 11: A sum of the nuclear potentials for ethanol molecule. Size of the computational box is given in bohr.

Black-Box Tensor Solver

The tensor-structured Hartree–Fock solver[46] based on factorized calculation of the two-electron integrals[44] includes efficient tensor implementation of the MP2 energy correction[45] scheme. Though it is yet implemented in Matlab, its performance in time and accuracy is compatible with the standard packages based on analytical evaluation of the two-electron integrals. Due to 1D complexity of all calculations, it enables 3D grids of the size 10^{15}, yielding mesh size of the order of atomic radii, 10^{-4} Å. That ensures high accuracy of calculations, which is controlled by the ε-ranks of tensor truncation.

The solver works in a black-box way: input the grid-based basis-functions and coordinates of nuclei in a molecule and start the program. Calculation of TEI for H_2O on grids $32\,768^3$ takes two minutes on a laptop. The time for TEI with $n^3 =$

131 072^3 for Alanine amino acid takes approximately one hour in Matlab, including incorporated density fitting.

Next examples compare the results from benchmark Molpro program with calculations by the tensor-based solver by using the same Gaussian basis, but now discretized on 3D Cartesian grid. For all molecules we use the "cc-pVDZ" Gaussian basis set. The core Hamiltonian part in these calculations is taken from Molpro. Note that since in the tensor solver the density fitting is included in "blind" calculation of TEI, it is not easy to compare the CPU times of our calculations with those in "ab initio" procedure in the standard programs, because the density fitting step there is usually considered as off-line pre-computing.

Fig. 12, top, shows convergence to the ground state energy E_0 in ab initio iterations for H_2O molecule (cc-pVDZ-41), where TEI is computed on the 3D grid of size 1 31 072^3. Fig. 12, bottom presents the zoom of the last 20 iterations. The ground state energy from Molpro (E_0 = 76.308 hartree) corresponding to the basis set cc-pVDZ-41 with 41 Gaussian-type basis functions is shown by the black dashed line.

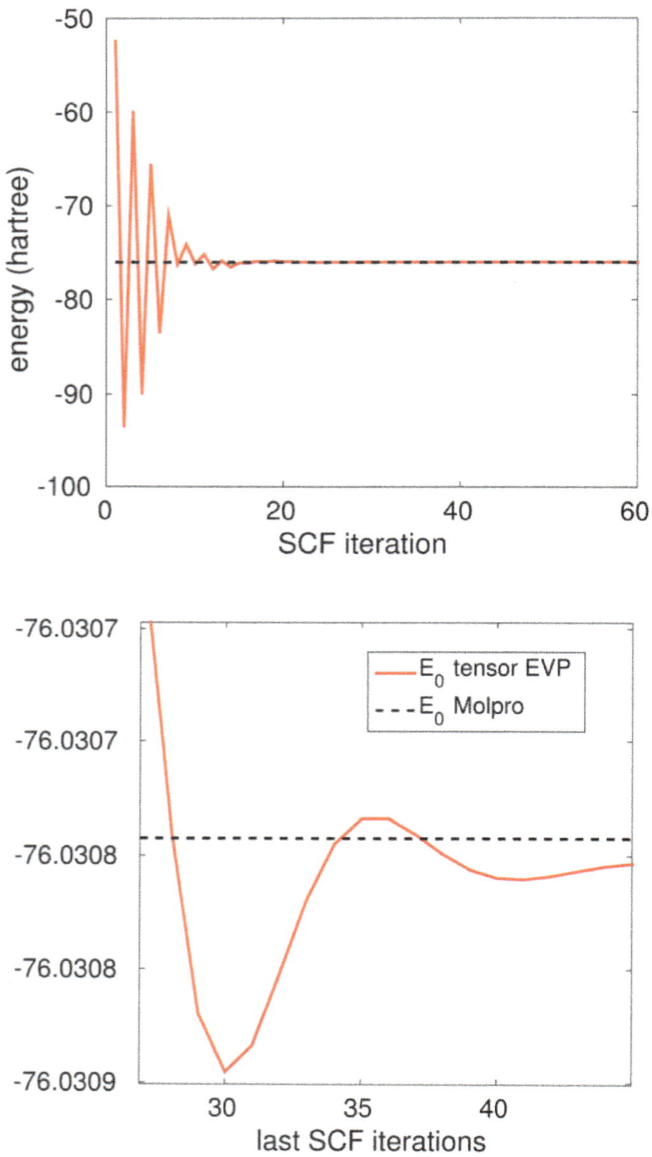

Figure 12: Top: convergence of the ground state energy to E0 in ab initio SCF Hartree–Fock iterations for H2O; bottom: zooming of final 20 iterations for H_2O.

Fig. 13 (top) shows the SCF EVP iteration history for Glycine amino acid (cc-pVDZ-170), where dashed line indicates convergence of the residual and the red solid line shows convergence of the error $\Delta E_{0,g}$ in ground state energy compared with the result from MOLPRO package[72] with the same basis set,

$\Delta E_{0,g} = |E_{0,g} - E_0|/|E_0|$.

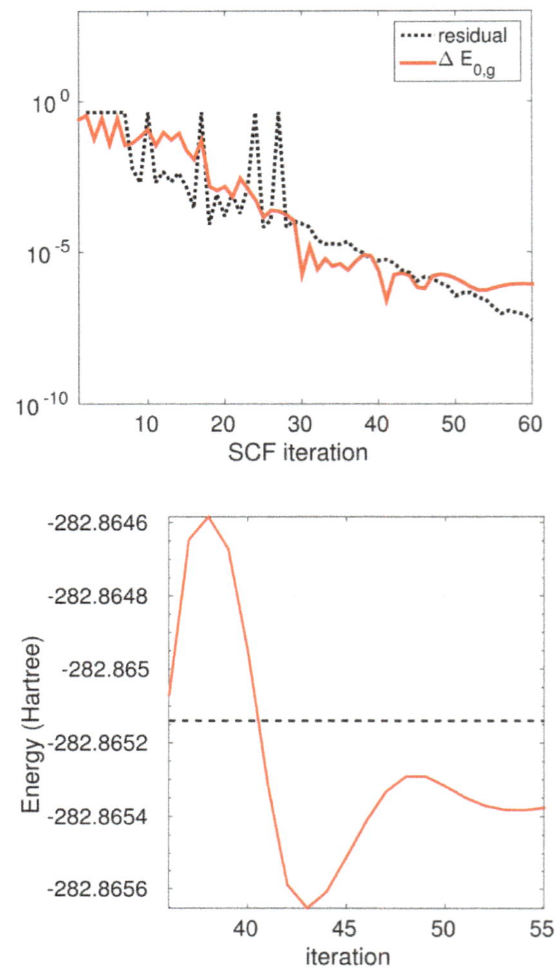

Figure 13: Top: convergence of the residual in ab initio Hartree–Fock iteration for glycine ($C_2H_5NO_2$) molecule. Red line shows the difference with E_0 from molpro at every iteration. Bottom: ground state energy $E_{0,g}$ (red solid line) at final 20 iterations for glycine molecule, dashed line shows E_0 from molpro.

Fig. 13 (bottom) illustrates the convergence of the ground state energy at final 20 iterations for Glycine, dashed line is the energy from Molpro.

The first row in Table 3 shows the number of orbitals and basis functions used in Hartree–Fock calculations for H_2O (cc-pVDZ-41), H_2O_2 (cc-pVDZ-68), and $C_2H_5NO_2$ (cc-pVDZ-170) molecules. Second row represents times (s) for one step of SCF iteration in ab initio solution of the Hartree Fock EVP. Next two rows show the relative difference in energy (13), for different grid sizes used in TEI calculations. This numerics demonstrates that in the case of fine enough spacial $n \times n \times n$-grids the accuracy in 7–8 digits (i.e. relative accuracy about 10^{-7}–10^{-8}) can be achieved for moderate size molecules up to small amino acids. All calculations are performed in the computational box of size $[-20,20]^3$ bohr^3. This tensor-based solver can be considered as the computational tool for trying the alternatives to Gaussian-type basis sets.

Table 3 Time (s) for one ab initio SCF EVP iteration and accuracy of ab initio solution with respect to grid-size in TEI calculations. Matlab on an Intel Xeon X5650

	H_2O	H_2O_2	$C_2H_5NO_2$
N_{orb}, N_b	5; 41	9; 68	20; 170
Time	0.35	0.55	6.0
$\Delta E_{0,g}$, $(65\,536^3)$	3.0×10^{-7}	8.0×10^{-8}	9.1×10^{-7}
$\Delta E_{0,g}$, $(131\,072^3)$	1.4×10^{-7}	3.9×10^{-8}	8.0×10^{-7}

FROM MP2 ENERGY CORRECTION TO EXCITED STATES

MP2 Correction Scheme by Using Tensor Formats

Given the set of Hartree–Fock molecular orbitals $\{C_p\}$ and the corresponding energies $\{\varepsilon_p\}$, $p = 1, 2,\ldots,N_b$, where $\{C_i\}$ and $\{C_a\}$ denote the occupied and virtual orbitals, respectively. First, one has to transform the TEI matrix $B = [b_{\mu\nu,\lambda\sigma}]$, corresponding to the initial AO basis set, to those represented in the molecular orbital (MO) basis,

$$V = \left[v_{ia,jb}\right] : \quad v_{iajb} = \sum_{\mu,\nu,\lambda,\sigma=1}^{N_b} C_{\mu i} C_{\nu a} C_{\lambda j} C_{\sigma b} b_{\mu\nu,\lambda\sigma}, \tag{14}$$

where $a, b \in J_v$, $i, j \in J_o$, and $J_o := \{1,\ldots,N_{orb}\}$, $J_v := \{N_{orb} + 1,\ldots,N_b\}$, with N_{orb} denoting the number of occupied orbitals. In the following, we shall use the notation

$N_v = N_b - N_{orb}$, $N_{ov} = N_{orb} N_v$.

The straightforward computation of the matrix V by above representation makes the dominating impact to the overall numerical cost of order $O(N_b^5)$. The method of complexity $O(N_b^4)$ based on the low-rank tensor decomposition of the matrix V was introduced in ref. 45. Indeed, it can be shown that the rank $R_B = O(N_b)$ approximation to the TEI matrix $B \approx LL^T$, with the $N\times R_B$ Cholesky factor L, allows to introduce the low-rank representation of the matrix V, (see ref. 6 and 45)

$V = L_V L_V^T$, $L_V \in R^{N_{ov} \times R_B}$,

and then reduce the asymptotic complexity of calculations to $O(N_b^4)$.

Given the tensor $V = [v_{iajb}]$, the second order MP2 perturbation to the HF energy is calculated by

$$E_{MP2} = -\sum_{a,b\in I_{vir}} \sum_{i,j\in I_{occ}} \frac{v_{iajb}\left(2v_{iajb} - v_{ibja}\right)}{\varepsilon_a + \varepsilon_b - \varepsilon_i - \varepsilon_j}, \tag{15}$$

where the real numbers ε_k, $k = 1,\ldots,N_b$, represent the HF eigenvalues.

Introducing the so-called doubles amplitude tensor **T**,

$$\mathbf{T} = \left[t_{iajb}\right] : \ t_{iajb} = \frac{\left(2v_{iajb} - v_{ibja}\right)}{\varepsilon_a + \varepsilon_b - \varepsilon_i - \varepsilon_j}, \quad a, b \in I_{vir};\ i, j \in I_{occ},$$

the MP2 perturbation takes the form of a simple scalar product of tensors,

$E_{MP2} = -\langle \mathbf{V},\mathbf{T}\rangle = -\langle \mathbf{V} \odot \mathbf{T}, \mathbf{1}\rangle$,

where **1** denotes the rank-1 all-ones tensor. Introducing the low ε-rank reciprocal "energy" tensor

$$\mathbf{E} = \left[e_{abij}\right] := \left[\frac{1}{\varepsilon_a + \varepsilon_b - \varepsilon_i - \varepsilon_j}\right], \quad a, b \in I_{vir};\ i, j \in I_{occ}, \tag{16}$$

and the partly transposed tensor (transposition in indices a and b)

$$\mathbf{V}' = \left[v_{iajb}{}'\right] := \left[v_{ibja}\right],$$

allows to decompose the doubles amplitude tensor **T** as follows

$$\mathbf{T} = \mathbf{T}^{(1)} + \mathbf{T}^{(2)} = 2\mathbf{V} \odot \mathbf{E} - \mathbf{V}' \odot \mathbf{E}. \tag{17}$$

Notice that the denominator in (15) remains strongly positive if $\varepsilon_a > 0$ for $a \in I_{vir}$ and $\varepsilon_i < 0$ for $i \in I_{occ}$. The latter condition (nonzero homo lumo gap) allows to prove the low ε-rank decomposition of the tensor **E**.[44,45]

Each term in the right-hand side in (17) can be treated separately by using rank-structured tensor decompositions of **V** and **E**, possibly combined with various symmetries and data sparsity. Numerical tests illustrating the tensor approach to the MP2 energy correction are presented in ref. 45.

Toward Low-Rank Approximation of the Bethe–Salpeter Equation for Calculation of Excitation Energies

One of the commonly used approaches for calculation of the excited states in molecules and solids, along with the time-dependent DFT, is based on the solution of the Bethe–Salpeter equation (BSE), see for example.[11,61] The BSE approach leads to the challenging computational task on the solution of the eigenvalue problem for determining the excitation energies ω_n, governed by a large fully populated matrix of size $O(N_{ov}^2) \approx O(N_b^2)$,

$$\begin{pmatrix} A & B \\ B^* & A^* \end{pmatrix} \begin{pmatrix} \mathbf{x}_n \\ \mathbf{y}_n \end{pmatrix} = \omega_n \begin{pmatrix} I & 0 \\ 0 & -I \end{pmatrix} \begin{pmatrix} \mathbf{x}_n \\ \mathbf{y}_n \end{pmatrix}, \qquad (18)$$

so that the computation of the entire spectrum is prohibitively expensive. Here the large matrix blocks of size $N_{ov} \times N_{ov}$ take a form

$$A = \Delta\varepsilon + V - \bar{W}, \quad B = V - \tilde{W},$$

where the diagonal "energy" matrix is defined by

$$\Delta\varepsilon = [\Delta\varepsilon_{ia,jb}] \in R^{N_{ov} \times N_{ov}}: \quad \Delta\varepsilon_{ia,jb} = (\varepsilon_a - \varepsilon_i)\delta_{ij}\delta_{ab},$$

while the matrices $\bar{W} = [\bar{w}_{ia,jb}]$ and $\tilde{W} = [\tilde{w}_{ia,jb}]$ are determined by permutation of the so-called static screened interaction matrix $W = [w_{ia,jb}]$, via $\bar{W}_{ia,jb} = w_{ij,ab}$, and $[\tilde{w}_{ia,jb}] = [w_{ib,aj}]$, respectively. In turn, the forth order tensor $\mathbf{W} = [w_{iajb}]$ is constructed by certain linear transformations of the tensor $\mathbf{V} = [v_{iajb}]$, see ref. 11 and 61.

A number of numerical methods for structured eigenvalue problems have been discussed in the literature.[4,15,51,55]

The tensor approach to the solution of the partial BSE eigenvalue problem for eqn (18) proposed in ref. 6 suggests to compute the reduced basis set by solving the simplified eigenvalue problem via the low-rank plus diagonal approximation to the matrix blocks A and B, and then solve spectral problem for the subsequent Galerkin projection of the initial system (18) to this reduced basis. This procedure relies entirely on multiplication of the simplified BSE matrix with vectors.

It was demonstrated on the examples of moderate size molecules[6] that a small reduced basis set, obtained by separable approximation with the rank parameters of about several tens, allows to reveal several lowest excitation energies and respective excited states with the accuracy about 0.1–0.02 eV.

Fig. 14 illustrates the BSE energy spectrum of the NH_3 molecule (based on HF calculations with cc-pDVZ-48 GTO basis) for the lowest $N_{red} = 30$ eigenvalues vs. the rank truncation parameter $\varepsilon = 0.6$ and 0.1, where the ranks of V and the BSE matrix block W are 4, 5 and 28, 30, respectively. For the choice $\varepsilon = 0.6$ and $\varepsilon = 0.1$, the error in the 1st (lowest) eigenvalue for the solution of the problem in reduced basis is about 0.11 eV and 0.025 eV, correspondingly. The CPU time in the laptop Matlab implementation of each example is about 5 s.

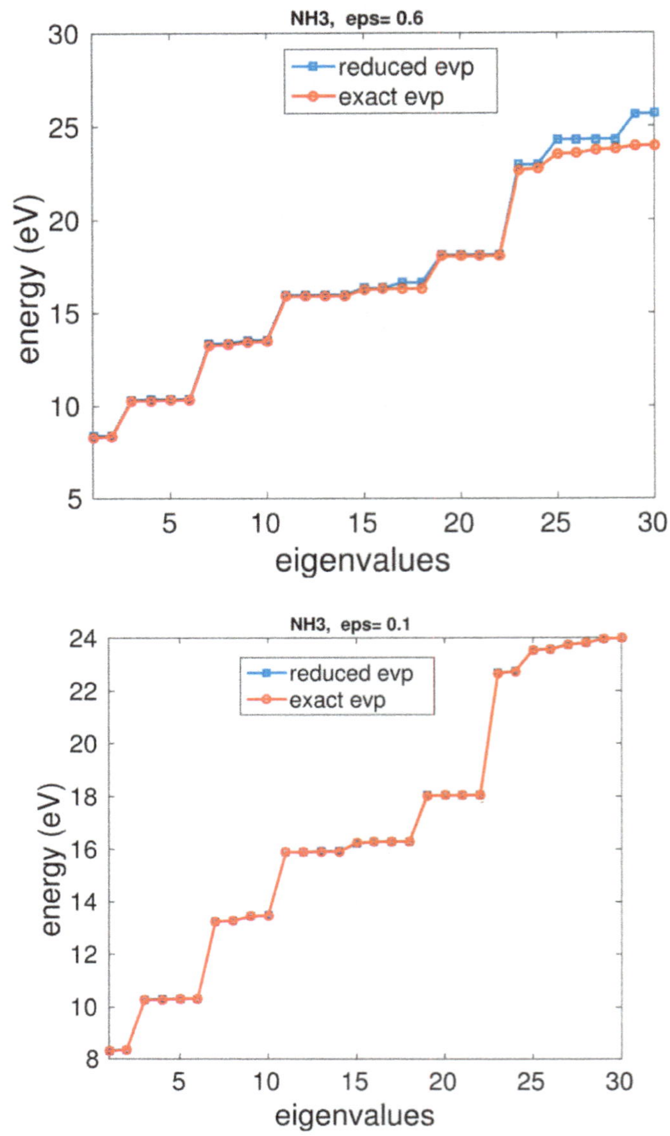

Figure 14: Comparison of $m_0 = 30$ lower eigenvalues for the reduced and exact BSE systems for NH_3 molecule: $\varepsilon = 0.6$, top; $\varepsilon = 0.1$, bottom.

TENSOR APPROACH TO SIMULATION OF LARGE CRYSTALLINE CLUSTERS

In this section, we briefly discuss the generalization of the tensor-based Hartree–Fock solver to the case of large lattice structured and periodic systems[47,48] arising in the numerical modeling of crystalline, metallic and polymer type compounds.

Fast Tensor Calculation of a Lattice Sum of Interaction Potentials

One of the challenges in the numerical treatment of large molecular systems is the summation of long-range potentials allocated on large 3D lattices.[23,52,53,58] The conventional Ewald summation techniques based on a separate evaluation of contributions from the short- and long-range parts of the interaction potential exhibit $O(L^3 \log L)$ complexity scaling for a cubic $L \times L \times L$ 3D lattice.

In the contrary, the main idea of the novel tensor summation method introduced in ref. 47 and 49 suggests to benefit from the low-rank tensor decomposition of the generating kernel approximated on the fine n × n × n representation grid in the 3D computational box Ω_L. This allows to completely decouple the 3D sum into the three independent 1D summations, thus reducing drastically the numerical expenses. The resultant potential sum, which now requires only O(n) storage and O(nL) computational demands, is represented by a few assembled canonical/Tucker n-vectors of complicated shape (see Fig. 15), where n is the univariate grid size for a cubic 3D lattice.

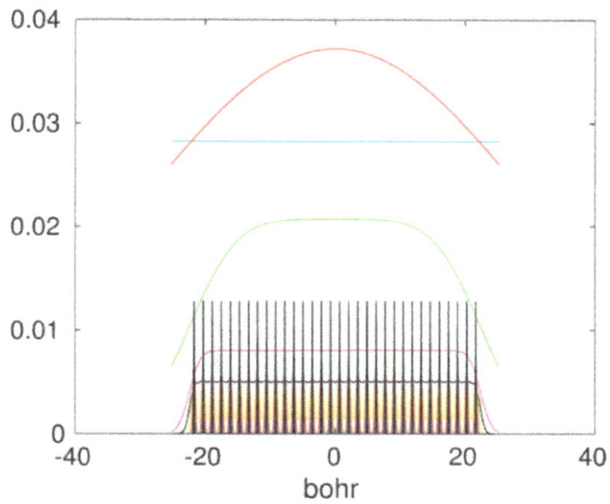

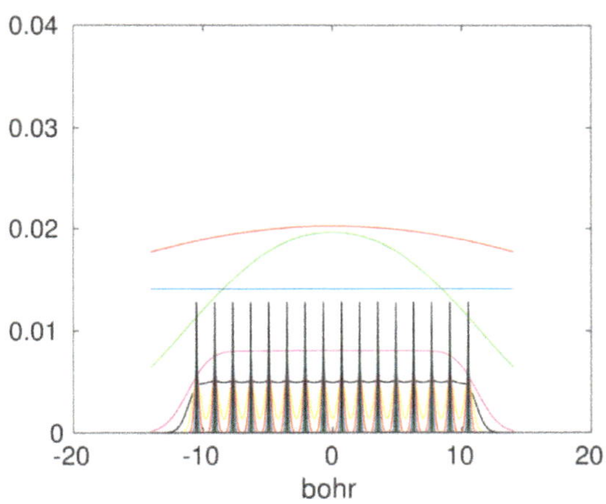

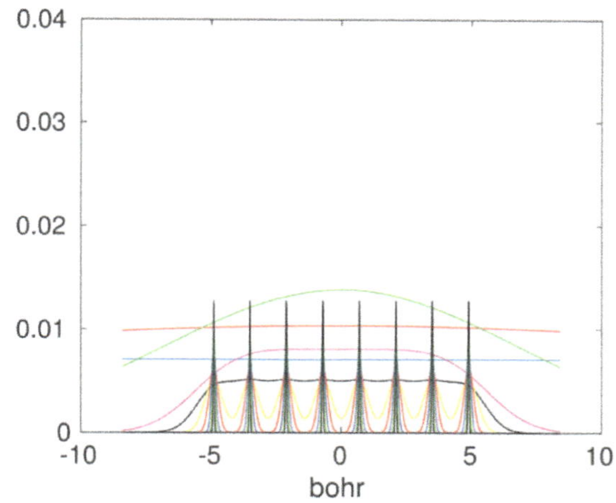

Figure 15: Assembled x-, y- and z-axis canonical vectors for a cluster of 32 × 16 × 8 hydrogen atoms.

For ease of exposition, we consider the electrostatic nuclear potential of a single hydrogen atom, $V_c(x) = \frac{Z}{\| x \|}$. Define the scaled unit cell, $\Omega_0 = [0,b]^3$, of size b × b × b and consider a sum of a finite number of interaction potentials in a symmetric computational box

$$\Omega_L = B \times B \times B, \quad \text{with} \quad B = b\left[-\frac{L}{2}, \frac{L}{2}\right], \quad L = 2L_0 \in \mathbb{N},$$

consisting of a union of L × L × L unit cells $\Omega_{\mathbf{k}}$, obtained from Ω_0 by a shift along the lattice vector b**k**, where **k** = $(k_1, k_2, k_3) \in Z^3$, such that kl ∈ K: = $K_- \cup K_+$ for l = 1, 2, 3 with $\mathcal{K}_- := \left\{-1, \ldots, -\frac{L}{2}\right\}$ and $\mathcal{K}_+ := \left\{0, 1, \ldots, \frac{L}{2} - 1\right\}$. Hence, we have $\Omega_L = \cup_{k1,k2,k3 \in K} \Omega_{\mathbf{k}} \in R^3$.

Recall that b = nh, where h > 0 is the fine mesh size that is the same for all spatial variables, and n is the number of grid points for each variable. We also define the accompanying domain $\widetilde{\Omega}_L$ obtained by scaling of Ω_L with the factor of 2, $\widetilde{\Omega}_L = 2\Omega_L$, and, similar to (31), introduce the respective rank-R reference (master) tensor

$$\widetilde{\mathbf{P}} = \sum_{q=1}^{R} \widetilde{\mathbf{p}}_q^{(1)} \otimes \widetilde{\mathbf{p}}_q^{(2)} \otimes \widetilde{\mathbf{p}}_q^{(3)} \in \mathbb{R}^{2n \times 2n \times 2n}, \tag{19}$$

approximating $\frac{1}{\| x \|}$ in $\widetilde{\Omega}_L$ on a 2n × 2n × 2n representation grid with mesh size h.

Let us consider a sum of single Coulomb potentials on a L × L × L lattice,

$$V_{c_L}(x) = \sum_{k_1,k_2,k_3 \in \mathcal{K}} \frac{Z_1}{\|x - a_1(k_1,k_2,k_3)\|}, \quad x \in \Omega_L \in \mathbb{R}^3. \tag{20}$$

The assembled tensor approach applies to the potentials defined on n × n × n 3D Cartesian grid. It reduces the sum over a rectangular 3D lattice, $\mathbf{P}_{cL} \in R^{n \times n \times n}$,

$$\mathbf{P}_{c_L} = \sum_{\mathbf{k} \in \mathcal{K}^3} \mathcal{W}_{\nu(\mathbf{k})} \widetilde{\mathbf{P}} = \sum_{k_1,k_2,k_3 \in \mathcal{K}} \sum_{q=1}^{R} \mathcal{W}_{(\mathbf{k})} \left(\widetilde{\mathbf{p}}_q^{(1)} \otimes \widetilde{\mathbf{p}}_q^{(2)} \otimes \widetilde{\mathbf{p}}_q^{(3)} \right),$$

to the summation of directional vectors for the canonical decomposition of shifted single Newton kernels,[47]

$$\mathbf{P}_{c_L} = \sum_{q=1}^{R} \left(\sum_{k_1 \in \mathcal{K}} \mathcal{W}_{(k_1)} \widetilde{\mathbf{p}}_q^{(1)} \right) \otimes \left(\sum_{k_2 \in \mathcal{K}} \mathcal{W}_{(k_2)} \widetilde{\mathbf{p}}_q^{(2)} \right) \otimes \left(\sum_{k_3 \in \mathcal{K}} \mathcal{W}_{(k_3)} \widetilde{\mathbf{p}}_q^{(3)} \right), \tag{21}$$

where $W_{\nu(\mathbf{k})} = W_{k1} \otimes W_{k2} \otimes W_{k3}$ is the shift-and-windowing (onto Ω_L) separable transform along the **k**-grid. Remarkably that the rank of the resulting sum is the same as for the R-term canonical reference tensor (19) representing the single Newton kernel.

The numerical cost and storage size are bounded by O(RLn) and O(Rn), respectively, where $n = n_0 L$, and n_0 is the grid size in the unit cell.

Fig. 15 presents the assembled canonical vectors for the cluster of 32 × 16 × 8 Hydrogen nuclei, with a distance 1.4 bohr between nuclei. The resulting potential is shown in Fig. 16. Size of the computational box is 50.4 × 28.0 × 16.8 bohr^3. The corresponding 3D Cartesian grid size is 9216 × 5120 × 3072. Here the empty interval between the lattice and the boundary of the computational box equals to 3 bohr.

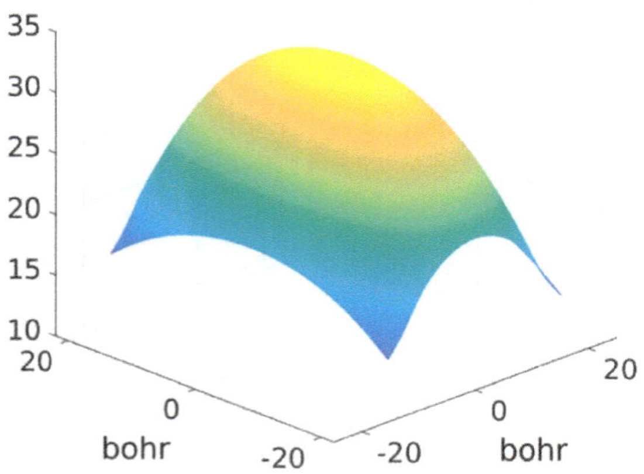

Figure 16: Assembled canonical sum of the Newton potentials for the cluster of 32 × 16 × 8 hydrogen atoms.

The next table represents CPU times for the lattice summation of the Newton kernels over L × L × L cubic box, with very fine n × n × n representation grid.

L^3	4096	32 768	262 144	2 097 152
Time (s)	1.8	0.8	3.1	15.8
3D grid size, n^3	5632^3	9728^3	$17\,920^3$	$34\,304^3$

The summation method in the canonical format was extended to the Tucker tensors which allows the principal generalization of this techniques to the case of rather complicated lattices with defects (Theorem 3.2, ref. 49), so that the resulting sum takes a form

$$\mathbf{T}_{c_L} = \sum_{m_1=1}^{r_1} \sum_{m_2=1}^{r_2} \sum_{m_3=1}^{r_3} b_{m_1,m_2,m_3} \left(\sum_{k_1 \in \mathcal{K}} \mathcal{W}_{(k_1)} \widetilde{\mathbf{t}}^{(1)}_{m_1} \right),$$

$$\otimes \left(\sum_{k_2 \in \mathcal{K}} \mathcal{W}_{(k_2)} \widetilde{\mathbf{t}}^{(2)}_{m_2} \right) \otimes \left(\sum_{k_3 \in \mathcal{K}} \mathcal{W}_{(k_3)} \widetilde{\mathbf{t}}^{(3)}_{m_3} \right)$$

where $\widetilde{\mathbf{t}}^{(\ell)}_{m_\ell}$, l = 1, 2, 3, represents the Tucker vectors of the rank-**r** master tensor approximating the Newton kernel $\frac{1}{\|x\|}$ in $\widetilde{\Omega}_L$ on a 2n × 2n × 2n representation grid.

In the case of defected lattices the increasing rank of the final sum of Tucker tensors can be reduced by the stable ALS based algorithms applicable to the Tucker tensors (see ref. 48). The particular examples of the lattice geometries suitable for our approach are presented in Fig. 17, see ref. 48 for more detailed discussion.

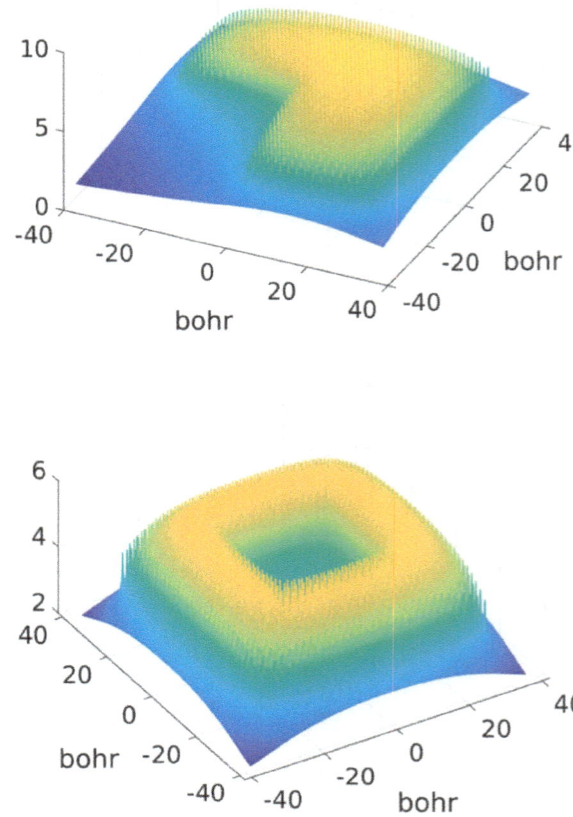

Figure 17: Assembled canonical sum of the Coulomb potentials on the L-shaped (top) and O-shaped (bottom) sub-lattices of a 32 × 32 × 1 lattice.

For the reduced Hartree–Fock equation, where the Fock operator is confined to the core Hamiltonian, the tensor-structured block-circulant representation of the Fock matrix was introduced[48] that allows the special low-rank approximation of the matrix blocks. This opens the way for the numerical treatment of large eigenvalue problems with structured matrices arising in the solution of the Hartree–Fock equation for large crystalline systems with defects.

Interaction Energy of the Coulomb Lattice Sums

Given the nuclear charges $\{Z_k\}$, centered at points x_k, $\mathbf{k} \in K^3$, located on a finite L × L × L lattice $\mathcal{L}_L = \{x_k\}$ with the step-size b, the interaction energy of the total electrostatic potential of these charges is defined by the lattice sum

$$E_L = \frac{1}{2} \sum_{\mathbf{k},\mathbf{j} \in \mathcal{K}, \mathbf{k} \neq \mathbf{j}} \frac{Z_{\mathbf{k}} Z_{\mathbf{j}}}{\|x_{\mathbf{j}} - x_{\mathbf{k}}\|}, \quad \textit{i.e. } \text{for } \|x_{\mathbf{j}} - x_{\mathbf{k}}\| \geq b. \tag{22}$$

The fast and accurate computation of electrostatic interaction energy (as well as the related forces and stresses) is one of the difficult tasks in computer modeling of macromolecular structures like finite crystals, and biological systems.

The tensor summation scheme (21) can be directly applied to this computational problem. In what follows we show that (22) can be treated as a particular case of the previous scheme served for calculation of (20) on a fine spacial grid. For this discussion, we assume that all charges are equal, $Z_{\mathbf{k}} = Z$.

First, notice that the rank-R reference tensor $h^{-3}\tilde{\mathbf{P}}$ defined in (19) approximates with high accuracy $O(h^2)$ the (and its shifted version) Coulomb potential $\frac{1}{\|x\|}$ in $\widetilde{\Omega}_L$ (for $\|x\| \geq b$ that is required for the energy expression) on the fine $2n \times 2n \times 2n$ representation grid with mesh size h. Likewise, the tensor $h^{-3}\mathbf{P}_{cL}$ approximates the potential sum $V_{cL}(x)$ on the same fine representation grid including the lattice points $x_{\mathbf{k}}$.

We propose to evaluate the energy expression (22) by using tensor sums as in (21), but now applied to a small sub-tensor of the rank-R canonical reference tensor $\tilde{\mathbf{P}}$, that is $\tilde{\mathbf{P}}_L := [\tilde{\mathbf{P}}_{|x\mathbf{k}}] \in \mathbb{R}^{2L\times 2L\times 2L}$, obtained by tracing of $\tilde{\mathbf{P}}$ at the accompanying lattice of the double size $2L \times 2L \times 2L$, i.e. $\widetilde{\mathscr{L}}_L = \{x_{\mathbf{k}}\} \cup \{x_{\mathbf{k}'}\} \in \widetilde{\Omega}_L$. Here $\tilde{\mathbf{P}}_{|x\mathbf{k}}$ denotes the tensor entry corresponding to the **k**-th lattice point designating the atomic center $x_{\mathbf{k}}$. We are interested in the computation of the rank-R tensor $\widehat{\mathbf{P}}_{c_L} = [\mathbf{P}_{c_L|x_{\mathbf{k}}}]_{\mathbf{k}\in\mathscr{K}} \in \mathbb{R}^{L\times L\times L}$, where $\mathbf{P}_{c_L}|x_{\mathbf{k}}$ denotes the tensor entry corresponding to the **k**-th lattice point on $\mathscr{L}_L$. The tensor $\widehat{\mathbf{P}}_{c_L}$ can be computed at the expense $O(L^2)$ by

$$\widehat{\mathbf{P}}_{c_L} = \sum_{q=1}^{R}\left(\sum_{k_1\in\mathscr{K}} \mathscr{W}_{(k_1)}\widetilde{\mathbf{p}}^{(1)}_{L,q} \otimes \sum_{k_2\in\mathscr{K}} \mathscr{W}_{(k_2)}\widetilde{\mathbf{p}}^{(2)}_{L,q} \otimes \sum_{k_3\in\mathscr{K}} \mathscr{W}_{(k_3)}\widetilde{\mathbf{p}}^{(3)}_{L,q}\right).$$

This leads to the representation of the energy sum (22) with accuracy $O(h^2)$ in a form

$$E_{L,T} = \frac{Z^2 h^{-3}}{2}\left(\left\langle \widehat{\mathbf{P}}_{c_L}, 1\right\rangle - \sum_{\mathbf{k}\in\mathscr{K}} \mathbf{P}_{|x_{\mathbf{k}}=0}\right),$$

where the first term in brackets represents the full canonical tensor lattice sum restricted to the **k**-grid composing the lattice $\mathscr{L}_L$, while the second term introduces the correction at singular points $x_j - x_{\mathbf{k}} = 0$. Here $\mathbf{1} \in \mathbb{R}^{L\times L\times L}$ is the all-ones tensor. By using the rank-1 tensor $\mathbf{P}_{0L} = \mathbf{P}_{|x\mathbf{k}=0}\mathbf{1}$, the correction term can be represented by a simple tensor operation

$$\sum_{\mathbf{k}\in\mathscr{K}} \mathbf{P}_{|x_{\mathbf{k}}=0} = \langle \mathbf{P}_{0L}, \mathbf{1}\rangle.$$

Finally, the interaction energy E_L allows the approximate representation

$$E_L \approx E_{L,T} = \frac{Z^2 h^{-3}}{2}\left(\left\langle \widehat{\mathbf{P}}_{c_L}, \mathbf{1}\right\rangle - \langle \mathbf{P}_{0L}, \mathbf{1}\rangle\right), \qquad (23)$$

that can be implemented in $O(L^2) \ll L^3 \log L$ complexity by tensor operations with the rank-R canonical tensors in $\mathbb{R}^{L\times L\times L}$.

Table 4 illustrates the performance of the algorithm described above. We compare the exact value computed by (22) with the approximate tensor representation in (23) computed on the fine representation grid with $n = n_0 L$, $n_0 = 128$. We consider the lattice systems composed of Hydrogen atoms with interatomic distance 2.0 bohr. The geometric size of the largest 3D lattice with 256^3 potentials is of the order of $(2.0\cdot 256 + 6.0)^3$ bohr^3,§ which is more than 20^3 cubic nanometers. Here T_{full} means CPU time (sec) for direct summation of by (22) of the order of L^6, where the number in brackets shows $\sqrt{T_{\text{full}}}$ of the order of $O(L^3)$.

Table 4 Comparison of times for the full (T_{full}) and tensor-based (T_{tens}) calculation of the interaction energy sum for the lattice electrostatic potentials. Matlab on a 2 Intel Xeon Hexa-Core/2677

L^3	T_{full}, $(O(L^3))$	T_{tens}	$E_{L,T}$	abs. err.
24^3	37, (6.1)	1.2	3.7×10^6	2×10^{-8}
32^3	250, (15.8)	1.5	1.5×10^7	1.5×10^{-9}
48^3	3374, (58.8)	2.8	1.12×10^8	0
64^3	—	5.7	5.0×10^8	—
128^3	—	13.5	1.6×10^{10}	—
256^3	—	68.2	5.2×10^{11}	—

The presented approach for fast calculation of the interaction energy can extended to the case of non-uniform rectangular lattices and, under certain assumptions, to the case of non-equal nuclear charges $Z_{\mathbf{k}}$. Moreover, it applies to many other types of spherically symmetric interaction potentials, for example, to shielded Coulomb interaction or van der Waals attraction sums corresponding to the distance function $\|x\|^{-2}$ and $\|x\|^{-6}$, respectively.

CONCLUSIONS

The goal of this paper is to attract interest of the specialists in computational quantum chemistry to recent results and open questions of the grid-based tensor approach in electronic structure calculations. Here we focus mostly on the description of main mathematical and algorithmic aspects of the tensor decomposition schemes and demonstrate their benefits in some applications.

The scope of applications which can be regarded as consistent, ranges from the Hartree–Fock energy estimates for moderate size molecules, including calculation of TEI tensor with incorporated algebraic density fitting, to calculation of the excitation energies for molecules, and up to a unique superfast method for calculating the lattice potential sums and the interaction energy of long range potentials on a lattice in a finite volume. Tensor approach allows to treat above problems using moderate computational facilities. All numerics given in the paper presents implementations in Matlab.

The described numerical tools are not restricted to the applications presented here, but can be applied to various hard computational problems in (post) Hartree–Fock calculations related to accurate evaluation of multidimensional integrals, and efficient storage and manipulations with large multivariate data arrays.

The presented method for summation of long-range potentials with sub-linear computational cost does not have analogues in what is used so far in computational quantum chemistry and can have a good future. For example, it can be useful in modeling of large finite molecular structures like nanostructures or quantum dots, where the periodic approach may be inconsistent. Calculating a sum of several millions of lattice potentials on fine 3D grids takes only several seconds in Matlab on a laptop.[47] This method gives also the unique possibility to present the summed 3D lattice potential in the whole computational region with very high accuracy. Integration and differentiation of this 3D potential can be easily performed on representation grid due to 1D computational costs.

Tensor approach is now being evolved for modeling the electronic structure of finite crystalline-type molecular systems. We hope that tensor numerical methods will have a good future in solving challenging multidimensional problems of computational quantum chemistry.

REFERENCES

1. J. Almlöf, in Direct methods in electronic structure theory, ed. D. R. Yarkony, Modern Electronic Structure Theory, World Scientific, Singapore, 1995, vol. II, pp. 110–151.
2. M. Bachmayr, Adaptive low-rank wavelet methods and applications to two-electron Schrödinger equations, PhD dissertation, RWTH Aachen, 2012.
3. P. Baudin, J. Marin, I. G. Cuesta and A. M. S. de Meras, Calculation of excitation energies from the CC2 linear response theory using Cholesky decomposition, J. Chem. Phys., 2014,140, 104111.
4. P. Benner, H. Faßbender and M. Stoll, Solving Large-Scale Quadratic Eigenvalue Problems with Hamiltonian Eigenstructure using a Structure-Preserving Krylov Subspace Method, Electronic Transactions on Numerical Analysis, 2008, 29, 212–229.
5. P. Benner and T. Mach, The LR Cholesky Algorithm for Symmetric Hierarchical Matrices, Linear Algebra and its Applications, 2013, 439(4), 1150–1166.
6. P. Benner, V. Khoromskaia and B. N. Khoromskij, A reduced basis approach for calculation of the Bethe–Salpeter excitation energies using low-rank tensor factorizations, 2015, Preprint arXiv:1505.02696.
7. C. Bertoglio and B. N. Khoromskij, Low-rank quadrature-based tensor approximation of the Galerkin projected Newton/Yukawa kernels, Comput. Phys. Commun., 2012, 183(4), 904–912.
8. F. A. Bischoff, R. J. Harrison and E. F. Valeev, Computing many-body wave functions with guaranteed precision: The first-order Møller-Plesset wave function for the ground state of helium atom, J. Chem. Phys., 2012, 137, 104103.
9. S. F. Boys, G. B. Cook, C. M. Reeves and I. Shavitt, Automatic Fundamental Calculations of Molecular Structure, Nature, 1956, 178, 1207–1209.
10. D. Braess, Asymptotics for the Approximation of Wave Functions by Exponential-Sums, J. Approx. Theory, 1995, 83, 93–103.
11. M. E. Casida, in Recent Advances in Density Functional Methods, Part I, ed. D. P. Chong, World Scientific, Singapoure, 1995, p. 155.
12. A. Cichocki and Sh. Amari, Adaptive Blind Signal and Image Processing: Learning Algorithms and Applications, Wiley, 2002.
13. W. Dahmen, S. Proessdorf and R. Schneider, Multiscale methods for pseudo-differential equations on manifolds, Wavelet analysis and its applications, 5, Academic Press, 1995.
14. S. Dolgov, Tensor product methods in numerical simulation of high-dimensional dynamical problems, PhD thesis, University of Leipzig, 2014, http://nbn-resolving.de/urn:nbn:de:bsz:15-qucosa-151129.
15. S. Dolgov, B. N. Khoromskij, D. Savostyanov and I. Oseledets, Computation of extreme eigenvalues in higher dimensions using block tensor train format, Comput. Phys. Commun., 2014, 185(4), 1207–1216.
16. L. De Lathauwer, B. De Moor and J. Vandewalle, A multilinear singular value decomposition, SIAM J. Matrix Anal. Appl., 2000, 21, 1253–1278.
17. T. H. Dunning Jr, Gaussian basis sets for use in correlated molecular calculations. I. The atoms boron through neon and hydrogen, J. Chem. Phys., 1989, 90, 1007–1023.
18. A. Durdek, S. R. Jensen, J. Juselius, P. Wind, T. Flå and L. Frediani, Adaptive order polynomial algorithm in a multi-wavelet representation scheme, Appl. Num. Math., 2014, 92, 40–53.
19. P. P. Ewald, Die Berechnung optische und elektrostatischer Gitterpotentiale, Ann. Phys., 1921, 64, 253.
20. L. Frediani, E. Fossgaard, T. Flå and K. Ruud, Fully adaptive algorithms for multivariate integral equations using the non-standard form and multiwavelets with applications to the Poisson and bound-state Helmholtz

kernels in three dimensions, Mol. Phys., 2013, 111(9–11), 1143–1160.

21. L. Genovese, A. Neelov, S. Goedecker, T. Deutsch, S. A. Ghasemi, A. Willand, D. Caliste, O. Zilberberg, M. Rayson, A. Bergman and R. Schneider, Daubechies wavelets as a basis set for density functional pseudopotential calculations, J. Chem. Phys., 2008, 129, 014109.
22. L. Grasedyck, D. Kressner and C. Tobler, A literature survey of low-rank tensor approximation techniques, 2013, arXiv:1302.7121v1.
23. L. Greengard and V. Rochlin, A fast algorithm for particle simulations, J. Comput. Phys., 1987, 73, 325.
24. M. Griebel and J. Hamaekers, Sparse grids for the Schroedinger equation, M2AN, 2007, 41, 215–247.
25. W. Hackbusch and B. N. Khoromskij, Low-rank Kronecker product approximation to multi-dimensional nonlocal operators, Part I. Separable approximation of multi-variate functions, Computing, 2006, 76, 177–20.
26. W. Hackbusch and S. Kühn, A new scheme for the tensor representation, J. Fourier Anal. Appl., 2009, 15, 706–722.
27. W. Hackbusch and R. Schneider, in Tensor Spaces and Hierarchical Tensor Representations, ed. S. Dahlke and W. Dahmenet al., Lecture Notes in Computer Science and Engineering, 102, Springer, 2014.
28. R. J. Harrison, G. I. Fann, T. Yanai, Z. Gan and G. Beylkin, Multiresolution quantum chemistry: Basic theory and initial applications, J. Chem. Phys., 2004, 121(23), 11587–11598.
29. T. Helgaker, P. Jørgensen and J. Olsen, Molecular Electronic-Structure Theory, Wiley, New York, 1999.
30. N. Higham, in Analysis of the Cholesky decomposition of a semi-definite matrix, ed. M. G. Cox and S. J. Hammarling, Reliable Numerical Computations, Oxford University Press, Oxford, 1990, pp. 161–185.
31. F. L. Hitchcock, The expression of a tensor or a polyadic as a sum of products, J. Math. Phys., 1927, 6, 164–189.
32. E. G. Hohenstein, R. M. Parrish and T. J. Martinez, Tensor hypercontraction density fitting, Quartic scaling second- and third-order Møller-Plesset perturbation theory, J. Chem. Phys., 2012, 137, 044103.
33. V. Kazeev and B. N. Khoromskij, Explicit low-rank QTT representation of Laplace operator and its inverse, SIAM J. Matrix Anal. Appl., 2012, 33(3), 742–758.
34. B. N. Khoromskij, O(d log N)-Quantics Approximation of N-d Tensors in High-Dimensional Numerical Modeling, J. Const. Approx., 2011, 34(2), 257–289, Preprint 55/2009, MPI MiS, Leipzig 2009, http://www.mis.mpg.de/publications/preprints/2009/prepr2009-55.html.
35. B. N. Khoromskij, Fast and Accurate Tensor Approximation of Multivariate Convolution with Linear Scaling in Dimension, J. Comput. Appl. Math., 2010, 234, 3122–3139.
36. B. N. Khoromskij, Structured Rank-($r_1,\ldots,r_d$) Decomposition of Function-related Tensors in R^d, Comput. Method. Appl. Math., 2006, 6(2), 194–220.
37. B. N. Khoromskij, in Tensor Numerical Methods for High-dimensional PDEs: Basic Theory and Initial Applications, ESAIM: Proceedings and Surveys, ed. N. Champagnat, T. Leliévre and A. Nouy, 2015, vol. 48, pp. 1–28, DOI:10.1051/proc/201448001.
38. B. N. Khoromskij and V. Khoromskaia, Low Rank Tucker-Type Tensor Approximation to Classical Potentials, Cent. Eur. J. Math., 2007, 5(3), 523–550.
39. B. N. Khoromskij and V. Khoromskaia, Multigrid Accelerated Tensor Approximation of Function Related Multi-dimensional Arrays, SIAM J. Sci. Comput., 2009, 31(4), 3002–3026.
40. V. Khoromskaia, Computation of the Hartree–Fock Exchange in the Tensor-structured Format, Comput. Method. Appl. Math., 2010, 10(2), 1–16.
41. V. Khoromskaia, Numerical Solution of the Hartree–Fock Equation by Multilevel Tensor-structured methods, PhD dissertation, TU Berlin, 2010, http://opus.kobv.de/tuberlin/volltexte/2011/2948/.
42. B. N. Khoromskij, V. Khoromskaia and H.-J. Flad, Numerical Solution of the Hartree–Fock Equation in Multilevel Tensor-structured Format, SIAM J. Sci. Comput., 2011, 33(1), 45–65.
43. V. Khoromskaia, D. Andrae and B. N. Khoromskij, Fast and accurate 3D tensor calculation of the Fock operator in a general basis, Comput. Phys. Commun., 2012, 183, 2392–2404.
44. V. Khoromskaia, B. N. Khoromskij and R. Schneider, Tensor-structured factorized calculation of two-electron integrals in a general basis, SIAM J. Sci. Comput., 2013, 35(2), A987–A1010.
45. V. Khoromskaia and B. N. Khoromskij, Møller-Plesset (MP2) Energy Correction Using Tensor Factorizations of the Grid-based Two-electron Integrals, Comput. Phys. Commun., 2014, 185, 2–10.
46. V. Khoromskaia, Black Box Hartree–Fock Solver by the Tensor Numerical Methods, Comput. Method. Appl. Math., 2014, 14(1), 89–111.
47. V. Khoromskaia and B. N. Khoromskij, Grid-based lattice summation of electrostatic potentials by assembled canonical tensors, Comput. Phys. Commun., 2014, 185, 3162–3174.
48. V. Khoromskaia and B. N. Khoromskij, Tensor Approach to Linearized Hartree–Fock Equation for Lattice-type and Periodic Systems, 2014, Preprint 62/2014, MIS MPI, arXiv:1408.3839, submitted.
49. V. Khoromskaia and B. N. Khoromskij, Tucker tensor method for fast grid-based summation of long-range potentials on 3D lattices with defects, 2014, Preprint arXiv:14.1994.
50. T. G. Kolda and B. W. Bader, Tensor Decompositions and Applications, SIAM Rev., 2009, 51(3), 455–500.
51. D. Kressner, Numerical methods and software for general and structured eigenvalue problems, PhD dissertation, TU Berlin, 2004.
52. K. N. Kudin and G. E. Scuseria, Revisiting infinite lattice sums with the periodic Fast Multipole Method, J. Chem. Phys., 2004, 121, 2886–2890.

53. S. A. Losilla, D. Sundholm and J. Juselius, The direct approach to gravitation and electrostatics method for periodic systems, J. Chem. Phys., 2010, 132(2), 024102.

54. Ch. Lubich, On Variational Approximations in Quantum Molecular Dynamics, Math. Comput., 2005, 74, 765–779.

55. L. Lin, Y. Saad and C. Yang, Approximating spectral densities of large matrices, 2014, Preprint arXiv:1308.5467v2.

56. I. V. Oseledets, Approximation of $2^d \times 2^d$ matrices using tensor decomposition, SIAM J. Matrix Anal. Appl., 2010, 31(4), 2130–2145.

57. I. V. Oseledets and E. E. Tyrtyshnikov, Breaking the Curse of Dimensionality, or How to Use SVD in Many Dimensions, SIAM J. Sci. Comput., 2009, 31(5), 3744–3759.

58. C. Pisani, M. Schütz, S. Casassa, D. Usvyat, L. Maschio, M. Lorenz and A. Erba, CRYSCOR: a program for the post-Hartree–Fock treatment of periodic systems, Phys. Chem. Chem. Phys., 2012, 14, 7615–7628.

59. R. Parrish, E. G. Hohenstein, T. J. Martinez and C. D. Sherrill, Discrete variable representation in electronic structure theory: Quadrature grids for least-squares tensor hypercontraction, J. Chem. Phys., 2013, 138, 194107.

60. P. Pulay, Improved SCF convergence acceleration, J. Comput. Chem., 1982, 3, 556–560.

61. E. Ribolini, J. Toulouse and A. Savin, in Electronic excitation energies of molecular systems from the Bethe–Salpeter equation: Example of H_2 molecule, ed. S. Ghosh and P. Chattaraj, Concepts and Methods in Modern Theoretical Chemistry, Electronic Structure and Reactivity, 2013, vol. 1, p. 367.

62. U. Schollwöck, The density-matrix renormalization group in the age of matrix product states, Ann. Phys., 2011, 326(1), 96–192.

63. H. Sekino, Y. Maeda, T. Yanai and R. J. Harrison, Basis set limit Hartree Fock and density functional theory response property evaluation by multiresolution multiwavelet basis, J. Chem. Phys., 2008, 129, 034111.

64. A. Smilde, R. Bro and P. Geladi, Multi-way Analysis, Wiley, 2004.

65. F. Stenger, Numerical methods based on Sinc and analytic functions, Springer-Verlag, Heidelberg, 1993.

66. D. Sundholm, P. Pyykkö and L. Laaksonen, Two-dimensional Fully Numerical Molecular Calculations. X. Hartree–Fock Results for He_2, Li_{12}, Be_2, HF, OH^-, N_2, CO, BF, NO^+, and CN^-, Mol. Phys., 1985, 56, 1411–1418 .

67. A. Szabo ans N. Ostlund, Modern Quantum Chemistry, Dover Publication, New York, 1996.

68. L. R. Tucker, Some mathematical notes on three-mode factor analysis, Psychometrika, 1966, 31, 279–311.

69. F. Verstraete, D. Porras and J. I. Cirac, DMRG and periodic boundary conditions: A quantum information perspective, Phys. Rev. Lett., 2004, 93(22), 227205.

70. G. Vidal, Efficient classical simulation of slightly entangled quantum computations, Phys. Rev. Lett., 2003, 91(14), 147902.

71. H. Wang and M. Thoss, Multilayer formulation of the multiconfiguration time-dependent Hartree theory, J. Chem. Phys., 2003, 119, 1289–1299.

72. H.-J. Werner and P. J. Knowleset al., MOLPRO, version 2002.10, a package of ab initio programs for electronic structure calculations.

73. S. R. White, Density-matrix algorithms for quantum renormalization groups, Phys. Rev. B: Condens. Matter Mater. Phys., 1993, 48(14), 10345–10356.

74. H. Yserentant, Regularity and Approximability of Electronic Wave Functions, Lecture Notes in Mathematics series, Springer-Verlag, 2010

CHAPTER 2

APPLICATION OF HARTREE-FOCK METHOD FOR MODELING OF BIOACTIVE MOLECULES USING SAR AND QSPR

Cleydson B. R. Santos[1,2,3], Cleison C. Lobato[1], Francinaldo S. Braga[1], Sílvia S. S. Morais[4], Cesar F. Santos[1], Caio P. Fernandes[1,3], Davi S. B. Brasil[1,5], Lorane I. S. Hage-Melim[1], Williams J. C. Macêdo[1,2], José C. T. Carvalho[1,2,3]

[1]Laboratory of Modeling and Computational Chemistry, Federal University of Amapá, Macapá, Brazil

[2]Postgraduate Program in Biotechnology and Biodiversity-Network BIONORTE, Campus Universitário Marco Zero, Macapá, Brazil

[3]Laboratory of Drug Research, School of Pharmaceutical Sciences, Federal University of Amapá, Macapá, Brazil

[4]Laboratory of Statistical and Computational Modeling, University of the State of Amapá, Macapá, Brazil

[5]Institute of Technology, Federal University of Pará, Belém, Brazil

ABSTRACT

The central importance of quantum chemistry is to obtain solutions of the Schrödinger equation for the accurate determination of the properties of atomic and molecular systems that occurred from the calculation of wave functions accurate for many diatomic and polyatomic molecules, using Self Consistent Field method (SCF). The application of quantum chemical methods in the study and planning of bioactive compounds has become a common practice nowadays. From the point of view of planning it is important to note, when it comes to the use of molecular modeling, a collective term that refers to methods and theoretical modeling and computational techniques to mimic the behavior of molecules, not intend to reach a bioactive molecule simply through the use of computer programs. The choice of method for energy minimization depends on factors related to the size of the molecule, parameters of availability, stored data and computational resources. Molecular models generated by the computer are the result of mathematical equations that estimate the positions and properties of the electrons and nuclei, the calculations exploit experimentally, the characteristics of a structure, providing a new perspective on the molecule. In this work we show that studies of Highest Occupied Molecular Orbital Energy (HOMO), Low Unoccupied Molecular Orbital Energy (LUMO) and Map of molecular electrostatic potential (MEP) using HatreeFock method with different basis sets (HF/3-21G*, HF/3-21G**, HF/6-31G, HF/6-31G*, HF/ 6-31G** and HF/6-311G), that are of great importance in modern chemistry, biochemistry, molecular biology, and other fields of knowledge of health sciences. In order to obtain a significant correlation, it is essential that the descriptors are used appropriately. Thus, the quantum chemical calculations are an attractive source of new molecular descriptors that can, in principle, express all the geometrical and electronic properties of molecules and their interactions with biological receptor.

INTRODUCTION

Brazil has the largest genetic diversity in plant species in the world, however it is estimated that less than 10% were evaluated as their biological characteristics and less than 5% were submitted the detailed phytochemical studies. Despite a recent increase in research in this area, plants still constitute a source relatively underused and potentially very valuable for discovery of new biologically active substances [1] [2] . The bioactive compounds present in the vegetable

Citation: Santos, C. , Lobato, C. , Braga, F. , Morais, S. , Santos, C. , Fernandes, C. , Brasil, D. , Hage-Melim, L. , Macêdo, W. and Carvalho, J. (2014) Application of Hartree-Fock Method for Modeling of Bioactive Molecules Using SAR and QSPR. Computational Molecular Bioscience, 4, 1-24. doi: 10.4236/cmb.2014.41001.

kingdom have important functions and biological actions and may be considered as promoters of human health. We have already recognized the association between intake of fruits and vegetables and decreased risk of development various disorder chronic degenerative, such as cancer, inflammation, cardiovascular disease, cataracts, macular degeneration and other being carotenoids and phenolic compounds, some of the groups of bioactive compounds to which are attributed such actions. The seeds and extracts of annatto (Bixa orellana L.) are used as colorants in food, pharmaceutical and cosmetic industries due to the predominant presence of the carotenoid bixin [3]. The efficiency of biological activity of bioactive compounds of vegetable origin depends on their structure and concentration. In turn, the amount of these substances in vegetables is largely influenced by genetic factors and environmental conditions, in addition of the degree of maturation and plant variety, among other things. It is known also that the biological capacity is influenced by the substrate used in the assay, by the solvent and by extraction technique used, as well as the binomial time-temperature. Regarding the organic solvents, methanol, by managing to extract high amount of bioactive compounds, it has been touted as the most effective [4] -[7] .

The Montrichardia linifera (Arruda) Schott, Araceae family, popularly known as "aninga", is an aquatic macrophyte that forms large clonal populations along the rivers and streams of the Amazon. Those bordering consider this poisonous plant because its sap causes skin burns and eye contact can cause blindness [8] . However, paradoxically, it is widely used in traditional Amazonian medicine, mainly due to the healing properties of sap and of the juice of this plant, which has already been mentioned in the literature since the nineteenth century for the treatment of wounds and ulcers [9] -[12] , leading to the hypothesis that this species may contain biologically active substances. However, very little is known about their chemical composition and biological activities that eventually may be useful also against human infections caused by parasites, which are serious problems in tropical and subtropical in developing countries, despite the discovery of new anti-protozoal drugs [13] . Between antimalarial compounds isolated from plants, artemisinin is one of the most important discoveries nowadays [14] - [16]. Artemisinin (or Qinghaosu, QHS, Figure 1) represents the most relevant advance in the treatment of malarial disease for the last 20 years [17]. Artemisinin is a sesquiterpene lactone with an endoperoxide group, which has been used in traditional Chinese medicine for many centuries as a natural product for fever and malarial treatment. This drug was isolated by Chinese chemists in the early 1970s from the ancient Artemisia annua L. Nowadays, artemisinin and derivatives are widely used around the world because of their potent antimalarial activity, fast action, and low toxicity. As a result, artemisinin and its derivatives have become recognized as a new generation of antimalarial drugs [18].

Pinheiro, Ferreira and Romero (2001) techniques combined quantum chemical (Hartree-Fock 3-21G) and multivariate analyzes methods (PCA, HCA, KNN and SIMCA) to study and propose diidroartemisinin derivatives. Through the technique PCA and HCA, seven (7) descriptors that were responsible for the classification of compounds into two distinct classes were selected, and with construction of qualitative models KNN and SIMCA two (2) compounds of a set of twelve (12) test predicted as of high activity were proposed [19] .

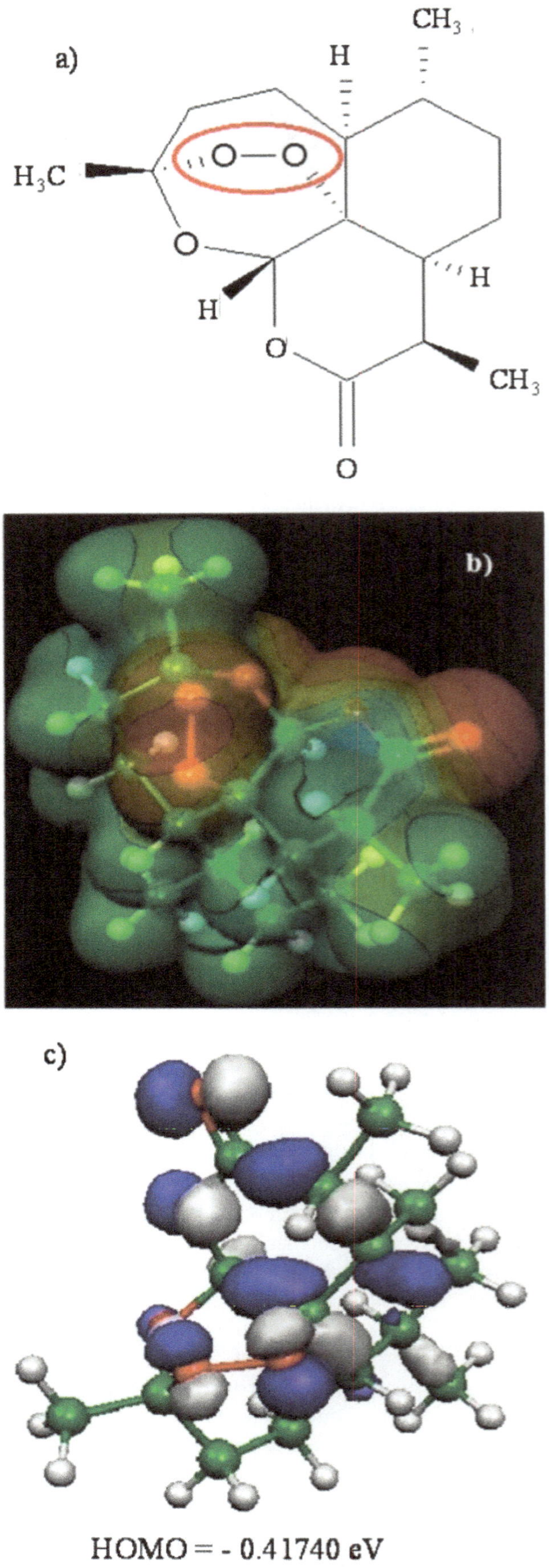

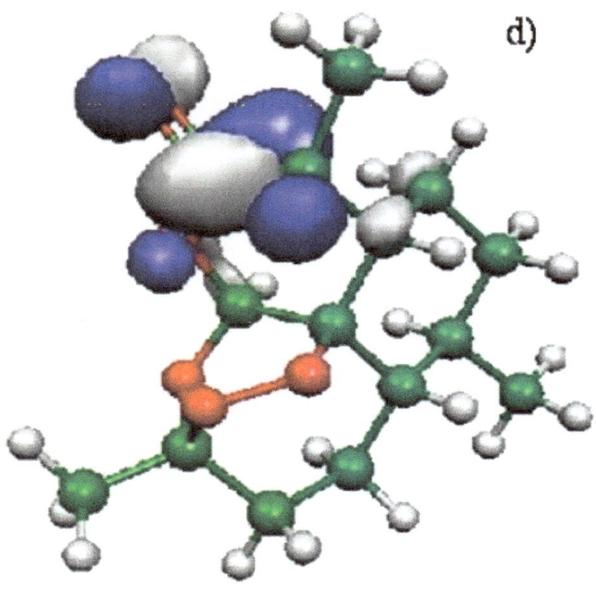

Figure 1. (a) The artemisinin structure and the region essential for expression of the biological activity (pharmacophore) was visualized using ChemSketch 12.00 program [23]. (b) Map of molecular electrostatic potential (MEP). ((c) and (d)) Orbital's energy Homo and Lumo. The MEP, Homo and Lumo were visualized and calculated using HartreeFock (HF) method and HF/6-31G** basis set by the Molekel program [24].

Artemisinin derivatives with antimalarial activity against Plasmodium falciparum, which is resistant to mefloquine, were studied using quantum chemical methods (HF/6-31G*) and the partial least-squares (PLS) method. Three main components explained 89.55% of the total variance, with $Q^2 = 0.83$ and $R^2 = 0.92$. From a set of 10 proposed artemisinin derivatives (artemisinin derivatives with unknown antimalarial activity against Plasmodium falciparum), a novel compound was produced with superior antimalarial activity compared with the compounds previously described in the literature [20]. Cardoso et al. (2008) studied artemisinin and some of its derivatives with activity against D-6 strains of Plasmodium falciparum using the HF/3-21G method. To verify the reliability of the geometry obtained, Cardoso et al. compared the structural parameters of the artemisinin trioxane ring with theoretical and experimental values from the literature. MEP was used in an attempt to identify key features of the compounds that are necessary for their activities, and they use those to propose new artemisinin derivatives [21]. There is however, the need for discovery of new antimalarial drugs using techniques of quantum chemistry which has been extensively used in the pharmaceutical industry, where predicting the probable activity of a drug from molecular orbital calculations is much less expensive than manufacturing it in order to perform expensive tests [22].

THE IMPORTANCE AND APPLICATION OF THE METHODS OF QUANTUM CHEMISTRY TO MOLECULAR MODELING

The central importance of quantum chemistry is to obtain solutions of the Schrödinger equation for the accurate determination of the properties of atomic and molecular systems. Within the area of physical chemistry, quantum chemistry is used to calculate various thermodynamic properties such as entropy, heat capacity of gases, for the interpretation of molecular spectra, calculations of bond length, bond angles, dipole moment and the understanding of intermolecular forces. In organic quantum chemistry can estimate the stability of molecules, calculate the properties of intermediate reactions, reproduce the aromaticity of organic compound and simulate spectra of nuclear magnetic resonance (NMR). In analytical chemistry is used for the interpretation of intensities of the spectral lines. In the area of inorganic chemistry is used for the ligand field theory, where predictions are made, and properties justify complex ions of transition metals. In Biochemistry of quantum chemistry calculations are used to study the conformation of biological molecules such as enzyme-substrate binding and solvation of organic molecules [25] -[28] .

The application of quantum chemical methods in the study and planning of bioactive compounds has become a common practice nowadays. From the point of view of planning, it is important to note that, when speaking of the use of molecular modeling (a collective term that refers to theoretical methods and computational techniques to model or mimic the behavior of molecules), it is not intended to reach a bioactive molecule simply through the use of computer programs. The development process of these molecules due to their complexity, necessarily involves multidisciplinary approach, which employs a large set of computational methods of systematic way form facilitate and optimize the development process of bioactive compounds, on a constant exchange of information with groups of chemical synthesis and evaluation of the activity of these compounds. These computational methods can be used as tools of rational design of bioactive compounds, so called because it is guided by a rational hypothesis about the mechanism of action of these compounds. The action of bioactive molecules is a very complex phenomenon, but one of the paradigms of medicinal chemistry is that these molecules are related to their effects or interactions to chemical reactions with macromolecular structures present in living systems, proteins, in its large majority. Cases these proteins are cell receptors, bioactive molecules are classified as receptor agonists or antagonists, in the case of enzymes, these molecules act as enzyme inhibitors [29] [30] .

The Origin of Quantum Chemistry Methods

The history and debates related to the so called "semiempirical method" of London-Eyring-Polanyi are reported on a fascinating work of Nye [31]. The approach of Eyring and Polanyi, aiming to merge theory with experimental results to construct potential energy surfaces, showed that it is possible to gain insights into the mechanisms of adiabatic reactions, leading to important concepts related to the dynamics of chemical reactions, such as transition state and activated complex. Already at that time, there were debates comparing the approaches based on first principles with the semiempirical. In 1933 James and Coolidge [32] published an ab initio calculation when obtained the binding energy of H_2 molecule with an accuracy of 98%. This calculation took a year to be done, but its success increased confidence of authors that then criticized the semiempirical method as a "happy cancellation of errors" that occurs for not taking into account "terms of considerable importance" [32].

This expression "cancellation of errors" remains to this day in the imagination of researchers, although it is not possible to identify precisely that are these errors, and mainly, how and why they if would cancel. Polanyi in 1937 counter-argued: "Personally, I do not attach importance to an exact agreement between theory and experiment at this stage, but I believe that the theory [referring to the semiempirical method] can claim to provide a reasonable description of the mechanism of reactions chemical that would otherwise remain obscured" [33] .

In 1941 Hirschfelder presented the calculation of at least one hundred activation energies for several reactions using the method of London-Eyring-Polanyi, showing that only in some cases the values did not agree with the experimental, with an error of 10 kcal/mol [34] . Hirschfelder defended the semiempirical method as "a method sufficiently flexible so that it could be made consistent with any set of chemical facts, maintaining consistent with the basic principles of quantum mechanics". According to Nye [31] , in this statement Hirschfelder has correctly identified that an ad hoc character is inherent to semi-empirical approaches. This ad hoc character involves risk, since it may well be possible to force an agreement between the semi-empirical and experimental model, even if the model is based on an erroneous theory. Such risk is even greater if the model is used for extrapolation. The final triumph of the first principles approach to the reaction $H + H_2$ ® $H_2 + H$, occur only came in 2003 when Mielke and colleagues [35] presented a new constants of experimental and theoretical speeds. The theoretical calculations were obtained using quantum dynamics in an exact adiabatic potential surface including Born-Oppenheimer corrections. The experiment and theory now agree in the range of 167 K to 2112 K, within experimental error, causing the problem can be solved now.

The "semiempirical" expression was first used in theoretical chemistry in 1931 by Michael Polanyi (1891-1976) and Henry Eyring (1901-1981) [36] [37] in their attempt to combine thermodynamics, chemical kinetics, quantum mechanics and the theory of binding electrons valence. In these 82 years, semiempirical method of London-Eyring-Polanyi to construct potential energy surfaces, provided a work tool useful to get insights about how the physical and chemical processes occur, and to motivate the development of new techniques and experiments to study the temporary combination of atoms called transition state as in the works of Ahmed Zewail. The novelty of the visual maps of potential energy surfaces and the language of potential wells and barriers of activation has also become an important pedagogical tool for chemistry. In contrast, first-principles calculations of quantum dynamics, with equivalent accuracy to that of Mielke and colleagues [35] , remain prohibitive, almost impossible to be performed for the vast majority of chemical reactions of practical interest. Undoubtedly: the ab initio methods, based entirely on first principles eventually prevail. However, if the reaction of the hydrogen atom with a hydrogen molecule could be regarded as definitely settled 10 years ago, what about all the other chemical and biochemical reactions? [38]

HARTREE-FOCK EQUATIONS

A development of great importance in quantum chemistry occurred from calculation functions accurate wave diatomic and for many polyatomic molecules, by the method of self-consistent field developed by Douglas Hartree [39] . In his theory Hartree considered the wave function as being formed by an antisymmetric linear combination of products of spin-orbitals. Then the wave function Hartree-Fock for atoms or molecules obeying the Pauli exclusion principle, should be written as a product antisimetrizado spin-orbitals, called Slater determinant, that in the formula restricted and normalized to a system of closed layers containing 2m electrons, is given by:

$$\Psi(1,2,2m) = \frac{1}{\sqrt{2m!}} \begin{vmatrix} \Phi_1(1) & \bar{\Phi}_1(1) & \cdots & \bar{\Phi}_m(1) \\ \Phi_1(2) & \bar{\Phi}_1(2) & \cdots & \bar{\Phi}_m(2) \\ \vdots & \vdots & \ddots & \vdots \\ \Phi_1(2n) & \bar{\Phi}_1(2m) & \cdots & \bar{\Phi}_m(2m) \end{vmatrix} \tag{1}$$

in which each spin-orbital F_i (x,a) is a product of a spatial function j_i (x) and a spin function a or b

$$\Phi_i(x,\alpha) = \varphi_i(x)\alpha(i), \bar{\Phi}_i(x,\beta) = \varphi_i(x)\beta(i) \tag{2}$$

The functions of an electron j_i (x) are known as "orbitals", a term proposed by Mulliken [40], which is the quantum mechanical analogue of the classical orbit [41] . It is considered that the orbitals are orthonormal

$$\langle \varphi_i(1) | \varphi_j(1) \rangle = \delta_{ij}, \delta_{ij} = \begin{cases} 0 \text{ se } i \neq j \\ 1 \text{ se } i = j \end{cases} \tag{3}$$

All elements forming a given column of Slater determinant involve the same spin-orbital already all elements in the same row involve the same electrons. When the configuration of an atom or molecule layer is closed this is represented by a single determinant, unlike a configuration layer to open, since in this case it is used a sum of Slater determinant. The analysis made the following refers the systems configurations closed layer. Open layer settings will not be discussed here, viewed the complexity of the formulas. Aiming to show a simple format for the expressions, it is used the system of atomic units in the equations.

The expression refers to the total electronic energy Hartree-Fock E for a system of electrons n nuclei in fundamental state is given by the theorem of the variational method.

$$E = \frac{\langle \Psi | \hat{H} | \Psi \rangle}{\langle \Psi | \Psi \rangle} \tag{4}$$

where Y comes to be wave function in the form of Slater determinant and $\hat{H}$ is the Hamiltonian operator without considering the coordinates spin system and non-relativistic. In the case of molecules that in our case the operator $\hat{H}$ is the Hamiltonian purely electronic which is obtained after the separation of Born-Oppernheimer [42] that written in atomic units has the form:

$$\hat{H} = -\frac{1}{2}\sum_i \nabla_i^2 - \sum_\alpha \sum_i \frac{Z_A}{r_{\alpha i}} + \sum_{i<} \sum_j \frac{1}{r_{ij}} \tag{5}$$

The first term in Equation (5) is the operator for the kinetic energy of the electrons, the second term comes to be potential energy of attraction between electrons and nuclei, and the last term refers to the potential energy of repulsion between the electrons, Z_A is the charge of the nucleus A, r_{ai} is the distance of the electron i the nucleus a and r_{ij} is the distance of the electron i to the electron j.

The operator $\hat{H}$ can be separated into two other operators

$$\hat{H} = \hat{H}_1 + \hat{H}_2 \tag{6}$$

The operator $\hat{H}_1$ is relative to the kinetic energy of the electrons and the potential energy of interaction of the electrons with the nuclei:

$$\hat{H}_1 = \sum_{i}^{2m} \hat{h}(i) \tag{7}$$

The operator $\hat{h}(i)$ is the corresponding Hamiltonian to the motion of an electron in the field generated by the nuclei only

$$\hat{h}(i) = -\frac{1}{2}\nabla_i^2 - \sum_{\alpha} \frac{Z_A}{r_{\alpha j}} \tag{8}$$

The operator $\hat{H}_2$ is associated with the interelectronic repulsion

$$\hat{H}_2 = \sum_{i} \sum_{< j} \frac{1}{r_{ij}} \tag{9}$$

Utilizing the wave function (1) and the Hamiltonian (5) in the expression for the energy functional (4), has been as a result the expression for the total electronic energy of a system represented by a wave function in the form of Slater determinant.

$$E = 2\sum_{i=1} h_{ii} + \sum_{i} \sum_{j} \left(2J_{ij} - K_{ij}\right) \tag{10}$$

This relationship has reduced the integration of many electrons the set of three dimensional integrals (h_{ij}) and six (J_{ij}) and (K_{ij}), written of following form:

$$h_{ii} = \left\langle \varphi_i(1) \middle| \hat{h}(1) \middle| \varphi_i(1) \right\rangle \tag{11}$$

$$J_{ij} = \left\langle \varphi_i(1)\varphi_j(2) \middle| \frac{1}{r_{12}} \middle| \varphi_i(1)\varphi_j(2) \right\rangle \tag{12}$$

$$K_{ij} = \left\langle \varphi_i(1)\varphi_j(2) \middle| \frac{1}{r_{12}} \middle| \varphi_j(1)\varphi_i(2) \right\rangle \tag{13}$$

The integral of an electron h_{ii} represents the sum of the kinetic energy of an electron in the orbitalj$_i$(l) and its potential integral due to the action of the nuclei. The Coulomb integrals J_{ij} and of exchange K_{ij} are associated to interaction between an electron in the orbital j$_i$(x) and another in the orbital j$_i$(x). In classical mechanics the integral Coulomb is the interaction energy between two distributions of charges, already integral exchange does not have analogue in classical mechanics, and their presence arises due to anti-symmetry of the wave function [41]. Varying E in Equation (4) for each orbital and making dE = 0 while keeping orthonormality of the wave function arrives at the 1-electron equations that define the Hartree-Fock method

$$\hat{F}(1)\varphi_i(1) = \varepsilon_i\varphi_i(1), i = 1, 2, \cdots, m \tag{14}$$

where Fock operator $\hat{F}(1)$ is given by

$$\hat{F}(1) = \hat{h}(1) + \sum_{j}^{m}\left[2\hat{J}_i(1) - \hat{K}_j(1)\right] \tag{15}$$

where $\hat{J}_i$ and $\hat{K}_j$ are respectively operators of Coulomb and exchange

$$\hat{J}_i(1)\varphi_i(1) = \left[\int \varphi_j^*(2)\frac{1}{r_{12}}\varphi_j(2)\,\mathrm{d}\tau_2\right]\varphi_i(1) \tag{16}$$

$$\hat{K}_j(1)\varphi_i(1) = \left[\int \varphi_j^*(2)\frac{1}{r_{12}}\varphi_i(2)\,\mathrm{d}\tau_2\right]\varphi_j(1) \tag{17}$$

The Hartree-Fock equations are solved in general by an interactive procedure (SCF). At the end of the procedure, ε_i are the eigenvalues monoeletrônicos HF system. Each ε_i is often called orbital energy and is interpreted as being the energy of an electron in the orbital j$_i$, resulting from its kinetic energy, energy of attraction with the nuclei and their energy exchange and repulsion due to all other electrons in their charge density $|\varphi_i|^2$.

The Equation (14) provides

$$\varepsilon_i = \left\langle \varphi_i(1) \middle| F(1) \middle| \varphi_i(1) \right\rangle \tag{18}$$

Substituting Equation (15) in (18) we can relate the orbital energies with the integrals (11), (12) and (13).

$$\varepsilon_i = h_{ii} + \sum_{j=1}^{m}\left(2J_{ij} - K_{ij}\right) \tag{19}$$

It may be also find the total electron energy E from the orbital energies. But this total energy is not simply equal the sum of the energies of an electron. This fact is due to the sum of the energies of an electron including twice each electron-electron interaction, i.e., repulsion between electrons 1 and 2 contributes to the energy of an electron associated with both electrons. Thus it follows that the second term in the equation below corrects this problem

$$E = 2\sum_{i}^{m}\varepsilon_i + \sum_{i}^{m}\sum_{j}^{m}\left(2J_{ij} - J_{ij}\right) \tag{20}$$

Thus substituting (19) in (20) have

$$E = \sum_{i=1}^{m}\left[2h_{ii} + \sum_{j=1}^{m}\left(2J_{ij} - K_{ij}\right)\right] \tag{21}$$

which equals

$$E = \sum_{i=1}^{m}\left(\varepsilon_i - h_{ii}\right) \tag{22}$$

with h_{ii} being the same Equation (11).

Hartree-Fock-Roothaan Method

The problem of solving the Hartree-Fock equations for atoms and molecules is due to the fact of absence of central symmetry. Therefore it is necessary to use approximations for the best orbital. Thus for systems containing many electrons, an approximated way of revolver the Hartree-Fock equations consists of expand the orbital Hartree-Fock j$_i$(x) on a linear combination of K basis functions c_m, in accordance proposed by Roothaan [43] , and this method denominated of Hartree-Fock-Roothaan (HFR) or also Molecular-Orbital Linear Combination Atomic Orbitals (LCAO-MO). Thus the orbital can be expanded in a linear combination as follows.

$$\varphi_i = \sum_{\mu}^{K} c_{\mu i} \chi_{\mu} \tag{23}$$

where the c_{mi} are the coefficient of expansion, which will be treated as variational parameters and basis functions c_m are Slater type atomic orbitals or Gaussian type. To accurately represent the orbitals, functions should form a complete set. However, this requires an infinite number of such functions. What you should use is actually a finite number of basis functions. Orbitals should obey the orthonormality condition. Otherwise you can make a linear transformation thereby making them orthonormal

$$\sum_{\mu\nu} c^*_{\mu i} c_{\mu j} S_{\mu\nu} = \delta_{ij} \tag{24}$$

where S_{mv} is the overlap integral

$$S_{\mu\nu} = \langle \chi_\mu(i) | \chi_\nu(i) \rangle \tag{25}$$

For the molecules j_i (x) are molecular orbitals and c_m are the atomic orbitals. Thus there is a significant improvement in computational calculations when the orbital functions are expanded in terms of a finite set of basis functions. The integro-differential equations are then transformed into algebraic equations for the expansion coefficients [41]. The total electronic energy is obtained when Equation (23) is replaced in (21).

$$E = 2\sum_{i}^{ocup} \sum_{\mu\nu} c^*_{\mu i} c_{\nu i} h_{\mu\nu} + \sum_{ij}^{ocup} \sum_{\mu\nu\lambda\sigma} c^*_{\mu i} c^*_{\lambda j} c_{\nu i} c_{\sigma j} \left[2\langle \mu\lambda | \nu\sigma \rangle - \langle \mu\nu | \lambda\sigma \rangle \right] \tag{26}$$

where

$$h_{\mu\nu} = \langle \chi_\mu(1) | \hat{h}(1) | \chi_\nu(1) \rangle \tag{27}$$

with $\hat{h}(1)$ given by Equation (8) by replacing i by 1.

The integrals $\langle \mu\lambda | \nu\sigma \rangle$ and $\langle \mu\nu | \lambda\sigma \rangle$ are integral interaction two electrons are represented by

$$\langle \mu\lambda | \nu\sigma \rangle = \int\int \chi^*_\mu(1) \chi^*_\lambda(2) \left| \frac{1}{r_{12}} \right| \chi_\nu(1) \chi_\sigma(2) d\tau_1 d\tau_2 \tag{28}$$

$$\langle \mu\nu | \lambda\sigma \rangle = \int\int \chi^*_\mu(1) \chi^*_\nu(2) \left| \frac{1}{r_{12}} \right| \chi_\lambda(1) \chi_\sigma(2) d\tau_1 d\tau_2 \tag{29}$$

The best coefficients c_{mi} are determined by varying the electron energy given by Equation (26) with respect to them, obeying the condition of orthonormality. Thus we arrive at the equations of Hartree-Fock-Roothaan

$$\sum_{\nu} c_{\nu i} \left\{ h_{\mu\nu} + \sum_{j}^{ocup} \sum_{\nu\sigma} c^*_{\nu j} c_{\sigma j} \left[2\langle \mu\lambda | \nu\sigma \rangle - \langle \mu\sigma | \lambda\sigma \rangle \right] \right\} = \sum_{j} \sum_{\nu} \varepsilon_{ij} c_{\nu j} S_{\mu\nu} \tag{30}$$

where the operator Fock defined as

$$F_{\mu\nu} = h_{\mu\nu} + \sum_{j}^{ocup} \sum_{\nu\sigma} c_{\nu j} c_{\sigma j} \left[2\langle \mu\lambda | \nu\sigma \rangle - \langle \mu\nu | \lambda\sigma \rangle \right] \tag{31}$$

We can replace the Equation (31) in (30) and thus obtain the equations of Hartree-Fock-Roothaan the following notation

$$\sum_{\nu} c_{\nu i} F_{\mu\nu} - \sum_{j} \sum_{\nu} \varepsilon_{ij} c_{\nu j} S_{\mu\nu} = 0, \mu = 1, 2, \cdots, k \tag{32}$$

Applying a unitary transform on Equation (32) to diagonalize the matrix ε, has been

$$\sum_{\nu} (F_{\mu\nu} - \varepsilon_i S_{\mu\nu}) c_{\nu i} = 0, \mu = 1, 2, \cdots, k \tag{33}$$

Equation (33) can be written in matrix form in this way has been

$$FC = SCE \tag{34}$$

The equations of Hartree-Fock-Roothaan are resolved in the same way that the equations of Hartree-Fock, i.e., by an interactive process (SCF-MO-LCAO). Those obtained at the end of the procedure are the eigenvalues of HFR system.

Hartree-Fock Limit

In Hartree-Fock method, is used a electronic wave function composed only by a Slater determinant, thereby providing only approximate description of exact wave function because this cannot be described by a single Slater determinant. The exact solution of the Schrödinger equation would not be obtained in case the spatial orbitals j_i (x) be expanded into a linear combination of basis functions c_m, although the larger and more complete is the set of those functions greater the degree of flexibility in expansion for spin-orbital and lowest expected value for energy. Larger bases sets decrease the HF energy up to a certain limit. This limit is the lowest that can be obtained from a wave function of a determinant and it is called of Hartree-Fock limit. Even so, as a variational method, this limit energy E_{HF} will still be above the exact nonrelativistic energy E_{ex} (NR) due to the energy of electron correlation E_{corr}

$$E_{ex}(NR) = E_{HF} + E_{corr} \quad (35)$$

The SCF method is a valid approach, but generates errors in energy because it describes of approximate mode the interactions between the electrons. Should be also consider the instantaneous interactions between the electrons. The moves of the electrons are correlated with each other, i.e., there is a correlation in the different positions of the electron and must be taken into account.

The exact value of the energy E including electron correlation and relativistic effects E_{rel} is given by:

$$E = E_{HF} + E_{corr} + E_{rel} \quad (36)$$

APPLICATION OF BASIS SETS SEPARATE VALENCE USED IN THE MOLECULAR MODELING

To perform atomic and molecular calculations, determines a basis set formed by mathematical functions. These bases consist of a linear combination of the wave functions of an electron in terms of a finite number of base functions, which contains a set of parameters to be optimized. The denomination of the base sets depends on various types of basis functions, of the number of those employed functions in the expansion of monoeletrônicas functions (orbital) and of characteristics of the parameters to be optimized. Thus it is of fundamental importance the careful choice these functions when you want to get accurate results. The main basis sets separate valence available in computational chemistry programs for performing calculations of molecular properties are presented below:

Slater Type Functions

The atomic systems containing only one electron, hydrogenic atoms, have as a solution of the Schrödinger nonrelativistic equation, a function of type

$$\Psi_{nlm}(r,\theta,\varphi) = R_{nl}(r)Y_{lm}^{(\theta,\varphi)} \quad (37)$$

where n, l and m are the principal quantum numbers, orbital angular momentum and magnetic, R_{nl} (r)is the radial part and $Y_{lm}^{(\theta,\varphi)}$ is the angular part denominated spherical harmonics. The radial part has the form

$$R_{nl}(r) = N_{nl}\left(\frac{2Zr}{na}\right)^{l}\exp\left(-\frac{Zr}{na}\right)L_{n+1}^{2l+1}\left(\frac{2Zr}{na}\right) \quad (38)$$

where N_{nl} comes to be a normalization factor, Z is nuclear charge, r is radius, a is a constant and $L_{n+1}^{2l+1}\left(\frac{2Zr}{na}\right)$ represents in a generic form a polynomial called associated Laguerre polynomial. The angular part is equal to

$$Y_{lm}^{(\theta,\varphi)} = N_{lm}P_l^{(m)}(\cos\theta)e^{im\phi} \quad (39)$$

being N_{lm} a normalization factor and $P_l^{(m)}$ the associated Legendre polynomial.

For more that hydrogenic orbitals are orthogonal they do not form a complete set of continuous functions. They have limited application because many of the required integrals in molecular orbital calculations are somewhat hard to calculated, especially when you have high values of the principal quantum number because of the complexity of the polynomial in r. Then Slater (1930) proposed a form analytical simpler for the radial function, then introducing functions such as Slater (Slater Type Functions-STFs) [42].

$$R_{nl}(r) = \frac{[2\zeta/a_0]^{n+1/2}}{[(2n)!]^{1/2}}r^{n-1}\exp(-\zeta r/a_0) \quad (40)$$

Since the orbital exponent is written as

$$\zeta = \frac{Z-s}{n^*} \quad (41)$$

where s is a constant related to the shielding effect of the electrons of the internal layers of atoms and n^* is an effective principal quantum number. Therefore the general shape of Slater type functions can be written as follows, which is similar to the equation.

$$\Psi_{nlm}(\zeta,r,\theta,\varphi) = [(2n)!]^{1/2}(2\zeta)^{n+1/2}r^{n-1}e^{-\zeta P}Y_{lm}^{(\theta,\varphi)} \quad (42)$$

Slater evaluated through empirical rules to select of parameters s and n^*, thereby aiming at a good approximation for the atomic orbitals best of this type. The values of z were determined by the variational method by Clementi (1963 and 1967) for neutral atoms to the ruthenium in their fundamental states, using for this the SCF method [44] [45]. The exponents z are positive numbers and adjustable in the methods of calculation. These exponents determine the size of the orbital, thereby large exponents characterize dense orbital and small exponents characterize diffuse orbitals.

The so-called Slater type orbitals (STOs) are formed by the product of the angular part of Equation (39), by the product of the radial part of Equation (40). The STOs generate reasonable representations of atomic orbitals. However, as they replace the polynomial in r in the hydrogenic orbitals by a simple power r, do not have the proper number of radial nodes, just not well represent the internal part of an orbital.

One of the limitations is that they are not of mode some orthogonal, as much as it can be corrected using a set of STOs orthogonalized. In addition the use of these orbitals in molecular SCF calculations makes the integrals multicenter, that involve interactions between electrons, becomes somewhat difficult to be solved numerically, thereby increasing the computational time.

Minimum Base Sets

Minimum base sets or base "single-zeta" is set containing a single function to represent each occupied atomic orbital of different quantum numbers n and l of electron configuration. Thus, this set presents a reduced computational time to make these calculations, so can be used in calculations involving large molecules which are our case. In the calculations where a minimum base set is used are used in general Slater type orbitals. Due to the small size of the base, minimum base sets generate results only form qualitative of the properties. Nevertheless the calculations of the electronic structure were until around 1960 all performed in terms of minimum base. The first semiempirical methods more used were based in a set of minimum base STOs [46] [47].

Gaussian Type Functions

Contracted Gaussian Functions

The use of STOs in electronic structure calculations generates computational problems due to the emergence of multicenter integrals, although these orbitals well describe the functional behavior of molecular orbitals. In order to simplify the calculation of multicenter integrals, Boys (1950) proposed for calculations involving molecules using Gaussian type functions (GTFs) [48]

$$R_{nl}(r) = \left[\frac{2^{2n+3/2}}{(2n-1)!\sqrt{\pi}}\right]^{1/2} \alpha^{2n+1/4} r^{n-1} \exp\left(-\alpha r^2\right) \tag{43}$$

where n and l are the main quantum numbers and orbital angular momentum and a is the orbital exponent which is a variational parameter. The use of GTFs in calculations of electronic interaction integrals presents an enormous advantage, because the product two of these functions entries in different centers is equivalent to a single function centered on a new center [49] , thereby there is a reduction of multicenter integrals to integrals calculated in terms of functions centered on the same point.

The functions of the type Gaussian have a poor behavior of atomic orbitals in relation to Slater type functions, the reason is that the GTFs do not have the "cusp" in the region near the nucleus. Thus the GTFs have a behavior different functional that seen for molecular orbitals, so it is necessary to use two (2) to five (5) GTFs to represent adequately each STO. As the number of electronic interaction integrals originated using of a basis set dimension m increases in the order of m^4, fastest speed and simplify the calculations of the electronic interaction integrals in terms of GTFs compensates the large number of integrals to be calculated when comparing using STFs.

The solution of the equations of SCF method in the calculation HFR is also a process using an enormous computational time, it is also proportional to the fourth power of the number of base functions. However, the number of cycles of the interactive process SCF method increases with the number of coefficients to be optimized. Thus the use of GTFs contracted originated from linear combinations of Gaussian primitive is generally more adequate [47] [50] [51] .

The contracted basis functions can be chosen in order to resemble STOs, atomic orbitals HF or any other set of functions. A type of GTFs contracted is set STO-NG. This nomenclature comes being a set of GTFs contracted that describes a STF through N GTFs primitives used in contraction. Each STO is then approximated as a linear combination of N GTFs, in which the coefficients and exponents contracted are chosen so that the basis function approach is an STF. Thus the computational time decreases, but the results generated by the set of contracted GTFs are not good.

The STOs ns and np are approximated by their respective functions

$$\varphi_{ns}(r) = \sum_{k=1}^{m} d_{ns,k} g_{1s}\left(\alpha_{nk,} r\right), n = 1,2,3 \tag{44}$$

$$\varphi_{np}(r) = \sum_{k=1}^{m} d_{np,k} g_{2p}\left(\alpha_{nk,} r\right), n = 1,2,3 \tag{45}$$

where g_{1s} and g_{2p} are GTFs, a_{nk} are gaussian exponents and $d_{ns,k}$ and $d_{np,k}$ are coefficients of the expansion, where the last three parameters are determined by the least squares method [51] .

Basis Sets Separate Valence

In order to obtain better results, many studies have been made aimed at finding basis sets with this capability. In recent years, besides the basis sets previously described, other sets are being used in electronic structure calculations. Among these, there are basis sets separate valence [42].

Basis sets separate valence are extended sets GTFs contracted, where the most usual are listed below: 4-31G, 3-21G, 6-31G e 6-311G. In such sets are used two functions for the valence orbitals and only one for the orbitals of the internal layer, may each of these functions is a linear combination or not primitive gaussian. The occurrence of this is due the internal layer contribute little to the chemical properties of interest. As the functions of the internal layers not duplicate it generates effects on total energy, however represents little it comes of dipole moments, ionization potentials of the valence, charge density, dissociation energy and other chemical properties. For example, the base 4-31G each atomic orbital of the internal layers is described by only one GTF contracted which is formed by the combination of four linear (4) GTFs primitive. For each atomic orbital of the valence shell there are two basic functions, being that of them is a GTF contracted from the linear combination of three (3) GTFs primitive, describing the internal part of valence orbital, since the other is a single GTF primitive, which describes the outer part of the valence orbital. Considering the first row atoms (Li to F) we have:

$$\varphi_{1s}(r) = \sum_{k=1}^{4} d_{k,1s} g_{1s}\left(\alpha_{k,1s}, r\right) \tag{46}$$

$$\varphi'_{2s}(r) = \sum_{k=1}^{3} d'_{k,2s} g_{1s}\left(\alpha'_{k,2sp}, r\right) \tag{47}$$

$$\varphi''_{2s}(r) = g_{1s}\left(\alpha''_{2sp}, r\right) \tag{48}$$

$$\varphi'_{2p}(r) = \sum_{k=1}^{3} d'_{k,2p} g_{2p}\left(\alpha'_{k,2sp}, r\right) \tag{49}$$

$$\varphi''_{2p}(r) = g_{2sp}\left(\alpha''_{2sp}, p\right) \tag{50}$$

For the hydrogen atom, which has no internal layer, it has been:

$$\varphi'_{1s}(r) = \sum_{k=1}^{3} d'_{k,1s} g_{1s}\left(\alpha'_{k,1s}, r\right) \tag{51}$$

where the functions which appear with (') are internal functions while appearing with (") are external functions. Similarly to the base structure previously described is the basis 6-31G. Only now each orbital of internal layers is represented by a GTF contracted originated from linear combination of six (6) GTFs primitive. In the case of the base 6-311G, includes a GTF primitive the set 6-31G to represent a new layer of the outer valence.

Polarization Function

All basis sets discussed so far have a peculiar characteristic, i.e., they comprise functions restricted to be centered in the nuclei. However, there is evidence showing that the atomic orbitals distort or polarize when they form a molecule. For this reason one should take into consideration the possibility of non-uniform displacement of electric charges outside the atomic nucleus, i.e. the polarization. Thus it is possible to obtain a better description of the changes as well as also deformations in the atomic orbitals within the molecule.

One way to consider such a polarization is introduced on the basis in question functions wherein the values of l (quantum number of the orbital angular momentum) larger than those of fundamental state of a given atom. In this type of functions gives the name of polarization functions. For the hydrogen atom of the description fundamental state uses only functions s. The functions p, d, ... centered on H, for molecular calculate, are considered as functions of polarization. Generally, with the inclusion of polarization functions in the molecular basis there is a higher possibility of obtaining better results for many of the properties of chemical interest, such as dissociation energy and dipole moments. In real terms, it is noted that it is not satisfactory to include polarization functions of symmetry d and f together with the basis sets s and p small, that is, the polarization functions should only be added when working with basis sets said saturated [52] .

Of the bases valence separate with polarization functions most commonly used in molecular calculations are the STO-3G*, 3-1G*, 6-31G*, 6-31G**, 6-311G* and 6-311G** [53] . The base 6-31G* and 6-31G** are formed by the inclusion of polarization functions to the base 6-31G, being that the base 6-31G* is constructed by adding a set of five (5) polarization functions of type GTFs of symmetry d to the base 6-31G, for each different atom of hydrogen and helium, already the basis 6-31G** is constructed by adding a set of three (3) polarization functions of type GTFs of symmetry p to the set 6-31G*, for each hydrogen atom.

Diffuse Functions

The use of valence basis set with polarization function in calculations involving anions does not generate good results, because the electron cloud of the anionic systems possess a tendency to expansion. Thus it is necessary to also add appropriate diffuse functions because they allow a greater occupation of space by the orbital region. The importance of diffuse functions in calculations of transition metals is due to the metal atoms present orbital type d and these have diffuse characteristics. So it becomes necessary adding diffuse functions to basis function associated to configuration of the neutral metal atom in order to obtain a better description of the metal complex. The great importance of diffuse functions is due to the fact that they better describe the farthest molecular orbital of the nuclei [54] .

COMPUTATIONAL CHEMISTRY SOFTWARE FOR CALCULATING MOLECULAR PROPERTIES

Several computational chemistry softwares make available calculations of molecular properties through semiempirical and Hatree-Fock methods. We can cite: MOPAC2002 [55] , VAMP [56] , MNDO97 [57] , Gaussian98 [58] , Spartan [59] , GAMESS [60] , Chem3D [61] , CAche [55] , HyperChem [62] , ZINDO [63] , GEOMOS [64] . Most of the great advances obtained in the research of semi-empirical methods were essentially incorporated on a series of programs entitled MOPAC [62]. MOPAC should be understood as an abbreviation of Molecular Orbital PACkage. Initiated in 1983, on group of Prof. M. Dewar, with the participation of Prof. J. J., P. Stewart, the MOPAC project currently has 30 years, and the main landmarks of this story are:

- 1983: MOPAC ® Geometry optimization, vibrational analysis and transition states.
- 1984: MOPAC 2 ® Solution of errors.
- 1985: MOPAC 3 ® AM1 method, calculation of polymers and CI (Configuration interaction).
- 1987: MOPAC 4 ® Solution of errors.
- 1988: MOPAC 5 ® PM3 method and way of reaction.
- 1989: MOPAC 6 ® Repair errors, calculation of electrostatic potential and extent of PM3 for 28 elements.
- 1993: MOPAC 7/93 ® Model solvent, nonlinear optical properties and analysis of symmetry.
- 1997: MOPAC 97 ® MNDO-d method 9) 1999: MOPAC2000 ® Dynamic memory allocation, large molecules, solids, transition metals, improving the search algorithm of the transition state.
- 2001: MOPAC2002 ® Improvement of the package solvate, all atoms of the main group, PM5 method and parameterization for more transition metals.
- 2004: MOPAC2005 ® Calculation of complex Eu(III) [65] and Th(IV) [66] by Sparkle Model.
- 2006: MOPAC 7.1 ® A translation of MOPAC 6 for Fortran 90 is available on the Internet, having incorporated the Sparkle/AM1 model for calculating all trivalent lanthanide ions [67] ;
- 2007: MOPAC2007 ® An application of MOPAC7.1 incorporating new methods of RM1, PM6, Sparkle/AM1 and Sparkle/PM3. This program is available free of charge to the academy and for applications nonprofit [67] .

Several programs two-dimensional of design of molecules are available and easy to use, as ChemWindow, Isis Draw, ChemDraw [68] and Chem3D [69] . They allow the preparation of figures and diagrams with desired quality and accuracy and facilitate the documentation and scientific communication. The software ChemSketch 12.00 [23] is an advanced design that provides chemical molecular properties, optimization and 3D visualization, ability to name the molecules, as IUPAC, and still has a large database of chemical structures and laboratory materials. The software automatically calculates the valence of each atom and restricts the construction of the molecule based on the octet rule, unless instructed to do this restriction. Then is possible to request the construction of 3D spatial form of the species studied, which triggers another window where the academic can rotate tridimensionally the species studied, in addition to observing these species in different visualizations with possibility to visualize bonds and spatial arrangement of species prominently in each of these representations. The design and visualization of 3D drugs, with steric factors relevant to biological activity, are important for analysis of the size, volume and shape of the molecules [70] . The Molekel is a free software multiplatform molecular visualization. It was originally developed at the University of Geneva by Flükiger in the 1990s for Silicon Graphics computers. In 1998, Stefan Portmann took responsibility and released version 3.0. The version 4.0 was almost one version of the platform independent. Other developments lead version 4.3, before Stefan Portman moved and stopped developing the codes. In 2006, the Swiss National Supercomputing Centre (CSCS) restarted the project and version 5.0 was released on December 21 of the same year [24] .

To obtain the conformational analysis and energy minimization, we can cite: the program Chem3D [69] widely used in studies with this objective [70] . However, other programs like Molecular Modeling Pro [71] , ChemSite (ChemSW) [72] , Alchemy, Sybyl, ChemX, cache and WebLab Viewer are also available.

In the area of molecular modeling, graphics construction and projects of drugs, the program Hyperchem [73] for being a tool specializing in 3D structures of interest to the medical, pharmaceutical and organic chemistry. The program lets you design complicated molecules. This software is also an alternative in the field of spectroscopy, which besides the ability to simulate a priori by the NMR spectra quantum methods, contains a database of approximately 10^4 molecules applicable to macromolecules as well as small molecules. The software also includes animations, and quantum chemical calculations and molecular mechanics.

The choice of method for energy minimization depends on factors related to the size of the molecule, parameters of availability and stored data and computational resources. Molecular models generated by the computer are the result of mathematical equations that estimate the positions and properties of the electrons and nuclei, the calculations exploit experimentally, the characteristics of a structure, providing a new perspective on the molecule [74] .

In quantum chemistry the softwares more used are Gaussian and GaussView that uses the laws of quantum mechanics to predict the energies, structures and properties and vibrational frequency of molecular systems [75] . The GaussView 5.0 is a program that can work on Windows and responsible for building the structures under study, by viewing these as well as for generating the input of the species under study for the program calculations–Gaussian 03W. This includes an advanced molecular modeler, which can be used for construction and molecular dimensions of the three test [75] .

The Gaussian 03W is a program that can work on Windows and Linux that performs computations used in the study of reaction mechanisms, equilibrium geometries of neutral molecules, radicals and ions, and the determination of physicochemical parameters. Appreciates structure, reactivity, thermodynamic properties, energy barriers (transition states), conformational analysis, employing the optimization of molecules and theoretical calculations of vibrational spectra. From the optimization is obtained the most appropriate structure to the molecule, whereas the lengths and bond angles and power stabilization calculated by E (RB = HF - LYP) ua [75] .

APPLICATION OF HARTREE-FOCK METHOD USING COMPUTATIONAL TOOLS

Obtaining molecular properties depends of method and basis sets, and represents a means of chemical information contained in the molecular structure of the compound studied. The structure-activity relationships (SAR) represent a core aspect of medicinal chemistry. The fact that a small change in structure leads to a small change in biological activity and allows chemists to rationalize substitutions at specifics positions, giving them the freedom to modify a molecule to improve various properties such as lipophilicity, bioavailability, and so on without sacrificing potency (to a large extent). From the modelers' perspective, the principle of similar structures having similar activities [76] is a cornerstone of quantitative structure-activity relationship (QSAR) modeling [77] [78]. This information is transformed and encoded for lots of problems chemical, pharmacological and toxicological studies on the relationship between structure-activity, quantitative structure-activity and structure-property (SAR, QSAR and QSPR) [79] -[81]

Ferreira et al. [82] studied artemisinin and 18 derivatives with antimalarial activity against W-2 strains of Plasmodium falciparum through quantum chemistry and multivariate analysis. The geometry optimization of structures was realized using the Hartree-Fock method and the 3-21G** basis set. Maps of molecular electrostatic potencial and docking molecular were used to investigate the interaction between the ligands and the receptor (Heme).

Santos et al. [83] performed studies using Hartree-Fock method and the 6-31G** basis set were employed to calculate the molecular properties of artemisinin and 20 derivatives with antimalarial activity. Maps of molecular electrostatic potential (MEPs) and molecular docking were used to investigate the interaction between ligands and the receptor (heme). Principal component analysis and hierarchical cluster analysis were employed to select the most important descriptors related to activity. The correlation between biological activity and molecular properties was obtained using the partial least squares and principal component regression methods. The regression PLS and PCR models built in this study were also used to predict the antimalarial activity of 30 new artemisinin compounds with unknown activity. The models obtained showed not only statistical significance but also predictive ability. The significant molecular descriptors related to the compounds with antimalarial activity were the hydration energy (HE), the charge on the O11 oxygen atom (QO11), the torsion angle O1-O2-Fe-N2 (D2) and the maximum rate of R/

Sanderson Electronegativity (RTe+). These variables led to a physical and structural explanation of the molecular properties that should be selected for when designing new ligands to be used as antimalarial agents.

Case Study on Aspirin

Aspirin introduced in 1899 was one of the first drugs developed and is still one of the most widely used. Estimated 20 billion aspirin tablets are consumed each year in the United States. Originally planned to ease the pain and relieve sore muscles, proved to be a highly complex drug with the power and unexpected limitations. It turned out that it reduces the incidence of heart attacks and is effective in reducing the incidence of Alzheimer's disease and cancer of the digestive tract. At the same time, however, aspirin attacking the stomach lining, causing bleeding or even ulcers, and usually cause intestinal problems [84] . One of the forms of action of aspirin is blocking an enzyme (a type of protein) called COX-2, which promotes inflammation, pain and fever. Unfortunately, it also interferes with COX-1 a correlate enzyme that produces essential hormones to the health of stomach and kidney. An analgesic and anti-inflammatory agent is efficient in COX-1. Figure 2 is shown the structure of aspirin which acts transferring part of its molecule known as acetyl group, for COX-2, disabling it. This drug-receptor interaction is irreversible nature due to the formation of a covalent bond resulting from the nucleophilic attack of the hydroxyl group of the amino acid $serina_{530}$ to the electrophilic acetyl grouping present in aspirin [85] .

A substitute for aspirin (new drug) has to keep this aspect of the molecule, and the replacement should the general format and size of the molecule in such a way it clicks into place in molecular target the same manner that aspirin.

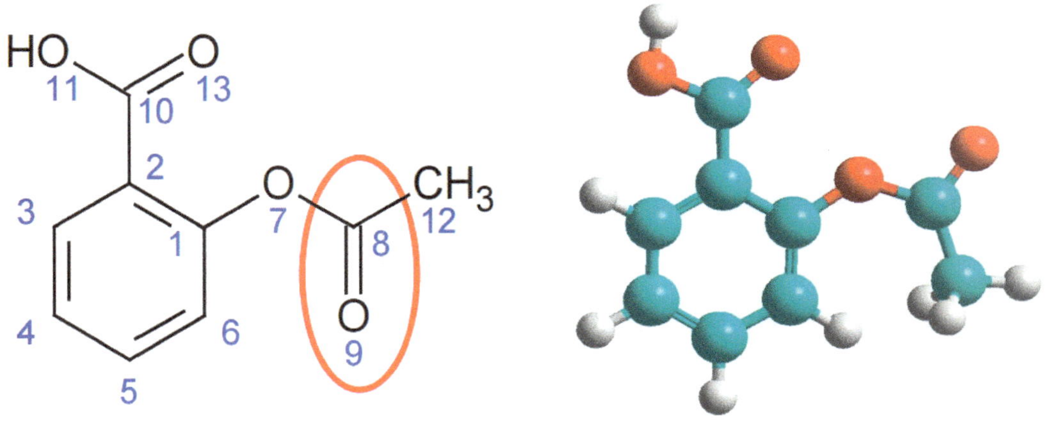

Figure 2. Aspirin (structure) and region essential for the expression of biological (pharmacophore) was visualized using ChemSketch 12.00 program [23] and HyperChem Release 6.02 [73] .

The use of maps of electrostatic potential (MEP) is a computational tool that aids in the process of recognition of one molecule by another, as interactions of types receptor drugs and enzyme substrates, because being by their potential that chemical species interact with other in biological recognition process. The electronic parameters are one of the main factors governing drug-receptor interaction, in this sense; the MEP can be considered an alternative approach in order to understand the electrostatic contribution of this drug and its new derivatives for biological activity [86] . In the construction of the MEP are necessary three steps: construction of the surface density of the molecule, construction of the electrostatic surface potential and applying colors the surface obtained to designate potential values. One of the frequent topics of theoretical chemistry is research to improve methods to elucidate the behavior of molecules and other reactive chemical species. Among the numerous existing reactivity indices the molecular electrostatic potential V(r) that is generated around a molecule by its nuclei and electrons, is known for being a real physical property can be determined experimentally by diffraction methods, as well as computationally [87] .

In Figure 3 are shown the MEPs of aspirin with Hatree-Fock method in different basis sets (HF/3-21G*, HF/ 3-21G**, HF/6-31G, HF/6-31G*, HF/6-31G** and HF/6-311G). In this figure we can observe that in the HF/6-31G** method, showed the lowest positive electrostatic potential (region blue color) of 0.06715 au (atomic unit). However, HF/6-311G method, showed higher positive electrostatic potential 0.07607 au. The variation between them was ±0.00892au (HF/6-311G and HF/6-31G**); Also in this figure we observed that the HF/6-311G method, showed the lowest negative electrostatic potential (region red color) equal to −0.10235 au. However, HF/3-21G** method showed higher negative electrostatic potential of −0.09071 au. The variation between them was ±0.01164 au (HF/6-311G and HF/3-21G**). Thus, the presence of a negative potential surface in carbonyl oxygen atom of the acetyl group is nucleophilic (affinity for positive nuclei), while the carbonyl carbon is electrophilic (electron affinity) in accordance with the literature [85] . Thus, we have shown that the use of different basis sets present values of different electrostatic potentials for the same case study. Therefore, obtaining molecular properties will depend on the method and basis set to reproduce experimental data with greater accuracy. In this case, the MEP of aspirin in different basis set was used to evaluate the key features of aspirin from qualitative comparisons in the region of the acetyl group, which according to Bernardinelli et al. [88] the geometric form of the electrostatic potential is similar for all active compounds. However, new derivatives of aspirin must have some structural similarity in terms of their electrostatic potentials that allow one to be recognized by the other, with similar biological activities [87] [89] [90] .

Structure-activity relationship (SAR) indicates molecular structure modifications that increase the drug effectiveness.

In general, reports show that these modifications are made throughout small changes in the leading compound structure, followed by trials in laboratory to quantify the variations in the biological activity due to changes in the molecular structure [91] .

The quantum-chemical descriptors widely used in SAR, QSAR and QSPR studies are related to the energy of the frontier orbitals (HOMO and LUMO). The reason for this is related to the fact that these properties provide information about the character electron donor and/or electron-acceptor and a compound thus forming a charge transfer complex (CTC) [92] . The energy of Highest Occupied Molecular Orbital Energy (HOMO) and Lowest Unoccupied Molecular Orbital Energy (LUMO) are quantum-chemical descriptors, which play an important role in chemical reactions and the formation of many complex charge transfer [93] . In Figure 4 is shown the fron-

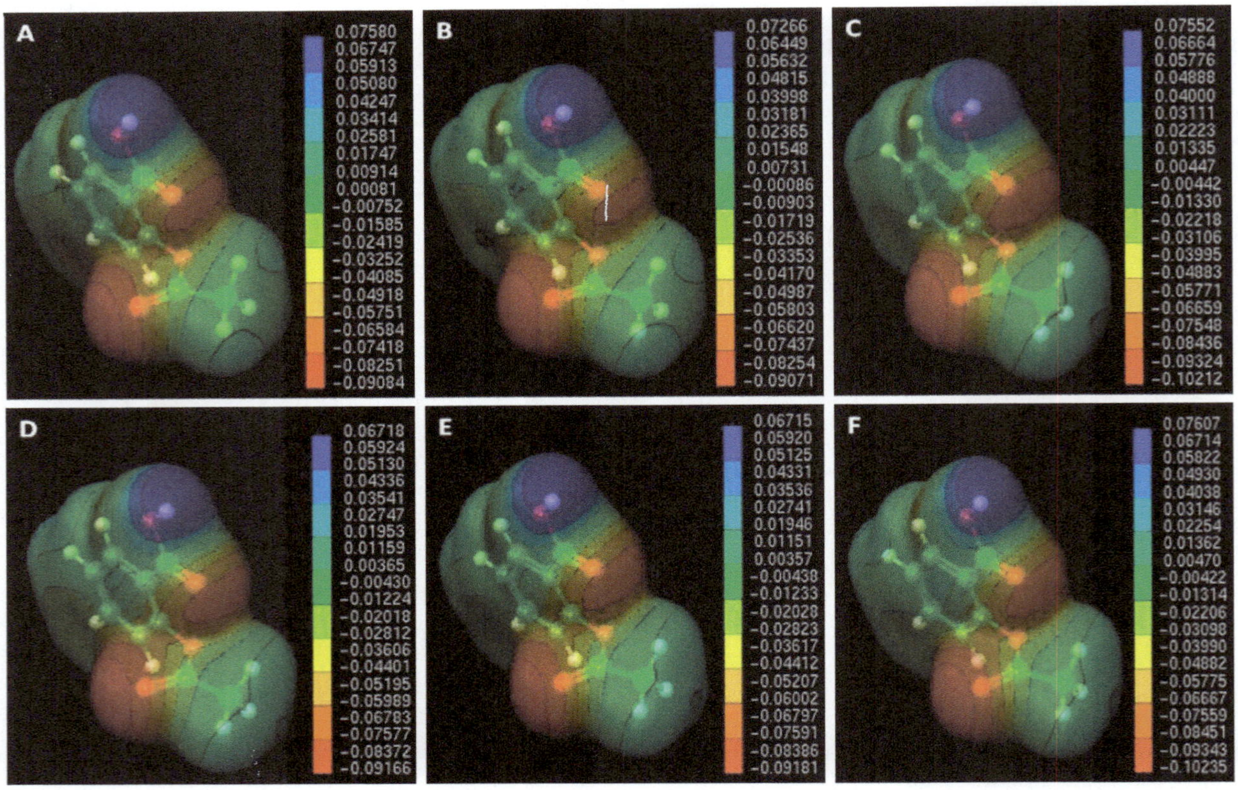

Figure 3. Molecular Electrostatic Potential Map of aspirin in different basis sets (A – HF/3-21 G*, B – HF/3- 21G**, C – HF/6-31G, D – HF 6-31G*, E – HF 6-31G**, F – HF 6-311G). The MEPs were visualized and calculated by the Molekel program [24] .

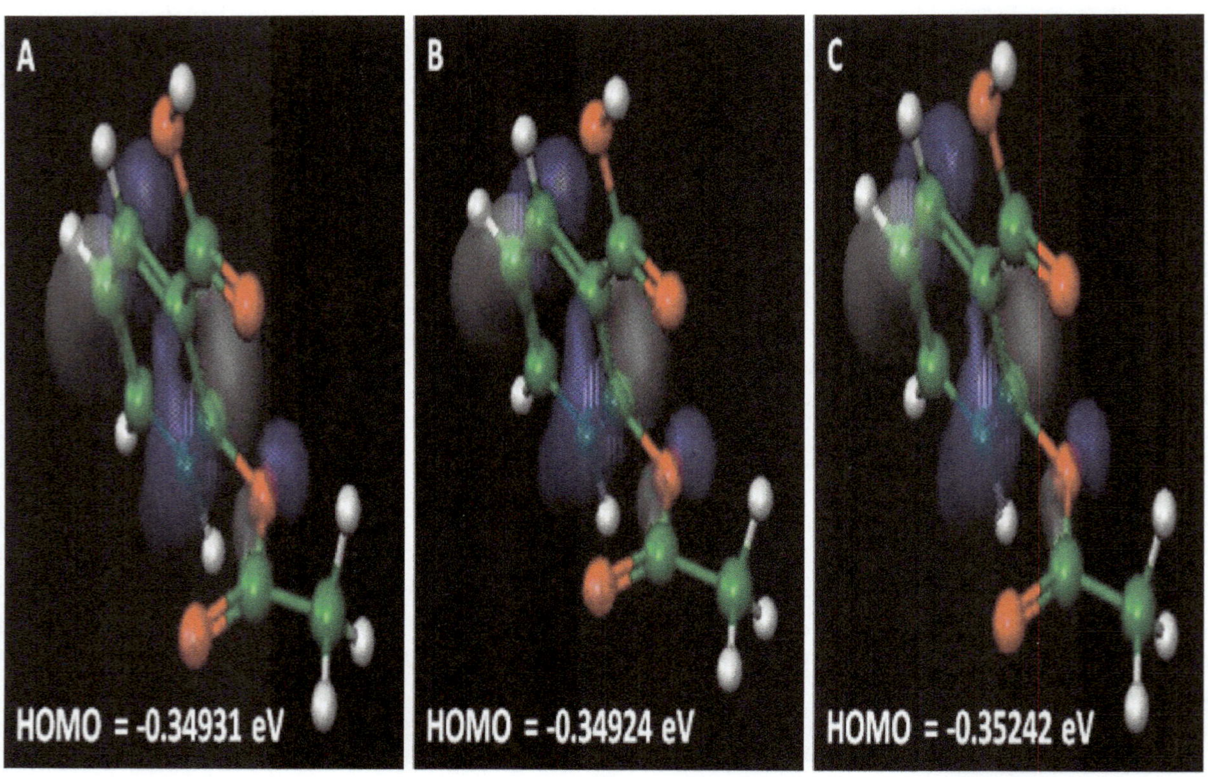

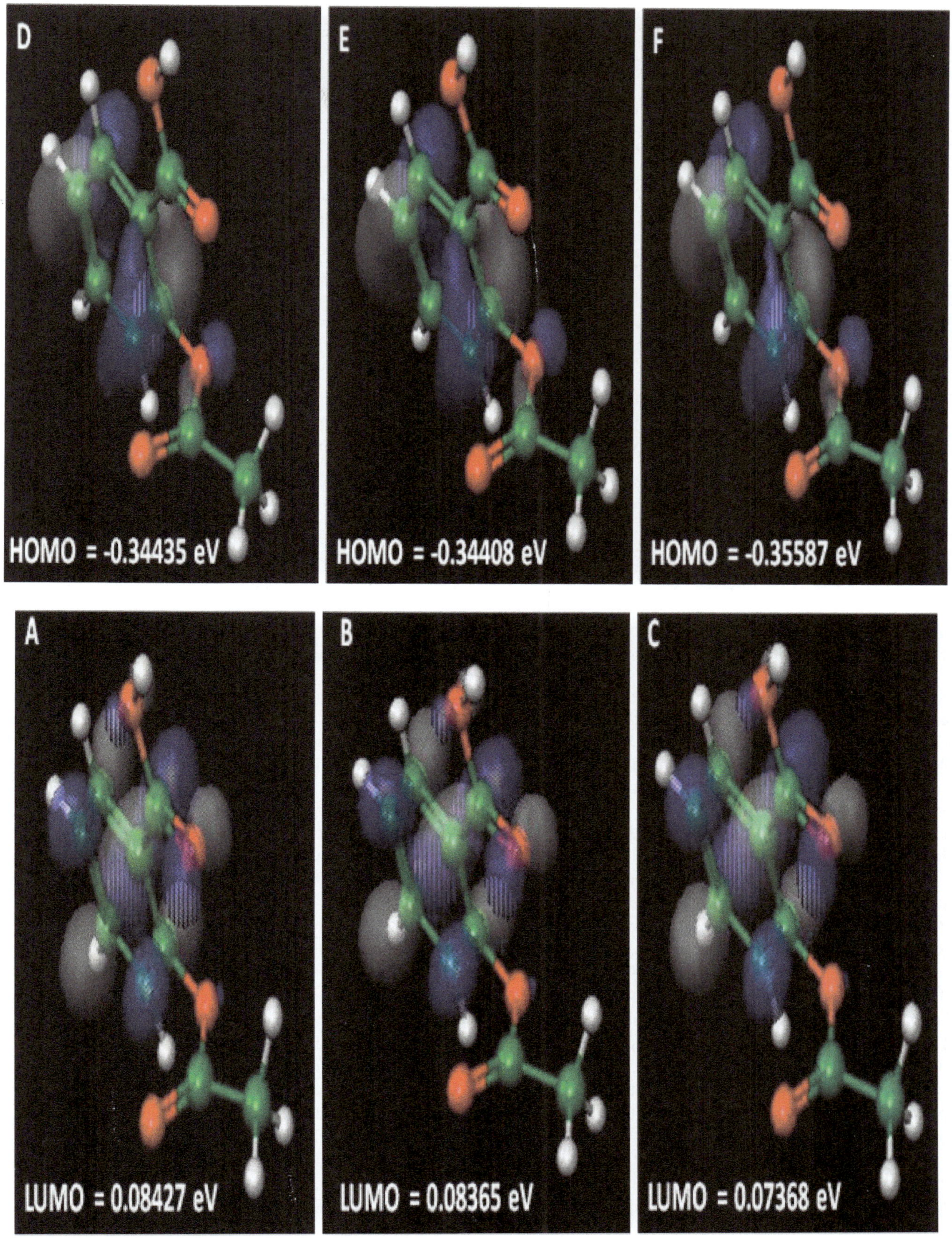
D
HOMO = -0.34435 eV
E
HOMO = -0.34408 eV
F
HOMO = -0.35587 eV
A
LUMO = 0.08427 eV
B
LUMO = 0.08365 eV
C
LUMO = 0.07368 eV

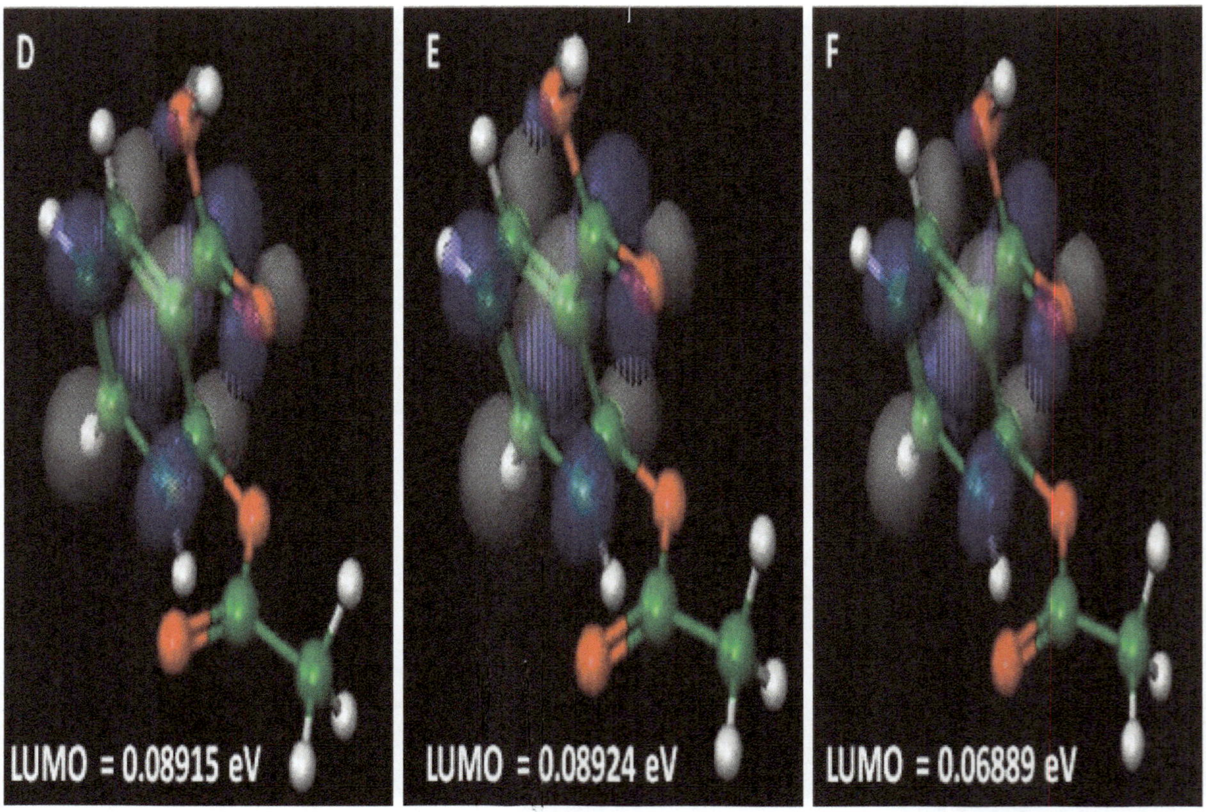

Figure 4. Orbital's energy Homo and Lumo of aspirin in different basis sets (A – HF/3-21 G*, B – HF/ 3-21G**, C – HF/6-31G, D – HF 6-31G*, E – HF 6-31G**, F – HF 6-311G). The Homo and Lumo were visualized and calculated by the Molekel program [24] .

tiers orbital (HOMO and LUMO) with the respective values of energy (eV), using Hatree-Fock method at different basis sets (HF/3-21G*, HF/3-21G**, HF/6-31G, HF/6-31G*, HF/6-31G** and HF/6-311G) for aspirin, and this figure we note the delimited region for the HOMO orbital which measures the electron-donor character of aspirin, and the LUMO which measures the electron-acceptor character. From these definitions, two important features can be observed: the higher the energy of the HOMO greater electron-donating ability, which can be observed in HF/6-31G** with a value of HOMO = −0.34408 eV, and the lower the energy of the LUMO will be lowest resistance to accept electrons that can be noticed in HF/6-311G with a value of LUMO = 0.06889 eV. In this figure the HOMO is located in the region around the benzene ring and the acetyl group, when the substituents are made in the aromatic ring or acetyl group, depending on the substituent will have a high electron density, such as the carbonyls, amines or amides, the more pronounced HOMO region is strongly influenced to perform electronic stereo secondary effects, which may compromise the pharmacological activity of the compound. The energy of HOMO is directly related to the ionization potential of the compound and characterizes the ability of the molecule to perform nucleophilic attacks. Also in Figure 4, the LUMO is located in close region to the benzene ring and of the carboxylic group, so the energy of LUMO is directly related to the electron affinity, characterized by the susceptibility of the compound in relation to attacks by nucleophiles [94] .

The energies of HOMO and LUMO have been used for decades as indices of chemical reactivity and are commonly correlated with other indices, such as electron affinity and ionization potential [95] -[99] . The difference between the energies of the HOMO-LUMO orbitals (gap) is an important indicator of molecular stability. Molecules with low gap value are generally reactive, while molecules with higher gap value indicate high stability of the molecule, in the sense of low reactivity in chemical reactions [100] . In Table 1 are shown some molecular properties obtained with Hartree-Fock method in different basis sets (HF/3-21G*, HF/3-21G**, HF/6-31G, HF/6-31G*, HF/6-31G** e HF/6-311G) and the correlation matrix of Pearson [101] -[103] , considering how an independent variant the total energy (Etotal) of aspirin, and other properties as the dependent variable, this treatment was performed using Stastica 6.2 program [104] . In this table it is observed that the correlation between the molecular properties of aspirin is less than or equal to 0.96890 (LUMO), while the correlation between the molecular properties and the total energy is less than or equal to 0.85448 (QC8), which represents the charge on the carbon atom 8, see Figure 2. Among these properties obtained those that had greater relevance to building a model QSPR in function of the total energy were the volume (−0.78677), hydration energy (−0.45775), GAP (−0.43419) and QC_8(0.85448). Therefore, we can represent the QSPR model according to the values of the statistical parameters in Equation (52) below. The statistical quality [105] of the regression equations was gauged by parameters like correlation coefficient (r) or squared correlation coefficient (r^2), explained variance (R_A^2, i.e., adjusted R^2), standard error of estimate (SEE), and variance ratio (F) [106] -[108] . The better regression models were selected on the basis of the higher r, F value (a statistic of assessing the overall significance) and the lower SEE.

$$E_{\text{total}} = 286.1621 - 1.1850(V) + 8.7851(HE) + 615.3683(GAP) + 39.8875(QC_8) \quad (52)$$

This type of treatment shows that QSPR studies are important to obtain of molecular properties that take into account different aspects of chemical information, this information can be through experiments or theoretical calculations simple counting, consider the entire molecule, fragments or functional groups, knowledge of the 3D structure of the molecule or molecular graphics his or her simply formula, information defined by scalar values, vectors or scalar fields [109] .

In the discovery of a superaspirin Jacob et al. (2012) [110] synthesized by conjugation aspirin-glucose (Figure 5) in order to study the solubility in water and anticancer activity as compared to aspirin and evaluated that aspirin-glucose was seven times more soluble in water than aspirin, and about 8 to 9 times more active in inhibiting cell growth than aspirin in its anticancer activity in cell culture breast, pancreatic, and cell lines prostate, while the activity was similar in a line benign non-cancer cells. According to the computer calculations performed to aspirin and aspirin-glucose in HF/6-31G** verified that the anticancer activity of aspirin-glucose was increased by the fact of presenting high value GAP = −0.43232 eV, having greater molecular stability and low chemical reactivity, being that the value of GAP to aspirin in HF/6-31G** was −0.43332 eV, having a variation of GAP = $\pm 10^{-3}$ eV (between aspirin and aspirin-glucose).

In Figure 5 is shown the map of electrostatic potential for aspirin-glucose, and observed the region of negative electrostatic potential, characterized by red color, which is the region essential for the expression of biological activity (pharmacophore). Therefore, these results are directly related with the conclusion that Jacob et al. indicated that the hydrolysis of the aspirin-glucose in human serum is at a relatively slower rate compared to aspirin where there has been significant anticancer activity at the doses studied under the experimental conditions.

Table 1. Molecular properties obtained at different basis sets of aspirin and Pearson correlation matrix

Method and basis set	Total energy (u.a)	Surface Area (Grid) (Å²)	Volume (Å³)	Hydration Energy (kcal/mol)	HOMO (eV)	LUMO (eV)	GAP (eV)	**QC_8 (C)**	**QO_9(C)**	**C8-O9 (Å)**
HF/3-21G*	−641.3663	345.70	527.07	-8.00	−0.34931	0.08427	−0.43358	0.856	−0.584	1.19329
HF/3-21G**	−641.4183	345.02	526.10	-8.04	−0.34924	0.08365	−0.43289	0.825	−0.585	1.19312
HF/6-31G	−644.6781	347.46	530.81	-8.04	−0.35242	0.07368	−0.42610	0.778	−0.541	1.20473
HF/6-31 G*	−644.9649	345.45	528.40	-7.89	−0.34435	0.08915	−0.43350	0.783	−0.544	1.18114
HF/6-31G**	−644.9837	345.33	528.31	-7.89	−0.34408	0.08924	−0.43332	0.772	−0.545	1.18127
HF/6-311	−644.8264	345.99	529.85	-8.03	−0.35587	0.06889	−0.42476	0.723	−0.490	1.20201
Total energy	1.00000									
Surface Area (Grid)	−0.36056	1.00000								
Volume	−0.78677	0.83299	1.00000							
Hydration Energy	−0.45775	−0.42636	−0.09622	1.00000						
HOMO	−0.05826	−0.53886	−0.44601	0.86149	1.00000					
LUMO	0.18092	−0.65181	−0.64085	0.75742	0.96890	1.00000				
GAP	−0.43419	0.72982	0.81092	−0.58584	−0.86310	−0.96123	1.00000			
QC_8	0.85448	−0.27889	−0.73507	−0.09622	0.33329	0.52087	−0.69165	1.00000		
QO_9	0.34835	0.28008	0.03220	−0.61938	−0.54970	−0.45390	0.31382	0.18617	1.00000	
C_8-O_9	0.10478	0.71376	0.48478	−0.91231	−0.94599	−0.92477	0.83323	−0.18682	0.55693	1.00000

a)

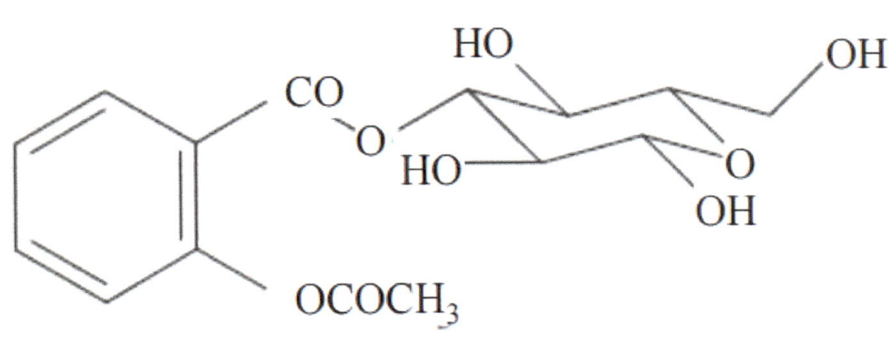

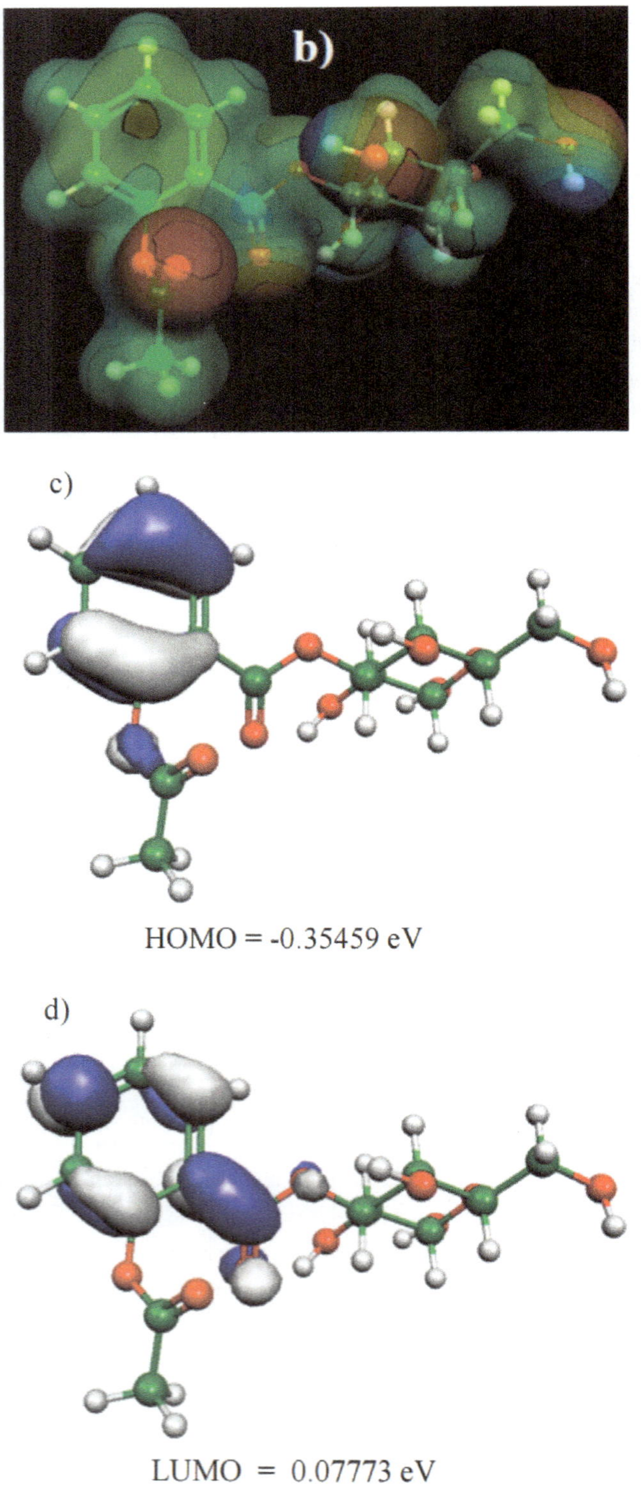

Figure 5. (a) The glucose-aspirin structure was visualized using ChemSketch 12.00 program [23] . (b) Map of molecular electrostatic potential (MEP). ((c) and (d)) Orbital's energy Homo and Lumo. The MEP, Homo and Lumo were visualized and calculated using Hartree-Fock (HF) method and HF/6-31G** basis set by the Molekel program [24] .

The anticancer activity in vitro was much stronger for aspirin-glucose compared to aspirin in cancer cell lines. Therefore, further studies are needed among them we can mention the use of maps of electrostatic potential as an indicator of site of chemical reactivity to confirm this finding in a system in vivo. The high solubility of the conjugated glucose aids the development of a form of injection of aspirin.

This new derivative can be considered in the future a "superaspirin" that must pass safety tests for long periods before being placed on pharmacy shelves, but the time to replace aspirin and other non-steroidal anti-inflammatory drugs.

FINAL CONSIDERATIONS

The computational chemistry is considered as one of the greatest intellectual realizations of the twentieth century, and has been the conceptual basis that allows the understanding of chemistry of a way much deeper than that existing before age 20, epoch wherein were launched the bases of quantum theory. The impact of this theory in chemistry can be verified for its practical implications in various fields such as spectroscopy, electron microscopy, molecular modeling, among others.

The molecular modeling provides important information for the process of drug discovery, because it facilitates obtaining of specific molecular properties of a molecule that can influence the interaction with the receptor and biological activity.

The Hartree-Fock method provides a quantitative prediction of high quality for a wide variety of chemical and biological systems, but the calculations are time consuming and high computational cost. A resource commonly employed is optimizing geometry with a basis set simpler, and then perform calculations "Single Point" with a basis set most complete allowing the determination of energy and other molecular properties of a system, using a base of more sophisticated calculation.

The application of the Hartree-Fock method depends on the basis set and of system to be studied to obtain different molecular properties related to the biological activity of bioactive molecules. The computer calculations should describe the necessary characteristics related to the experimental data of the molecule or set of molecules under study, for that molecular modeling can be represented of effective and/or efficient form. However, these descriptors are not completely universal, because they are dependent on the structures and systems studied. Even though it is based on a minimum of energy, the calculated descriptors have values very close to their respective empirical values and indicate trends electronic systems under study. The aspects highlighted in this work are evident, therefore, the quantum-chemical descriptors have a wide range of applications in SAR, QSAR and QSPR studies, as well as in many areas of integration of fundamental knowledge of Organic Chemistry, Biochemistry, Molecular Biology, Pharmacology and Pharmaceutical Chemistry.

ACKNOWLEDGEMENTS

We gratefully acknowledge the financial support of the Brazilian Agency National Council of Scientific and Technological Development (CNPq Proc. 306676/2010-9). The authors would like to thank the Postgraduate Program in Biotechnology and Biodiversity—Network BIONORTE and to the Laboratory of Modeling and Computational Chemistry, Federal University of Amapá for the computational support.

REFERENCES

1. Luna, J.S., Santos, A.F., Lima, M.R.F., Omena, M.C., Mendonça, F.A.C., Bieber, L.W. and Sant'Ana, A.E.G. (2005) A Study of the Larvicidal and Molluscicidal Activities of Some Medicinal Plants from Northeast Brazil. Journal of Ethnopharmacology, 97, 199-206. http://dx.doi.org/10.1016/j.jep.2004.10.004
2. Amarante, C.B., Müller, A.H., Póvoa, M.M. and Dolabela, M.F. (2011) Estudo Fitoquímico Biomonitorado Pelos Ensaios de Toxicidade Frente à Artemia Salina e de Atividade Antiplasmódica do Caule de Aninga (Montrichardia linifera). Acta Amazônica, 41, 431-434. http://dx.doi.org/10.1590/S0044-59672011000300015
3. Robards, K., Prenzler, P.D., Tucker, G., Swatsitang, P. and Glover, W. (2003) Phenolic Compounds and Their Role in Oxidative Processes in Fruits. Food Chemistry, 66, 401-436.http://dx.doi.org/10.1016/S0308-8146(99)00093-X
4. Moure, A., Cruz, J., Franco, D., Dominguez, J., Sineiro, J., Dominguez, H., Nnunez, M. and Parajó, J. (2001) Natural Antioxidants from Residual Sources. Food Chemistry, 72, 145-171. http://dx.doi.org/10.1016/S0308-8146(00)00223-5
5. Huang, D., Ou, B. and Prior, R.L. (2005) The Chemistry behind Antioxidant Capacity Assays. Journal of Agricultural and Food Chemistry, 53, 1841-1856.http://dx.doi.org/10.1021/jf030723c
6. Zibadi, S., Farid, R., Moriguchi, S., Lu, Y., Foo, L., Tehrani, P., Lreich, J. and Watson, R. (2007) Oral Administration of Purple Passion Fruit Peel Extract Attenuates Blood Pressure in Female Spontaneously Hypertensive Rats and Humans. Nutrition Research, 27, 408-416. http://dx.doi.org/10.1016/j.nutres.2007.05.004
7. Sun, J., Chu, Y.F., Wu, X. and Liu, R.H. (2002) Antioxidant and Antiproliferative Activities of Common Fruits. Journal of Agricultural and Food Chemistry, 50, 7449-7454.http://dx.doi.org/10.1021/jf0207530
8. Amarante, C.B., Silva, J.C.F., Solano, F.A.R., Nascimento, L.D., Moraes, L.G., Silva, G.F. and Uno, W.S. (2009) Spectrometric Study of the Leaves of Aninga (Montrichardia linifera) Collected from the Guama River, Campus of UFPA, Belém-PA. A Contribution to the Chemical Study of the Araceae Family. Revista Científica da UFPA, 7, 1-19.
9. D'Oliveira, H.V. (1854) System of Brazilian Vegetable Materia Medica. Casa de Eduardo & Henrique Laemmert, Rio de Janeiro.
10. Moreira, N.J. (1862) Dictionary of Brazilian Medicinal Plants. Typographia do Correio Mercantil, Rio de Janeiro.
11. Catsro, J.M. (1878) Purgatives Brazil's Indigenous. Faculdade de Medicina do Rio de Janeiro.
12. Fenner, R., Betti, A.H., Mentz, L.A. and Rates, S.M.K. (2006) Plantas Utilizadas na Medicina Popular Brasileira com Potencial Atividade Antifúngica. Revista Brasileira de Ciências Farmacêuticas, 42, 369-394. http://dx.doi.org/10.1590/S1516-93322006000300007
13. Moo-Puc, R., Robledo, D. and Freile-Pelegrin, Y. (2008) Evaluation of Selected Tropical Seaweeds for in Vitro AntiTrichomonal Activity. Journal of Ethnopharmacology, 120, 92-97. http://dx.doi.org/10.1016/j.jep.2008.07.035
14. Osorio, D.E.J., Montoya, P.G.L. and Arango, A.G.J. (2006) Alkaloidal Natural Products with Antiprotozoal Activity. Vitae, 13, 61-84.
15. Vikas, D.K., Vishweshwar, R. and Lakshmi, N.M. (2000) Current Status of Artemisinin and Its Derivatives as Antimalarial Drugs. Life Science, 66, 279-300.
16. Amarante, C.B., Silva, J.C.F., Müller, R.C.S. and Müller, A.H. (2011) Avaliação da Composição Mineral Do Chá da Folha Senescente de Montrichardia linifera (Arruda) Schott (Araceae) por Espectrometria de Absorção Atômica com Chama (FAAS). Química Nova, 34, 419-423. http://dx.doi.org/10.1590/S0100-40422011000300010
17. Meshnick, S.R., Jefford, C.W., Posner, G.H., Avery, M.A. and Peters, W. (1996) Second-Generation Antimalarial Endoperoxides. Parasitology Today, 12, 79-82.http://dx.doi.org/10.1016/0169-4758(96)80660-0
18. Klayman, D.L. (1985) Qinghaosu (Artemisinin): An Antima-Larial Drug from China. Science, 228, 1049-1055. http://dx.doi.org/10.1126/science.3887571
19. Pinheiro, J.C., Ferreira, M.M.C. and Romero, O.A.S. (2001) Antimalarial Activity of Dihydroartemisinin Derivatives against P. falciparum Resistant to Mefloquine: A Quantum Chemical and Multivariate Study. Journal of Molecular Structure: THEOCHEM, 572, 35-44.http://dx.doi.org/10.1016/S0166-1280(01)00522-X
20. Pinheiro, J.C., Kiralj, R., Ferreira, M.M.C. and Romero, O.A.S. (2003) Artemisinin Derivatives with Antimalarial Activity against Plasmodium falciparum Designed with the Aid of Quantum Chemical and Partial Least Squares Methods. QSAR & Combinatorial Science, 22, 830-842. http://dx.doi.org/10.1002/qsar.200330829
21. Cardoso, F.J.B., Figueiredo, A.F., Lobato, M.S., Miranda, R.M., Almeida, R.C.O. and Pinheiro, J.C. (2008) A Study on Antimalarial Artemisinin Derivatives Using MEP Maps and Multivariate QSAR. Journal of Molecular Modeling, 14, 39-48.http://dx.doi.org/10.1007/s00894-007-0249-9
22. Atkins, P. and Jones, L. (2002) Chemical Principles: The Quest for Insight. Freeman, New York.
23. ACD/Chemsketch Freware, version 12.00. (2010) Advanced Chemistry Development, Inc, Toronto.
24. Flukiger, P. (2002) Molekel 4.2. Swiss Center for Scientific Computing, Mano.
25. Dyer, J.R. (1965) Applications of Absorption Spectroscopy of Organic Compounds. Prentice-Hall, New Jersey.
26. Silverstein, R.M., Webster, F.X. and Kiemle, D. (2005) Spectrometric Identification of Organics Compounds. Jhon Wiley & Sons, New Jersey.
27. Bunge, A.V. (1979) Introdução a Química Quântica. Edgard Blücher, São Paulo.

28. Souza, A.A. and Farias, R.F. (2007) Elementos de Química Quântica. Átomo, Campinas.
29. Sant'Anna, C.M.R. (2009) Métodos de Modelagem Molecular Para Estudo e Planejamento de Compostos Bioativos: Uma Introdução. Revista Virtual de Química, 1, 49-57.
30. Barreiro, E.J. and. Fraga, C.A.M. (2001) Química Medicinal, as Bases Moleculares da Ação dos Fármacos. Artmed, Porto Alegre.
31. Nye, M.J. (2007) Working Tools for Theoretical Chemistry: Polanyi, Eyring, and Debates over the "Semiempirical Method". Journal of Computational Chemistry, 28, 98-108.http://dx.doi.org/10.1002/jcc.20527
32. James, H.M. and Coolidge, A.S. (1933) The Ground State of the Hydrogen Molecule. The Journal of Chemical Physics, 1, 825. http://dx.doi.org/10.1063/1.1749252
33. Marcus, R.A. (1964) Generalization of the Activated Complex Theory of Reaction Rates. I. Quantum Mechanical Treatment. The Journal of Chemical Physics, 41, 2614-2623.http://dx.doi.org/10.1063/1.1726329
34. Hirschfelder, J.O. (1941) Semi-Empirical Calculations of Activation Energies. The Journal of Chemical Physics, 9, 645. http://dx.doi.org/10.1063/1.1750966
35. Mielke, S.L., Peterson, K.A., Schwenke, D.W., Garrett, B.C., Truhlar, D.G., Michael, J.V., Su, M.C. and Sutherland, J.W. (2003) $H+H_2$ Thermal Reaction: A Convergence of Theory and Experiment. Physical Review Letters, 91, 63201- 63204.http://dx.doi.org/10.1103/PhysRevLett.91.063201
36. Atkinson, R.C. and Suppes, P. (1968) Program in Computer-Assisted Intruction: Final Report. In: Simonson, M.R. and Thompson, A., Eds., Educational Computing Foundations, Prentice Hall, Upper Saddle River.
37. Nye, M.J. (2000) Instruments and Experimentation in the History of Chemistry. MIT Press, Cambridge.
38. Morgon, N.H. and Coutinho, K. (2007) Métodos de Química Teórica e Modelagem Molecular. Livraria da Física, São Paulo.
39. Hartree, D.R. (1928) The Wave Mechanics of an Atom with a Non-Coulomb Central Field. Part I. Theory and Methods. Mathematical Proceedings of the Cambridge Philosophical Society, 24, 89-110. http://dx.doi.org/10.1017/S0305004100011919
40. Mulliken, R.S. (1932) Electronic Structures of Polyatomic Molecules and Valence. II. General Considerations. Physical Review, 41, 49-71.http://dx.doi.org/10.1103/PhysRev.41.49
41. Blinder, S.M. (1965) Basic Concepts of Self-Consistent-Field Theory. American Journal of Physics, 33, 431-520. http://dx.doi.org/10.1119/1.1971665
42. Levine, I.N. (1991) Quantum Chemistry. Prentice-Hall, New York.
43. Roothaan, C.C.J. (1951) New Developments in Molecular Orbital Theory. Reviews of Modern Physics, 23, 69-89. http://dx.doi.org/10.1103/RevModPhys.23.69
44. Clementi, E. and Raimondi, D.L. (1963) Atomic Screening Constants from SCF Functions. The Journal of Chemical Physics, 38, 2686-2689. http://dx.doi.org/10.1063/1.1733573
45. Clementi, E., Raimondi, D.L. and Reinhardt, W.P. (1967) Constantes Triagem Atômicas de SCF Funções. II. Átomos com 37-86 Elétrons. The Journal of Chemical Physics, 47, 1300-1307. http://dx.doi.org/10.1063/1.1712084
46. Hoffmann, R., (1963) An Extended Huckel Theory. I. Hydrocarbons. The Journal of Chemical Physics, 39, 1397-1412. http://dx.doi.org/10.1063/1.1734456
47. Pople, J.A. and Beveridge, D.L. (1970) Aproximate Molecular Orbital Theory. McGraw-Hill, New York.
48. Boys, S.F. (1950) Electronic Wave Functions. I. A General Method of Calculation for the Stationary States of Any Molecular System. Proceedings of the Royal Society A, 200, 542-554. http://dx.doi.org/10.1098/rspa.1950.0036
49. Shavitt, I. (1963) The Gaussian Function in Calculations of Statistical Mechanics and Quantum Mechanics. In: Alder, B., Fernbach, S. and Rotenberg, M., Eds., Methods in Computational Physics, Academic Press, New York, 1-45.
50. Hehre, W.J., Stewart, R.F. and Pople, J.A. (1969) Self-Consistent Molecular-Orbital Methods. I. Use of Gaussian Expansions of Slater-Type Atomic Orbitals. The Journal of Chemical Physics, 51, 2657-2664. http://dx.doi.org/10.1063/1.1672392
51. Stewart, R.F. (1969) Small Gaussian Expansions of Atomic Orbitals. The Journal of Chemical Physics, 50, 2485-2495. http://dx.doi.org/10.1063/1.1671406
52. Mulliken, R.S. and Liu, B. (1971) Self-Consistent-Field Wave Functions of P2 and PO, and the Role of d Functions in Chemical Bonding and of s-p Hybridization in N2 and P2. Journal of the American Chemical Society, 93, 6738-6744.http://dx.doi.org/10.1021/ja00754a004
53. Hehre, W. (1986) Ab Initio Molecular Orbital Theory. John Wiley & Sons, New York, 678-682.
54. Santos, C.B.R., Lobato, C.C., Vieira, J.B., Brasil, D.S.B., Brito, A.U., Macêdo, W.J.C., Carvalho, J.C.T. and Pinheiro, J.C. (2013) Evaluation of Quantum Chemical Methods and Basis Sets Applied in the Molecular Modeling of Artemisinin. Computational Molecular Bioscience, 3, 66-79. http://dx.doi.org/10.4236/cmb.2013.33009
55. Stewart, J.J.P. (1999) MOPAC2002. Fujitsu Limited, Tokyo.http://www.cachesoftware.com/
56. Clark, T. (2002) VAMP, Computer Cahemie Centrum. Friedrich-Alexander-Universität Erlangen-Nürnberg, Nägelsbachstra Be 25, D-91052, Erlangen.
57. Thiel, W. (1998) MNDO97. Organisch-Chemisches Institut, Universitaet Zuerich, Winterthurerstrasse 190, CH-8057, Zuerich.
58. Frisch, M.J., Trucks, G.W., Schlegel, H.B., Scuseria, G.E., Robb, M.A., Cheeseman, J.R., Zakrzewski, V.G., Montgomery Jr., J.A., Stratmann, R.E., Burant, J.C., Dapprich, S., Millam, J.M., Daniels, A.D., Kudin, K.N.,

Strain, M.C., Frakas, O., Tomasi, J., Barone, V., Cossi, M., Cammi, R., Mennucci, B., Pomelli, C., Adamo, C., Clifford, S., Ochterski, J., Pettersson, G.A., Ayala, P.Y., Cui, Q., Morokuma, K., Malick, D.K., Rabuck, A.D., Raghavachari, K., Foresman, J.B., Cioslowski, J., Ortiz, J.V., Baboul, A.G., Stefanov, B.B., Liu, G., Liashenko, A., Piskorz, P., Komaromi, I., Gomperts, R., Martin, R.L., Fox, D.J., Keith, T., Al-Laham, M.A., Peng, C.Y., Nanayakkara, A., Gonzalez, C., Challacombe, M., Gill, P.M.W., Johnson, B., Chen, W., Wong, M.W., Andres, J.L., Gonzalez, C., Head-Gordon, M., Replogle, E.S. and Pople, J.A. (1998) Gaussian 98, Revision A. 7, Gaussian, Inc., Pittsburgh.

59. SPARTAN, Wavefunction, Inc., 18401, Von Karman Avenue, Suite 370, Irvine CA 92612 USA.
60. Gordon, M., GAMESS, Gilman Hall, Iowa State University (Ames).
61. CambridgeSoft Corporation (1999) Chem3D. Cambridge.
62. Ostlund, N. (1993) HyperChem. Hypercube Inc., Gainesville
63. Zerner, M.C. (1998) ZINDO Manual. QTP, University Of Florida, Gainesville, 32611-38435.
64. Ranaldi, D., Hoggan, P.E., Cartier, A., Baka, K., Mornard, G., Thery, V., Loos, M., Mokrane, A. and Dillet, V. (2003) GEOMOS Manual. Laboratorie de Chimie Théorique, Université de Nancy I, Nancy.
65. Rocha, G.B., Freire, R.O., da Costa Jr., N.B., de Sá, G.F. and Simas, A.M. (2004) Sparkle Model for AM1 Calculation of Lanthanide Complexes: Improved Parameters for Europium. Inorganic Chemistry, 43, 2346-2354. http://dx.doi.org/10.1021/ic034882p
66. Young, D.C. (2001) Computational Chemistry—A Practical Guide for Applying Techniques to Real-World Problems. Cytoclonal Pharmaceutics Inc., John Wiley & Sons, Inc., New York.
67. Stewart, J.J.P. (2007) MOPAC2007, Stewart Computational Chemistry EUA.http://openmopac.net
68. Cambridge Soft Corporation (1998) CS ChemDraw 5.0, Chemical Strutucture Drawing Standard. Cambridge.
69. Cambridge Soft Corporation (1998) CS Chem3D 5.0, Molecular Modeling and Analysis. Cambridge.
70. Carvalho, I., Pupo, M.T., Borges, A.D.L. and Bernardes, L.S.C. (2003) Introdução A Modelagem Molecular de Fármacos no Curso Experimental de Química Farmacêutica. Quimica Nova, 26, 428-438. http://dx.doi.org/10.1590/S0100-40422003000300023
71. ChemSW®, Inc., (2001) Molecular Modeling ProTM4.0, Computational Chemistry Program. ChemSW®, Inc., Fairfield.
72. ChemSW® Inc., (2001) ChemSite: Interative 3D Molecular Modeling. ChemSW®, Inc., Fairfield.
73. Chemplus, Modular Extensions for HyperChem Release 6.02, Molecular Modeling for Windows. (2000) HyperClub, Inc, Gainesville.
74. Patrick, G.L. (2001) An Introduction to Medicinal Chemistry. Oxford University Press, Oxford.
75. Frisch, M.J., Trucks, G.W., Schlegel, H.B., Scuseria, G.E., Robb, M.A., Cheeseman, J.R., Montgomery Jr., J.A., Vreven, T., Kudin, K.N., Burant, J.C., Millam, J.M., Iyengar, S.S., Tomasi, J., Barone, V., Mennucci, B., Cossi, M., Scalmani, G., Rega, N., Petersson, G.A., Nakatsuji, H., Hada, M., Ehara, M., Toyota, K., Fukuda, R., Hasegawa, J., Ishida, M., Nakajima, T., Honda, Y., Kitao, O., Nakai, H., Klene, M., Li, X., Knox, J.E., Hratchian, H.P., Cross, J.B., Adamo, C., Jaramillo, J., Gomperts, R., Stratmann, R.E., Yazyev, O., Austin, A.J., Cammi, R., Pomelli, C., Ochterski, J.W., Ayala, P.Y., Morokuma, K., Voth, G.A., Salvador, P., Dannenberg, J.J., Zakrzewski, V.G., Dapprich, S., Daniels, A.D., Strain, M.C., Farkas, O., Malick, D.K., Rabuck, A.D., Raghavachari, K., Foresman, J.B., Ortiz, J.V., Cui, Q., Baboul, A.G., Clifford, S., Cioslowski, J., Stefanov, B.B., Liu, G., Liashenko, A., Piskorz, P., Komaromi, I., Martin, R.L., Fox, D.J., Keith, T., Al-Laham, M.A., Peng, C.Y., Nanayakkara, A., Challacombe, M., Gill, P.M.W., Johnson, B., Chen, W., Wong, M.W., Gonzalez, C. and Pople, J.A. (2004) Gaussian 03, Revision C.02. Gaussian, Inc., Wallingford.
76. Johnson, M. and Maggiora, G. (1990) Concepts and Applications of Molecular Similarity. John Wiley & Sons, New York.
77. Leach, A. (2001) Molecular Modelling: Principles and Applications. Prentice Hall, Englewood Cliffs.
78. Martin, Y. (2010) Quantitative Drug Design: A Critical Introduction. 2nd Edition, CRC Press, Boca Raton. http://dx.doi.org/10.1201/9781420071009
79. Todeschini, R. and Gramatica, P. (1997) WHIM-3D 3.3. Milano.
80. Todeschini, R. and Gramatica, P. (1998) New 3D Molecular Descriptors: The WHIM and QSAR Applications. Perspectives in Drug Discovery and Design, 9-11, 355-380.http://dx.doi.org/10.1023/A:1027284627085
81. Guha, R. (2012) Exploring Structure-Activity Data Using the Landscape Paradigm. Wiley Interdisciplinary Reviews: Computational Molecular Science, 2, 829-841. http://dx.doi.org/10.1002/wcms.1087
82. Ferreira, J.E.V., Figueiredo, A.F., Barbosa, J.P., Cristino, M.G.G., Macêdo, W.J.C., Silva, O.P.P., Malheiros, B.V., Serra, R.T.A. and Pinheiro, J.C. (2010) A Study of New Antimalarial Artemisinins through Molecular Modeling and Multivariate Analysis. Journal of the Serbian Chemical Society, 75, 1533-1548.http://dx.doi.org/10.2298/JSC100126124F
83. Santos, C.B.R., Vieira, J.B., Lobato, C.C., Hage-Melim, L.I.S., Souto, R.N.P., Lima, C.S., Costa, E.V.M., Brasil, D.S.B., Macêdo, W.J.C. and Carvalho, J.C.T. (2014) A SAR and QSAR Study of New Artemisinin Compounds with Antimalarial Activity. Molecules, 19, 367-399. http://dx.doi.org/10.3390/molecules19010367
84. Flower, R. (2003) What Are All the Things that Aspirin Does? British Medical Journal, 327, 572-573. http://dx.doi.org/10.1136/bmj.327.7415.572
85. Fraga, C.A.M. (2001) Razões da Atividade Biológica: Interações Microe Biomacromoléculas. Química Nova na Escola, São Paulo-SP, 33-42.

86. Santos, C.B.R., Lobato, C.C., Sousa, M.A.C., Macêdo, W.J.C. and Carvalho, J.C.T. (2014) Molecular Modeling: Origin, Fundamental Concepts and Applications Using Structure-Activity Relationship and Quantitative Structure-Activity Relationship. Reviews in Theoretical Science, 2, 91-115. http://dx.doi.org/10.1166/rits.2014.1016

87. Politzer, P. and Murray, J.S. (2002) The Fundamental Nature and Role of the Electrostatic Potential in Atoms and Molecules. Theoretical Chemistry Accounts, 108, 134-142.http://dx.doi.org/10.1007/s00214-002-0363-9

88. Bernardinelli, G., Jefford, C.W., Maric, D., Thomson, C. and Weber, J. (1994) Computational Studies of the Structures and Molecules. International Journal Quantum Chemistry, 52, 117-131. http://dx.doi.org/10.1002/qua.560520710

89. Murray, J.S. and Politzer, P. (2011) The Electrostatic Potential: An Overview. Wiley Interdisciplinary Reviews: Computional Molecular Science, 1, 153-163. http://dx.doi.org/10.1002/wcms.19

90. Galabov, B., Ilieva, S., Koleva, G., Allen, W.D., Schaefer III, H.F. and Schleyer, P.V.R. (2013) Structure-Reactivity Relationships for Aromatic Molecules: Electrostatic Potentials at Nuclei and Electrophile Affinity Indices. Wiley Interdisciplinary Reviews: Computational Molecular Science, 3, 37-55. http://dx.doi.org/10.1002/wcms.1112

91. Martins, G.R., Napolitano, H.B., Camargo, L.T.F.M. and Camargo, A.J. (2012) Structure-Activity Relationship Study of Rutaecarpine Analogous Active against Central Nervous System Cancer. Journal of the Brazilian Chemical Society, 23, 2183-2190.http://dx.doi.org/10.1590/S0103-50532013005000004

92. Honório, K.M. and Silva, A.B.F. (2003) An AM1 Study on the Electron-Donating and Electron-Accepting Character of Biomolecules. International Journal of Quantum Chemistry, 95, 126-132. http://dx.doi.org/10.1002/qua.10661

93. Hirao, H. (2011) Correlation Diagram Approach as a Tool for Interpreting Chemistry: An Introductory Overview. Wiley Interdisciplinary Reviews: Computational Molecular Science, 1, 337-349. http://dx.doi.org/10.1002/wcms.20

94. Grant, G.H. and Richards, W.G. (1996) Computational Chemistry. Oxford Science Publications, Oxford.

95. Contreras, R., Domingo, L.R., Andrés, J., Pérez, P. and Tapia, O. (1999) Nonlocal (Pair Site) Reactivity from SecondOrder Static Density Response Function: Gasand Solution-Phase Reactivity of the Acetaldehyde Enolate as a Test Case. The Journal of Physical Chemistry A, 103, 1291-1436. http://dx.doi.org/10.1021/jp982801i

96. Clare, B.W. (1995) The Relationship of Charge Transfer Complexes to Frontier Orbital Energies in QSAR. Journal of Molecular Structure: Theochem, 331, 63-78. http://dx.doi.org/10.1016/0166-1280(94)03783-H

97. Clare, B.W. (1995) Charge Transfer Complexes and Frontier Orbital Energies in QSAR: A Congeneric Series of Electron Acceptors. Journal of Molecular Structure: Theochem, 337, 139-150. http://dx.doi.org/10.1016/0166-1280(95)04135-S

98. Clare, B.W. (1994) Frontier Orbital Energies in Quantitative Structure-Activity Relationships: A Comparison of Quantum Chemical Methods. Theoretical Chemistry Accounts, 87, 415-430. http://dx.doi.org/10.1007/BF01127805

99. Heaton, C.A., Miller, A.K. and Powell, R.L. (2001) Predicting the Reactivity of Fluorinated Compounds with Copper Using Semi-Empirical Calculations. Journal of Fluorine Chemistry, 107, 1-3. http://dx.doi.org/10.1016/S0022-1139(00)00324-9

100. Zhang, G. and Musgrave, C.B. (2007) Comparison of DFT Methods for Molecular Orbital Eigenvalue Calculations. The Journal of Physical Chemistry, 111, 1554-1561.http://dx.doi.org/10.1021/jp061633o

101. Comelli, N.C., Duchowicz, P.R., Lobayan, R.M., Jubert, A.H. and Castro, E.A. (2012) QSPR Study of Valproic Acid and Its Functionalized Derivates. Molecular Informatics, 31, 181-188. http://dx.doi.org/10.1002/minf.201100119

102. Tropsha, A. (2010) Best Practices for Qsar Model Development, Validation and Exploitation. Molecular Informatics, 29, 476-488.http://dx.doi.org/10.1002/minf.201000061

103. Barbosa, J.P., Ferreira, J.E.V., Figueiredo, A.F., Almeida, R.C.O., Silva, O.P.P., Carvalho, J.R.C., Cristino, M.G.G., Pinheiro, J.C., Vieira, J.L.F. and Serra, R.T.A. (2011) Molecular Modeling and Chemometric Study of Anticancer Derivatives of Artemisinin. Journal of the Serbian Chemical Society, 76, 1263-1282. http://dx.doi.org/10.2298/JSC111227111B

104. STATISTICA (Data Analysis Software System). (2004) Version 6.1. StatSoft, Inc., Tulsa.http://www.statsoft.com

105. Snedecor, G.W. and Cochran, W.G. (1967) Statistical Methods. Oxford and IBH, New Delhi, 381-418.

106. Chatterjee, S., Hadi, A.S. and Price, B. (2000) Regression Analysis by Examples. 3rd Edition, Wiley VCH, New York.

107. Diudea, M.V. (2000) QSPR/QSAR Studies for Molecular Descriptors. Nova Science, Huntingdon, New York.

108. Debnath, B., Gayen, S., Bhattacharya, S., Samanta, S. and Jha, T. (2003) QSAR Study on Some Pyrodoacridine Ascididemin Analogues as Anti-Tumor Agents. Bioorganic & Medicinal Chemistry, 11, 5493-5499. http://dx.doi.org/10.1016/j.bmc.2003.09.022

109. Santos, C.B.R., Vieira, J.B., Formigosa, A.S., Costa, E.V.M., Pinheiro, M.T., Silva, J.O., Macêdo, W.J.C. and Carvalho, J.C.T. (2014) Validation of Computational Methods Applied in Molecular Modeling of Artemisinin with Antimalarial Activity. Journal of Computational and Theoretical Nanoscience, 11, 1-9.

CHAPTER 3 REVIEW ON DFT AB INITIO CALCULATIONS OF SCALAR COUPLING CONSTANTS

Ibon Alkorta and José Elguero

Instituto de Química Médica, C.S.I.C., Juan de la Cierva, 3, E-28006 Madrid

ABSTRACT

The present review summarizes the information available on the ab initio calculations of spin-spin nuclear coupling constants through hydrogen bonds or in van der Waals complexes. It also reports the sources of experimental data on $^{nh}J_{XY}$ scalar couplings.

INTRODUCTION

We intend to summarize the present knowledge about the ab initio calculations of coupling constants in non-bonded systems, also called supermolecules. Bonded means here, covalently bonded, because most supermolecules are formed by fragments linked together by hydrogen bonds (HBs). We will start with a brief summary of the theory of scalar coupling constants followed by the ab initio approach to these magnitudes. Then, the three main sections will be devoted to "Hydrogen bonded molecules: couplings through a hydrogen bond", to "Hydrogen bonded molecules: couplings through a dihydrogen bond" and to "Van der Waals complexes".

According to the IUPAC recommendations [1], ^{n}J should be used for nuclear spin-spin coupling constants through n bonds (usually given in frequency units) Parentheses may be used (for example) to indicate the species of nuclei coupled, e.g. J(^{13}C,^{1}H) or, additionally, the coupling path, e.g. J(POCF). Where no ambiguity arises, the elements involved can be, alternatively, given as subscripts, e.g. J_{CH}. The nucleus of higher mass should be given first. ^{n}K, is the reduced nuclear spin-spin coupling constant defined as $K_{jk} = 4\pi^2 J_{jk}/h\gamma_j\gamma_k$, [2] where h is the Planck constant and γ_i,γ_j are the magnetogyric ratios of i and j nuclei. Reduced coupling constants are useful when comparing coupling constants for different isotopes. K is given in N A^{-2} m^{-3}, the only problem is that due to the small value of h (6.626176 x 10^{-34} J Hz^{-1}), K is very large and usually given in 10^{19} to 10^{21} units. For instance, $^{1}J_{CH}$ in methane is equal to 125 Hz, since γ_{1H} = 26.7510 * 10^{7} and γ_{13C} = 6.7283*10^{7} rad T^{-1} s^{-1}, $^{1}K_{CH}$ = 4.141*10^{20} N A^{-2} m^{-3}. The use of reduced coupling constants avoids some misunderstandings when comparing couplings between different nuclei, for instance, the fact that $^{2h}J_{NH}$ (9 Hz) is larger than $^{1h}J_{NN}$ (3 Hz) has been described as counterintuitive [3]. The corresponding reduced couplings are $^{2h}K_{NH}$ = 0.74 • 10^{20} and $^{1h}K_{NN}$ = 2.43 • 10^{20} N A^{-2} m^{-3}.

Couplings through hydrogen bonds are usually named ^{nh}J (other authors used ^{hn}J, a unification by the IUPAC of the notation seems highly desirable) where h stands for hydrogen bond and n is the number of bonds, including the HB, that separates the coupled nuclei.

Citation: Alkorta, I.; Elguero, J. Review on DFT and ab initio Calculations of Scalar Coupling Constants. Int. J. Mol. Sci. 2003, 4, 64-92. DOI: 10.3390/i4030064

The Theory of Scalar Spin-Spin Coupling Constants

Magnetic phenomena are included consistently in the Dirac relativistic equation. However, nonrelativistic limit (Pauli) is usually sufficient for interpretation of NMR experiments [4] if heavy atoms are excluded. Quantum mechanically, the scalar coupling can be defined as the mixed second derivative of the energy of the molecule with respect to the magnetic moments of two nuclei:

$J_{IS,xy} = \partial^2 E/\partial \mu_{I,x} \partial \mu_{S,y}$

It can be subdivided into four distinct components:

- – Fermi contact (FC), H_{FC}.
- – Paramagnetic spin-orbit (PSO), H_{PSO}.
- – Diamagnetic spin-orbit (DSO), H_{DSO}.
- – Spin-dipolar (SD), H_{SD}.

$$H = H_{FC} + H_{PSO} + H_{DSO} + H_{SD} \quad (1)$$

In general, the most important contribution to the scalar coupling is the Fermi contact term, the exception being coupling constants involving fluorine atoms, in particular $^{19}F \cdots ^{19}F$ couplings. This term relies on the probability of finding an electron at the site of the two coupled nuclei (this term rigorously disappears when there is no overlap between atoms). Concerning the FC term, it is expected that σ electrons will play a very significant role since these are the only electrons that do not have nodes at the nuclear sites. Unlike π or δ orbitals, only σ orbitals have a nonzero value at the site of the nucleus. However, the FC coupling is also transmitted through the π-electronic system owing to exchange interactions between the σ- and π-electronic systems (the FC spin information «leaks» into the p-electronic system). In fact, σ orbitals are characterized by a cusp at the site of the nucleus. Thus, it can be seen that, unlike shielding, scalar coupling calculations will be especially sensitive to the shape of the electronic wave functions near the nucleus. This gives rise to the difficulty in using Gaussian functions; the problem stems from the difference between the cusp of a Gaussian orbital and that of an STO one. Thus, scalar coupling calculations are often much more demanding than chemical shifts.

In addition to the direct contact term (an electron actually visiting both nuclei), there is a magnetic interaction through space between the spin-orbit momentum of an electron and that of a nucleus. These two terms are very similar to the ones seen in defining shielding except that both sources of perturbation are nuclear magnetic moments (there is no external magnetic field). Many authors point out that there is a compensation between the PSO and DSO terms.

There is also the spin dipole-dipole interaction between electrons and nuclei. This interaction also does not require the Fermi-contact term. And lastly, there is a mixed Fermi contact, spin dipolar interaction where one interaction is between an electron visiting the site of one of the nuclei and interacting with the other nuclei through space. This interaction is primarily responsible for the observed anisotropy of the coupling tensor.

Ab initio Calculations of Scalar Coupling Constants

Concerning the examples found in the literature, they can be classified into three main groups:

1a. Methods based on DFT (Density Functional Theory).

Type DFT-FPT (Pople's Finite Perturbation Theory): Gaussian set of programs.

Type SOS (Sum-over-States)-DFPT (Density Finite Perturbation Theory): deMon Package (Salahub, Malkin, Malkina). This method only allows calculating the FC, PSO and DSO terms, not the SD term. Some authors, like Bartlett (see later on), stress that these methods are not recommended for atoms with lone pairs.

Type CPDTF-KS (Kohn-Sham operator) used by Cremer and Peralta [5,6].

2a. Methods based on MCSCF (Multiconfiguration Self Consistent Field): Dalton package (Ruud, Helgaker, see [7] for a detailed discussion about Hartree-Fock and post-Hartree-Fock calculations).

3a. Methods based on EOM (Equation of Motion)-CCSD (Coupled Cluster Singles and Doubles): ACES package (Gauss, Bartlett, Perera, del Bene).

To these methods correspond different types of functional and basis sets. Here are some representative examples:

1b. Functionals like VWN, PWP, B3PW91, B3LYP and their unrestricted versions and basis sets like IGLO-III, 6-311G**, 6-31G**.

2b. Basis sets like IGLO-III, aug-cc-pVDZ, etc.

3b. Basis sets like aug-cc-pV5Z, aug-cc-pVTZ, etc.

When increasing the distance, the "through-space" couplings decay roughly as an exponential. The results reported by several authors show a Morse curve appearance (quick decrease, minimum value, very small asymptotic increase). For instance, the ^{129}Xe-^{131}Xe coupling, as calculated by Salsbury and Harris [8] shows this behavior: $^1J_{Xe \cdots Xe} = [1-\exp(-0.41\,(dXe \cdots Xe - 9.8)]^2$, $r^2 = 0.99$ (Eq. 1), $d_{Xe \cdots Xe}$ being the distance in atomic units (see Figure 1).

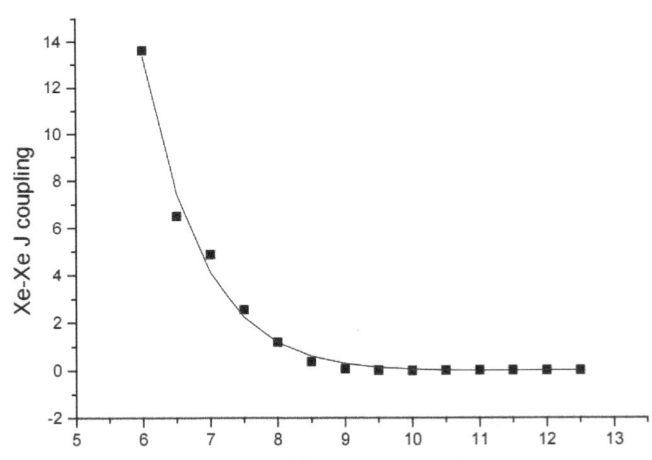

Figure 1. Morse fitting of Salsbury and Harris data (two first points removed)

Hydrogen Bonded Molecules: Couplings through a Hydrogen Bond

The hydrogen-bond-induced changes of the intramolecular spin-spin coupling constants will not be discussed systematically (for instance, $^{1}J_{NH}$ before and after the NH group was involved in an HB). Any hydrogen bond (HB) of the A–H···B type can adopt three limit situations (we will use IMHB for intramolecular hydrogen bonds): classical, shared and proton transfer. In cases were the proton is shared between A and B the distinction between $^{1}J_{AH}$ and $^{1h}J_{B\cdots H}$ disappears; finally, when the proton is completely transferred, the situation again becomes "classical", with $^{1}J_{BH}$ and $^{1h}J_{A\cdots H}$.

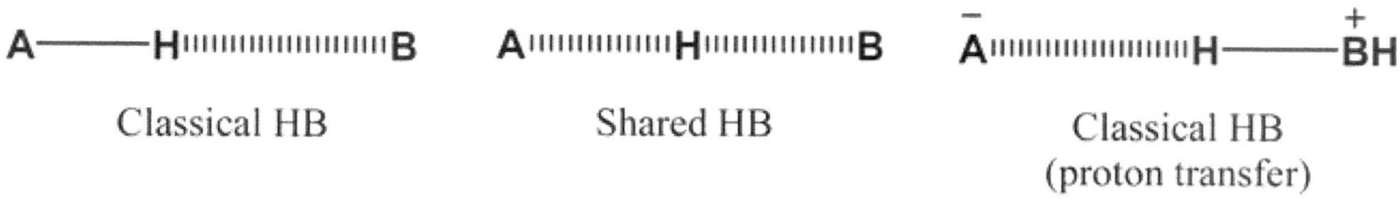

Traditional AH···B HBs have normal A···B distances, A–H slightly elongated (relative to the monomer); similarly the ion-pair, B^{+}–H···A, have A···B distances comparable to the preceding case, B^{+}–H slightly elongated (relative to the isolated B^{+}–H cation). The proton-shared structure is characterized by a short A···B distance and an elongation of the A···H and B···H distances.

The contribution of the Bartlett, Perera and Del Bene's group to this topic is so significant that we have decided to divide this section in three parts: a) Other groups [9,10,11,12 ,13,14,15,16,17,18,19,20,21,22,23,24]; b) Bartlett-Perera-Del Bene [25,26,27,28,29,30,31,32,33,34,35,36]; c) Data analysis [37,38,39,40,41].

a) Other groups. Three publications by Pecul, Sadlej and Leszczynski are noteworthy for the quality of the calculations. In the first one, Pecul and Sadlej [9] described their calculation on water dimer. In the first part they use the conventional case of water monomer (Table 1) to compare different approaches to the calculation of spin-spin coupling constants (they were dominated by the Fermi contact and paramagnetic spin-orbital terms).

Table 1. The nuclear spin-spin coupling constants (Hz) in the water monomer

		$^{1}J_{OH}$	$^{2}J_{HH}$
Experimental			
	cyclohexane	–78.7	—
	nitromethane	–80.6	–7.3
Calculated			
	MCSCF (RAS4/HIII)	–77.2	–12.6
	MCSCF (CAS/HIIIa)	–75.2	–9.3
	EOM-CCSD[a]	–74.9	–10.8

[a] From the literature (different authors) [9].

These results show that MCSCF (included in the DALTON program) when used with a basis set of very high quality are comparable to EOM-CCSD calculations. For the water dimer **1**, at the RAS level, they found $^{2h}J_{O1\text{-}O2} = 1.6$, $^{1h}J_{O1\text{-}H5} = 4.4$, $^{3h}J_{O1\text{-}H6} = 0.5$, $^{3h}J_{O2\text{-}H3} = 0.2$, $^{2h}J_{H3\text{-}H5} = 0.3$ and $^{4h}J_{H3\text{-}H6} = –0.08$ Hz.

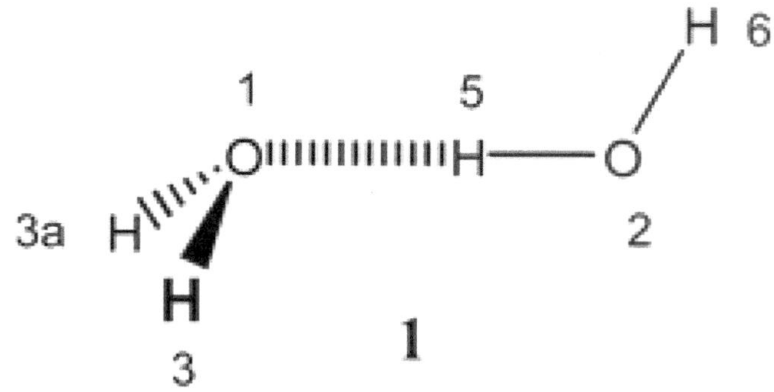

1

Pecul, Leszczynski and Sadlej published two papers on scalar coupling constants in hydrogen bonded systems. In the first one [10], they studied X-H···O complexes [CH_2O–H_2O **2**, C_2H_2–H_2O **3**, CH_3OH–H_2O **4** and $(HCOOH)_2$ **5**]. The calculations were carried out at the MCSCF level and the main conclusions are: $^{1h}J_{OH}$ correlates with the strength of the HB (expressed in terms of the interaction energy); the intermolecular $^{1h}J_{OH}$ couplings are determined primarily by the FC term (these couplings are similar to those through covalent bonds, what distinguishes them is their magnitude and, more surprisingly, their sign); the $^{2h}J_{XY}$ couplings through the X-H···Y HB are also dominated by the FC term. In the formic acid dimer **5**, the $^{2h}J_{OO}$ coupling attains 7.1 Hz.

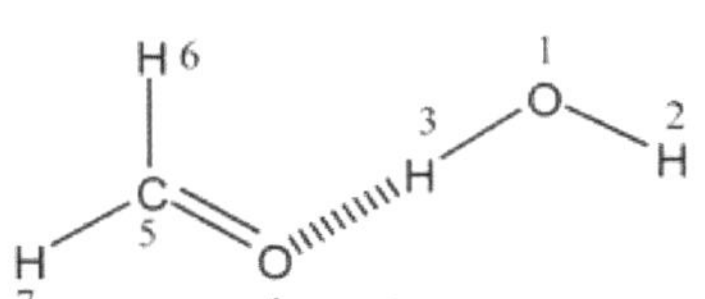

2

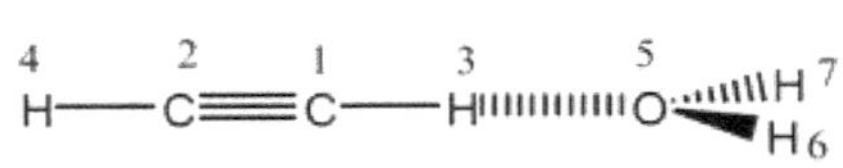

3

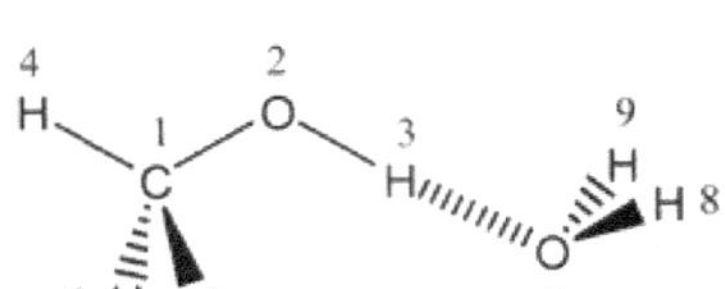

4

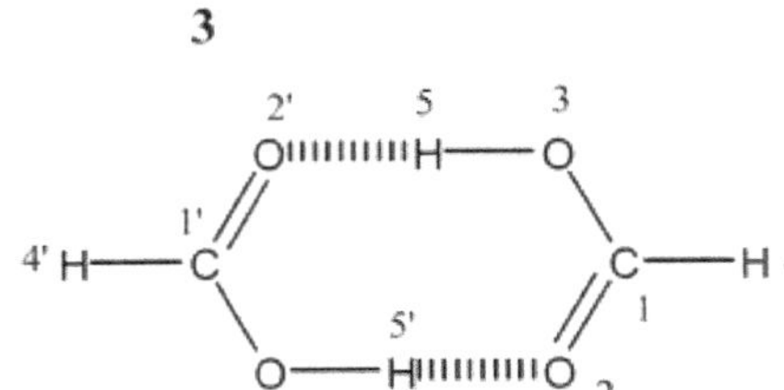

5

The second paper [11] deals with the important case of N-H···O=C (this last atom is designed C' in most publications following the common use in peptides) and N-H···N=C systems. Thus, formamide dimer (Fa-Fa) **6** and formamide-formamidine dimer (Fa-Fi) **7** have been studied at the MCSCF level. Complex **7** should present a large $^{2h}J_{NN}$ coupling of 8.3 Hz. Another interesting conclusion is that $^{2h}J_{CH}$ couplings are positive for N–H···N=C bonds and negative for N–H···O=C bonds. The authors have examined the variation of the couplings with the ON distance, for instance, there is a fast exponential decay of $^{2h}J_{NN'}$.

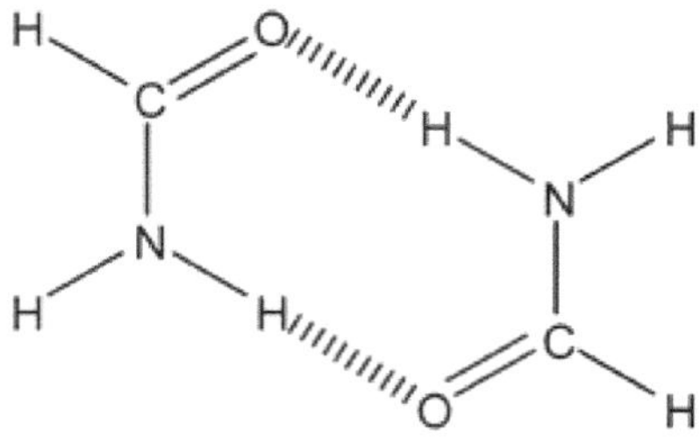

6

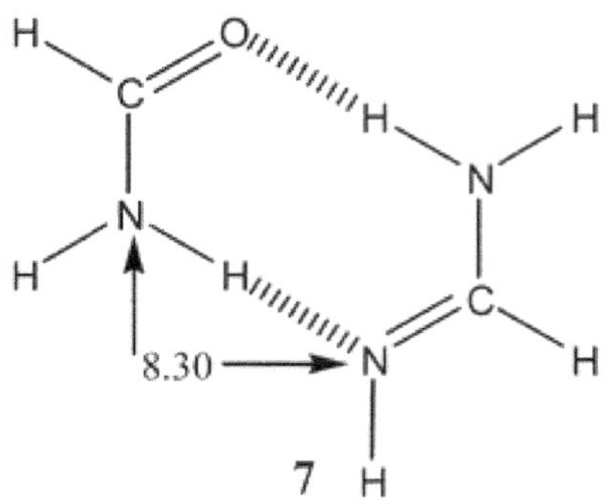

7

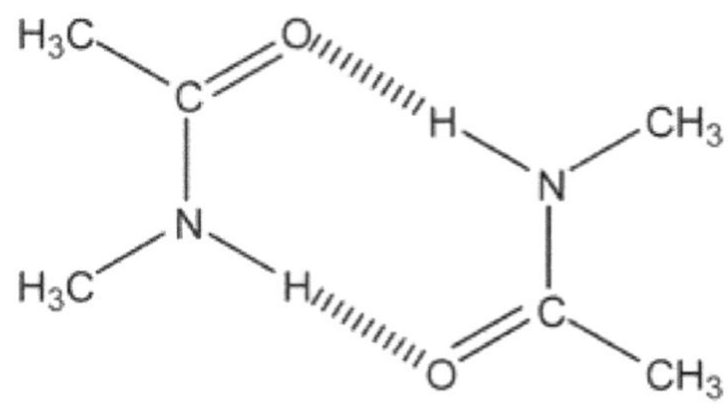

8

Scheurer and Brüschweiler [12] have studied the N-methylacetamide dimer **8** as a model of the trans-hydrogen $^{3h}J_{NC'}$ coupling involved in polypeptide N–H···O=C' hydrogen bonds (trans here does not mean a trans disposition but that the coupling occurs through the HB). Using the DFT-based methodology of Malkin, Malkina and Salahub (SOS-DFTP), they explored the double dependence of $^{3h}J_{NC'}$ with the NO distance and the out-of-plane tilt angle and they showed that there exists a linear correlation between $^{3h}J_{NC'}$ and the isotropic H^N chemical shift. They computed also the $^{2h}J_{NN}$ couplings in guanine/cytosine (5.36 Hz), uracil/adenine (7.65 Hz) and thymine/adenine pairs (6.45 Hz).

A second paper by Czernek and Brüschweiler [13] is related to the experimental observation of $^{2h}J(^{31}P-^{1}H)$ and $^{3h}J(^{31}P-^{15}N)$ in protein-nucleotide complexes (see Appendix, Table 7). Using the same methodological approach, they calculated two complex fragments **9** and **10** with different N–H···O=P geometries.

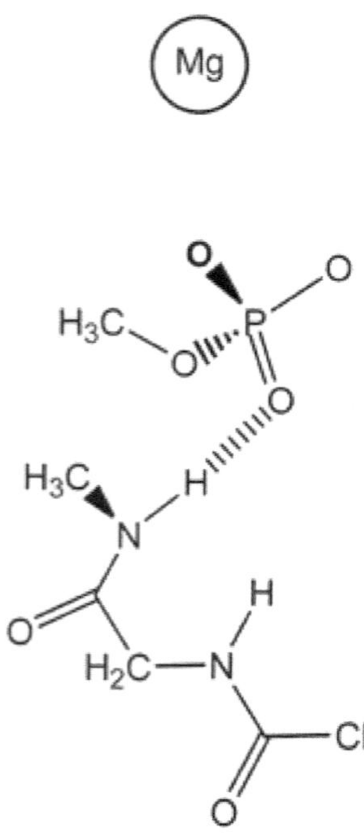

9

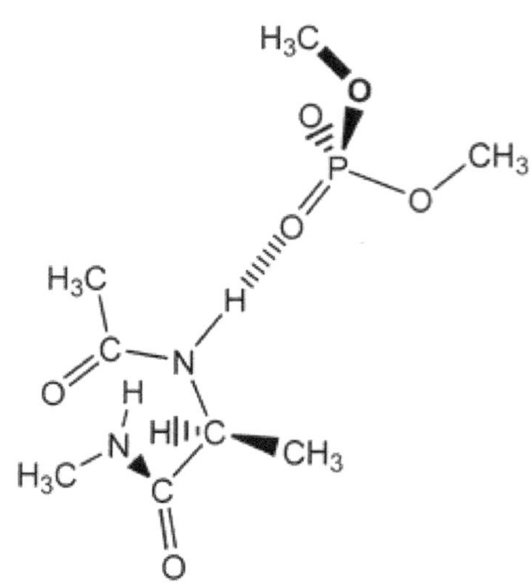

10

Fragment **9** is a model of Gly15 together with $Mg[PO_4(CH_3)]$ and fragment **10** is a model of Ala18 together with $[PO_4(CH_3)_2]^-$. They devised a two-dimensional surface of the form:

$$^{nh}J_{PX}(r_{NO}, \alpha_{POH}) = a_0 e^{-a_1 r_{NO}} \{\cos(\alpha_{POH} - \alpha_{offset})\}^{a_2} \quad (2)$$

Each value of $^{nh}J_{PX}$ (for a pair of N···O distance and P–O···H angle) depends on a series of parameters that have been calculated. For instance, $a_0 = -9628.6$ Hz, $a_1 = 2.7232$ Å^{-1}, $a_2 = 2.4761$, $\alpha_{offset} = 3.1853$ rad for $^{3h}J_{PN}$. They proposed to use equation (2) in a similar way to the Karplus equation.

It will be interesting to know whether the type of equation known as the Karplus equation is followed in supramolecular complexes. For instance, if the dependence of $^{3}J_{FF}$ in compound **11** has a counterpart in $^{4h}J_{FF}$ in complex **12**.

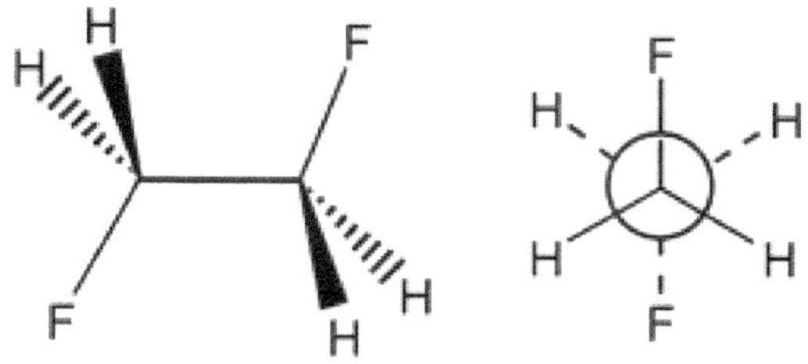

11

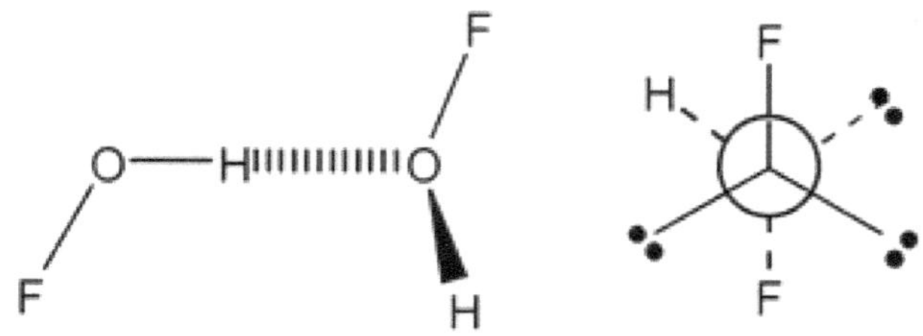

12

Barfield's contribution is important having published four papers between 1999 and 2002 [14,15,16,17,18]. In his first paper, published with Grzesiek, two DNA triplets **13** (T•A-T) and **14** (C^+•G–C) were studied theoretically (DFT-FTP// UB3PW91/6-311G**) to be compared with their experimental results (Table 7 of the Appendix) [14,15]. In these systems, two trans-hydrogen bond $^{2h}J_{NN}$ and $^{1h}J_{HN}$ scalar couplings were calculated. Interesting correlations were found between the chemical shift of the proton of the HB and $^{1h}J_{HN}$ (positive), $^{2h}J_{NN}$ (positive) and $^{1}J_{NH}$ (negative), both between calculated values (Eqs. 3 and 4) and between experimental values (Eq. 5).

$$^{1}J_{HN} = -91.9 + 1.01\ \delta^{1}H,\ r^2 = 0.984 \quad (3)$$

$$^{2h}J_{NN} = -9.4 + 1.17\ \delta^{1}H,\ r^2 = 0.9999 \quad (4)$$

$$^{2h}J_{NN} = -9.6 + 1.27\ \delta^{1}H,\ r^2 = 0.861 \quad (5)$$

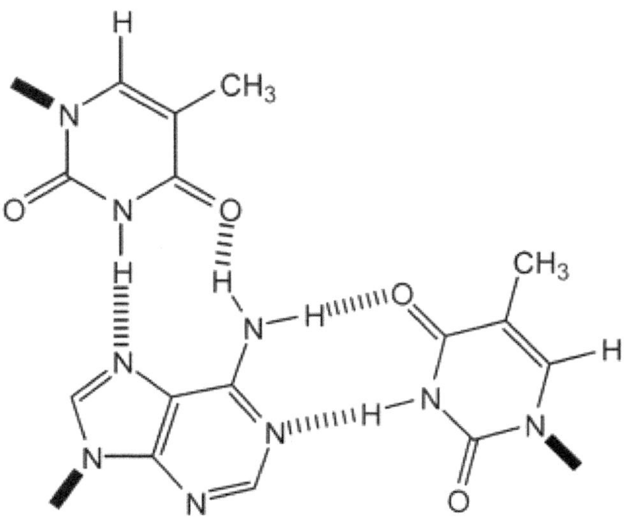

13

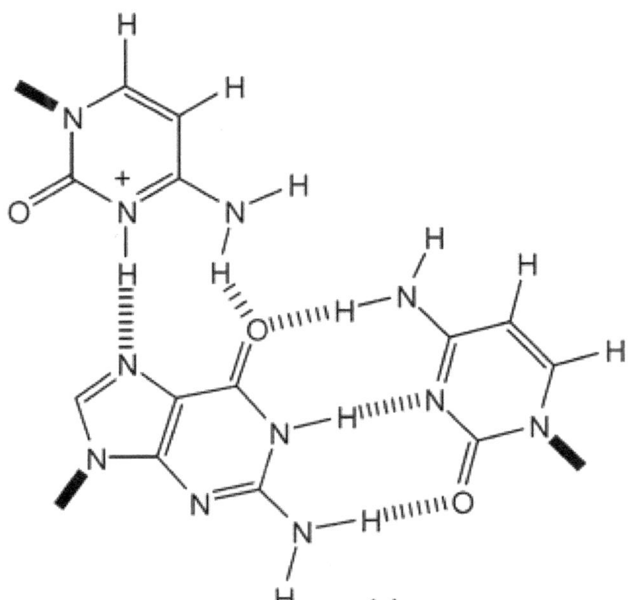

14

Barfield, O'Leary et al. measured (see Table 7, Appendix) $^{2h}J_{HH}$ couplings in 1,3 and 1,4-diols of favorable conformation and computed (DFT/FTP) these couplings for different geometries of the fragment O–H···O–H with acceptable results [16,17]. In his last publication, Barfield addressed the problem of proteins, and specifically, the notorious $^{3h}J_{NC'}$ coupling [18]. He used as a model, formamide dimer, but instead of a cyclic dimer, like **6**, he used a more realistic open-dimer **15**.

He found that $^{3h}J_{NC'}$ depends on r_{OH}, θ_1, θ_2 and the torsion angle ρ about the C=O bond (H···O=C›–N). Several equations were devised in this very important publication. Those relating $^{3h}J_{NC'}$ with $\cos^2\rho$ bear a clear similarity with the Karplus relationship.

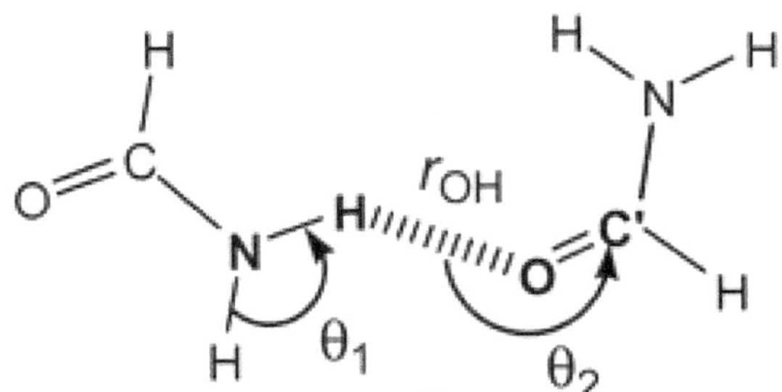

15

The authors of one of the most widely used methods for calculating coupling constants, Malkin and Malkina, collaborated with Limbach in the study of the experimental findings of these last authors (reported in Table 7). Thanks to experiments at very low temperatures (between 110 and 150 K) using freons as solvents, Golubev, Limbach and their coworkers measured ^{1}H-^{19}F coupling constants in fluoride ion/ hydrogen fluoride complexes (see Scheme 1) [19].

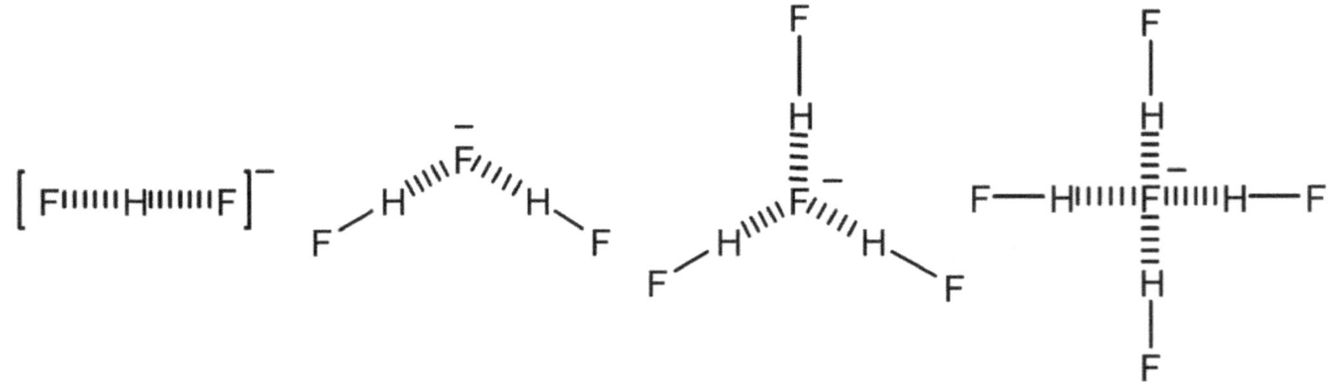

Scheme 1. Species formed by mixing HF and tetrabutylammonium fluoride.

They measured or estimated all the possible coupling constants that were also calculated at the MCLR-CAS level [19]. This collection of data is still remarkable for the wealth of information it contains being completely apart from peptides, nucleic acids or their combination.

The second publication is of a more theoretical nature [20]. The authors used the famous Steiner-Limbach diagrams, where geometrical (such as q_2) and NMR properties (such as δ^1H of the H in the HB, ^nJ and ^{nh}J) were represented against q_1:

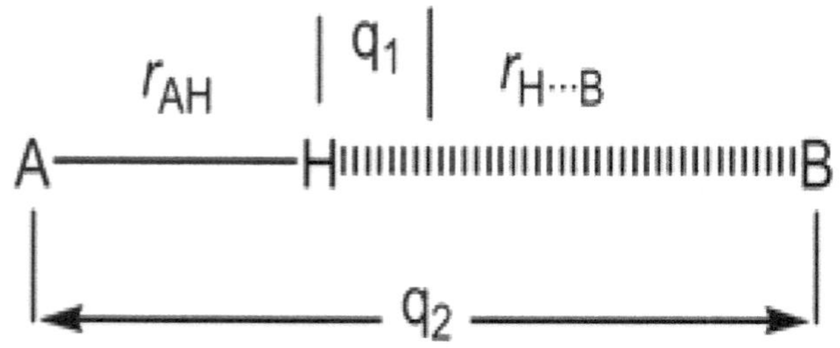

$$q_1 = (r_{AH} - r_{H\cdots B})/2 \quad (6)$$

$$q_2 = r_{AH} + r_{H\cdots B} \quad (7)$$

The cases of $[F(HF)_n]^-$, n = 1-4, discussed above, of [C≡N…N≡;C]$^-$, formamidine dimer **16** were calculated. For instance, we report below the calculated value (in Hz) of $^{2h}J_{NN}$ of **16**.

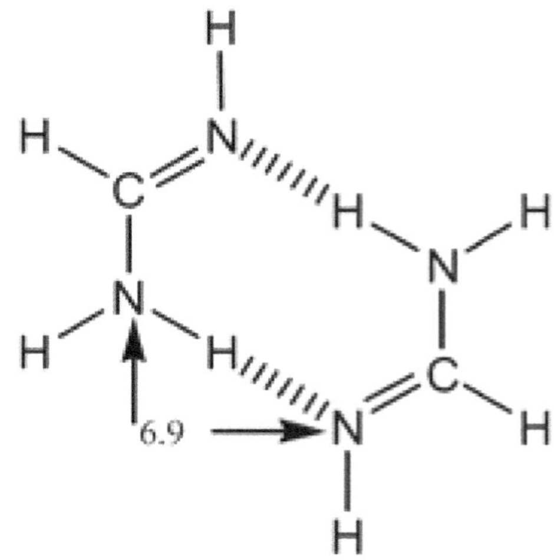

16

Although not reporting coupling constants, a third paper by Limbach is worth quoting because it describes the influence of an electric field on the geometry and properties of A–H···B complexes [21].

Bagno has examined the case of formamide dimer **15** in relation with the case of ubiquitin (see Table 7) [22]. He has built up the coupling surface of $^{3h}J_{NC'}$ in function of the N···O distance and the dihedral angle. His conclusions, based on FTP/DFT calculations are similar to those of Barfield [18], which were obtained two years later.

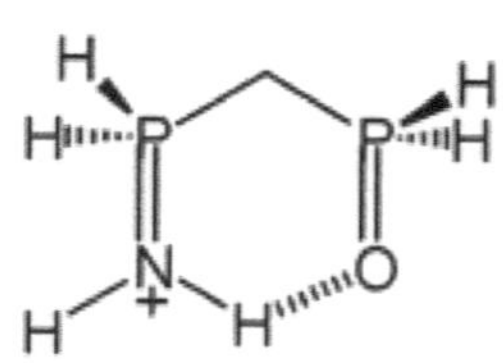

17a (C_s)
$\phi_1 = \phi_2 = 0°$

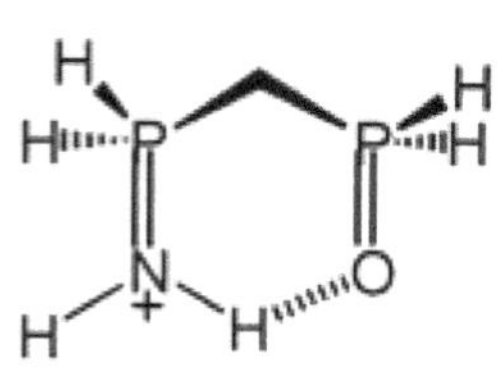

17b (C_1)
$\phi_1 = 62.0°, \phi_2 = -1.5°$

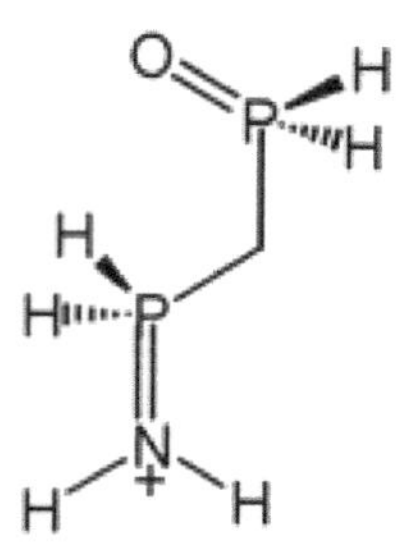

18 (C_s)
$\phi_1 = 180°, \phi_2 = 0°$

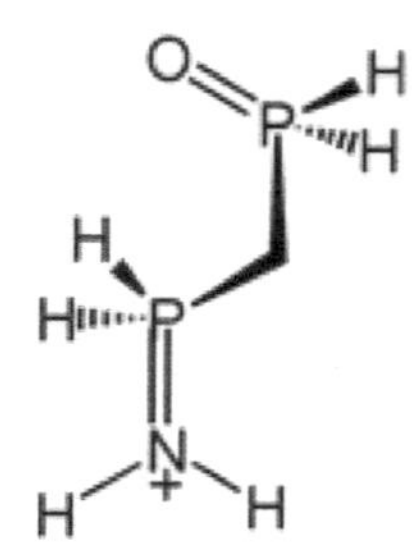

19 (C_1)
$\phi_1 = 172.3°, \phi_2 = -3.6°$

Finally, we have calculated the coupling constants of structures **17-19** in an attempt to determine whether the coupling of 1.5 Hz between ^{15}N and ^{31}P (not directly bonded) is transmitted through the skeleton ($^4J_{NP}$) or through the HB ($^{3h}J_{NP}$) [23]. Only the Fermi contact term was calculated using the FTP/B3LYP/aug-cc-pVTZ approach. The conclusion is that the main transmission pathway is through the covalent bonds.

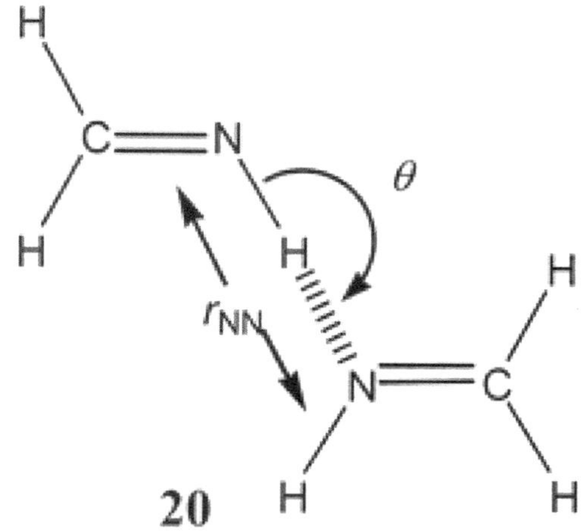

20

Bryce and Wasylishen have modeled the $^{2h}J_{NN}$ in nucleic acid base pairs using methylene-imine dimer **20** in function of hydrogen bond geometry (r_{NN} and θ) [24]. The variation of $^{2h}J_{NN}$ as a function of r_{NN} is fit to an exponential decay [r_{NN} = 3.36 – 0.37 ln ($^{2h}J_{NN}$)]. The influence of θ is less pronounced.

b) Bartlett, Perera, Del Bene›s contribution [25,26,27,28,29,30,31,32,33,34,35,36].

Remember that all papers by this group are based on high-level EOM-CCSD calculations. Perera and Bartlett [25] initiated their systematic exploration of scalar couplings through hydrogen-bonds inspired by Limbach-Malkina›s paper on $[FHF]^-$and $[F(HF)_n]^-$ (n = 1-4) [20]. These authors criticize the use of DFT methods when dealing with fluorine atoms "The MCLR results reported for $[FHF]^-$ by Limbach et al. are unrealistic". The estimated value for $^{2h}J_{FF}$ is > 200 Hz, the MCLR calculated value is –133 Hz and the EOM-CCSD value is + 225 Hz. We will see in the following discussion that ^{19}F NMR properties are amongst the most difficult to calculate.

In the second publication of the same authors with Del Bene [26], they examined the case of the "low-barrier" hydrogen bonds, a topic of general interest for its biological implications, trying to find out whether predicted NMR coupling constants across hydrogen bonds can be used as fingerprints for specifying hydrogen bond types. They calculated a certain number of ^{17}O–^{17}O, ^{15}N–^{15}N and ^{15}N–^{17}O systems (Table 2). Amongst their main conclusions are that for these nuclei, the FC term dominates, that the relationship between $^{2h}J_{OO}$ (Hz) and d_{OO} (Å) shows an exponential decay, and that $^{2h}J_{NN}$ for d_{NN} = 2.9 Å is 7.2 Hz close to the experimental findings (see Table 7).

Table 2. NMR total and FC coupling constants (Hz) for A–H–B systems

Complex	AB	HB type[a]	d_{AB} (Å)	Fermi-contact (Hz)	J_{total} (Hz)
$O_2H_5^+$	OO	PS	2.38	39.92	39.54
$O_2H_3^-$	OO	PS	2.44	16.28	17.96
O_2H_4	OO	T	2.91	1.47	1.29
$^+H_2OH:NCH$	ON	PS	2.47	34.07	34.12
$HOH:NC^-$	ON	T	2.82	6.62	6.55
$CNH:OH_2$	NO	T	2.84	8.57	8.51
HOH:NCH	ON	T	3.13	1.14	1.16
$^+HCNH:NCH$	NN	PS	2.52	32.46	32.52
$CNH:NC^-$	NN	PS	2.58	21.47	21.52
CNH:NCH	NN	T	3.00	5.60	5.62

[a] PS = Proton-shared; T = Traditional.

Then, Del Bene and Jordan examined the case of the complex between hydrogen chloride and ammonia that Del Bene had already studied in what concerns its geometry, energy and vibrational aspects [27]. To the $Cl–H\cdots NH_3$ complex they applied an external electric field along the HB to modify the character of the HB, an approach already used by Limbach and ourselves [21]. When the field increases, the system moves from traditional, to shared, to ion-pair. We report their most significant results in Table 3. They criticized the use by Limbach and ourselves of the q coordinates concluding that they are only useful for symmetric systems (which is not correct), that the FC term dominates and showing that $^{2h}J_{15N\text{-}35Cl}$ is negative.

Table 3. Equilibrium Cl···N distances, proton chemical shifts and $^{2h}J_{15N\text{-}35Cl}$ coupling constants as functions of external field strength

Field (a.u.)	$d_{Cl\cdots N}$ **(Å)**	δ **(ppm)**	$^{2h}J_{15N\text{-}35Cl}$ **(Hz)**
0.0000	3.080	10.0	–5.7
0.0010	3.056	10.8	–6.1
0.0025	3.019	12.0	–6.9
0.0040	2.975	13.6	–8.0
0.0055	2.832	20.9	–11.8
0.0100	2.896	19.1	–9.1
0.0150	3.004	16.4	–6.7

Contrary to most authors, they prefer graphical representations to statistical models. Their graphs clearly showed that their seven data belong to two families, the first set corresponds to 0.0000-0.0040 a.u. fields and the second one to 0.0055-0.0150 a.u. fields. These sets can be adjusted to linear correlations, although they probably correspond to the two branches of an exponential [21] (eqs. 8-13).

- First set (traditional Cl–H···N HB):

$d_{N\cdots Cl} = (3.082\pm0.002) - (26.2\pm0.9)$ field, $n = 4$, $r^2 = 0.998$ (8)

$\delta^1H = (8.93\pm0.11) + (893\pm47)$ field, $n = 4$, $r^2 = 0.994$ (9)

$^{2h}J_{Cl\cdots N} = -(5.60\pm0.11) - (576\pm49)$ field, $n = 4$, $r^2 = 0.986$ (10)

- Second set (ion-pair $Cl^-\cdots H\text{–}N^+$ HB):

$d_{N\cdots Cl} = (2.73\pm0.02) + (18\pm2)$ field, $n = 3$, $r^2 = 0.987$ (11)

$\delta^1H = (23.6\pm0.4) - (475\pm40)$ field, $n = 3$, $r^2 = 0.993$

$^{2h}J_{Cl\cdots N} = -(14.6\pm0.3) + (536\pm35)$ field, $n = 3$, $r^2 = 0.996$ (13)

The proton shared situation is very unstable [21] and should be found in the 0.0040-0.0055 au region.

The next contribution by Del Bene and Bartlett was devoted to the $^{2h}J_{NN}$ coupling present in some $^{15}N\text{–}H\cdots{}^{15}N$ fragment [28] (see Table 7). Here again, the FC term dominates, other terms being negligible. Their results are reported in Table 4 (note that is $^{2h}J_{NN}$ positive).

21

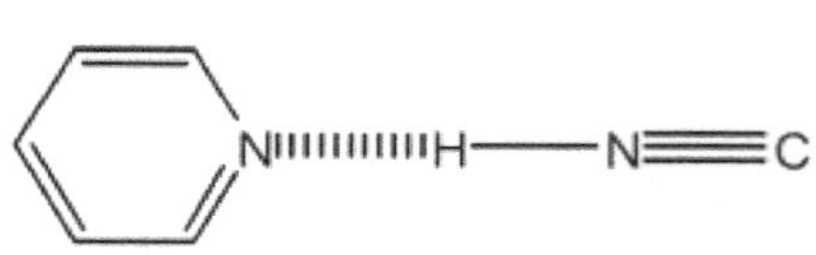

23

22

Table 4: Equilibrium N···N distances, symmetry and $^{2h}J_{15N\text{-}15N}$ coupling constants

Complex	**symmetry**	$d_{N\cdots N}$ **(Å)**	$^{2h}J_{15N\text{-}15N}$ **(Hz)**
pyrrole:NCH (**21**)	C_{2v}	3.16	3.0
CNH:NCH (**22**)	$C_{\infty v}$	3.00	5.5
$CNH:NH_3$	C_{3v}	2.85	8.7
CNH:NCLi	$C_{\infty v}$	2.83	9.6
CNH:pyridine (**23**)	C_{2v}	2.79	10.7

Reference [29] reports the calculations carried out on the problem of a long range coupling constant present in iminophosphorane-substituted proton-sponges **24** (see Table 7). Using as a model structure **25**, the results reported in Table 5 were found (only the FC term was computed).

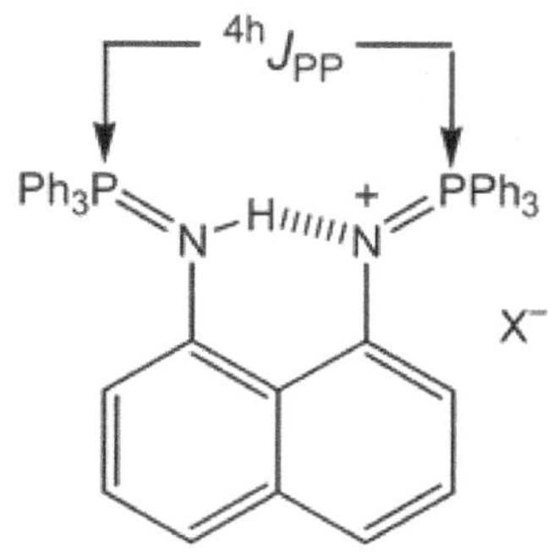

24

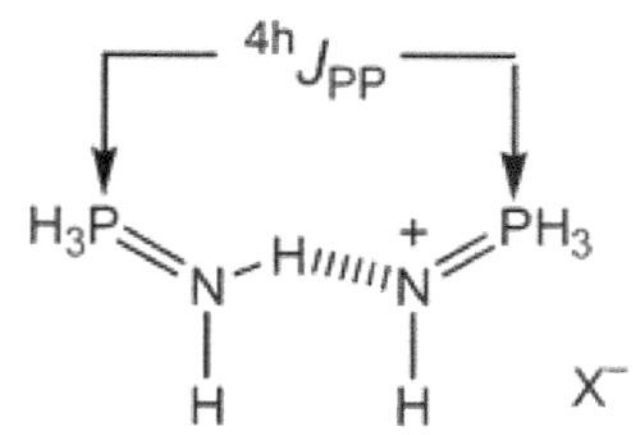

25

Table 5: N···N and P···P distances and computed $^{4h}J_{31P\text{-}31P}$ coupling constants in model **25**

N···N distance (Å)	P···P distance (Å)	$^{4h}J_{31P\text{-}31P}$ (Hz)
2.53	4.79	14.3
2.83	5.09	8.1
3.03	5.29	5.9
3.33	5.59	3.9
3.53	5.79	3.0

These data can be analyzed statistically leading to eqs. (14) and (15):

$d_{P\cdots P} = (2.27\pm0.02) + (0.997\pm0.003)\ d_{N\cdots N}$, n = 5, $r^2 = 1.000$ (14)

$\ln\ ^{4h}J_{31P\text{-}31P} = (6.96\pm0.10) - (4.66\pm0.09)\ \ln d_{N\cdots N}$, n = 5, $r^2 = 0.999$ (15)

Del Bene, Parera and Bartlett used in another contribution the reduced coupling constants K to correct some erroneous statements about the relative importance of scalar couplings [30]; for the N–H⁺–N and O–H⁺–O complexes $^{2h}K_{X\cdots X} > {}^{1}K_{X\text{-}H} > {}^{1h}K_{X\cdots X}$. Jordan, Del Bene et al. devoted another paper to the N–H···N system [31]. They used the pyrrole:hydrogen cyanide (**21**) and the hydrogen isocyanide:hydrogen cyanide (**22**) systems and, as perturbation, an electric field or chemical substitution (fluorine and Be^{2+} on the pyrrole, Li, Na, S^- and O^- instead of H in NCH). Some of their results are reported in Table 6.

Table 6: Chemical shifts of the hydrogen-bonded proton and $^{2h}J_{15N\text{-}15N}$ coupling constants as a function of field strength

Field (a.u	δ^1H (ppm)	$^{2h}J_{15N\text{-}15N}$ (Hz)
0.0000	1.9	6.4
0.0040	2.4	7.3
0.0100	3.2	8.8
0.0150	4.1	10.3
0.0200	5.2	12.3
0.0225	6.0	13.7
0.0250	7.1	15.5

In all cases, proton-shared and ion-pair N_a–H–N_b HBs are unlikely to be formed in neutral complexes. These authors introduced an important representation to understand the nature and properties of hydrogen-bonded systems, plotting N_a–H (Å) against N_b–H (Å) and representing the square of the ground-state vibrational wave function superimposed on the potential energy surface.

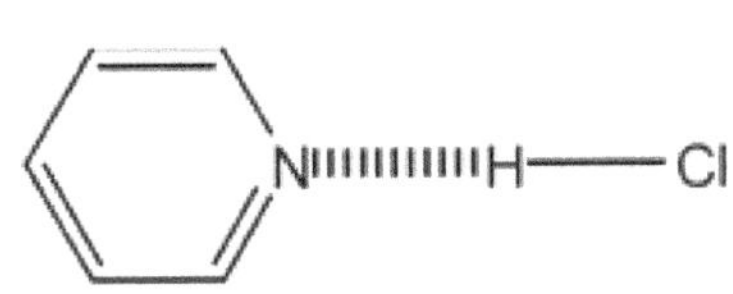

26

The next problem they tackled concerns the hydrogen chloride:pyridine complex (**26**) [32]. Pyridine is a very good

HBA because the positive charge can be delocalized over the ring. In this example, increasing the field, they succeeded to attain the proton-shared and afterwards the ion-pair structures. They carried out the calculations using an isotropic external electric field (analogous to the Onsager's model), optimizing the structure for each field (under some restrictions) and then calculating the coupling constants with the field turned off (turning on the external field during the calculations increases the absolute value of $^{2h}J_{35Cl\cdots15N}$). Their calculations are in agreement with Limbach's results on temperature effects o HBs: decreasing the temperature corresponds to an increase of the field and the traditional system evolves to proton-shared and then to ion pair.

Del Bene, Jordan, Parera and Bartlett [33] returned to the problem of $^{2h}J_{FF}$ in $[FHF]^-$ and $[FDF]^-$ that was their first publication on scalar couplings across HBs [25]. They examined the possible effects of zero-point vibrational effects and of vibrational averaging, to conclude that both have essentially no effect on $^{2h}J_{FF}$. They confirmed that, although the FC term dominates, the PSO term is also important. Using Limbach's estimation of 220 Hz for $^{2h}J_{FF}$ in $[FHF]^-$ for comparison, they reported for $[FHF]^-$ 212.7 Hz and for $[FDF]^-$ 223.1 Hz. This is an important result that allows a new entrance to isotope effects.

Two papers [34,35] can be consulted to get an overview of the preceding series of papers. One dealing with ^{15}N,^{15}N spin-spin coupling constants across N–H–N and N–H$^+$–N hydrogen bonds and entitled "Can coupling constants provide reliable estimates of N–N distances in biomolecules?" (with a positive answer both for the distance and the angular variations) and the other with "what a difference a decade makes: progress in ab initio studies of the hydrogen bond".

The last paper until now reports calculations of $^{3h}J_{15N\text{-}31P}$ spin-spin coupling constants across N-H···O-P hydrogen bonded complexes [36]. This work is similar in subject but different in methodology to that of Czernek and Brüschweiler [12]. Instead of complex structures **9** and **10**, equation-of-motion coupled cluster singles and doubles (EOM-CCSD) calculations were performed in model cationic and anionic complexes including NH_4^+:OPH, NH_4^+:OPH_3, NH_3:$^-O_2PH_2$, NFH_2:$^-O_2PH_2$, and NF_2H:$^-O_2PH_2$. Three-bond coupling constants can be appreciable when the phosphorous is P(V), but are negligible with P(III). $^{3h}J_{N\text{-}P}$ values in complexes with cyclic or open structures are less than 1 Hz, a consequence of the nonlinear arrangement of N, H, O, and P atoms. For complexes with these structures, $^{3h}J_{N\text{-}P}$ may not be experimentally measurable. In contrast, complexes in which the N, H, O, and P atoms are collinear or nearly collinear have larger values of $^{3h}J_{N\text{-}P}$, even though the N-P distances are longer than N-P distances in cyclic and open structures. In linear complexes, $^{3h}J_{N\text{-}P}$ is dominated by the Fermi-contact term, which is distance dependent. Therefore, N-P (and hydrogen-bonding N-O) distances in these complexes can be determined from experimentally measured ^{15}N-^{31}P coupling constants.

c) Data analysis. Peralta, Contreras et al. analyzed using NBOs (Natural Bonding Orbitals) the C-H···O interaction in NCH···OH_2 and NCH···O=CH_2 **27** and their effect on nuclear magnetic shielding constants [37]. In a subsequent paper, they expanded the same methodology to the dissection of fluorine–fluorine couplings (see below, **31**). They proposed a partition of the Fermi contact term as a sum of contributions of core orbitals, non-bonding electron pairs and bonding orbitals. As expected the coupling constant through the space are governed by the non-bonding electron pair term in all the cases studied [38].

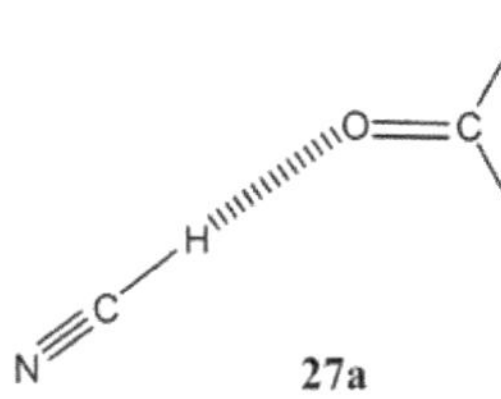

27a

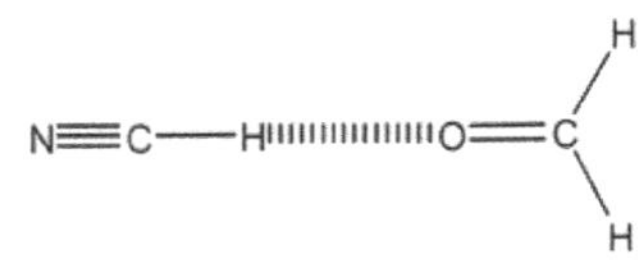

27b

Weinhold, the creator of NBO analysis, applied his method first to an example of covalent bonds ($^3J_{HH}$ couplings in ethane) [39] and then to $^{1h}J_{NH}$ and $^{2h}J_{NN}$ in DNA A-T base pair [3]. The contributions of the orbitals involved in the HB to the coupling constants are able to explain the differences observed between the $^{2h}J_{NN}$ and $^{1h}J_{NH}$ [3]. Arnold, Oldfield et al. used DFT theory to calculate and analyze through-space ^{19}F-^{19}F coupling constants [40] and found an exponential relationship between the coupling constant value and the internuclear distance. The same authors compared the $^{nh}J_{NC'}$ in proteins with the electron density properties obtained using Bader's Theory of Atoms in Molecules (AIM) [41]. They found an exponential relationship between the mutual penetration of the electronic clouds and the coupling constant values.

Hydrogen Bonded Molecules: Couplings through a Dihydrogen Bond

This fascinating topic has been covered recently [42], we have replaced the letter h of hydrogen bonds by the letter d of dihydrogen bonds. EOM-CCSD calculations were performed to determine one-bond ($^{1d}J_{H\text{-}H}$) and three-bond ($^{3d}J_{X\text{-}M}$) spin-spin coupling constants across X–H···H–M dihydrogen bonds for complexes with ^{13}C-^1H, ^{15}N-^1H, and ^{17}O-^1H proton-donor groups and proton-acceptor metal hydrides ^7Li-^1H and ^{23}Na-^1H. Unlike two-bond spin-spin coupling constants across N-H-N, N-H-O, O-H-O, and Cl-H-N hydrogen bonds, which are determined solely by the Fermi-contact term (see above), $^{1d}J_{H\text{-}H}$ receives non-negligible contributions from the PSO and DSO terms. However, these terms tend to cancel, so that the curve for the distance-dependence of $^{1d}J_{H\text{-}H}$ is determined by the distance-dependence of the Fermi-contact term. The value of $^{1d}J_{H\text{-}H}$ is dependent on the nature of the proton-donor and proton acceptor, and the relative orientation of the bonded pair. Hence, it would be difficult to extract structural information from experimentally measured coupling constants without a model tailored specifically for the experimental complex. $^{3d}J_{C\text{-}Li}$ values for the equilibrium structures of seven linear complexes stabilized by C-H···H-Li bonds are dependent on C-Li distances, and are also sensitive to structural changes which remove any one of these four atoms from the dihydrogen bond. $^{3d}J_{O\text{-}M}$ for the complexes HOH:HLi and HOH:HNa exhibit unusual behavior as a function of the O-M distance, increasing with increasing distance through a change of sign, reaching a maximum, and then subsequently decreasing.

Van der Waals Complexes

The number of publications devoted to the study of scalar couplings in van der Waals complexes is much more reduced than those concerning HBs. The reason is fairly simple: there are no experimental results! Therefore, the only existing papers reported explorations of hypothetical situations, although their conclusions are important: i) the transmission of the scalar nuclear spin-spin coupling is a property not restricted to the HBs; ii) the existence of the nuclear spin-spin coupling transmitted through an interaction is not an evidence of the covalent character of this interaction.

Salsbury and Harris [8] examined the case of the xenon dimer. Exploiting the fact that xenon has two isotopes with spins ≠ 0 (^{129}Xe, spin I = 1/2, and ^{131}Xe, spin I = 3/2), they have calculated the FC term for the Xe_2 and for the Xe-^{1}H complexes. In the regions of interest J(Xe-H) is microHz and J(^{129}Xe-^{131}Xe) is milliHz.

More recently, a similar study was carried out on He_2 (^{3}He, spin 1/2). Obviously, this coupling is not accessible experimentally. At the energy minimum (5.6 a.u. = 2.96 Å), the coupling should amount to 1.3 Hz [43]. All the terms were calculated at the EOM-CCSD level, but only FC is important. The same author with Sadlej and Leszczynski [44] has calculated that the HF···CH_4 van der Waals dimer should have a J_{HF} ≈ 4 Hz. In this same paper they reported calculations of the [FHF]$^-$ anion (Scheme 1) and the F–H···F–H dimer. Bryce and Wasylishen extended these calculations to HF···CH_3F (in two conformations) [45]; they found that the isotropic coupling constants are dominated by the Fermi-contact mechanism.

At lower level (DFT, only FC, PSO and DSO terms) but much more complex structures, Bagno, Saielli and Scorrano [46] carried out calculations on three van der Waals complexes: methane-methane dimer (not reported), methane-benzene **28** and benzene-benzene (T-shaped) **29**.

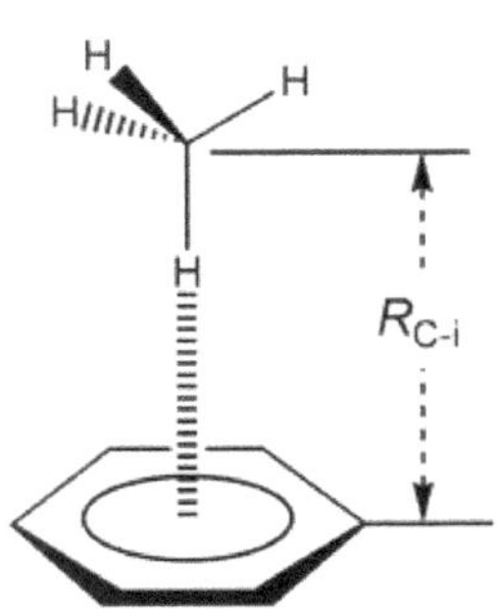

28

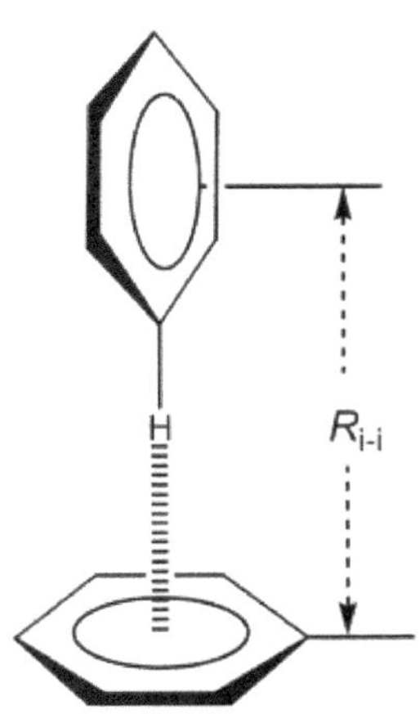

29

The ^{1}H-^{1}H couplings are very small, for all purposes null, but the ^{1}H-^{13}C should be in the 0.2-0.3 Hz range (for equilibrium geometries, R_{C-i} = 3.685 Å and R_{i-i} = 4.90 Å) and therefore experimentally measurable. However, these model systems are not suitable for an experimental verification due to their too-large exchange rate compared to their small couplings. A better candidate would be the ^{15}N-^{19}F coupling in the complex CH_3-CN···C_6F_6 **30** [47].

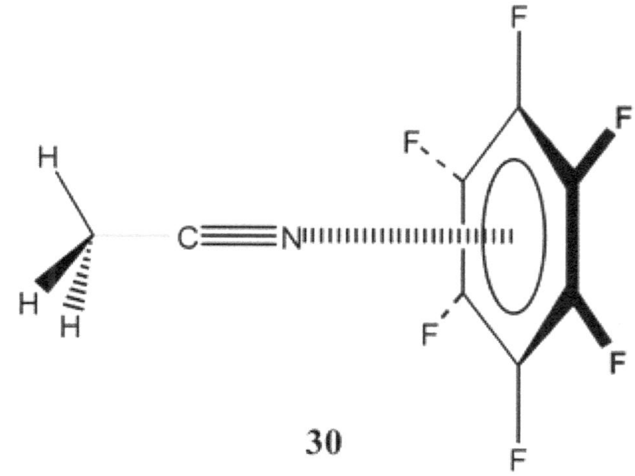

30

Bagno, Saielli and Scorrano [48] have examined a large set of van der Waals complexes (CH_4/CH_4, CH_4/C_2H_4, C_2H_4/C_2H_4, CH_4/C_6H_6 **28**, C_6H_6/C_6H_6 **29**) as well as some more realistic and complex systems in search of type **2** interactions (calixarene/CH_3CN, the Tröger's balance of Wilcox, etc.).

This is probably the best place to describe the attempts of two groups to analyze the ^{19}F/^{19}F nuclear spin-spin coupling through-space. Both used, as a model, two hydrogen fluoride molecules, Peralta, Contreras et al. [38] in disposition **31** and Mallory et al. [50] in disposition **32**.

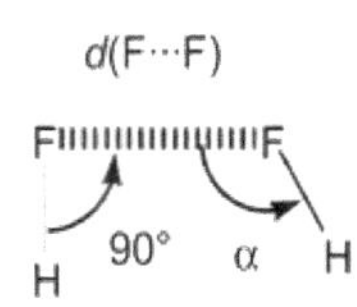

31

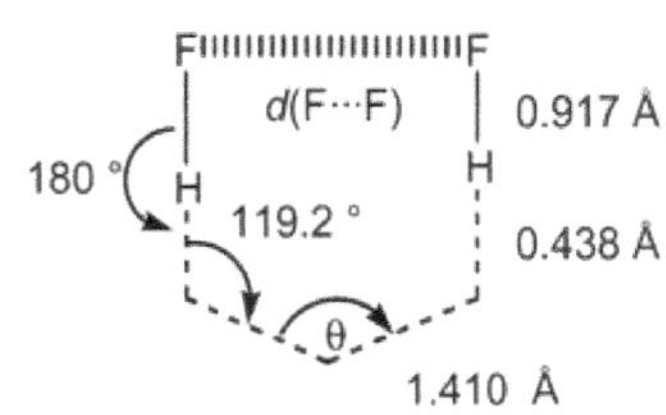

32

The second authors did not carry out calculations on ^{19}F^{19}F couplings, and concluded only that there is an exponential relationship between J_{FF} and d(F···F) [50]. Peralta, Contreras and Snyder have carried out a natural bond order (NBO) dissection of J_{FF} [38].

CONCLUSIONS

We are, at the end of 2002, in an interesting and unstable situation: the calculation of scalar couplings shows the same impetus and problems as some ten years ago did the calculation of nuclear shieldings, being still highly demanding on computational effort. The authors face a difficult election: either to considerably simplify the problem or to lower the level of the method to the point to obtain dubious results. Even with a considerable simplification, there are problems accessible only to people having access to big supercomputational centers.

It is expected than in the next ten years, development in software and hardware will allow some fascinating problems to be tackled. For instance, other interactions deserved to be explored like inverse HBs [51,52], three-centered (bifurcated) HBs [53], charge-transfer bonds [54], etc. One of the main objectives that should be approached in the next years is the calculation of coupling constants including heavy atoms (therefore, including relativistic corrections) because the structure of many fundamental problems in chemistry (catalysis) and biochemistry involve these atoms.

We hope that this review will encourage other physicists, chemists, biochemists, molecular biologists and NMR spectroscopists to endeavor the calculation of spin-spin scalar coupling constants.

APPENDIX

We feel useful to summarize in Table 7 the experimentally measured nuclear spin-spin coupling constants through a hydrogen bond, because they are the test used for many theoretical works [55,56,57,58,59,60,61,62,63,64,65,66,67, 68,69,70,71,72,73,74,75,76,77,78,79,80,81,82,83,84,85,86 ,87,88,89,90,91,92,93,94,95,96,97,98,99]. Two cases have to be distinguished: intermolecular and intramolecular HBs (these last we will name as IMHBs). The first class is without ambiguity but not so the second one. We consider coupling constants through HBs between different aminoacids of the same protein, due to folding, as intermolecular coupling constants. Couplings observed in IMHBs can be transmitted through the HB or through the covalent skeleton. The transmission through the skeleton could represent up to a 100% of the observed coupling. There is no experimental possibility to separate both mechanisms and even not a theoretical partition method available. These couplings through IMHBs have been ascribed to ^{nh}Js with the qualitative argument that, in similar compounds, the skeleton is unable to transmit the spin information. Some of them are illustrated in Scheme 2.

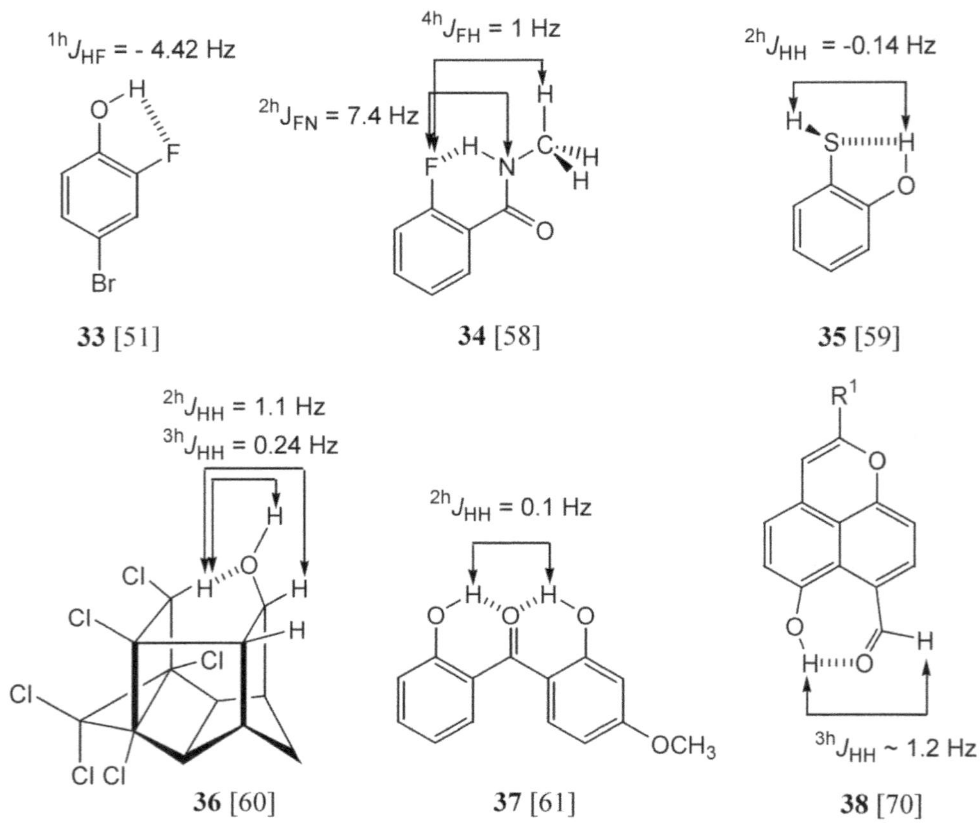

Scheme 2. Some examples of scalar couplings through intramolecular HBs (IMHB).

Amongst the intermolecular HBs, the most common situations are summarized in Scheme 3. They are found both in nucleic acids and in proteins (in this last case, the atoms of the second aminoacid residue are noted with a dash).

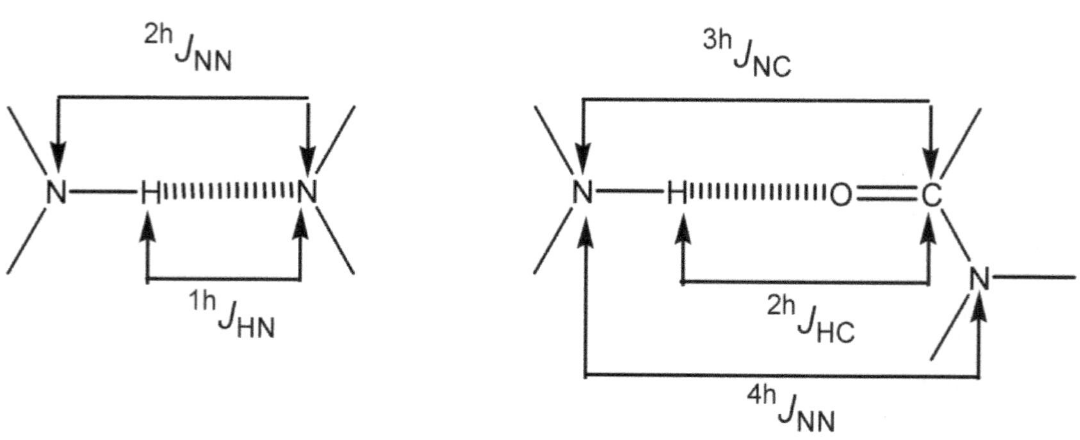

Scheme 3. The most frequent intermolecular couplings.

Especially noteworthy is reference [91]; there, for the first time, a coupling constant through a hydrogen bond was measured in the solid state [a $^{2h}J(^{15}N\text{-}H\cdots{}^{15}N)$]: this opens extraordinary expectations for the study of ^{nh}Js in intermolecular simple systems where there is not a single example.

Table 7. Experimental coupling constants through hydrogen bonds

Compound	Couplings measured	HB	Ref.
	^{1h}Js		
F···H	$^{1h}J_{F\cdots H}$	IMHB	[55]
[F···H···F]⁻	$^{1h}J_{F\cdots H}$	inter	[19]
[F···H···N(collidine)]	$^{1h}J_{F\cdots H}$	inter	[56,57]
[F···H···N(collidine)]	$^{1h}J_{F\cdots H}$	inter	[56,57]
[RCO_2^-···H···N(pyridine)]	$^{1h}J_{F\cdots H}$	inter	[58]
[Pyridine···H+···pyridine]	$^{1h}J_{F\cdots H}$	inter	[59]
DNA, RNA	$^{1h}J_{F\cdots H}$	inter	[14,15,16,60,61,62,63,64,65,66,67]
Nucleic acids	$^{1h}J_{F\cdots H}$	inter	[68,69,70,71,72]
Fulvenes	$^{1h}J_{F\cdots H}$	IMHB	[73,74,75]
	^{2h}Js		
[F···H···F]–	$^{2h}J_{F\text{-}H\cdots F}$	inter	[19]
[F···H···N(collidine)]	$^{2h}J_{F\text{-}H\cdots N}$	inter	[56,57]
N-H···F	$^{2h}J_{N\text{-}H\cdots F}$	IMHB	[76]
H-S···H	$^{2h}J_{H\text{-}S\cdots H}$	IMHB	[77]
H-O···H	$^{2h}J_{H\text{-}O\cdots H}$	IMHB	[16,78]
H···O···H	$^{2h}J_{H\cdots O\cdots H}$	IMHB	[79]
Proteins	$^{2h}J_{C=O\cdots H}$ (N-H···O=C)	inter	[80,81,82,83,84,85,86,87,88,89,90]
DNA, RNA	$^{2h}J_{N\text{-}H\cdots N}$	inter	[14,15,16,60,61,62,63,64,65,66,67]
Nucleic acids	$^{2h}J_{N\text{-}H\cdots N}$	inter	[68,69,70,71,72]
Apomyoglobin	$^{2h}J_{N\text{-}H\cdots N}$ (imidazole···H-imidazolium)	inter	[91]
Proton sponges	$^{2h}J_{N\text{-}H\cdots N}$	IMHB	[92]
Fulvenes	$^{2h}J_{N\text{-}H\cdots N}$	IMHB	[73,74,75,93]
Protein/DNA	$^{2h}J_{P=O\cdots H}$ (P=O···H-N)	inter	[94]
Flavoprotein	$^{2h}J_{P\text{-}O\cdots H}$ (P=O···H-N, P=O···H-O)	inter	[95]
Metalloprotein	$^{2h}J_{Cd\cdots H}$ (N-H···S-Cd)	inter	[96,97]
Metalloprotein	$^{2h}J_{Hg\cdots H}$ (N-H···S-Hg)	inter	[97]
	^{3h}Js		
H-C=O···H-O	$^{3h}J_{H\text{-}C=O\cdots H}$	IMHB	[98]
Galactal	$^{3h}J_{H\text{-}O\cdots H}$ (O-H···O-H)	IMHB	[99]
H-C-O...H	$^{3h}J_{H\text{-}C=O\cdots H}$	IMHB	[78]
Proteins	$^{3h}J_{N\text{-}H\cdots O=C'}$	inter	[80,81,82,83,84,85,86,87,88,89,90]
Nucleic acids	$^{3h}J_{N\text{-}H\cdots O=C'}$	inter	[68,69,70,71,72]
Protein/DNA	$^{3h}J_{N\text{-}H\cdots O=P}$	inter	[68,69,70,71,72]
	^{4h}Js		
H-C-N-H···F	$^{4h}J_{H\text{-}C\text{-}N\text{-}H\cdots F}$	IMHB	[76]
Proton sponges	$^{4h}J_{P=N^+\text{-}H\cdots N=P}$	IMHB	[100,101]
Nucleotides	$^{4h}J_{H\text{-}C\text{-}N\cdots H\text{-}N}$	inter [a]	[102,103]

[a] Observation of the interaction but J was not measured.

REFERENCES

1. Harris, R. K.; Kowalewski, J.; De Menezes, S. C. Parameters and symbols for use in Nuclear Magnetic Resonance. Pure Appl. Chem. 1997, 69, 2489–2495.
2. Raynes, W. T. A proposal concerning the units of the reduced nuclear spin-spin coupling constant. Magn. Reson. Chem.1992, 30, 686.
3. Wilkens, S. J.; Westler, W. M.; Weinhold, F.; Markley, J. L. Trans-hydrogen-bond $^{h2}J_{NN}$ and $^{h1}J_{NH}$ couplings in the DNA A-T base pair: Natural Bond Orbital analysis. J. Am. Chem. Soc. 2002, 124, 1190–1191.
4. Bour, P.; Budesinsky, M. Sum-over-states of the nuclear spin-spin coupling constants. J. Chem. Phys. 1999, 110, 2836–2843.
5. Sychrovsky, V.; Gräfenstein, J.; Cremer, D. Nuclear magnetic resonance spin-spin coupling constants from coupled perturbed density functional theory. J. Chem. Phys. 2000, 113, 3530–3547.
6. Peralta, J. E.; Barone, V.; Contreras, R. H.; Zaccari, D. G.; Snyder, J. P. Through-bond and through-space J_{FF} spin-spin coupling in peridifluoronaphthalenes: Accurate DFT evaluation of the four contributions. J. Am. Chem. Soc. 2001, 123, 9162–9163.
7. Helgaker, T.; Jaszunski, M.; Ruud, K. Ab initio methods for the calculation of NMR shielding and indirect spin-spin-coupling constants. Chem. Rev. 1999, 99, 293–352.
8. Salsbury, F. R.; Harris, R. A. Estimation of the Fermi contact contribution to the xenon-hydrogen and xenon-xenon spin-spin coupling constants. Mol. Phys. 1998, 94, 307–312.
9. Pecul, M.; Sadlej, J. The nuclear spin-spin coupling constants in the water dimer. Chem. Phys. Lett. 1999, 308, 486–494.
10. Pecul, M.; Leszczynski, J.; Sadlej, J. Comprehensive ab initio studies of nuclear magnetic resonance shielding and coupling constants in XH...O hydrogen-bonded complexes of simple organic molecules. J. Chem. Phys. 2000, 112, 7930–7938.
11. Pecul, M.; Leszczynski, J.; Sadlej, J. The shielding constants and scalar couplings in N-H...O:C and N-H... N:C hydrogen bonded systems: An ab initio MO study. J. Phys. Chem. A 2000, 104, 8105–8113.
12. Scheurer, C.; Brüschweiler, R. Quantum-chemical characterization of nuclear spin-spin couplings across hydrogen bonds. J. Am. Chem. Soc. 1999, 121, 8661–8662.
13. Czernek, J.; Brüschweiler, R. Geometric dependence of $^{3h}J(^{31}P\text{-}^{15}N)$ and $^{2h}J(^{31}P\text{-}^{1}H)$ scalar couplings in protein-nucleotide complexes. J. Am. Chem. Soc. 2001, 123, 11079–11080.
14. Dingley, A. J.; Masse, J. E.; Peterson, R. D.; Barfield, M.; Feigon, J.; Grzesiek, S. Internucleotide scalar couplings across hydrogen Bonds in Watson-Crick and Hoogsteen base pairs of a DNA triplex. J. Am. Chem. Soc. 1999, 121, 6019–6027.
15. Barfield, M.; Dingley, A. J.; Feigon, J.; Grzesiek, S. A DFT study of the interresidue dependencies of scalar J-coupling and magnetic shielding in the hydrogen-bonding regions of a DNA triplex. J. Am. Chem. Soc. 2001, 123, 4014–4022.
16. Fierman, M.; Nelson, A.; Khan, S. I.; Barfield, M.; O'Leary, D. J. Scalar Coupling Across the Hydrogen Bond in 1,3- and 1,4-Diols. Org. Lett. 2000, 2, 2077–2080.
17. Barfield, M.; Bergset, J. M.; O'Leary, D. J. Density functional theory studies of transannular $^{1}H\text{-}^{1}H$ J-coupling in half-cage alcohols and rigid 1,3- and 1,4-diols: Conformational dependencies and implications for intramolecular hydrogen bond detection in carbohydrates. Magn. Reson. Chem. 2001, 39, S115–S125.
18. Barfield, M. Structural Dependencies of interresidue scalar coupling $^{h3}J_{NC'}$ and donor ^{1}H chemical shifts in the hydrogen bonding regions of proteins. J. Am. Chem. Soc. 2002, 124, 4158–4168.
19. Shenderovich, I. G.; Smirnov, S. N.; Denisov, G. S.; Gindin, V. A.; Golubev, N. S.; Dunger, A.; Reibke, R.; Kirpekar, S.; Malkina, O. L.; Limbach, H.-H. Nuclear magnetic resonance of hydrogen-bonded clusters between F^- and $(HF)_n$. Experiment and theory. Ber. Bunsen-Phys. Ges. 1998, 102, 422–428.
20. Benedict, H.; Shenderovich, I. G.; Malkina, O. L.; Malkin, V. G.; Denisov, G. S.; Golubev, N. S.; Limbach, H.-H. Nuclear scalar spin-spin couplings and geometries of hydrogen bonds. J. Am. Chem. Soc. 2000, 122, 1979–1988.
21. Ramos, M.; Alkorta, I.; Elguero, J.; Golubev, N. S.; Denisov, G. S.; Benedict, H.; Limbach, H.-H. Theoretical study of the infulence of electric fields on hydrogen-bonded acid-base complexes. J. Phys. Chem. A 1997, 101, 9791–9800.
22. Bagno, A. Quantum chemical modeling of through-hydrogen bond spin-spin coupling in amides and ubiquitin.Chem.-Eur. J. 2000, 6, 2925–2930.
23. López-Leonardo, C.; Alajarín, M.; Llamas-Lorente, P.; Bautista, D.; Jimeno, M. L.; Alkorta, I.; Elguero, J. The molecular structure and NMR properties of P-phosphinoylmethyl aminophosphonium salts. Struct. Chem. submitted.
24. Bryce, D. L.; Wasylishen, R. E. Modeling $^{2h}J_{iso}(N,N)$ in nucleic acid base pairs: Ab initio characterization of the $^{2h}J(N,N)$ tensor in the methyleneimine dimer a a function of hydrogen bond geometry. J. Biomol. NMR 2001, 19, 371–375.
25. Perera, S. A.; Bartlett, R. J. NMR spin-spin coupling constants for hydrogen bonds of $[F(HF)_n]^-$, n = 1-4, clusters. J. Am. Chem. Soc. 2000, 122, 1231–1232.
26. Del Bene, J. E.; Perera, S. A.; Bartlett, R. J. Predicted NMR coupling constants across hydrogen bonds: A fingerprint for specifying hydrogen bond type? J. Am. Chem. Soc. 2000, 122, 3560–3561.
27. Del Bene, J. E.; Jordan, M. J. T. Vibrational spectroscopic and NMR properties of hydrogen-bonded complexes. Do they tell us the same thing? J. Am. Chem. Soc. 2000, 122, 4794–4797.
28. Del Bene, J. E.; Bartlett, R. J. N-N spin-spin coupling constants $[^{2h}J(^{15}N\text{-}^{15}N)]$ across N-H...N hydrogen bonds

in neutral complexes. To what extent does the bonding at the nitrogens influence $^{2h}J_{N-N}$? J. Am. Chem. Soc. 2000, 122, 10480–10481.

29. Del Bene, J.; Perera, S. A.; Bartlett, R. J.; Alkorta, I.; Elguero, J. $^{4h}J(^{31}P-^{31}P)$ coupling constants through N-H$^+$–H hydrogen bonds: A Comparison of computed ab initio and experimental data. J. Phys. Chem. A 2000, 104, 7165–7166.

30. Del Bene, J. E.; Perera, S. A.; Bartlett, R. J. What parameters determine N-N and O-O coupling constants ($^{2h}J_{X-X}$) across X-H$^+$–X hydrogen bonds? J. Phys. Chem. A 2001, 105, 930–934.

31. Toh, J. S.-S.; Jordan, M. J. T.; Husowitz, B. C.; Del Bene, J. E. Can Proton-Shared or ion-pair N–H–N hydrogen bonds be produced in uncharged complexes? A systematic ab initio study of the structures and selected NMR and IR properties of complexes with N–H–N hydrogen bonds. J. Phys. Chem. A 2001, 105, 10906–10914.

32. Chapman, K.; Crittenden, D.; Bevitt, J.; Jordan, M. J. T.; Del Bene, J. E. Relating environmental effects and structures, IR, and NMR properties of hydrogen-bonded complexes: ClH:Pyridine. J. Phys. Chem. A 2001, 105, 5442–5449.

33. Del Bene, J. E.; Jordan, M. J. T.; Perera, S. A.; Bartlett, R. J. Vibrational effects on the F-F spin-spin coupling constant ($^{2h}J_{F-F}$) in FHF$^-$ and FDF$^-$. J. Phys. Chem. A 2001, 105, 8399–8402.

34. Del Bene, J. E.; Perera, S. A.; Bartlett, R. J. ^{15}N,^{15}N spin-spin coupling constants across N–H–N and N–H$^+$–N hydrogen bonds: Can coupling constants provide reliable estimates of N–N distances in biomolecules? Magn. Reson. Chem. 2001,39, S109–S114.

35. Del Bene, J. E.; Jordan, M. J. T. What a difference a decade makes: progress in ab initio studies of the hydrogen bond.THEOCHEM 2001, 573, 11–23.

36. Del Bene, J. E.; Perera, S. A.; Bartlett, R. J.; Elguero, J.; Alkorta, I.; López-Leonardo, C.; Alajarín, M. $^{3h}J(^{15}N-^{31}P)$ spin-spin coupling constants across N-H···O-P hydrogen bonds. J. Am. Chem. Soc 2002, 124, 6393.

37. Peralta, J. E.; Ruiz de Azua, M. C.; Contreras, R. H. Natural bond orbitals analysis of C-H···O interactions in NCH/H_2O and NCH/OCH_2, and their effect on nuclear magnetic shielding constants. THEOCHEM 1999, 491, 23–31.

38. Peralta, J. E.; Contreras, R. H.; Snyder, J. P. Natural bond orbital dissection of fluorine-fluorine through-space NMR coupling (JF,F) in polycyclic organic molecules. Chem. Commun. 2000, 2025–2026.

39. Wilkens, S. J.; Westler, W. M.; Markley, J. L.; Weinhold, F. Natural J-coupling analysis: Interpretation of scalar J-couplings in terms of Natural Bond Orbitals. J. Am. Chem. Soc. 2001, 123, 12026–12036.

40. Arnold, W. D.; Mao, J.; Sun, H.; Oldfield, E. Computation of through-space $^{19}F-^{19}F$ scalar couplings via Density Functional Theory. J. Am. Chem. Soc. 2000, 122, 12164–12168.

41. Arnold, W. D.; Oldfield, E. The chemical nature of hydrogen bonding in proteins via NMR: J-couplings, chemical shifts, and AIM Theory. J. Am. Chem. Soc. 2000, 122, 12835–12841.

42. Del Bene, J. E.; Perera, S. A.; Bartlett, R. J.; Alkorta, I.; Elguero, J.; Mó, O.; Yáñez, M. One-bond ($^{1d}J_{H-H}$) and three-bond ($^{3d}J_{X-M}$) spin-spin coupling constants across X-H···H-M dihydrogen bonds. J. Am. Chem. Soc. submitted.

43. Pecul, M. The nuclear spin-spin coupling constant in He_2. J. Chem. Phys. 2000, 113, 10835–10836.

44. Pecul, M.; Sadlej, J.; Leszczynski, J. The $^{19}F-^{1}H$ coupling constants transmitted through covalent, hydrogen bond, and van der Waals interactions. J. Chem. Phys. 2001, 115, 5498–5506.

45. Bryce, D. L.; Wasylishen, R. E. Ab initio characterization of through-space indirect nuclear spin-spin coupling tensors for fluorine-X (X = F, C, H) spin pairs. J. Mol. Struct. 2002, 602-603, 463–472.

46. Bagno, A.; Saielli, G.; Scorrano, G. DFT calculation of intermolecular nuclear spin-spin coupling in van der Waals dimers. Angew. Chem., Int. Ed. 2001, 40, 2532–2534.

47. Alkorta, I.; Rozas, I.; Jimeno, M. L.; Elguero, J. A theoretical and experimental study of the interaction of C_6F_6 with electron donors. Struct. Chem. 2001, 12, 459–464.

48. Bagno, A.; Saielli, G.; Scorrano, G. Through-space spin-spin couplings in van der Waals dimers and CH/π interacting systems. An ab initio study. Chem. Eur. J. 2002, 8, 2047–2056.

49. Bagno, A.; Saielli, G.; Scorrano, G. Substituent effects on the through-space nuclear magnetic spin-spin coupling in van der Waals dimers. ARKIVOC 2002. in press.

50. Mallory, F. B.; Mallory, C. W.; Butler, K. E.; Lewis, M. B.; Xia, A. Q.; Luzik, E. D.; Fredenburgh, L. E.; Ramanjulu, M. M.; Van, Q. N.; Francl, M. M.; Freed, D. A.; Wray, C. C.; Hann, C.; Nerz-Stormes, M.; Carroll, P. J.; Chirlian, L. E. Nuclear spin-spin coupling via nonbonded interactions. 8. The distance dependence of through-space fluorine-fluorine coupling. J. Am. Chem. Soc. 2000, 122, 4108–4116.

51. Alkorta, I.; Rozas, I.; Elguero, J. Non-conventional hydrogen bonds. Chem. Soc. Rev. 1998, 27, 163–170.

52. Rozas, I.; Alkorta, I.; Elguero, J. Inverse hydrogen-bonded complexes. J. Phys. Chem. A 1997, 101, 4236–4244.

53. Rozas, I.; Alkorta, I. Elguero, Bifurcated hydrogen bonds: three-centered interactions. J. Phys. Chem. A 1998, 102, 9925–9932.

54. Alkorta, I.; Rozas, I.; Elguero, J. Charge-transfer complexes between dihalogen compounds and electron donors. J. Phys. Chem. A 1998, 102, 9278–9285.

55. Rowbotham, J. B.; Smith, M.; Schaeffer, T. Stereospecific spin-spin coupling between hydroxyl protons and ^{19}F nuclei in o-, m- and p-fluorophenyl derivatives. "Through-space" interactions via the hydrogen bond. Can. J. Chem. 1975, 53, 986–992.

56. Golubev, N. S.; Shenderovich, I. G.; Smirnov, S. N.; Denisov, G. S.; Limbach, H.-H. Nuclear scalar spin-spin coupling reveals novel properties of low-barrier hydrogen bonds in a polar environment. Chem.-Eur. J. 1999, 5, 492–497.

57. Shenderovich, I. G.; Burtsev, A. P.; Denisov, G. S.; Golubev, N. S.; Limbach, H.-H. Influence of the temperature-dependent dielectric constant on the H/D isotope effects on the NMR chemical shifts and the hydrogen bond geometry of the collidine-HF complex in $CDCl_3/CDClF_2$ solution. Magn. Reson. Chem. 2001, 39, S91–S99.

58. Smirnov, S. N.; Golubev, N. S.; Denisov, G. S.; Benedict, H.; Schah-Mohammedi, P.; Limbach, H.-H. Hydrogen/deuterium isotope effects on the NMR chemical shifts and geometries of intermolecular low-barrier hydrogen-bonded complexes. J. Am. Chem. Soc. 1996, 118, 4094–4101.

59. Schah-Mohammedi, P.; Shenderovich, I. G.; Detering, C.; Limbach, H.-H.; Tolstoy, P. M.; Smirnov, S. N.; Denisov, G. S.; Golubev, N. S. Hydrogen/deuterium isotope effects on the NMR chemical shifts and symmetry of homoconjugated hydrogen-bonded ions in polar solution. J. Am. Chem. Soc. 2000, 122, 12878–12879.

60. Dingley, A. J.; Grzesiek, S. Direct observation of hydrogen bonds in nucleic acid base pairs by internucleotide $^{2}J_{NN}$couplings. J. Am. Chem. Soc. 1998, 120, 8293–8297.

61. Pervushin, K.; Ono, A.; Fernández, C.; Szyperski, T.; Kainosho, M.; Wütrich, K. NMR scalar couplings across Watson-Crick base pair hydrogen bonds in DNA observed by transverse relaxation-optimized spectroscopy. Proc. Natl. Acad. Sci. USA 1998, 95, 14147–14151.

62. Wöhnert, J.; Dingley, A. J.; Stoldt, M.; Görlach, M.; Grzesiek, S.; Brown, L. R. Direct identification of NH···N hydrogen bonds in non-canonical base pairs of RNA by NMR spectroscopy. Nucleic Acid. Res. 1999, 27, 3104–3110.

63. Majumdar, A.; Kettani, A.; Skripkin, E. Observation and measurement of internucleotide $^{2}J_{NN}$ coupling constants between ^{15}N nuclei with widely separated chemical shifts. J. Biomol. NMR 1999, 14, 67–70.

64. Hennig, M.; Williamson, J. R. Detection of N-H···N hydrogen bonding in RNA via scalar couplings in the absence of observable imino proton resonances. Nucleic Acids Res. 2000, 28, 1585–1593.

65. Pervushin, K.; Fernández, C.; Riek, R.; Ono, A.; Kainosho, M.; Wütrich, K. Determination of $^{h2}J_{NN}$ and $^{h1}J_{HN}$ coupling constants across Watson-Crick base pairs in the Antennapedia homeodomain-DNA complex using TROSY. J. Biomol. NMR 2000, 16, 39–46.

66. Kojima, C.; Ono, A.; Kainosho, M. Studies of physicochemical properties of N-HYN hydrogen bonds in DNA, using selective ^{15}N-labeling and direct ^{15}N 1D NMR. J. Biomol. NMR 2000, 18, 269–277.

67. Bytchenkoff, D.; Chiarparin, E.; Früh, D.; Rüdisser, S.; Bodenhausen, G. Temperature dependence of internucleotide nitrogen-nitrogen scalar couplings in RNA. Magn. Reson. Chem. 2002, 40, 377–379.

68. Dingley, A. J.; Masse, J. E.; Feigon, J.; Grzesiek, S. Characterization of the hydrogen bond network in guanosine quartets by internucleotide $^{3h}J_{NC'}$ and $^{2h}J_{NN}$ scalar couplings. J. Biomol. NMR 2000, 16, 279–289.

69. Kettani, A.; Gorin, A.; Majumdar, A.; Hermann, T.; Skripkin, E.; Zhao, H.; Jones, R.; Patel, D. J. A dimeric DNA interface stabilized by stacked A·(G·G·G·G)·A hexads and coordinated monovalent cations. J. Mol. Biol. 2000, 297, 627–644.

70. Pervushin, K. The use of TROSY for detection and suppression of conformational exchange NMR line broadening in biological macromolecules. J. Biomol. NMR 2001, 20, 275–285.

71. Ishikawa, R.; Kojima, C.; Ono, A.; Kainosho, M. Developing model systems for the NMR study of substituent effects on the N–H···N hydrogen bond in duplex DNA. Magn. Reson. Chem. 2001, 39, S159–S165.

72. Majumdar, A. Applications of $^{1}H_a$-$^{1}H_b$ correlated $^{2h}J(N,N)$ spectroscopy for identifying H_a-C-N_a···H_d-N_d hydrogen bonds in nucleic acids. Magn. Reson. Chem. 2001, 39, S166–S170.

73. Claramunt, R. M.; Sanz, D.; Alarcón, S. H.; Pérez Torralba, M.; Elguero, J.; Foces-Foces, C.; Pietrzak, M.; Langer, U.; Limbach, H.-H. 6-Aminofulvene-1-aldimine: A model molecule for the study of intramolecular hydrogen bonds.Angew. Chem. Int. Ed. 2001, 40, 420–423.

74. Pietrzak, M.; Limbach, H.-H.; Pérez-Torralba, M.; Sanz, D.; Claramunt, R. M.; Elguero, J. Scalar coupling constants across the intramolecular NHN hydrogen bond of symmetrically and non-symmetrically substituted 6-aminofulvene-1-aldimines. Magn. Reson. Chem. 2001, 39, S100–S108.

75. Sanz, D.; Pérez Torralba, M.; Alarcón, S. H.; Claramunt, R. M.; Foces-Foces, C.; Elguero, J. Tautomerism in the solid state and in solution of a series of 6-aminofulvene-1-aldimines. J. Org. Chem. 2002, 67, 1462–1471.

76. Rae, I. D.; Weigold, J. A.; Contreras, R. H.; Biekofsky, R. R. Analysis of long-range through-space couplings via an intramolecular hydrogen bond. Magn. Reson. Chem. 1993, 31, 836–840.

77. Schaefer, T.; Wildman, T. A.; Salman, S. R. The perpendicular conformation of 2-hydroxythiophenol. Intramolecular hydrogen bonding to specific lone pair. J. Am. Chem. Soc. 1980, 102, 107–110.

78. Anet, F. A. L.; Bourn, A. J. R.; Carter, P.; Winstein, S. Effect of steric compression on coupling constants. J. Am. Chem. Soc. 1965, 87, 5249–5250.

79. Schaeffer, T.; Chum, K. Spin-spin coupling between proximate protons on neighbouring aromatic rings and between hydroxyl protons in 2,2'-dihydroxy-4-methoxybenzophenone. Coupling through the hydrogen bond. Can. J. Chem.1976, 54, 2231–2242.

80. Cordier, F.; Grzesiek, S. Direct observation of hydrogen bonds in proteins by interresidue $^{3h}J_{NC'}$ scalar couplings. J. Am. Chem. Soc. 1999, 121, 1601–1602.

81. Cornilescu, G.; Hu, J.-S.; Bax, A. Identification of the hydrogen bonding network in a protein by scalar couplings. J. Am. Chem. Soc. 1999, 121, 2949–2950.
82. Cordier, F.; Rogowski, M.; Grzesiek, S.; Bax, A. Observation of through-hydrogen-bond $^{2h}J_{HC'}$ in a perdeuterated protein. J. Magn. Reson. 1999, 140, 510–512.
83. Cornilescu, G.; Ramirez, B. E.; Franck, M. K.; Clore, G. M.; Gronenborn, A. M.; Bax, A. Correlation between $^{3h}J_{NC'}$ and hydrogen bond length in proteins. J. Am. Chem. Soc. 1999, 121, 6275–6279.
84. Wang, Y.-X.; Jacob, J.; Cordier, F.; Wingfield, P.; Stahl, S. J.; Lee-Huang, S.; Torchia, D.; Grzesiek, S.; Bax, A. Measurement of $^{3h}J_{NC'}$ connectivities across hydrogen bonds in a 30 kDa protein. J. Biomol. NMR 1999, 14, 181–184.
85. Meissner, A.; Sørensen, O. W. New techniques for the measurement of C'N and C'H^{N}J coupling constants across hydrogen bonds in proteins. J. Magn. Reson. 2000, 143, 387–390.
86. Meissner, A.; Sørensen, O. W. ^{3h}J Coupling between C^a and H^N across hydrogen bonds in proteins. J. Magn. Reson.2000, 143, 431–434.
87. Cordier, F.; Wang, C.; Grzesiek, S.; Nicholson, L. K. Ligand-induced strain in hydrogen bonds of the c-Src SH3 domain detected by NMR. J. Mol. Biol. 2000, 304, 497–505.
88. Li, H.; Yamada, H.; Akasaka, K.; Gronenborn, A. M. Pressure alters electronic orbital overlap in hydrogen bonds. J. Biomol. NMR 2000, 18, 207–216.
89. Jaravine, V. A.; Alexandrescu, A. T.; Grzesiek, S. Observation of the closing of individual hydrogen bonds during TFE-induced helix formation in a peptide. Protein Sci. 2001, 10, 943–950.
90. Alexandrescu, A. T.; Snyder, D. R.; Abilgaard, F. NMR of hydrogen bonding in cold shock protein A and an analysis of crystallographic resolution on comparisons of hydrogen bond lengths. Protein Sci. 2001, 10, 1856–1868.
91. Hennig, M.; Geierstanger, B. H. Direct detection of a histidine–histidine side chain hydrogen bond important for folding of apomyoglobin. J. Am. Chem. Soc. 1999, 121, 5123–5126.
92. Pietrzak, M.; Wehling, H.-H.; Limbach, N. S.; Golubev, C.; López, R. M.; Claramunt, J. ^{13}C detected scalar nitrogen-nitrogen couplings across the intramolecular symmetric NHN hydrogen bond of proton sponges. J. Am. Chem. Soc2001, 123, 4338–4339.
93. Brown, S. P.; Pérez-Torralba, M.; Sanz, D.; Claramunt, R. M.; Emsley, L. The Direct Detection of a Hydrogen Bond in the Solid State by NMR through the Observation of a Hydrogen-Bond Mediated ^{15}N-^{15}N J Coupling. J. Am. Chem. Soc.2002, 124, 1152–1153.
94. Mishima, M.; Hatanaka, M.; Yokoyama, S.; Ikegami, T.; Wälchli, M.; Ito, Y.; Shirakawa, M. Intermolecular ^{31}P-^{15}N and^{31}P-^{1}H scalar couplings across hydrogen bonds formed between a protein and a nucleotide. J. Am. Chem. Soc. 2000,122, 5883–5884.
95. Löhr, F.; Mayhew, S. G.; Rütterjans, H. Detection of scalar couplings across NH···OP and OH···OP hydrogen bonds in a flavoprotein. J. Am. Chem. Soc. 2000, 122, 9289–9295.
96. Blake, P. R.; Park, J. B.; Adams, M. W. W.; Summers, M. F. Novel observation of NH···S(Cys) hydrogen-bond-mediated scalar coupling in cadmium-113 substituted rubredoxin from Pyrococcus furiosus. J. Am. Chem. Soc. 1992, 114, 4931–4933.
97. Blake, P. R.; Lee, B.; Summers, M. F.; Adams, M. W. W.; Park, J. B.; Zhou, Z. H.; Bax, A. Quantitative measurement of small through-hydrogen-bond and 'through-space' proton-cadmium-113 and proton-mercury-199 J couplings in metal-substituted rubredoxin from Pyrococcus furiosus. J. Biomol. NMR 1992, 2, 527–533.
98. Platzer, N.; Buisson, J.-P.; Demerseman, P. Complete ^{1}H and ^{13}C NMR spectral characterization of 1,6-dioxapyrene and related compounds. An unusual coupling interaction through hydrogen bond in three precursors. J. Heterocyl. Chem. 1992, 29, 1149–1153.
99. Kwon, O.; Danishefsky, S. J. Synthesis of asialo GM_1. New insights in the application of sulfonamidoglycosylation in oligosaccharide assembly: subtle proximity effects in the stereochemical governance of glycosidation. J. Am. Chem. Soc. 1998, 120, 1588–1599.
100. Laynez, J.; Menéndez, M.; Saiz, J. L.; Llamas-Saiz, A. L.; Foces-Foces, C.; Elguero, J.; Molina, P.; Alajarín, M.; Vidal, A. Iminophosphorane-substituted proton sponges. Part 4. Comparison of X-ray molecular structures with solution properties (pKa, ^{1}H and ^{13}C NMR spectroscopy). J. Chem. Soc., Perkin Trans. 2 1993, 709.
101. Elguero, J.; Fruchier, A.; Jimeno, M. L.; Molina, P. An experimental study of the ^{4h}J(^{31}P-^{31}P) coupling constant and the^{12}C/^{13}C isotope effect on ^{31}P in an iminophosphorane-substituted proton sponge. J. Chem. Res. (S) 2002, 34–36.
102. Majumdar, A.; Kettani, A.; Skripkin, E.; Ptel, D. J. Observation of internucleotide NH···N hydrogen bonds in the absence of directly detectable protons. J. Biomol. NMR 1999, 15, 207–211.
103. Liu, A.; Majumdar, A.; Hu, W.; Kettani, A.; Skripkin, E.; Patel, D. J. NMR detection of N-H···O=C hydrogen bonds in^{13}C,^{15}N-Labeled Nucleic Acids. J. Am. Chem. Soc. 2000, 122, 3206–3210.
104. Density Functional Methods in Chemistry (Lebanowski, A.; Andzelm, J. Eds.). Springer: N.Y. , 1991.
105. Modern Density Functional Theory: A Tool for Chemists (Seminario, J. M.; Politzer, P. Eds.). Elsevier: Amsterdam, 1995; Vol. 2.
106. Theoretical Treatments of Hydrogen Bonding (Hadzi, D. Ed.). 1997.
107. Electronic Density Functional Theory: Recent Progress and New Directions (Dobson, J. F.; Vignale, G.; Das, M. P., Eds.). Plenum Press: N.Y., 1998.
108. Contreras, R. H.; Peralta, J. E. Angular dependence of spin-spin coupling constants. Prog. Nucl. Magn. Reson. Spectrosc.2000, 37, 321–425.

109. Contreras, R. H.; Peralta, J. E.; Giribet, C. G.; Ruiz de Azua, M. C.; Facelli, J. C. Advances in theoretical and physical aspects of spin-spin coupling constants. Annu. Rep. NMR Spectrosc. 2000, 41, 55–184.

110. Pervushin, K. Impact of transverse relaxation optimized spectroscopy (TROSY) on NMR as a technique in structural biology. Q. Rev. Biophys. 2000, 33, 161–197.

111. Koch, W.; Holthausen, M. C. A Chemist's Guide to Density Functional Theory, 2nd Edition edWiley-VICH, 2001.

112. Grzesiek, S.; Cordier, F.; Dingley, A. J. Scalar couplings across hydrogen bonds. Methods Enzymol. 2001, 338, 111–133.

113. Dingley, A. J.; Cordier, F.; Grzesiek, S. An introduction to hydrogen bond scalar couplings. Concepts Magn. Reson. 2001,13, 103–127.

114. Majumdar, A.; Patel, D. J. Identifying Hydrogen Bond Alignments in Multistranded DNA Architectures by NMR. Acc. Chem. Res 2002, 35, 1–11.

CHAPTER 4

EXTENSION OF THE INTERACTING QUANTUM ATOMS (IQA) APPROACH TO B3LYP LEVEL DENSITY FUNCTIONAL THEORY (DFT)

Peter Maxwell[ab], Ángel Martín Pendás[c] and Paul L. A. Popelier[ab]

[a]Manchester Institute of Biotechnology (MIB), 131 Princess Street, Manchester M1 7DN, UK

[b]School of Chemistry, University of Manchester, Oxford Road, Manchester M13 9PL, UK

[c]Departamento de Quimica Fisica y Analitica, Universidad de Oviedo, E-33006 Oviedo, Spain

ABSTRACT

An interaction between two atoms, bonded or non-bonded, consists of interatomic contributions: electrostatic energy, exchange energy and electronic correlation energy. Together with the intra-atomic energy of an atom, these contributions are the basic components of the Interacting Quantum Atom (IQA) energy decomposition scheme. Here, we investigate IQA's proper use in conjunction with an explicit implementation of the B3LYP functional. The recovery of the total molecular energy from the IQA components is emphasised, for the first time. A systematic study of three model systems of biological relevance, N-methylacetamide (NMA), the doubly capped tripeptide GlyGlyGly and an alloxan dimer, shows the stabilization effect of B3LYP on most of the interatomic exchange energies (V^{AB}_X) compared to their Hartree–Fock values. Diagrams of exchange energies versusinteratomic distance show the clustering of interactions, one cluster for each 1,n (n = 1 to 6 where the atoms are separated by n − 1 bonds). The positioning of some V^{AB}_X values outside their expected cluster marks interesting interactions.

INTRODUCTION

The spatial localization and partitioning of properties of matter is central to Chemistry. Fortunately, matter (at ambient conditions, and excluding metallic systems) allows itself to be described in terms of local features and behaviour, as well as recurring motifs. Indeed, molecular assemblies consist of molecules, which in turn consist of functional groups and largely transferable atoms (e.g. ref. 1–3). This locality and transferability enables Chemistry to classify, systematize and eventually explain and predict the phenomena it observes.

There are various alternatives to obtain this local information theoretically and computationally, most popularly by energy and electronic charge. Amongst the most used partitioning schemes are the "energy decomposition analysis" (EDA),[4,5] its variant the natural energy decomposition analysis (NEDA),[6] the natural bond orbital (NBO) analysis[7] or Quantum Chemical Topology (QCT),[8] which amongst its various segments incorporates the Quantum Theory of Atoms in Molecules (QTAIM) [9] and Interacting Quantum Atoms (IQA).[10] The latter approach is conceptually the most minimal of all energy partitioning schemes. IQA is parameter-free although computationally the most expensive method. However, very recently, IQA has gained popularity through its implementation in the efficient computer program AIMAll.[11] The IQA method is the subject of this paper, particularly, its use within the context of density functional theory (DFT).

Citation: Maxwel P, Pendas AM, Popelier PLA. Extension of the interacting quantum atoms (IQA) approach to B3LYP level density functional theory (DFT). Phys. Chem. Chem. Phys. 2016; Advance Article. DOI: 10.1039/c5cp07021j.

We also note again that IQA can operate on non-equilibrium geometries, and can thus go beyond the original QTAIM atomic energies, which are only valid at a stationary point (e.g. an equilibrium geometry).

QCT is a branch of theoretical chemistry that yields a wealth of calculated chemical information from the wavefunction of a molecule, molecular assembly, metallic system or ionic crystal. It uses the (mathematical) language of dynamical systems (e.g. attractor, critical point, separatrix, basin, gradient path) to obtain chemical insight and then ideally make predictions. It defines the so-called topological atom, which is a finite-volume object with a shape, which is specifically determined by the total system that this atom is part of topological atoms do not overlap nor leave gaps: they collectively exhaust space and their properties are additive. In other words, there is no need for corrections due to topological atoms intersecting, or leaving a gap between them (resulting in electron density not belonging to any one atom).

Each topological atom contributes its own intra-atomic energy to the total system's energy.[10] In addition to this contribution, and more importantly for this study, IQA also provides all inter-atomic energy contributions. When summed, the intra-atomic and inter-atomic energies (whether attractive or repulsive) account for the complete energy of a molecule or molecular complex. Being able to recover this complete energy is very important but has been ignored in some recent publications applying IQA. A more detailed description of the IQA approach can be found in the Theoretical background section.

IQA has been used to study various bonding patterns such as halogen bonding,[12] hydrogen bonding,[13,14] interelectronic exchange between electronegative atoms,[15] interactions of Zn(II) complexes,[16] interactions between organoselenium molecules and diiodine[17] and the interactions within halogentrinitromethanes.[18] In addition to interaction studies, investigations addressing steric repulsion and binding energies have been reported[19–23] including IQA's interpretation of chemical phenomena such as hyperconjugation.[24] Finally, charge transfer, chemical potentials and the nature of functional groups,[3] as well as the relationship between electronic exchange energy and molecular structure[25] have also been reported. Outside of aiding chemical interpretations, IQA has also been used in more technical studies, for example, detailing the relationship between various density partitions, atomic charge and the delocalization index[26] but also in rare and direct comparisons[20] with non-IQA methods such as EDA and NBO.

A practical limitation of IQA is its potential incompatibility with quantum mechanical methods that solve the Schrödinger equation. Currently, the application of IQA is limited to the following methods: Hartree–Fock (HF), several multiconfigurational expansions including Complete Active Space (CAS), Configuration Interaction with single and double excitations (CISD), and Full Configuration Interaction (FCI), together with Coupled Cluster with single and double excitations (CCSD). Neither perturbation theory nor standard density functional theory (DFT) methods provide well-defined a second-order reduced density matrix, and hence IQA cannot be applied to them. As a consequence, simply adding all IQA energy contributions for a B3LYP wavefunction, for example, does not at all return the original ab initio energy of the molecule or molecular system. However, this fact has not always prevented work[18,19,27–30] featuring alternative theory levels such as MP2 or B3LYP from being published. The aim of the current investigation is to identify and clarify what the inter-atomic exchange values represent, and whether they are a cause for concern. This aim will be reached for the most popular density functional to date: B3LYP.[31] Being able to recover the total molecular energy from its IQA energy components is important in force field design because energy cannot be spuriously created or suddenly go missing. In this sense, the recent extension of IQA to B3LYP wavefunctions benefits the development of the QCTFF force field.[32]

In using the B3LYP functional, the Kohn–Sham (KS) second-order density matrix must be estimated from the first-order density matrix and will therefore only be approximate (or "fictitious"[33]). This fact causes the exchange energy contribution to be incomplete, coming from the use of the KS orbitals for both the intra-atomic and inter-atomic contribution. Overall, here we will explicitly address the interpretation of the interatomic exchange energy produced by an IQA partitioning of a B3LYP wavefunction, compared with that of a Hartree–Fock (HF) wavefunction, for the same molecule.

Note that IQA is not the only method that has identified the use of KS orbitals as a concern before. The same concern surfaces in the calculation and interpretation of delocalization indices (DIs), also denoted δ(A,B). A delocalization index represents the number of electrons delocalized (i.e. exchanged or shared) between atoms A and B. Hence, it is not surprising that there is a relationship between δ(A,B) and the interatomic exchange energy between A and B. Indeed, using a binomial Taylor expansion one can prove[34,35] that the interatomic exchange energy is approximately equal to minus δ(A,B) divided by twice the internuclear distance. Thus, a large value for δ(A,B) corresponds to a large absolute value for the interatomic exchange energy, and hence energetic stabilization. In 2002, Poater et al.[36] reported the significance of having to use DFT-KS orbitals within a HF-formalism in order to obtain a numerical value at DFT level. They produced δ(A,B) values obtained from a HF treatment of a DFT wavefunction in order to gain useful chemical insight. However, as electron correlation was not fully considered, their values were consistently overestimated. Both δ(A,B) and the interatomic exchange–correlation energies can provide valuable information of chemical significance for polyatomic molecules.[25,37,38]

We will present a direct comparison of the exchange energies calculated from a HF wavefunction and calculated from a HF treatment of a B3LYP wavefunction, both at exactly the same molecular geometry. In order to examine a range of both intramolecular and intermolecular interactions between atoms, the following biologically relevant molecules are used as case studies: (i) N-methylacetamide (NMA), which serves as a well-studied prototype system for the ubiquitous peptide bond, (ii) triglycine (GlyGlyGly, with a peptide-bond cap at both termini) (TriGly), which is of relevance to the construction of our peptide force field QCTFF given the recent finding[39] that single amino acids are not sufficiently transferable towards proteins, and (iii) an alloxan dimer (Cambridge Structural Database (CSD) identifier ALOXAN), which is a remarkable crystal (for a reason explained just below) featuring upfront in an important review[40] on intermolecular interactions.

THEORETICAL BACKGROUND

Interacting Quantum Atoms (IQA)

Fig. 1 shows an example of a molecular system being partitioned into topological atoms. The picture shows a realm of atoms appearing as malleable boxes, or even "bubbles", negotiating energies and charges between them, and thereby forming a stable quantum system. As the nuclear positions change, so do these atoms' shapes and properties. It is clear that the topological atoms exhaust space: no overlap and no gaps. A remarkable property of crystalline alloxan (or pyrimidine-2,4,5,6-tetraone) is that hydrogen bonds do not appear in the crystal in spite of the two acidic NH entities for each molecule. Instead, crystalline alloxan features[40] an array of short, almost orthogonal intermolecular C═O···C═O contacts with distances of about 2.8 Å and C═O···C angles in the range between 155° and 163°. Two atomic interaction lines, each with a corresponding bond critical point, appear between the two monomeric units.

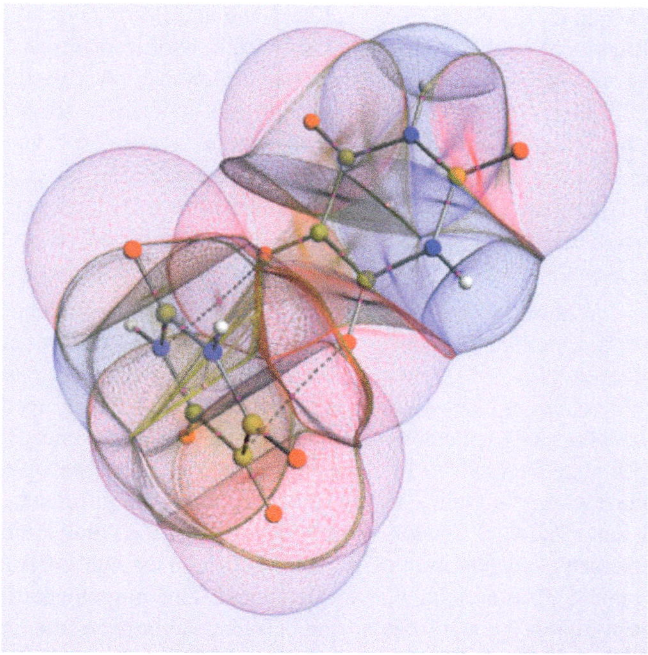

Figure 1: A representation of the topological atoms in the alloxan dimer. The configuration is directly taken from the Cambridge Structural Database, without geometry optimization, and a single-point energy conformation was obtained at B3LYP/6-31+G(d,p) level.

The IQA approach[10] follows the traditional QTAIM[9] formalism by partitioning a molecule into topological atoms. The IQA energies are calculated through the use of the first-order density matrix (non-diagonal) and the second-order (diagonal) density matrix, which allow the complete calculation of the Born–Oppenheimer energy of a molecule. To start, the intra-atomic and inter-atomic energies are given in eqn (1),

$$E_{\mathrm{IQA}} = \sum_{\mathrm{A}} E^{\mathrm{A}}_{\mathrm{IQA}} = \sum_{\mathrm{A}} E^{\mathrm{A}}_{\mathrm{intra}} + \frac{1}{2}\sum_{\mathrm{A}}\sum_{\mathrm{B}\neq\mathrm{A}} V^{\mathrm{AB}}_{\mathrm{inter}} \tag{1}$$

where E_{IQA} represents the molecular energy as obtained from an IQA partitioning, while E^{A}_{IQA} is the full atomic energy of atom A, E^{A}_{intra} is the intra-atomic energy of atom A (or "self-energy"), while V^{AB}_{inter} represents the interatomic interaction energy between atom A and B. The intra-atomic contribution, E^{A}_{intra}, is further partitioned into:

$$E^{A}_{intra} = T^{A} + V^{AA}_{ee} + V^{AA}_{en} \tag{2}$$

The intra-atomic energy comprises of the kinetic energy of the electrons (T^{A}), the electron–electron repulsive potential energy (V^{AA}_{ee}) and the electron–nucleus attractive potential energy (V^{AA}_{en}) within atom A.

The inter-atomic energy can be subdivided to give

$$V^{AB}_{inter} = V^{AB}_{nn} + V^{AB}_{en} + V^{AB}_{ne} + V^{AB}_{ee} \tag{3}$$

The inter-atomic energy comprises of the sum of four potential energies: the nucleus–nucleus repulsive energy (V^{AB}_{nn}), electron–nucleus attractive energy (V^{AB}_{en}), nucleus–electron attractive energy (V^{AB}_{ne}) and the electron–electron repulsive (V^{AB}_{ee}) energy. The final energy can be divided further into a Coulombic (V^{AB}_{Coul}) and an exchange–correlation (V^{AB}_{XC}) contribution according to eqn (4),

$$V^{AB}_{ee} = V^{AB}_{Coul} + V^{AB}_{X} + V^{AB}_{corr} = V^{AB}_{Coul} + V^{AB}_{XC} \tag{4}$$

Here V^{AB}_{XC} is a combination of exchange (V^{AB}_{X}) and correlation (V^{AB}_{corr}) potential energies.

The calculation of the electron–electron potential V^{AB}_{ee} requires integration over atomic volumes of the second-order density matrix, which is computationally very expensive because of the evaluation of a six-dimensional integral, that is, three dimensions for each atom. The V^{AB}_{ee} energy can be expressed according to Görling–Levy theory[41] using perturbation theory as follows:

$$V^{\mathrm{AB}}_{\mathrm{ee}} = \int_{\Omega_{\mathrm{A}}}\int_{\Omega_{\mathrm{B}}} \frac{\rho(\vec{r}_1)\rho(\vec{r}_2)}{\|\vec{r}_1 - \vec{r}_2\|} \mathrm{d}^3\mathbf{r}_1\mathrm{d}^3\mathbf{r}_2 - \int_{\Omega_{\mathrm{A}}}\int_{\Omega_{\mathrm{B}}} \frac{|\rho(\vec{r}_1;\vec{r}_2)|^2}{\|\vec{r}_1 - \vec{r}_2\|} \mathrm{d}^3\mathbf{r}_1\mathrm{d}^3\mathbf{r}_2 + V^{\mathrm{AB}}_{\mathrm{corr}} \tag{5}$$

The Coulombic term can be expressed as a functional of the electron density (first term of eqn (5)). However, the contributions from the exchange (second term) or correlation (third term) functions are more difficult as their analytical form remains unknown.[15] Moreover, for DFT functionals such as B3LYP, the exchange term uses components from both the first-order and second-order density matrices.[15] For these cases, the Kohn–Sham[42] second-order density matrix can be estimated from the first-order matrix, but will remain only approximate and differs from any analytical solution. Very recently, (dynamic electronic) correlation was incorporated[43] in IQA using closed shell coupled cluster theory. Compared to the exchange or Coulombic energies it is a minor quantity, at least for simple systems such as H_2, N_2, CO and H_2O.

Returning to the conventional HF-compatible IQA, a rearrangement of the inter-atomic energies allows an electrostatic contribution and a covalent contribution to be separated and defined:

$$V^{AB}_{elec} = V^{AB}_{nn} + V^{AB}_{en} + V^{AB}_{ne} + V^{AB}_{Coul} \tag{6}$$

$$V^{AB}_{XC} = V^{AB}_{X} + V^{AB}_{corr} \tag{7}$$

where V^{AB}_{elec} (or V^{AB}_{cl}) is the electrostatic contribution to the overall inter-atomic interaction energy between atoms A and B. The remaining exchange–correlation energy V^{AB}_{XC}, acts as a measurement of covalency between two atoms derived from the combination of inter-electron dependency and the Pauli Exclusion Principle. A final equation summarizes how interactions can then be typically analyzed through the combination of the electrostatic and covalent contributions to an interaction:

$$V^{AB}_{inter} = V^{AB}_{elec} + V^{AB}_{XC} \quad (8)$$

At this point the brief review of the traditional IQA approach concludes. We will now address how both the traditional and the B3LYP-compatible IQA partitioning have been implemented into the computer program AIMAll.[11,44]

First, a technical note is in place here. AIMAll slightly expands the IQA formalism explained above by additionally calculating so-called AA′ inter-atomic energies. Here, A′ represents every other atom in the system with the exclusion of atom A itself. Hence, A′ is equivalent to a summation over all atoms B that are not equal to A,

$$\sum_{B} V^{AB}_{inter} \equiv V^{AA'}_{inter} \quad (9)$$

Note that recovering the total interaction energy within a molecule, which is formally $\sum_{AB} V^{AB}_{inter}$, is more accurately calculated using $V^{AA'}_{inter}$ as an intermediate quantity rather than $\sum_{B} V^{AB}_{inter}$, where V^{AB}_{inter} is individually calculated for each atom pair.

In theory, summing over B or working with the A′ route should produce equal results but typically there are small differences in the energy values due to the nature of their separate calculation and associated errors (i.e. by grid quadrature or semi-analytical integration). In AIMAll, each of the interaction energy subcomponents (i.e. V^{AB}_{en}, V^{AB}_{ne} or V^{AB}_{ee}, see eqn (3)) can also be calculated in "AA′ mode". AIMAll uses the AA′ energies in the calculation of the total molecular energy instead of the summation over every interatomic AB interaction. Using AA′ energies has two advantages: (1) a faster complete calculation of the IQA molecular energy partitioning albeit with reduced chemical insight, and (2) a more accurate recovery of the ab initio energy.

Now we come to the actual incorporation of the explicit B3LYP functional into an IQA energy analysis. The strategy should be made clear upfront. The overall goal is to recover the total energy when using B3LYP, which has been glossed over in the literature so far. This goal is reached in this work by using the explicit functional only within a single atom, i.e. for the total atomic energy only. We will show that the functional cannot be used for an interatomic energy. Thus we have no choice but to adopt the Hartree–Fock-like expression for interatomic exchange energy but then using Kohn–Sham orbitals.

In the development of the B3LYP extension of IQA it is convenient to define energy contributions as total atomic energies,

$$V^{A}_{comp} = V^{AA}_{comp} + V^{AA'}_{comp} \quad (10)$$

where "comp" can be XC, ne, en, nn or Coul. The quantity V^{A}_{comp} should not be confused with the previously defined intra-atomic energy (or "self-energy"). This extra symbol is motivated by a future need (see below) to gather intra- and inter-atomic energy contribution, of a nature specified by "comp".

In previous versions of AIMAll, a B3LYP wavefunction would be treated identically to a HF wavefunction, which leads to a huge discrepancy between the total molecular energy and a summation of the various IQA energy components. In the older versions of AIMAll, both the intra-atomic and inter-atomic exchange energy is calculated from the HF exchange energy equation only:

$$E_X^{HF} = -\frac{1}{2}\sum_{i,j}\iint \psi_i^*(\mathbf{r}_1)\psi_j^*(\mathbf{r}_1)\frac{1}{r_{12}}\psi_i(\mathbf{r}_2)\psi_j(\mathbf{r}_2)d\mathbf{r}_1 d\mathbf{r}_2 \quad (11)$$

instead of from the (explicit) B3LYP exchange–correlation functional:[45]

$$E^{B3LYP}_{XC} = (1 - a_0)\, E^{LSDA}_{X} + a_0 E^{HF}_{X} + a_X \Delta E^{B88}_{X} + a_C E^{LYP}_{C} + (1 - a_C)\, E^{VWN}_{C} \quad (12)$$

where $a_0 = 0.20$, $a_X = 0.72$ and $a_C = 0.81$, and E^{LSDA}_X is the Local Spin Density Approximation (LSDA) standard local exchange functional, while E^{HF}_X is the HF exchange energy, and ΔE^{B88}_X is Becke's gradient correction to the Becke-88 exchange functional, E^{LYP}_C is Lee, Yang and Parr's correlation functional, and E^{VWN}_C is the Vosko–Wilk–Nusair local density approximation to the correlation functional. A "pseudo-DFT/B3LYP" compatible algorithm has quite recently been implemented in AIMAll in order to accurately recover a molecule's total ab initio energy. The qualifier "pseudo" refers to the ambiguity in the calculation of both interatomic and intra-atomic exchange. However, as explained below, a reasonable choice will be made to cope with this ambiguity.

Without extra thought or treatment, the introduction of a functional produces incorrect intra-atomic V^{AA}_{XC} and inter-atomic V^{AB}_{XC} energies (leading to incorrect V^{AA}_{ee} and V^{AB}_{ee} energies). As a result the total energy of the molecule is incorrectly calculated and one obtains a large energetic discrepancy. In order to obtain a correct result, the exchange–correlation functional needs to be explicitly incorporated. In a later AIMAll version (14.04.17 ref. 11), the total atomic exchange–correlation energy denoted V^{A}_{XC} (see eqn (10)) for the B3LYP functional was implemented. This implementation now allows the well-defined and correct calculation of the total V^{A}_{XC} component of V^{A}_{ee} for an atom A:

$$V^{A}_{ee} = V^{A}_{XC} + V^{A}_{elec} \quad (13)$$

In turn, the correct calculation of V^{A}_{ee} allows the complete recovery of an atomic energy E^{A}_{IQA} (see eqn (1)). When summing over all atoms, the ab initio energy is recovered from an IQA partitioning. It is important to keep in mind that, even with this implementation, standard functionals are constructed to include the interacting part of the kinetic energy, such that the atomic electron–electron repulsion in eqn (13) is inevitably contaminated by the residual kinetic energy term.

Interatomic exchange energies are chemically meaningful and therefore one must be able to calculate them, again in a DFT context. The division of the total atomic exchange–correlation into intra- (V^{AA}_{XC}) and inter- ($V^{AA'}_{XC}$ or V^{AB}_{XC}) components is actually ambiguous. This ambiguity limits the correct extraction of interatomic terms but a reasonable choice or decision can be made. An "amalgamated" approach has been introduced to tackle this remaining limitation. This approach involves first calculating the inter-atomic exchange–correlation contribution ($V^{AA'}_{XC}$ or V^{AB}_{XC}) via the pure Hartree–

Fock exchange equation only, but by inserting KS orbitals instead of HF orbitals,

$$V^{AB}_{X,HF(B3LYP)} = -\sum_{i,j}\int_A\int_B \frac{\psi_i^{KS}(\mathbf{r}_1)\,\psi_j^{KS}(\mathbf{r}_2)\,\psi_j^{KS}(\mathbf{r}_1)\,\psi_i^{KS}(\mathbf{r}_2)}{r_{12}} d\mathbf{r}_1 d\mathbf{r}_2 \quad (14)$$

Note that the subscript X (rather than XC) emphasizes that, formally, we extract only the exchange part from the exchange–correlation energy, using the HF framework.

Secondly, this approach calculates the intra-atomic exchange–correlation energy from the well-defined total atomic exchange–correlation, $V^{A}_{XC,B3LYP}$, and the HF-defined inter-atomic exchange, $V^{AA'}_{X,HF(B3LYP)}$, or

$$V^{AA}_{XC,amalgam} = V^{A}_{XC,B3LYP} - \frac{1}{2}V^{AA'}_{X,HF(B3LYP)} \quad (15)$$

where $V^{A}_{XC,B3LYP}$ is obtained by performing a three-dimensional integration over the volume of atom A, of E^{B3LYP}_{XC} defined in eqn (12). The calculation of $V^{A}_{XC,B3LYP}$ is possible because the E^{B3LYP}_{XC} functional obviously only depends on the electron density.

It is pivotal to appreciate the difference in treatment between inter- and intra-atomic exchange–correlation energies. Eqn (14) cannot be rewritten as a function of electron densities, and thus the B3LYP (which is a function of the electron density) cannot be introduced here. Thus, no interatomic exchange(–correlation) energy can be calculated from a 6D integral such as that in eqn (14) but then with the functional appearing in the integrand.

We also point out that a private communication with Dr Keith shows that $V^{A}_{XC,B3LYP}$ is a transferable quantity as demonstrated from a series of tests he conducted on H···H interactions in normal alkanes.

Finally, we note that in this proposed scheme, $V^{AB}_{X,HF(B3LYP)} = V^{BA}_{X,HF(B3LYP)}$, and that the total molecular exchange–correlation energy can be recovered as follows,

$$\sum_A V^{A}_{XC,B3LYP} = E^{molec}_{XC,B3LYP} \quad (16)$$

The recovery of the total molecular energy from the intra- and inter-atomic contributions is made explicit in the following equation,

$$E^{molec}_{IQA,B3LYP} = \sum_A E^{A}_{IQA,B3LYP} = \sum_A \Bigg[T^A + \frac{1}{2}V^A_{en} + \frac{1}{2}V^A_{ne} + \frac{1}{2}V^{AA'}_{nn} + \underbrace{V^{AA}_{Coul} + V^{AA}_{XC,amalgam} + \frac{1}{2}V^{AA'}_{Coul} + \frac{1}{2}V^{AA'}_{XC,HF(B3LYP)}}_{V^A_{ee,B3LYP}}\Bigg] \quad (17)$$

where the quantity on the left hand side represents the molecular energy as recovered from the B3LYP-IQA analysis, which is to be compared with the original (unpartitioned) energy of the B3LYP wave function.

In this investigation we will analyse the effect of only using the HF-defined exchange equation (eqn (11)) in the calculation of interatomic exchange energies derived from KS orbitals. Note that, with the exception of the total atomic B3LYP exchange–correlation functional energy, V^{A}_{XC}, all other intra-atomic and inter-atomic exchange–correlation energies are calculated incorporating only the HF exchange component without correlation. The appropriate notation will be applied throughout the analysis to maintain clarity. Subscript 'XC' is used for the total atomic exchange–correlation energy (V^{A}_{XC}) along with the hybrid intra-atomic exchange–correlation energy (V^{AA}_{XC}), and subscript 'X' is used for the inter-atomic exchange energy (V^{AB}_{X} or $V^{AA'}_{X}$).

Computational Methods

The program GaussView generates sensible geometries for NMA and TriGly, which were subsequently geometry-optimized with the GAUSSIAN09[46] program, in order to obtain a local energy minimum for NMA at HF/6-311G(d,p), and for TriGly at HF/6-31+G(d,p) level. The latter level of theory has been recommended[47] before for its favourable balance between accuracy and computational demand. Note that no frequency calculations were carried out because this work does not depend on the confirmation of molecular geometries being true minima. Subsequent wave functions and ab initio energies were calculated at HF, B3LYP and MP2 levels of theory (with the respective basis set) without re-optimization of the geometry. Note that the MP2 wave function did not feature in any IQA energy partitioning but was only used to obtain net atomic charges, necessary for proper quantifying of polarity (see Discussion, Section 4). The latter served in a discussion of bond polarity quantified at HF versus post-HF level. We emphasize that the molecular geometries were those obtained at HF level, and that they were never re-optimized at B3LYP or MP2 level, in order to guarantee a direct comparison of energy components. The geometry of the alloxan dimer was obtained directly from the CSD, without geometry optimization, and the corresponding wave functions were again calculated at HF, B3LYP and MP2 levels (with 6-31+G(d,p)).

The extensive analysis of NMA and TriGly is preceded by a succinct analysis of H_2 and LiH, at HF, B3LYP and full CI level using 6-311G (d,p). The latter level of theory is computationally extremely expensive and hence only possible for the smallest of molecules. Yet, this analysis is useful in setting the scene while making contact with the most advanced non-DFT post-Hartree–Fock method.

The AIMAll[11,44] software package (both versions 13.10.19 and 14.04.17) was used to perform the QTAIM analysis and the IQA partitioning calculations for the HF and B3LYP wavefunctions. AIMAll (v 13.10.19) was used for the IQA partitioning of the HF and B3LYP wavefunctions, via a HF-only IQA algorithm, to investigate and compare the V^{AB}_{X} values. The B3LYP wavefunctions were then analysed using AIMAll (v 14.04.17) to check if molecular energy is recovered when using the implementation of the B3LYP atomic exchange–correlation functional (V^{A}_{XC}) in the IQA algorithm. Note that both versions produce the same inter-atomic $V^{AB}_{X,B3LYP}$ values since both follow the HF exchange-only algorithm.

AIMAll was also used to calculate QTAIM atomic charges. All default parameters (e.g. "auto" quadrature grids

"Boaq" and "Briaq") were used within AIMAll, except the encompparameter (short for "Energy Components"), which was set to "4" in order to carry out a full IQA partitioning calculation.

AIMStudio, which is a component of the AIMAll package, and in-house software called IRIS were used for visualization.[48,49] Within AIMStudio, all default settings were used. Some non-default settings for IRIS were employed, i.e. wireframe for the surface and altered transparency.

Throughout the analysis no interaction energy has been doubly counted. Within the AIMAll format, all interaction energies are split into V^{AB} and V^{BA}, referring to any subscript (e.g. elec, inter, XC, X), and $V^{AB} \equiv V^{BA}$. Simply adding V^{AB}_{X} and V^{BA}_{X} gives the total exchange energy between atoms A and B, or $V^{AB}_{X,tot}$. For clarity, and to condense the data, only the $V^{AB}_{X,tot}$interaction will be reported in the figures. In line with the IQA formalism, $V^{AB}_{X,tot}$ will simply be reported as V^{AB}_{X}. For each system a table will be presented containing the quantitative analysis of the interaction energies. Within each table, the analysis is divided into sub-sections according to which interatomic exchange energy is more stable, $V^{AB}_{X,HF}$ or $V^{AB}_{X,B3LYP}$, where the subscript identifies the nature of the wavefunction used. Within each sub-section, maximum and mean difference percentages between the HF and B3LYP interatomic exchange energies will be presented. The total number of interaction cases for each 1,n interaction type within each sub-section is also included. Percentages were chosen for the analysis to allow a transferable measure that can be analysed, irrespectively of the magnitude of the energy value in question. Across the analysis, the interatomic exchange energies span a range of 10^{-6} kJ mol^{-1} to just over 10^{3} kJ mol^{-1}.

RESULTS

Materials

Before the extensive analysis of NMA and TriGly, the scene is set with a brief analysis of H_2 and LiH. NMA and TriGly are studied to investigate intra-molecular interactions in small and medium-sized systems. The alloxan dimer system adds another dimension by including intermolecular interactions. Fig. 2 shows the three systems; note that TriGly has 17 H atoms and 22 non-H atoms, while the alloxan dimer has 4 H atoms and 20 non-H atoms. Through individually analyzing every interaction in each of the three systems we aim to rationalize the difference between interatomic HF exchange energies, $V^{AB}_{X,HF}$, and interatomic energies obtained from a HF-treatment of a B3LYP wavefunction, denoted $V^{AB}_{X,B3LYP}$.

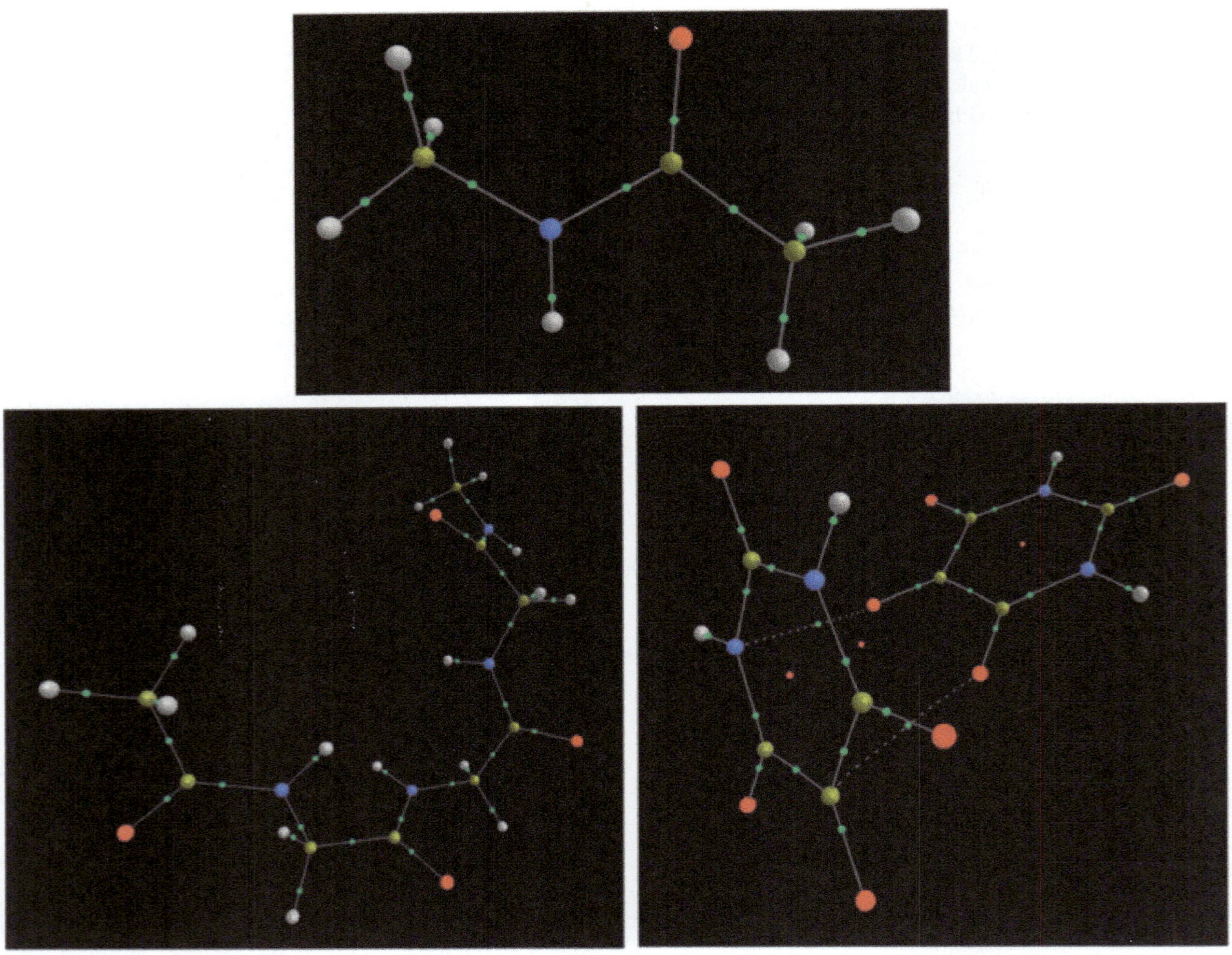

Figure 2: N-Methylacetamide (NMA) (top), TriGly (bottom left) and the alloxan dimer (bottom right). Green dots represent bond critical points (BCPs) and red dots represent ring critical points (RCPs). Images were produced using AIMStudio.

IQA Molecular Energy (E_{IQA}) *vs. ab initio* (E_{WFN})

Recent versions of the AIMAll program (version 14.04.17 and later) include the implementation of the explicit B3LYP exchange functional into the IQA partitioning. As a result, the IQA energies accurately recover the ab initio molecular energy (typically within <1 kJ mol^{-1}). Table 1 shows the IQA energy recovery of each system, when using the incorrect HF exchange (AIMAll version 13.10.19) and then the correct B3LYP exchange (14.04.17).

Table 1 IQA molecular energy (E_{IQA}), ab initio energy (E_{WFN}) and IQA residual error (ΔE) of NMA, TriGly and the alloxan dimer, at HF and B3LYP theory levels, using AIMAll versions 13.10.19 and 14.04.17. All energies are in kJ mol^{-1}. Note that, in the case of B3LYP, E_{IQA} (from AIMAll version 14.04.17) is the same energy as the left hand side of eqn (17) or$E^{molec}_{IQA,B3LYP}$

System	Theory level (and AIMAll version)	Energy		
		E_{WFN}	E_{IQA}	$\Delta E = E_{IQA} - E_{WFN}$
NMA				
	HF (v13.10.19)	−648691.11	−648691.03	−0.07
	B3LYP (v13.10.19)	−652690.04	−648588.54	−4101.50
	B3LYP (v14.04.17)	−652690.04	−652689.85	−0.19
TriGly				
	HF (v13.10.19)	−2277500.00	−2277500.95	0.95
	B3LYP (v13.10.19)	−2290953.80	−2277187.25	−13766.55
	B3LYP (v14.04.17)	−2290953.80	−2290953.39	−0.41
Alloxan dimer				
	HF (v13.10.19)	−2946006.19	−2946010.03	3.84
	B3LYP (v13.10.19)	−2961934.04	−2945606.78	−16327.26
	B3LYP (v14.04.17)	−2961934.04	−2961933.37	−0.67

Clearly, the ab initio energy of each molecular system is not recovered for B3LYP wavefunctions in AIMAll version 13.10.19 because the values for ΔE run in the thousands of kJ mol^{-1}. The ΔE values are also seen to grow with system size. This growing error is due to the increasing number of V^{A}_{ee} energies being incorrectly calculated, which we expect to be more dramatic for non-hydrogen atoms. Alloxan has 3 non-hydrogen atoms more than TriGly (in spite of having fewer atoms in total), which probably explains why its energy error is larger than that of TriGly.

When the explicit B3LYP V^{A}_{XC} functional is used in the IQA partitioning, the ab initio molecular energy is accurately recovered. For TriGly and the alloxan dimer, the B3LYP recovery error is much smaller than that of HF. An pitiable IQA energy recovery of ~3.8 kJ mol^{-1} was observed for the alloxan dimer, which is caused by poor integration errors (L(Ω) values) for 7 of the 8 carbon atoms in the dimer (see Table S1, ESI†). Extra integrations with non-default quadrature grids show that V^{AB}_{X} values are numerically stable, even when L(Ω) is quite large. Hence we deduce that another energy component, most likely the intra-atomic energy, is sensitive to the magnitude of L(Ω). Next, the breakdown of the interatomic exchange energies,V^{AB}_{X}, will be investigated for each system.

Interatomic Exchange Energies V^{AB}_{X}

Fig. S1–S9 contrast the HF and B3LYP HF-calculated V^{AB}_{X} energies for each interatomic interaction, together with the $V^{AB}_{X,HF} = V^{AB}_{X,B3LYP}$ bisector. The figures are grouped into a triplet for each molecular system. Within the grouped figures, the first figure always incorporates all interatomic interactions in the system (66, 528 and 276 for NMA, TriGly and alloxan, respectively). Typically, the first figures of each triplet are dominated by the strongly covalent backbone interactions within the molecule, corresponding to V^{AB}_{X} energies in the region of hundreds of kJ mol^{-1}. The second figure of each triplet only identifies V^{AB}_{X} energies within the range of −60 to −1 kJ mol^{-1} ('medium-range'). The third figure of each triplet only identifies V^{AB}_{X} energies within the range of −1 to 0 kJ mol^{-1} ('long-range'). These two energy ranges can be loosely translated to strong and weak interatomic interactions. Finally, Fig. 4 and 7 show a direct overlay of $|V^{AB}_{X}|$ values obtained at both HF and B3LYP level for the interactions in TriGly and the alloxan dimer, respectively. Fig. 6 is identical to Fig. S9 of the ESI,† and is shown in the main text for convenience and also due to its importance.

Small molecules and full CI

Table 2 compares three wave functions, in terms of interatomic exchange energy V_{XC} (in kJ mol^{-1}) and bond order δ, at Hartree–Fock (HF), B3LYP and full CI (FCI) level of theory, using the same basis set of 6-311G(d,p) for all (which is also used for NMA below).

Table 2 A comparison between Hartree–Fock (HF), B3LYP and full CI (FCI) wavefunctions in terms of interatomic exchange energy V_{XC} (in kJ mol^{-1}) and bond order δ. The molecules were geometry optimized only at HF level using at 6-311G(d,p) and the geometry kept constant for all three levels of theory. Note that V_{XC} is actually V_X for HF, and $V_{X,HF(B3LYP)}$ for B3LYP

Molecule	Measure	HF	B3LYP	FCI
H_2	V^{HH}_{XC}	−690.22	−687.08	−624.57
	δ_{HH}	1.00	1.00	0.85
LiH	V^{LiH}_{XC}	−92.14	−102.13	−103.02
	δ_{LiH}	0.19	0.22	0.21

In general, the results are in line with previously obtained insight. V_{XC} scales proportionally with δ, as expected. In covalent systems (such as H_2), the Coulomb correlation localizes the electrons in the atomic basins, which means that δ decreases and that the HF wavefunction is too delocalized. This is a general phenomenon that will be clearly and repeatedly recovered in NMA and TriGly. This exaggerated delocalization by HF is why the $|V_{XC}|$ value of H_2 is considerably smaller at the FCI level. In ionic molecules, such as LiH, the situation is reversed (although now the effect is much less pronounced). Here the mean field solution (i.e. HF) returns very well localized ions, while electron correlation transforms them somewhat back into neutral systems. This effect makes both δ and $|V_{XC}|$ increase. The effect of electron correlation on ions is to change their electron density just a little. B3LYP captures this effect nicely, and the B3LYP and FCI results are in very good agreement.

In particular, δ_{HH} must be equal to 1, exactly (1.000000), both at the HF and B3LYP levels. This value does only depend on the one-determinant structure of the (pseudo)-wavefunction. This also explains why B3LYP fails to provide a good V_{XC} value. Note that this means that other energy components are greatly affected by this inadequacy (kinetic, electron–nucleus, etc.) because the total B3LYP energy is close to the FCI one. The change in V_{XC} from HF to B3LYP is small since the real change caused by electron correlation in this system (as opposed to that found in LiH) is not in the electron density, but in the second-order density matrix. The HF, B3LYP, and FCI densities in H_2 are quite similar but it is the pair behavior that is different. This can only be taken into account using a multi-determinant expansion, which provides smaller δ and $|V_{XC}|$ values.

NMA

NMA contains 12 atoms and hence (12 − 1) × 12/2 = 66 unique interatomic interactions. NMA features regularly in biochemistry as a small and computationally inexpensive system that contains a peptide bond. NMA›s interactions span a range of interaction types from 1,2 to 1,6. To elaborate on this notation, a 1,2 interaction consists of two neighbouring atoms separated by a single 'typical' covalent chemical bond; a 1,3 interaction is separated by two bonds, a 1,4 interaction is separated by three bonds, while an 1,n interaction is separated by n − 1 bonds. Thus, NMA serves as a good example for a small variety of interaction ranges encompassing the so-called bonded (1,2; 1,3 and 1,4 interactions) and non-bonded interactions (1,5 and higher).

Fig. S1–S3 compare the HF and B3LYP V^{AB}_X values in NMA, along with a bisector. The V^{AB}_X energies are distributed both above and below the bisector, indicating that increased stability may occur at either HF or B3LYP level. Fig. S1 shows all 66 interactions, and at this grand scale, a clear stabilization is perceived for only 3 interactions, which are all 1,2. To see more detail, Fig. S2 and S3 zoom in on smaller energy windows, which are −60 to −1 kJ mol^{-1} and −1 to 0 kJ mol^{-1}, respectively. Fig. S2 shows the vast majority of 1,3 and 1,4 interactions. Energies are scattered around the bisector, with almost twice as many data points in the $|V^{AB}_{X,B3LYP}| > |V^{AB}_{X,HF}|$ half (which is hard to see, but clear from Table 3 below).

Table 3 Stabilization and destabilization of V^{AB}_X energies in going from HF to B3LYP for all 66 interatomic interactions in NMA. Maximum and mean absolute difference percentage (between and HF and B3LYP) sorted by interaction type "1,n" and by whether $|V^{AB}_{X,HF}| > |V^{AB}_{X,B3LYP}|$ or $|V^{AB}_{X,B3LYP}| > |V^{AB}_{X,HF}|$

	$\|V^{AB}_{X,HF}\| > \|V^{AB}_{X,B3LYP}\|$			$\|V^{AB}_{X,B3LYP}\| > \|V^{AB}_{X,HF}\|$		
	Destabilization caused by B3LYP			**Stabilization caused by B3LYP**		
1,*n*	**Maximum**	**Mean**	**No. interactions**	**Maximum**	**Mean**	**No. interactions**
1,2	1.4	1.1	7	15.2	9.6	4
1,3	14.4	6.4	12	54.8	19.9	6
1,4	2.8	2.8	1	62.4	28.9	15
1,5	0.0	0.0	0	57.8	34.8	12
1,6	16.1	10.5	2	358.1	103.8	7

In Fig. S3 (ESI), which shows only three 1,4 interactions and all 1,5 and 1,6 interactions, most data points now appear below the bisector, which means that B3LYP stabilizes these interactions compared to HF. This stabilization trend can also be seen as a 'shift' in the data points, in going from HF to B3LYP. This analysis benefits from more quantitative information about the distribution of the V^{AB}_{X} values at either side of the bisector. This information is provided in Table 3, which groups the V^{AB}_{X} values according to which side of the bisector they lie,i.e. either $|V^{AB}_{X,HF}| > |V^{AB}_{X,B3LYP}|$ (upper half in Fig. S1–S3, ESI) or $|V^{AB}_{X,B3LYP}| > |V^{AB}_{X,HF}|$ (lower half).

Table 3 shows only percentages, which characterize the degree to which the introduction of B3LYP changes the HF V^{AB}_{X} values. To understand how these percentages (maximum and mean) were calculated, it is instructive to focus on the 4 example interactions (N1–H8, N1–C9, N1–C2 and C2–O3) marked in red in Table S2 . These interactions also appear in Fig. S1 where they lie in the $|V^{AB}_{X,B3LYP}| > |V^{AB}_{X,HF}|$ region, that is, in the lower half, below the bisector. Table S2 makes clear that the largest relative energy shift, going from HF to B3LYP, is −111.8 kJ mol^{-1} (for N1–C2). Note that negative differences correspond to a stabilization caused by B3LYP. All B3LYP energies in this paper are always referred to HF, so it makes sense to work with the ratio of this energy difference to the original HF. Converting this ratio to a percentage, we define $100 \times |V^{AB}_{X,B3LYP} - V^{AB}_{X,HF}|/V^{AB}_{X,HF}$ as the "absolute difference percentage". The absolute difference percentage associated with C2–O3 then appears as 15.2% in Table 3 under the 1,2 entry referring to 4 interactions with $|V^{AB}_{X,B3LYP}| > |V^{AB}_{X,HF}|$. Table 3 also lists the mean difference percentage (i.e. 9.6%), where the mean represents an average over these 4 interactions.

Table 3 shows that the interactions predominately stabilized by B3LYP are the 1,4 interactions and higher. The 1,2 and 1,3 interactions are mostly destabilized by B3LYP, in a ratio of about 2 to 1. Secondly, when $|V^{AB}_{X,B3LYP}| > |V^{AB}_{X,HF}|$, the mean difference percentage increases monotonically with n in 1,n over the whole range of n. However, when $|V^{AB}_{X,HF}| > |V^{AB}_{X,B3LYP}|$, there is no apparent trend linking difference percentage to the interaction type.

Fig. 3 shows the shift in V^{AB}_{X} energies in going from HF to B3LYP for all 66 interactions in NMA. The logarithmic energy scale brings out the shift in terms of energy ratios. Fig. 3confirms, at a glance, that the larger n in an 1,n interaction, the more pronounced the relative energy stabilization caused by B3LYP. This effect is consistent with the behavior of the mean difference percentage in Table 3, for the stabilizing half on the right of this table. This effect may also be related to the fact that B3LYP is not an asymptotically correct functional, so it behaves pathologically at long-range.

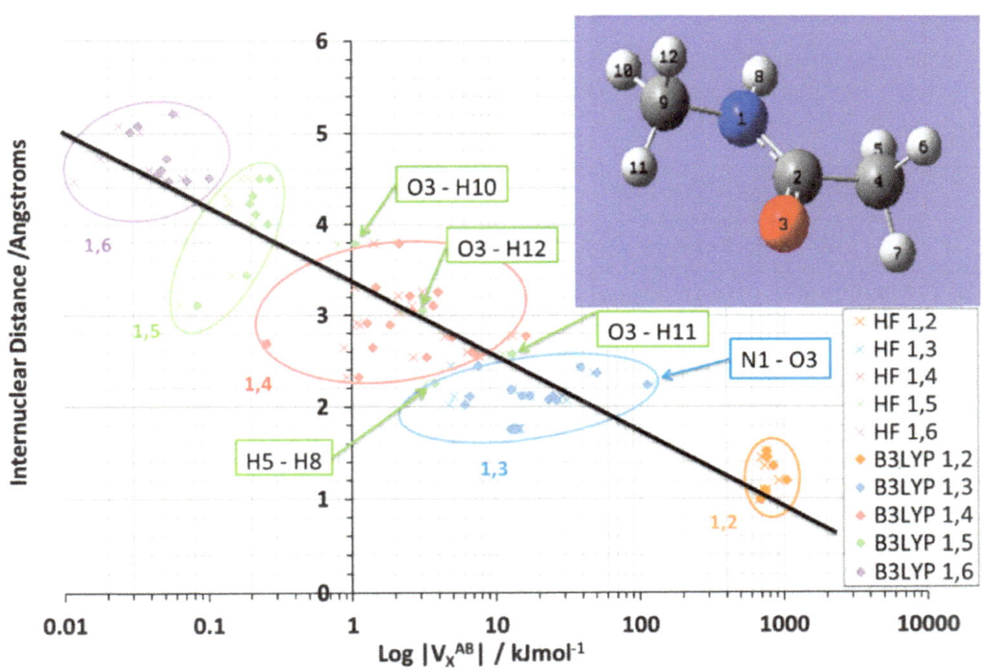

Figure 3: Logarithmic plot of $|V^{AB}_{X}|$ versus interatomic distance for all 66 NMA interactions at HF/6-311G(d,p) and B3LYP/6-311G(d,p) level of theory. Interaction types "1,n" have been encircled where possible, showing the clustering as well as the outliers.

It is pleasing to see that V^{AB}_{X} energies broadly fall apart into non-overlapping clusters, each cluster corresponding to an 1, n interaction, where n = 2, 3, 4, 5 or 6. The gap between the 1, 2 cluster and the 1,3 cluster is particularly pronounced, while the 1,3 and 1,4 clusters are contiguous. The overall behavior of V^{AB}_{X} energies as a function of internuclear distance is such is that the lowest values occur at the largest distances. The descending line in black captures this behavior coarsely, with a correlation coefficient r^2 of 0.85, referring to the B3LYP data only. This broadly respectable correlation does not attempt to spot any non-linearity in the data, nor does it detect a fine structure emerging by grouping V^{AB}_{X} values per interaction pair AB.

Closer inspection of the clusters reveals some energetic anomalies, that is, interactions appearing in unexpected clusters. For example, four 1,5 interactions (green) turn up far outside their cluster. Fig. 3 marks these as O3–H12, O3–H11, O3–H10 and H5–H8. The first two interactions of this set appear near the overall (B3LYP) correlation line (in black). Hence they are not anomalous in terms of how a V^{AB}_{X} energy varies with A···B distance. On the other hand, O3–H10 is anomalous, not just by connectivity "distance" (i.e. 1,5) but by actual (geometrical) distance. Indeed, this interaction is curiously strong for its relatively large internuclear distance. The reason for this observation is not clear. Inspired by earlier work[25] that showed that V^{H-H}_X values are anomalously large when part of a planar H–C–C–H arrangement in alkanes, it is tempting to invoke the near-planarity of the H10–C9–N1–C2–O3 fragment (where the H10–C9–N1–C2 dihedral angle is −163° (off by only 17° from 180°) and the C9–N1–C2–O3 dihedral angle is 4°). Finally, note that V^{N1-O3}_X is the strongest interaction in the 1,3 cluster, quite far to the right of the black line. This enhanced delocalization across the C–N peptide bond to the oxygen is reminiscent of the resonance canonical imparting double bond character to the CN interaction.

TriGly (GlyGlyGly)

There are 528 interatomic interactions in this oligopeptide (Fig. 2) ranging up to 1,15 interaction types. Ignoring a hydrogen atom each time, NMA occurs four times in GlyGlyGly, which therefore marks a major upscaling in the size of our analysis, while also establishing links between the NMA and GlyGlyGly.

Fig. S4–S6 again compare the HF and B3LYP V^{AB}_X values, but now in TriGly (counterpart of Fig. S1–S3, ESI†). Fig. S4 shows all 528 interactions, and at this grand scale, clear stabilization materializes for only a dozen or so interactions, which are all 1,2. Fig. S5 (−60 to −1 kJ mol^{-1}) shows the vast majority of 1,3 and 1,4 interactions, along with about a third of the 1,5 interactions. Energies are scattered around the bisector, with 2.6 times as many data points in the $|V^{AB}_{X,B3LYP}| > |V^{AB}_{X,HF}|$ half. Fig. S6 repeatedly shows a larger shift towards stabilization by B3LYP. The interactions from 1,4 to 1,8 follow trends identical to those observed for NMA, with greater stability favoured in the B3LYP wavefunction.

Table 4 shows that the interactions that B3LYP predominately stabilizes, are the 1,4 interactions and higher, as before for NMA, but this effect tails off at 1,13 and even reverts at 1,14. When lumped together, the 1,2 and 1,3 interactions are neither stabilized nor destabilized by B3LYP. In other words, 46 interactions are less stable under B3LYP while 40 are more so. As in the case of NMA, the mean difference percentage increases monotonically with n (in 1,n) up to 1,8 interactions, again when $|V^{AB}_{X,B3LYP}| > |V^{AB}_{X,HF}|$. Thus, in TriGly, the range of the monotonic trend observed up to the 1,6 interactions in NMA, is extended to 1,8. However, at this point, the mean difference percentage decreases (but not monotonically) until the 1,15 interactions are reached. When $|V^{AB}_{X,HF}| > |V^{AB}_{X,B3LYP}|$, there is still (as for NMA) no apparent trend linking difference percentage to the interaction type, throughout the whole range of interactions.

Table 4 Stabilization and destabilization of V^{AB}_X energies in going from HF to B3LYP for all 528 interatomic interactions in TriGly. Maximum and mean absolute difference percentage (between and HF and B3LYP) sorted by interaction type "1,n" and by whether $|V^{AB}_{X,HF}| > |V^{AB}_{X,B3LYP}|$ or $|V^{AB}_{X,B3LYP}| > |V^{AB}_{X,HF}|$

	$\|V^{AB}_{X,HF}\| > \|V^{AB}_{X,B3LYP}\|$			$\|V^{AB}_{X,B3LYP}\| > \|V^{AB}_{X,HF}\|$		
	Destabilization caused by B3LYP			**Stabilization caused by B3LYP**		
1,*n*	**Maximum**	**Mean**	**No. interactions**	**Maximum**	**Mean**	**No. interactions**
1,2	1.5	1.2	12	15.2	7.7	20
1,3	15.2	8.9	34	63.0	23.3	20
1,4	2.9	1.9	4	72.9	27.6	59
1,5	0.0	0.0	0	94.2	41.7	60
1,6	0.0	0.0	0	206.1	70.9	60
1,7	0.0	0.0	0	194.1	99.8	49
1,8	0.0	0.0	0	452.1	147.8	43
1,9	12.2	12.2	1	383.9	114.5	42
1,10	0.0	0.0	0	295.8	100.5	33
1,11	4.8	4.8	1	183.0	75.7	26
1,12	33.8	25.5	3	119.3	53.3	23
1,13	25.6	19.3	2	228.9	76.7	14
1,14	64.2	43.2	4	248.4	82.4	9
1,15	67.6	40.3	7	144.1	75.8	2

On a technical note, such large difference percentages for long-range 1,n interactions could be a cause for concern in the analysis above. However, the reliability of the values for V_X should not be called into question, as we now explain. Table S3 shows how a variation in the outer angular integration grid (within AIMAll) affects the V_X energies for the very high 1,n interactions for two example interactions in TriGly: C22···H27 (1,14) and H24···H28 (1,15). A change in the outer angular quadrature integration grid changes the integration error L(Ω) calculated for each atom. A computationally less expensive grid, for example, 'Low (Lebedev)' will typically[50] result in a higher absolute L(Ω) value for a given atom, compared to that of 'Sky High (Lebedev)'. Table S3 shows that the variation of the integration grid does cause a significant change in L(Ω) but importantly V_X remains very stable throughout. Such small changes in the values of V_X, compared to that of L(Ω), allow us to trust the V_X values, even for very high n in 1,n. This stability of V_X suggests that a large L(Ω) error affects another IQA energy component more, presumably E^A_{intra}.

Fig. 4 shows the effect on V^{AB}_X energies of introducing B3LYP, compared to the original HF values, for all 528 interactions in TriGly computed at the HF/6-31+G(d,p) and B3LYP/6-31+G(d,p) level of theory. As in Fig. 3, it again makes sense to express the overall behaviour of V^{AB}_X energies as a function of internuclear distance in a coarse, linear way. However, here we draw two lines, one for HF and one for B3LYP, instead of the single black line in Fig. 3. The descending line in black captures this behaviour, admittedly coarsely, with a correlation coefficient r^2 of 0.91 for both HF and B3LYP. It is clear that the cloud of B3LYP energies (in red) occurs right of the cloud of HF energies (in blue), which means that most interatomic interactions are indeed more stable at B3LYP level. It is interesting to again analyze anomalies in the same manner as in Fig. 3. It is best to truncate the analysis at 1,6 interactions because energies of about 1 J mol^{-1} or less are not really significant (unless hundreds of them are considered as a group and their energies simply added, as allowed by the additive nature of V_X).

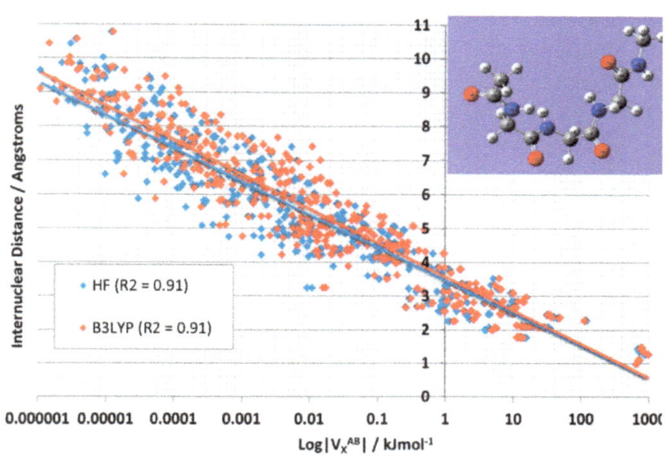

Figure 4: Logarithmic plot of V^{AB}_X versus interatomic distance for all 528 TriGly interactions at the HF/6-31+G(d,p) and B3LYP/6-31+G(d,p) level of theory.

Fig. 5 shows again the broad behavior seen before in NMA (Fig. 3): the further two atoms are apart, the less they exchange (i.e. reduced delocalization). In other words, well separated atoms have a small V^{AB}_X energy. This behavior is typical for saturated systems, such as this tripeptide, which lacks aromatic amino acids or other extensive electronically delocalized fragments. Secondly, as seen in NMA, the various 1,n interactions do cluster as well, albeit with more overlap compared to NMA. The most substantial overlap takes place between the 1,5 (green) and 1,4 (red) cluster. Such overlap is not surprising because TriGly is a more complex system, which contains a wider variety of interactions than found in NMA. In particular, because TriGly is curled (see inset) there are many instances where the actual (geometrical) distance is smaller than the distance based on bond connectivity (i.e. labelled by 1,n). This mismatch between geometrical and connectivity-based distance shows itself in 1,6 interactions where, for example, the H10···H19 distance is only 2.68 Å. The corresponding V^{AB}_X(B3LYP) energy is a respectable 0.4 kJ mol^{-1}, whereas the mean V^{AB}_X energy of all 1,6 interactions is at least an order of magnitude smaller.

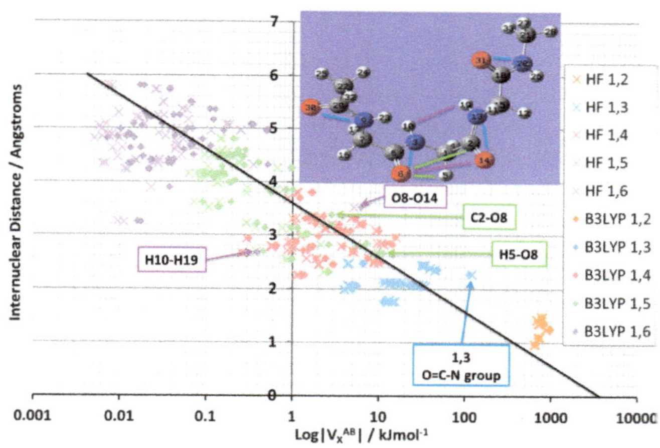

Figure 5: Logarithmic plot of $|V^{AB}_X|$ versus interatomic distance for TriGly interactions up to 1,6, at HF/6-31+G(d,p) and B3LYP/6-31+G(d,p) level of theory. Key outlying V_X values have been labelled in both the plot and the inset molecule image. The black line shows the overall correlation of the B3LYP energies, with a correlation coefficient r^2 of 0.91.

The next set of interactions of interest is the same as in NMA: chemically meaningful quantities such as V^{AB}_X are anticipated to be transferable. We expect the N···O interactions associated with the peptide bond to be very similar in energy value. QCT is known for providing highly transferable properties. Indeed, the $V^{N\cdots O}_X$ energy averaged across all four peptide bonds in TriGly and the one peptide bond in NMA, is −113.4 kJ mol^{-1} with a standard deviation of only 1.0 kJ mol^{-1} at B3LYP level. The standard deviation at HF level is only 0.2 kJ mol^{-1}, which is a remarkable testimony of transferability. As expected, the $V^{N\cdots O}_X$ energies again (see NMA, Fig. 3) appear at the strong end of the 1,3 cluster (Fig. 5, at the extreme right of blue data points, marked as "O= C–N group"). Finally, we mention that the 1,5 interaction H5···O8 also appears at the strong extreme of the 1,5 cluster, but energetically this interaction is less anomalous than any O···N interaction because $V^{H5\cdots O8}_X$ lies very near the black line. In contrast, the 1,6 interaction O8···O14 lies quite a bit off the black line, and marks an unusually strong stabilization of about 6 kJ mol^{-1} by "through-space" oxygen···oxygen delocalization. This observation sets the scene for exploring another type of interaction that received a great deal of attention in the last decade, called the n → π* interaction.

TriGly is an ideal system to make contact with the research theme of Raines and co-workers on so-called n → π* interactions in proteins.[51] The interaction occurs between protein backbone amides, and is characterized as arising from the delocalization of a lone pair of electrons (n, in fact, the p-rich lone pair) of an amide oxygen, to an antibonding orbital (π*) of the subsequent carbonyl group. This research group has elevated this type of interaction to one that may be as important as hydrogen bonding, which they characterize as an n_s → σ* interaction. Here, there is an s-rich lone pair of an amide oxygen delocalizing with the antibonding σ* orbital of the N–H bond of an amide bond, but now four amino residues further down the protein backbone. Although the n → π* interaction can continuously get stronger, Raines et al. propose that the interaction should unmistakably take place if the distance d between O_B and C_C in $C_A{=}O_B{\cdots}C_C{=}O_D$ is smaller than 3.2 Å, and if the angle θ, defined as $O_B{\cdots}C_C{=}O_D$, lies between 99° and 119°. Note that the quantum mechanical interpretation of Raines et al. is couched in the language of Natural Bond Analysis (NBO),[7] which suffers from practical issues of computational stability and lack of transparency. However, IQA is alternative candidate to express delocalization effects but now in an orbital invariant way, as well as computationally stable and transparently. In TriGly, the $C_A{=}O_B{\cdots}C_C{=}O_D$ system that comes nearest to obeying those geometrical criteria is $C_6{=}O_8{\cdots}C_2{=}O_{14}$, where the distance d is 3.4 Å and θ is 87°. It turns out that that the $V^{C2\cdots O8}_X$ energy is 2.8 kJ mol^{-1}. This energy is close to a value quoted by Raineset al., still in kcal mol^{-1}, and amounting to 0.5 in that energy unit, obtained after an NBO analysis of an α helix. Having said this, the 1,5 interaction C2···O8 (marked in Fig. 5), occupies a rather unremarkable place in the green cluster or with respect to the black line. A likely reason is that the $C_6{=}O_8{\cdots}C_2{=}O_{14}$ fragment falls out of the range proposed by Raines et al., strictly speaking.

Alloxan Dimer

This system is the first example of a van der Waals complex, which is of interest to the current analysis because of its surprising total lack of hydrogen bond as discovered by Bolton[52] in its stable crystalline structure. However, Dunitz and Schweizer ask if alloxan is really a "problematic" structure[53] although it has been regarded as such for about 40 years. They write that "It ("alloxan" red.) is held together mainly by whatever factors contribute to the cohesive energies… the C=O···C=O type of interaction would seem to play an important part". Unpublished work from our lab shows that the electrostatics between topological atoms plays an important role in the stability of the complex. Here we only discuss the exchange energy component. The investigation of the intermolecular interactions allows us to compare with the intra-molecular interactions, and present a complete analysis of the V^{AB}_X energies.

The alloxan monomer is a heterocyclic planar system. Within each alloxan monomer, intramolecular interactions only reach to the maximum value of 1,6. When added to the intermolecular interactions, a total interaction count of 24 × (24 − 1) = 276 is obtained, in the knowledge that each alloxan monomer consists of 12 atoms. In the current work, only one of the two known alloxan dimer crystal orientations[53] was investigated (see Fig. 2), the other being a dimer held by bifurcated hydrogen bonds. For the purpose of this analysis, a single dimer arrangement is sufficient. The monomers are roughly T-shaped (with a tilt), with two carbonyl groups of one monomer pointing towards the plane of the other monomer, approximately bisecting this plane.

Fig. S7–S9 again compare the HF and B3LYP V^{AB}_X values, but now in alloxan. Fig. S7 shows all 276 interactions, and at this grand scale, clear stabilization materializes for only a dozen or so interactions, which are all 1,2. The introduction of B3LYP renders the strongest interactions stronger. Fig. S8 (−60 to −1 kJ mol^{-1}) shows the vast majority of 1,3 and all but one of the 1,4 interactions, along with very few 1,5 interactions, as well as up to a quarter of inter-alloxan interactions. Energies are scattered around the bisector, with about 10 times as many data points in the $|V^{AB}_{X,B3LYP}| > |V^{AB}_{X,HF}|$ half. Fig. S9 repeatedly shows a larger shift towards stabilization by B3LYP. The interactions from 1,4 to 1,8 follow trends identical to those observed for NMA, with greater stability favored in the B3LYP wavefunction. Fig. 6 (which is identical to Fig. S9, ESI†) covers the range 0 to −1 kJ mol^{-1}, which corresponds to 1,5; 1,6 and the inter-alloxan interactions. Here, 83% interactions are stabilized by B3LYP. Table 5 confirms this observation and quantifies the number of interactions in either the $|V^{AB}_{X,B3LYP}| > |V^{AB}_{X,HF}|$ half or the $|V^{AB}_{X,B3LYP}| < |V^{AB}_{X,HF}|$ half.

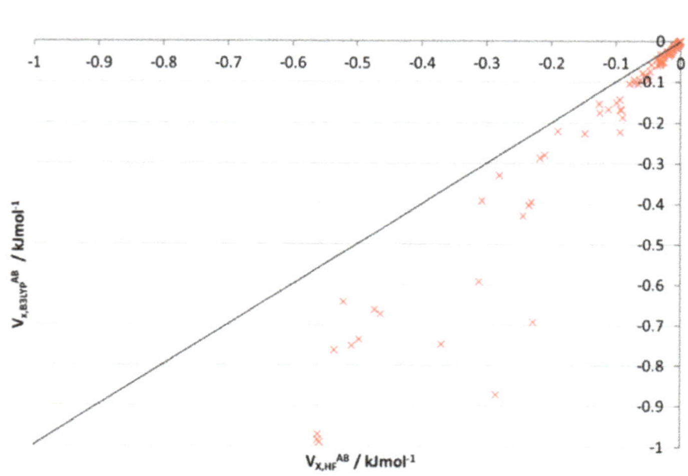

Figure 6: V^{AB}_X energies for long-range interactions (0 to −1 kJ mol^{-1}) in the alloxan dimer, from both HF and B3LYP wavefunctions. Note that this figure is repeated in the ESI† as Fig. S9 because there it is part of a triplet of plots on alloxan.

Table 5 Stabilization and destabilization of V^{AB}_{X} energies in going from HF to B3LYP for all 276 interatomic interactions in alloxan. Maximum and mean absolute difference percentage (between and HF and B3LYP) sorted by interaction type "1,n" and by whether $|V^{AB}_{X,HF}| > |V^{AB}_{X,B3LYP}|$ or $|V^{AB}_{X,B3LYP}| > |V^{AB}_{X,HF}|$. Just over half of the interactions (144 in total) are of intermolecular nature

	$|V^{AB}_{X,HF}| > |V^{AB}_{X,B3LYP}|$			$|V^{AB}_{X,B3LYP}| > |V^{AB}_{X,HF}|$		
	Destabilization caused by B3LYP			**Stabilization caused by B3LYP**		
1,*n*	**Maximum**	**Mean**	**No. interactions**	**Maximum**	**Mean**	**No. interactions**
1,2	0.3	0.2	4	16.0	12.0	20
1,3	21.2	9.4	14	56.0	26.6	22
1,4	0.0	0.0	0	76.3	38.3	42
1,5	0.0	0.0	0	107.3	77.4	24
1,6	0.0	0.0	0	206.7	136.2	6
Intermolecular	259.4	89.7	27	225.0	74.0	117

The predominately stabilizing trend caused by B3LYP also appears in purely intermolecular (i.e. alloxan···alloxan) interactions. In total, there are 144 of these interactions, which amounts to 52% of the complete total of 276 interactions. Table 5 shows these intermolecular interactions as a separate entry. There are now 27 cases for which $|V^{AB}_{X,HF}| > |V^{AB}_{X,B3LYP}|$ and 117 for which $|V^{AB}_{X,B3LYP}| > |V^{AB}_{X,HF}|$. The distribution of cases over these two stabilization regimes can be rationalized as follows. From the TriGly data we learnt that interactions are more stable at HF level if they are (very) long range (or 1,2 and 1,3 but this very short range is not relevant here). Hence, we expect the average internuclear distance for the 27 cases to be large. Indeed, it turns out to be 6.5 Å as shown in Table S6. The value of 6.5 Å falls in between 6.3 Å and 6.8 Å, which are the respective average internuclear distances for 1,8 and 1,9 interactions in TriGly.

Fig. 7 demonstrates the energetic stability that B3LYP brings to the vast majority of interactions in the alloxan dimer: the red cloud (B3LYP) is indeed predominately shifted to the right compared to the blue cloud (HF). This figure is the counterpart of Fig. 4, and again shows two lines, one for HF and one for B3LYP, with almost identical correlation coefficients r^2 of 0.85 and 0.84, respectively. Note that the x-axis is logarithmic hence $|V^{AB}_{X}|$ must be used. It is clear that the shift becomes more pronounced as the interatomic distance increases and the exchange energy decreases. The shift is more pronounced here than in TriGly (see Fig. 4).

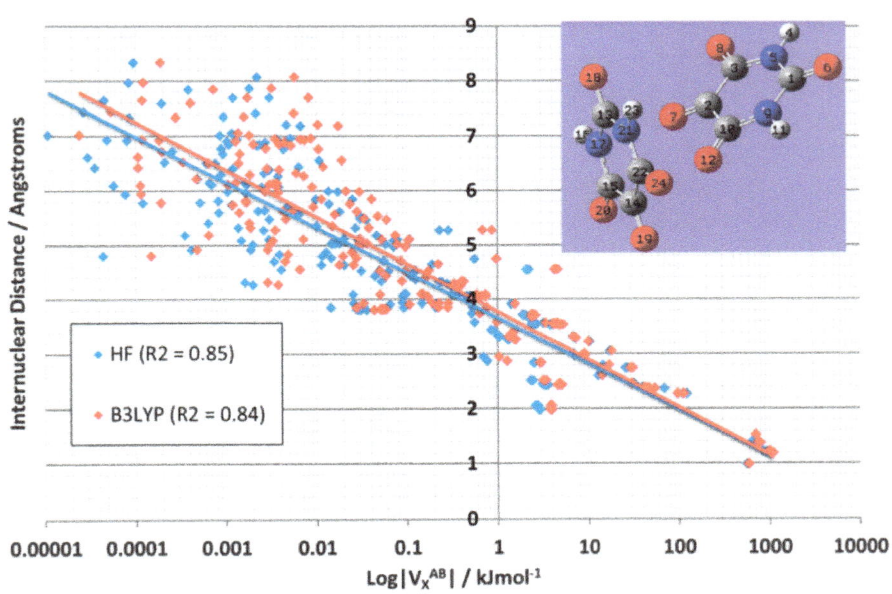

Figure 7: Logarithmic plot of V^{AB}_{X} versus interatomic distance for all 276 interactions in the alloxan dimer at HF/6-31+G(d,p) and B3LYP/6-31+G(d,p) level of theory.

Finally, Fig. 8 shows again the broad behaviour seen before in NMA and TriGly (Fig. 3 and 5) but now for the alloxan dimer. The black line represents the overall relationship between V^{AB}_X and internuclear distance. The corresponding correlation coefficient r^2 is 0.84. This time the clusters represents the exchange energy patterns of the alloxan monomers only, and show again minimal overlap. Four intramolecular O···O interactions stand out by their unusual strength: O7···O8, O7···O12 in one alloxan, and O19···O20, O19···O24 in the other. The average V^{OO}_X B3LYP energy is −30.8 kJ mol^{-1} with a pleasingly small standard deviation of 0.9 kJ mol^{-1}. These four O···O interactions appear outside of their 1,4 cluster. This observation is reminiscent of that made in TriGly where the 1,6 interaction O8···O14 appeared far outside its cluster.

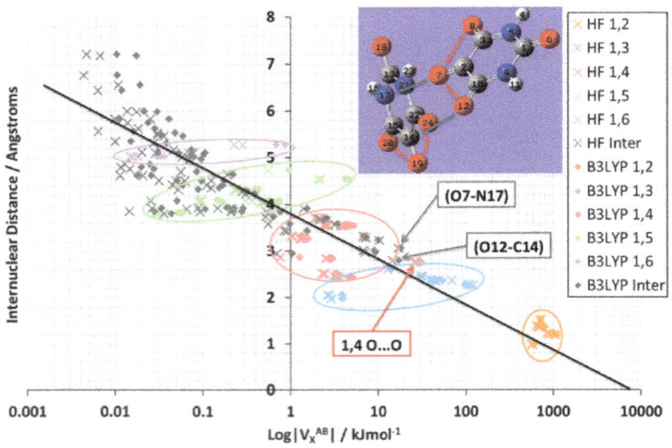

Figure 8: Logarithmic plot of $|V^{AB}_X|$ versus interatomic distance for alloxan interactions with $|V_{X,B3LYP}| > 0.01$ kJ mol^{-1}, at HF/6-31+G(d,p) and B3LYP/6-31+G(d,p) level of theory. Interaction types "1,n" have been encircled where possible, showing the clustering as well as the outliers. Bracketed interaction labels indicate the interaction also forms an AIL. The black line shows the overall correlation of the B3LYP energies, with a correlation coefficientr^2 of 0.84.

Fig. 8 also marks inter-alloxan interactions (in grey-brown) having values of $|V_{X,B3LYP}| > 0.01$ kJ mol^{-1}. These intermolecular interactions mainly populate the weaker end of the energy spectrum: they start in the utmost left region, which is weaker than the intramolecular 1,6 interactions (flat disk), and extend over the 1,5 interactions (green disk) well into the 1,4 interactions (red disk). In fact, two intermolecular interactions even outdo the 1,4 interactions in terms of strength: O7···N17 and O12···C14 (marked in Fig. 8, and ignoring the 4 intramolecular O···O interactions outside the 1,4 disk). They are the strongest intermolecular interactions, and remarkably correspond to the two interatomic interaction lines found between the two monomers. The O12···C14 interaction materialises between two C=O groups, and can hence be linked to Dunitz and Schweizer's comment on the important role of C=O···C=O type interactions played in the stability of the crystal. However, a convincing argument on the stability of the alloxan dimer can only be made when considering all IQA energy contributions.

DISCUSSION

The interatomic exchange energy results provide several points to consider more deeply. These points will be addressed in turn, starting with a brief comment on the comparison between delocalization indices and V^{AB}_X.

A systematic increase in energetic stability is seen between pairwise 1,4 bonded interactions and higher non-bonded interactions. This trend was observed unequivocally until ~1,14 interactions. However, interactions of 1,6 and higher, in general, had interatomic exchange energies that fell below the integration error of the atoms involved, eventually reaching values as little as ~1 × 10^{-6} kJ mol^{-1} in magnitude. Such interactions would be considered negligible for any application. Hence, up to ~1,6 interactions, served as an appropriate cut-off for interaction types.

A comparison can be made between the V^{AB}_X results presented and the trend previously observed in measuring pairwise electron delocalization, via the delocalization index (DI). The consistent overestimation of DI values, which has been observed before, matches the current results for non-bonded interactions (1,5 and higher). This overestimation provides evidence that some Coulomb electron–electron correlation (which is included in the one-electron density) is partly included in the pairwise exchange results, despite the HF-formalism not directly addressing it. This is a well-known phenomenon, which allows a simple explanation in terms of the different behavior of Kohn–Sham and Hartree–Fock orbitals. KS orbitals are more localized compared to HF orbitals, which is common in largely covalent interactions. This increased localization leads to smaller DIs. Because exchange energies are proportional to DIs, they will be smaller as well (in absolute value). The opposite will be true when KS orbitals become more strongly delocalized than their HF counterparts. In any case, the exact behavior will never fully be included without the electron-pair density being defined within the HF-formalism.

The 1,2 and 1,3 interactions are neither stabilized nor destabilized by B3LYP. To rationalize this fact, it helps to consider work of Poater et al.[36] regarding a similar analysis on δ(A,B) rather than on our V^{AB}_X values. From their study of ethane and diborane, one concludes that δ(A,B) values of the 1,2 and 1,3 interactions involved can either increase or decrease. Given the proportionality between $|V^{AB}_X|$ and δ(A,B), mentioned in the Introduction, we can conclude that the results of Poater et al. support our own findings, stated at the beginning of this paragraph. From this parallel behaviour of δ(A,B) and V^{AB}_X results, we conclude that some first-order density correlation must also being included in V^{AB}_X.

A more detailed analysis is also possible that allows for a physicochemical interpretation of the systematic differences found in our data beyond the general discussion already presented. Tables S2, S4 and S5 gather all the (1,n) n = 2,3,4 pairs for NMA, TriGly and the alloxan dimer, respectively, listing their $V^{AB}_{X,HF}$ and $V^{AB}_{X,B3LYP}$ values, their difference, and absolute difference percentages. In order to shed some light on them we take into account the major factors that influence V^{AB}_X. We will use orbital arguments, which are easier to grasp, although fully orbital invariant insights may also be given. At the HF or KS levels,

$$V_X^{AB} = -\sum_{i,j}\int_A\int_B \frac{\psi_i(\mathbf{r}_1)\psi_j(\mathbf{r}_2)\psi_j(\mathbf{r}_1)\psi_i(\mathbf{r}_2)}{r_{12}} d\mathbf{r}_1 d\mathbf{r}_2 \tag{18}$$

So the first, and probably most important factor in determining V^{AB}_X, is the degree of delocalization of the orbitals involved in a given AB interaction. It is easy to understand from eqn (18) that only orbitals delocalized between atoms A and B will contribute non-negligibly to V^{AB}_X. Inclusion of electron correlation effects, even at DFT level, tend to localize orbitals involved in direct covalent links, and to delocalize their long-range tails. For a given AB pair, other factors involve the size of the topological atoms, and the amount of electron charge contained in them. For instance, it is clear that in 1,2 interactions, a shift in the position of the interatomic surface (separating A and B) towards the ideal mid-partition will increase V^{AB}_X. In summary, homopolarity works in favour of covalency. Conversely, heteropolarity works against covalency. Similarly, V^{AB}_X is usually dominated by the diagonal i = j term in eqn (18), the so-called self-interaction contributions, which behave like products of densities. An increase in the electron population of an atom will necessarily increase these self-interaction terms, adding to a larger V^{AB}_X. There are other effects that the above expression (only valid for single-determinant expansions) will never capture. There are other effects that the above expression (only valid for single-determinant expansions) will never capture. At any standard HF or KS DFT level, a hydrogen molecule will be described by one Slater (pseudo)determinant consisting of a doubly occupied totally symmetric σ_g orbital, independently of the internuclear distance. Its DI will be exactly equal to 1, and its V^{AB}_X value will be very large, not vanishing at dissociation. This is the well-known single determinant dissociation problem of H_2 rephrased in terms of DI's or V^{AB}_X's. A proper wavefunction correlated method, as simple as a CAS [2,2] calculation, will mix at least two determinants, leading to a destructive interference that will make V^{AB}_X vanish at the large distance limit. We will now show how a combination of these effects rationalizes our findings.

It is useful to introduce net atomic charges (denoted q) into this discussion. Tables S7–S9 provide the net charges of all atoms in NMA, TriGly and the alloxan dimer, respectively. These data show that B3LYP produces net accurate charges that are very close to those generated by MP2, which models correlation effects independently from DFT.

As noticed above, our data favour a clearly biased distribution for 1,4 and higher interactions. When $|V^{AB}_{X,B3LYP}| > |V^{AB}_{X,HF}|$, the sum of the net charge on A and on B at the B3LYP level (i.e. q(A) + q(B)) is lower than the sum at the HF level across all 1,n interaction types. In other words, the total electron population contained in the B3LYP (topological) atoms is larger. This is in agreement with the comments on the role of self-interaction posed in the above paragraph, and larger atomic electron populations will be, in general, accompanied by larger $|V^{AB}_{X,B3LYP}|$ values, particularly when the implied extra electrons are delocalized between A and B, such that the product $\psi_i^2(\mathbf{r}_1)\psi_i^2(\mathbf{r}_2)$ contributes significantly to V_X (see eqn (18) and recall that we integrate electron 1 over A and electron 2 over B). Almost all the exceptions to this rule may also be rationalized. For instance, in the C2–C10 interaction in the alloxan dimer, q(A) + q(B) at B3LYP is lower, because these carbon atoms participate in very polar C=O interactions at the HF level. However, $|V^{AB}_{X,B3LYP}| < |V^{AB}_{X,HF}|$. In this case, the smaller net charge of the B3LYP carbon atoms is due to electrons provided by the neighbouring carbonylic oxygens, which do not affect the electrons participating in the C2–C10 interaction. The latter are still much more delocalized at the HF than at the B3LYP level, giving rise to a larger V_X in the former case, as the following paragraph explains.

Similarly, across all 3 systems, all the 1,2 bonded interactions that showed increased stability at the HF level were C–H or C–C interactions. In the case of C–H, the larger HF exchange stems from a spurious polarization giving rise to negatively charged hydrogen atoms, caused by a too delocalized C–H bonding orbital. However, B3LYP recovers the correct polarity by shifting the interatomic surface (or equivalently localizing the bonding orbital) and thereby decreasing the exchange. In the case of C–C, we face the well-known fact that too delocalized HF orbitals lead to exaggerated covalency, counteracted again by localization in the DFT description. We should add that it is a general phenomenon that inclusion of electron correlation considerably decreases both the DI and the exchange–correlation energy in 1,2 homoatomic bonds.

An analysis of the 1,3 interactions along the previous lines also explains the stability of the interactions at each level of theory. The 1,3 interactions that are more stable at HF, consisted of a wider variety of elements (or types of topological atoms, e.g. C···N, C···O, N···H) across the three case studies. Typically, at HF level these interactions typically exhibit a lower sum (not absolute value) of the net charge on A and on B. In some interactions (e.g. N···H), the polarity of the H atoms involved, changed according to whether the HF or B3LYP wavefunction was used. At HF, a number of H atoms had exaggerated charges and were often anionic. Some of these atoms exhibited a change in polarity, becoming cationic at B3LYP. The previous anion–anion behaviour would lead to an increase in the $|V^{AB}_X|$ at HF level. The small number of 1,4 cases that exhibited greater stability at HF were solely due to a decreased sum, q(A) +q(B), at this level. Typically, these cases were limited to N–H, N–N and H–H.

We should add, in passing, that obtaining intermolecular exchange contributions from KS orbitals is not new, because it is actually a standard procedure in the SAPT-DFT method originally proposed by Williams and Chabalowski,[54] and later improved by Hesselmann and Jansen.[55] This method has become an accurate, yet cheaper alternative to the very expensive symmetry-adapted perturbation-theory calculations.

Overall, this analysis validates the current expansion of IQA to B3LYP level of theory without reason for concern, in spite of the absence of an explicit second order reduced density matrix.

CONCLUSION

To the best of our knowledge, this study is the first to calculate and study interatomic exchange energies V^{AB}_X by using an explicit density functional, in this case that of B3LYP. As such, the IQA energy decomposition has been satisfactorily extended to DFT. Indeed, we recover the original (unpartitioned) energy of the system from the IQA energy contributions. The correct version (14.04.17 and higher), of the computer program AIMAll contains a pseudo-DFT IQA algorithm that makes this recovery possible within ~0.7 kJ mol^{-1} of the original ab initioB3LYP energy.

We studied the influence of B3LYP on V^{AB}_X, compared with the HF values, for three systems of biochemical interest:

N-methylacetamide (NMA), doubly capped tripeptide GlyGlyGly (TriGly) and an alloxan dimer. Over the three systems, a grand total of 870 V^{AB}_{X} energies for interactions ranging up to 1,15 interactions were carefully analyzed. The 1,6 and higher interactions, with V^{AB}_{X} energies corresponding to ~0.01 kJ mol^{-1}, were disregarded because they are negligible. The introduction of B3LYP invariably stabilized 1,4 and higher interactions, showing a consistent energy shift towards increased exchange energy, in terms of absolute value. Such a shift in stability is almost always due to an decrease in the sum of net atomic charges (q(A) + q(B)) caused by the use of Kohn–Sham orbitals.

However, for 1,2 and 1,3 bonded interactions, the V^{AB}_{X} energies were not always shifted in the same direction, i.e. sometimes the HF energies were more stable than the B3LYP energies. The exchange stability for each 1,2 interactions can be understood by considering a combination of the (de)localization of the KS or HF orbitals, the position of the interatomic surface and the sum of net atomic charges (inferring polarity) of the involved atoms. Through the orbital description, DFT methods have some correlation incorporated, resulting in more localized orbitals than HF. Only C–C and C–H interactions are the 1,2 interactions with greater V^{AB}_{X} energies using HF orbitals (over KS orbitals). Here, HF over-delocalized their atomic orbitals leading to greater exchange in the interaction. Contrastingly, the increased heteropolarity, experienced by some of the 1,2 interactions (e.g. N–H, N–C, C–O) through a HF orbital description, counteracts the lower sum of net atomic charges, leading to less interatomic exchange at HF but more so at B3LYP. All remaining 1,2 interactions had a reduced sum of net charges at B3LYP, resulting in greater interatomic exchange stability. The 1,3 interactions can be explained similarly to those of the 1,2 interactions.

Plots of V^{AB}_{X} versus internuclear distance shows that the vast majority of interactions emerge in well-defined clusters, each associated with a bond-connectivity 1, n. Similar A···B pairs yield very similar V^{AB}_{X} values, a hallmark of high transferability. Some anomalously strong interactions occur, shedding new light on patterns of exchange stability beyond that of the traditional covalent bonds.

In keeping with previous views and work on delocalization indices, HF-like B3LYP interatomic exchange energies (V^{AB}_{X}) also provide chemically relevant values, but now energies. This realization occurs despite only first-order (and not electron-pair) electron density correlation being accounted for. Overall, the extension of IQA to B3LYP can thus be justified.

SUPPLEMENTARY INFORMATION

Table S1: Integration errors (L(Ω)) for the alloxan dimer system calculated using the HF wave function. The 7 carbons (out of a maximum of 8) that have large integration errors are marked in red

ATOM	L (Ω) / kJmol^{-1}
o20	0.156
c15	0.278
c13	1.852
c1	-2.125
h23	0.084
c14	-3.095
o12	0.353
h11	0.083
h16	0.093
n17	-0.487
n21	0.321
o24	0.125
o19	0.141
n5	0.115
c22	-1.305
o18	0.140
c3	-1.658
h4	0.062
c10	-1.342
n9	-0.178
o6	0.090
c2	-3.685
o7	0.135
o8	0.114

Table S2: The exchange energies calculated at HF and B3LYP level for all 66 interatomic interactions in NMA. Shading highlights the interactions for which $|V_{X,HF}{}^{AB}| > |V_{X,B3LYP}{}^{AB}|$. Entries marked in red are used in the main text to discuss the example of calculating a percentage (utmost right column, |%|). The energy in green refers to the N...O interaction associated with the peptide bonds

Interaction	1,*n*	$V_{X,HF}$ / kJ/mol	$V_{X,B3LYP}$ / kJ/mol	ΔV_X (B3LYP-HF)	|%|
C4_H6	2	-741.176	-730.439	10.737	1.4
C4_H7	2	-743.571	-732.969	10.602	1.4
C4_H5	2	-751.265	-740.814	10.451	1.4
C9_H11	2	-733.722	-724.806	8.916	1.2
C9_H12	2	-736.765	-727.880	8.885	1.2
C9_H10	2	-740.246	-734.030	6.216	0.8
C2_C4	2	-759.252	-756.563	2.689	0.4
N1_H8	2	-675.847	-687.391	-11.545	1.7
N1_C9	2	-695.964	-761.652	-65.689	9.4
N1_C2	2	-733.292	-845.108	-111.816	15.2
C2_O3	2	-918.069	-1043.280	-125.212	12.0
N1_O3	3	-120.468	-111.981	8.487	7.0
N1_H10	3	-30.907	-26.442	4.465	14.4
N1_H12	3	-28.805	-24.743	4.063	14.1
N1_H11	3	-26.935	-23.241	3.694	13.7
H10_H12	3	-14.415	-13.638	0.777	5.4
H10_H11	3	-13.330	-12.662	0.668	5.0
H5_H6	3	-14.610	-13.981	0.629	4.3
H5_H7	3	-13.990	-13.474	0.516	3.7
H11_H12	3	-13.381	-12.880	0.501	3.7
H6_H7	3	-14.100	-13.663	0.437	3.1
C2_H5	3	-12.984	-12.730	0.253	1.9
C2_H6	3	-17.221	-17.185	0.036	0.2

N1_C4	3	-38.578	-38.676	-0.098	0.3
C2_H7	3	-15.004	-15.113	-0.110	0.7
C2_H8	3	-4.711	-6.109	-1.399	29.7
H8_C9	3	-5.123	-6.589	-1.465	28.6
O3_C4	3	-47.910	-50.335	-2.425	5.1
C2_C9	3	-4.877	-7.548	-2.671	54.8
N1_H5	4	-7.624	-7.411	0.213	2.8
H8_H12	4	-0.252	-0.256	-0.004	1.7
H8_H10	4	-0.913	-1.102	-0.189	20.7
H8_H11	4	-1.083	-1.279	-0.196	18.1
C2_H10	4	-1.044	-1.460	-0.416	39.9
N1_H7	4	-2.105	-2.524	-0.419	19.9
N1_H6	4	-2.140	-2.637	-0.497	23.2
C2_H11	4	-0.879	-1.381	-0.502	57.2
O3_H7	4	-6.316	-6.891	-0.575	9.1
O3_H6	4	-4.400	-4.993	-0.592	13.5
C4_H8	4	-2.666	-3.321	-0.655	24.6
C4_C9	4	-1.422	-2.127	-0.704	49.5
C2_H12	4	-1.131	-1.838	-0.706	62.4
O3_H5	4	-3.097	-3.963	-0.865	27.9
O3_H8	4	-2.611	-3.698	-1.086	41.6
O3_C9	4	-13.047	-16.163	-3.116	23.9

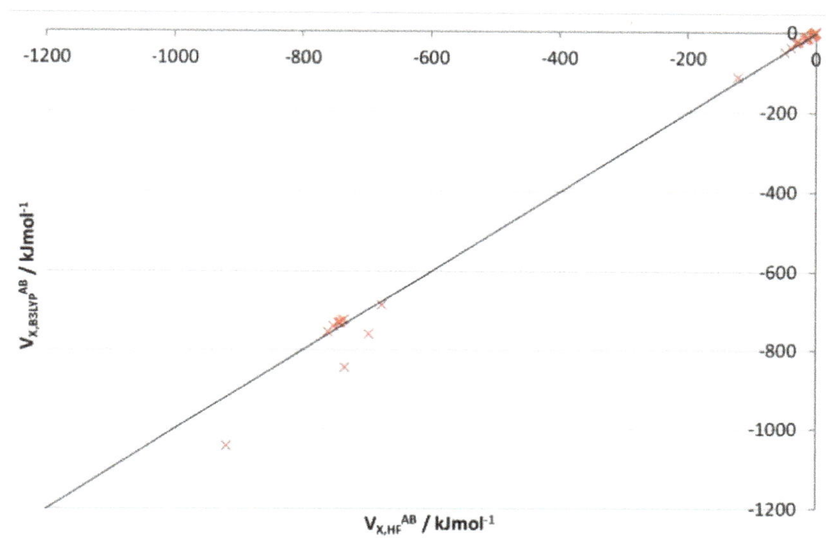

Figure S1: V_X^{AB} energies for all 66 interactions in NMA, from HF and B3LYP wavefunctions.

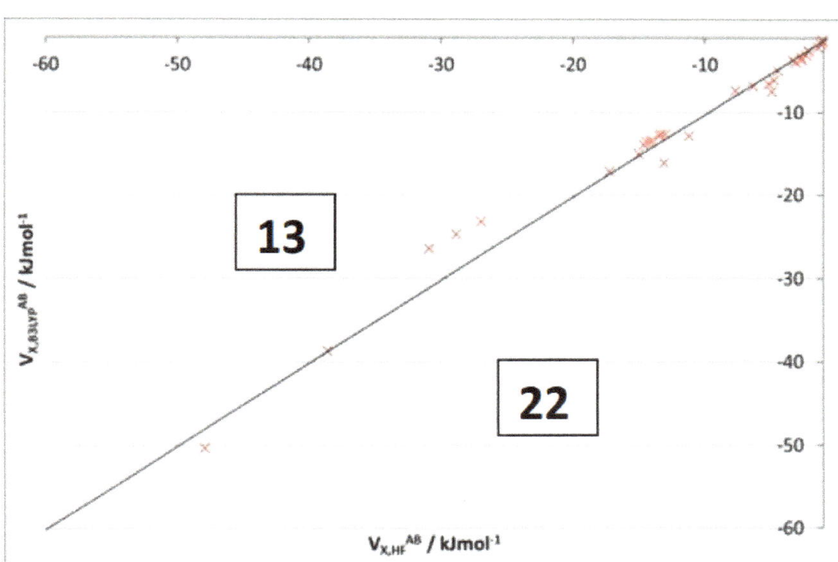

Figure S2: V_X^{AB} energies for medium-range interactions (-1 to -60 kJmol^{-1}) in NMA, from HF and B3LYP wavefunctions.

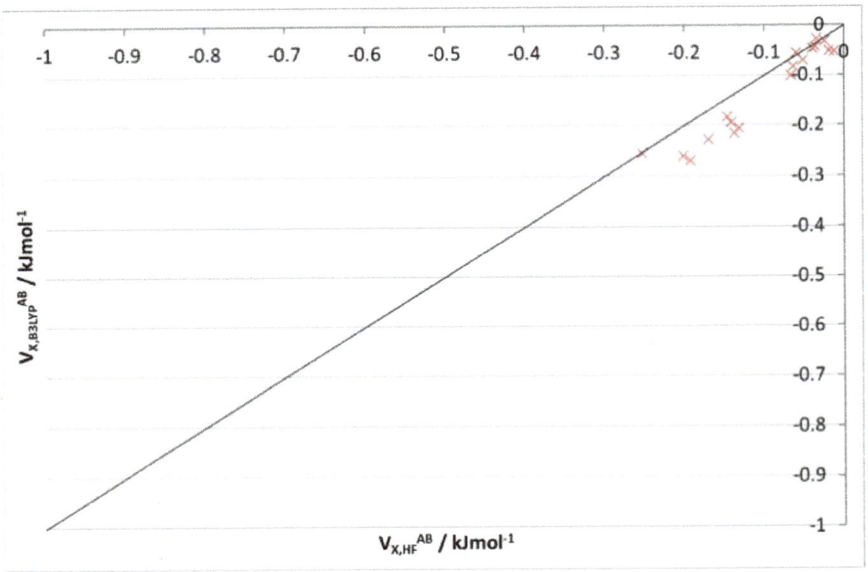

Figure S3: V_X^{AB} energies for long-range interactions (0 to -1 kJmol^{-1}) in NMA, from HF and B3LYP wavefunctions.

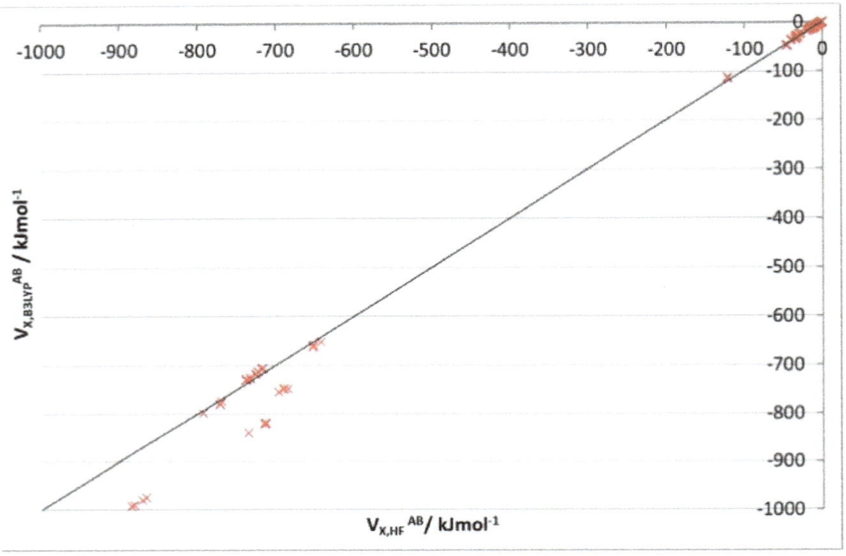

Figure S4: V_X^{AB} energies for all 528 interactions in TriGly, from both HF and B3LYP wavefunctions.

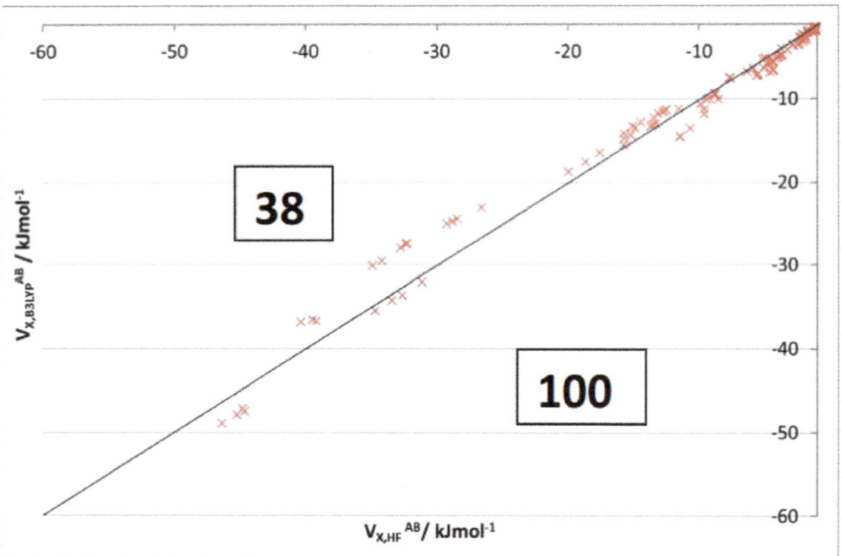

Figure S5: V_X^{AB} energies for medium-range interactions (-1 to -60 kJmol-1) in TriGly, for HF and B3LYP wavefunctions.

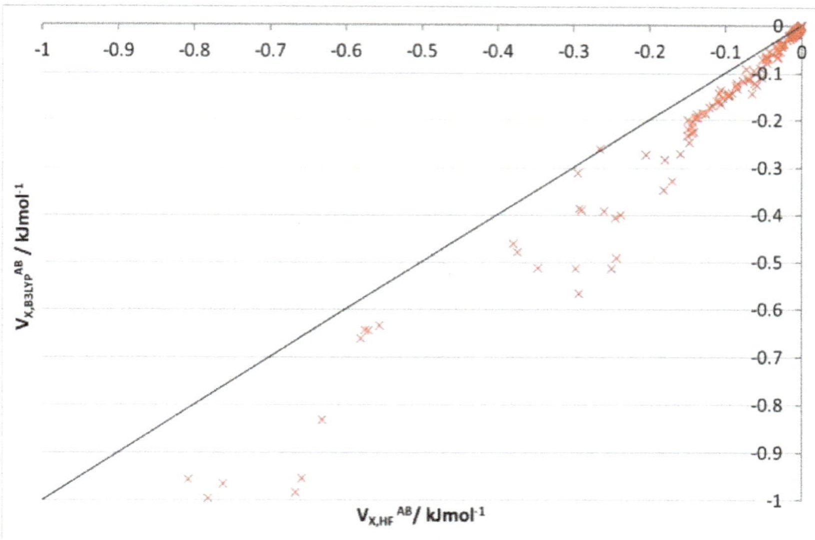

Figure S6: V_X^{AB} energies for long-range interactions (0 to -1 kJmol-1) in TriGly, for HF and B3LYP wavefunctions.

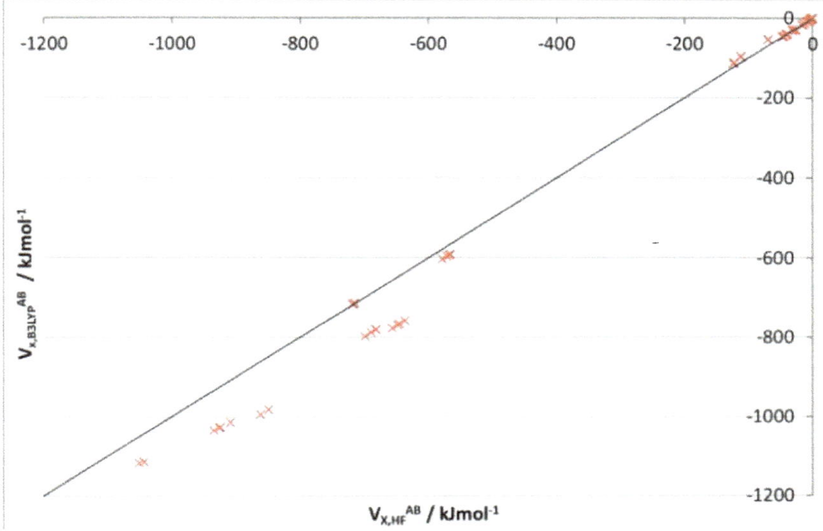

Figure S7: V_X^{AB} energies for all 276 interactions in the alloxan dimer, from HF and B3LYP wavefunctions.

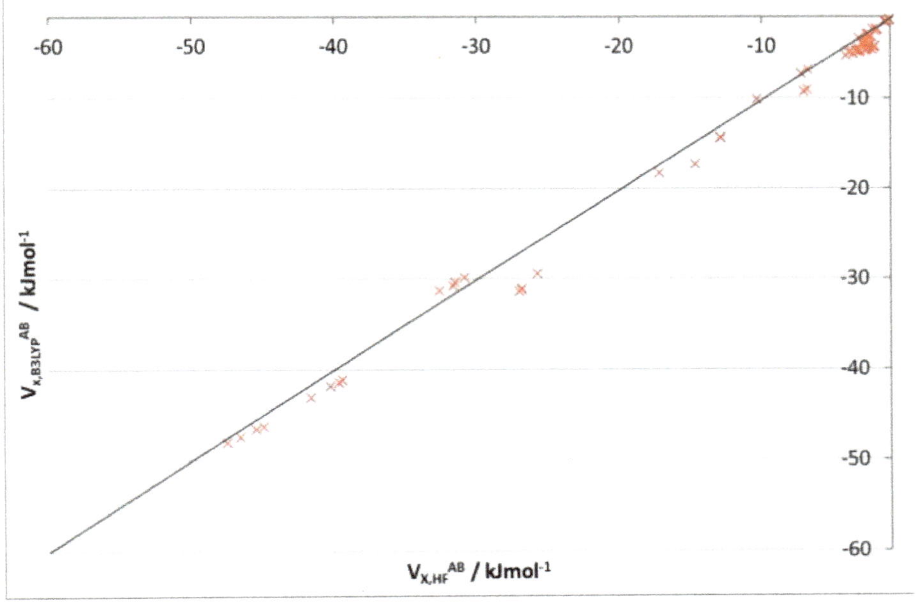

Figure S8: V_X^{AB} energies for medium-range interactions (-1 to -60 kJmol^{-1}) in the alloxan dimer, from HF and B3LYP wavefunctions.

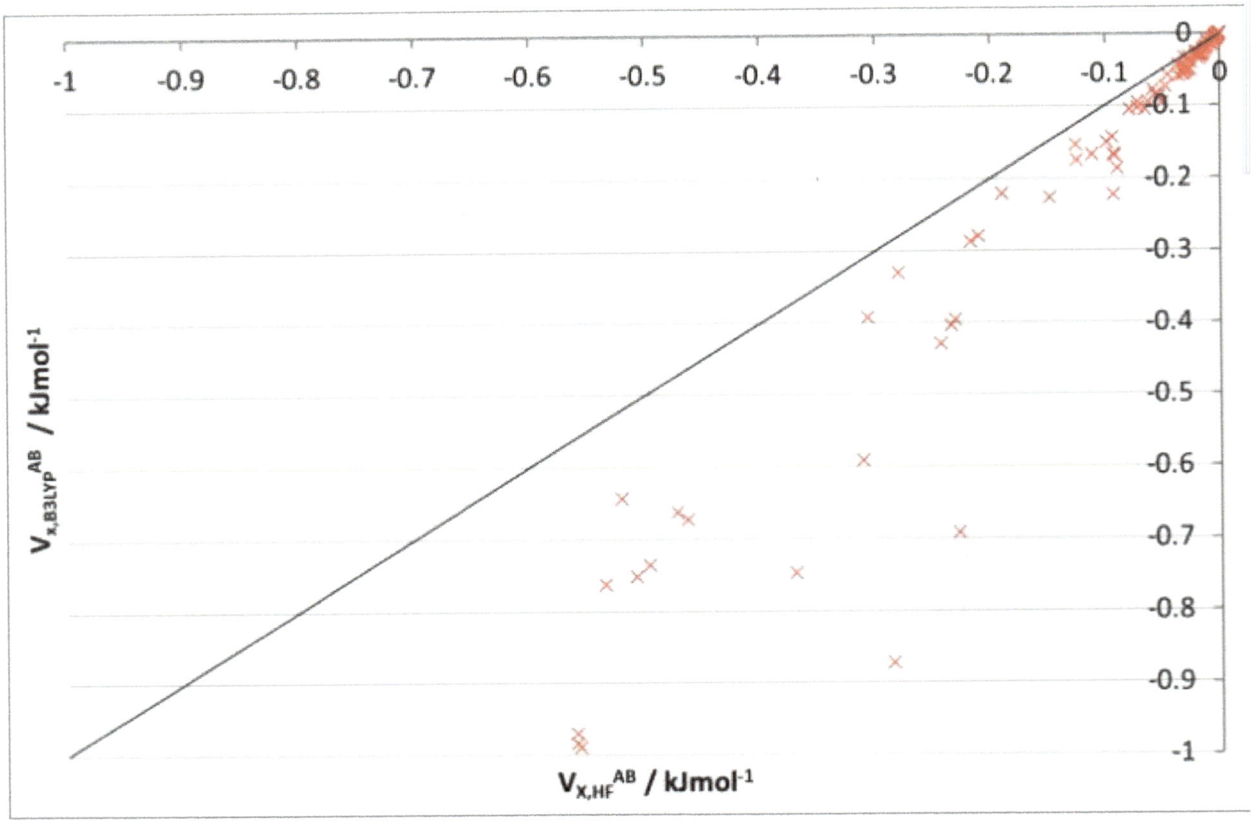

Figure S9: V_X^{AB} energies for long-range interactions (0 to -1 kJmol-1) in the alloxan dimer, from HF and B3LYP wavefunctions.

Table S3: Integration errors L(Ω) and corresponding V_X^{AB} energies calculated from five "outer angular" quadrature settings for two long-range interatomic interactions in TriGly: C22…H27 (1,14) and H23…H28 (1,15). All energies and integrations errors are in kJmol^{-1}

(A,B)	Boaq grid	VX(A,B) / kJmol^{-1}	L(Ω) A / kJmol^{-1}	L(Ω) B / kJmol^{-1}
C22_H27	Skyhigh	-6.93E-06	-0.2758	0.1675
	veryhigh	-6.91E-06	-0.1906	0.1557
	medium	-6.89E-06	-0.1875	0.1488
	low	-7.03E-06	-0.7195	0.1741
	auto	-6.91E-06	-0.1906	0.1557
H24_H28	Skyhigh	-4.97E-05	0.0994	0.1016
	veryhigh	-4.98E-05	0.1028	0.1030
	medium	-4.97E-05	0.1052	0.1014
	low	-4.97E-05	0.0981	0.1075
	auto	-4.98E-05	0.1028	0.1030

Table S4: Exchange energies, calculated at HF and B3LYP level, for all 1,2; 1,3 and 1,4 interatomic bonded interactions within the TriGly molecule. Shading highlights the interactions for which $|V_{X,HF}^{AB}| > |V_{X,B3LYP}^{AB}|$. The absolute percentage difference in the right column is defined as $100x|V_{X,B3LYP}^{AB} - V_{X,HF}^{AB}|/ V_{X,HF}^{AB}$. The energies in green refer to the N...O interactions associated with the peptide bonds

Interaction	1,*n*	$V_{X,HF}$ / kJ/mol	$V_{X,B3LYP}$ / kJ/mol	ΔV_X (B3LYP-HF)	|%|
H11_C15	2	-718.111	-707.345	10.766	1.5
C7_H16	2	-717.106	-706.835	10.271	1.4
C1_H5	2	-716.059	-706.319	9.740	1.4
H12_C15	2	-722.949	-713.466	9.483	1.3
C22_H32	2	-739.023	-729.908	9.115	1.2
C22_H25	2	-737.576	-728.595	8.981	1.2
C22_H24	2	-738.545	-729.711	8.834	1.2
C7_H17	2	-722.452	-713.669	8.782	1.2
C1_H4	2	-725.254	-717.753	7.501	1.0
C21_H27	2	-726.255	-718.901	7.354	1.0
C21_H33	2	-732.965	-725.863	7.102	1.0
C21_H28	2	-732.187	-727.780	4.407	0.6
C1_C2	2	-769.455	-774.591	-5.136	0.7
C22_C29	2	-792.726	-799.220	-6.494	0.8
N13_H19	2	-651.966	-659.569	-7.603	1.2
N20_H26	2	-651.848	-659.626	-7.778	1.2
C15_C18	2	-771.750	-779.557	-7.807	1.0
C6_C7	2	-770.986	-779.411	-8.425	1.1
N3_H10	2	-652.333	-662.159	-9.826	1.5
N9_H23	2	-642.046	-652.180	-10.134	1.6
N13_C15	2	-689.410	-747.143	-57.733	8.4
C7_N9	2	-696.471	-755.032	-58.561	8.4
C1_N3	2	-690.396	-749.521	-59.126	8.6
N20_C21	2	-684.412	-748.199	-63.787	9.3
C18_N20	2	-735.265	-840.287	-105.022	14.3
N3_C6	2	-713.698	-819.671	-105.973	14.8
C6_O8	2	-884.345	-992.693	-108.348	12.3
C2_N13	2	-713.683	-822.115	-108.432	15.2
N9_C29	2	-711.604	-820.053	-108.448	15.2
C2_O14	2	-880.945	-989.995	-109.049	12.4
C18_O31	2	-870.645	-980.148	-109.503	12.6
C29_O30	2	-865.326	-975.202	-109.876	12.7
N9_O30	**3**	**-120.818**	**-112.938**	**7.880**	**6.5**
N13_O14	**3**	**-120.830**	**-113.704**	**7.126**	**5.9**
N3_O8	**3**	**-120.862**	**-113.743**	**7.119**	**5.9**
N20_O31	**3**	**-120.964**	**-114.790**	**6.174**	**5.1**
N9_H17	3	-32.377	-27.446	4.931	15.2
N20_H33	3	-32.762	-27.928	4.833	14.8
H12_N13	3	-32.242	-27.411	4.832	15.0
N20_H28	3	-34.903	-30.111	4.791	13.7
N3_H4	3	-34.221	-29.555	4.665	13.6
N20_H27	3	-29.227	-25.030	4.197	14.4
H11_N13	3	-28.821	-24.733	4.087	14.2
N9_H16	3	-28.454	-24.403	4.051	14.2
N3_H5	3	-26.607	-23.082	3.526	13.3
C2_N3	3	-40.335	-36.869	3.466	8.6

C6_N9	3	-39.461	-36.588	2.873	7.3
N13_C18	3	-39.175	-36.673	2.501	6.4
H28_H33	3	-15.056	-13.286	1.771	11.8
H24_H32	3	-15.752	-14.153	1.599	10.1
H27_H28	3	-14.425	-12.838	1.587	11.0
H27_H33	3	-13.159	-11.805	1.354	10.3
H24_H25	3	-14.881	-13.537	1.344	9.0
H11_H12	3	-12.500	-11.294	1.206	9.6
H16_H17	3	-12.630	-11.445	1.185	9.4
H4_H5	3	-12.774	-11.592	1.183	9.3
H24_C29	3	-19.939	-18.790	1.149	5.8
H25_H32	3	-13.462	-12.321	1.142	8.5
C29_H32	3	-18.677	-17.589	1.089	5.8
C2_H4	3	-17.584	-16.526	1.058	6.0
C6_H17	3	-15.760	-14.775	0.985	6.3
H12_C18	3	-15.208	-14.356	0.852	5.6
H25_C29	3	-13.556	-13.099	0.457	3.4
C6_H16	3	-13.674	-13.285	0.389	2.8
H11_C18	3	-13.297	-12.980	0.317	2.4
C2_H5	3	-11.532	-11.279	0.253	2.2
C2_H11	3	-0.995	-1.623	-0.627	63.0
N9_C22	3	-34.695	-35.486	-0.790	2.3
C15_N20	3	-33.392	-34.245	-0.853	2.6
C1_N13	3	-31.098	-32.036	-0.938	3.0
C18_H26	3	-4.122	-5.125	-1.003	24.3
C2_H19	3	-3.942	-4.945	-1.003	25.4
C15_H19	3	-4.414	-5.435	-1.022	23.1
C6_H10	3	-3.952	-4.975	-1.023	25.9
N3_C7	3	-32.603	-33.629	-1.026	3.1
C7_H23	3	-4.390	-5.416	-1.027	23.4
C21_H26	3	-4.455	-5.534	-1.079	24.2
C1_H10	3	-4.619	-5.733	-1.114	24.1
C1_C6	3	-4.353	-6.455	-2.102	48.3
C7_C29	3	-4.361	-6.521	-2.160	49.5
C2_C15	3	-4.393	-6.617	-2.224	50.6
C18_C21	3	-4.573	-6.813	-2.240	49.0
C1_O14	3	-44.797	-47.192	-2.395	5.3
C22_O30	3	-46.347	-48.942	-2.595	5.6
C15_O31	3	-45.246	-47.945	-2.699	6.0
C7_O8	3	-44.650	-47.524	-2.874	6.4
H12_N20	4	-7.675	-7.456	0.219	2.9
N3_N13	4	-15.723	-15.541	0.182	1.2
N3_H17	4	-7.584	-7.449	0.135	1.8
H26_H33	4	-0.265	-0.260	0.004	1.7
C2_H10	4	-0.294	-0.311	-0.017	5.7
H4_N13	4	-5.175	-5.212	-0.036	0.7
N9_H24	4	-4.984	-5.049	-0.065	1.3
H12_H19	4	-0.574	-0.643	-0.069	12.1

H17_H23	4	-0.569	-0.644	-0.075	13.1
H16_H23	4	-0.556	-0.635	-0.079	14.2
H11_H19	4	-0.580	-0.662	-0.082	14.1
H5_N13	4	-1.695	-1.814	-0.118	7.0
H26_H27	4	-0.960	-1.119	-0.160	16.6
H5_H10	4	-0.885	-1.052	-0.167	18.9
N9_H25	4	-2.102	-2.285	-0.183	8.7
H26_H28	4	-1.109	-1.318	-0.209	18.9
N3_H16	4	-1.702	-1.912	-0.210	12.4
H11_N20	4	-1.755	-1.979	-0.224	12.8
H4_H10	4	-1.180	-1.411	-0.231	19.6
N9_H32	4	-3.691	-3.930	-0.239	6.5
C6_C29	4	-0.892	-1.217	-0.326	36.5
C2_C18	4	-0.890	-1.221	-0.330	37.1
N13_N20	4	-6.339	-6.731	-0.392	6.2
N3_N9	4	-5.880	-6.285	-0.405	6.9
H4_C6	4	-0.996	-1.407	-0.411	41.2
C18_H28	4	-1.098	-1.526	-0.428	39.0
C1_H19	4	-1.467	-1.896	-0.429	29.3
H5_C6	4	-0.862	-1.339	-0.477	55.3
C15_H26	4	-1.919	-2.419	-0.500	26.0
C18_H27	4	-0.985	-1.508	-0.522	53.0
C7_H10	4	-1.904	-2.438	-0.534	28.1
C1_C15	4	-1.109	-1.649	-0.541	48.8
C22_H23	4	-1.912	-2.464	-0.552	28.9
C1_C7	4	-1.121	-1.683	-0.562	50.1
C7_C22	4	-1.168	-1.732	-0.564	48.3
H11_O31	4	-8.800	-9.370	-0.570	6.5
C15_C21	4	-1.119	-1.689	-0.570	51.0
H16_C29	4	-0.975	-1.582	-0.606	62.2
O8_H16	4	-8.747	-9.362	-0.615	7.0
H17_C29	4	-1.067	-1.701	-0.635	59.5
H12_O31	4	-2.832	-3.485	-0.653	23.0
O8_H17	4	-2.883	-3.540	-0.657	22.8
C2_H12	4	-1.075	-1.741	-0.666	61.9
N13_O31	4	-9.218	-9.904	-0.686	7.4
O8_N9	4	-9.397	-10.103	-0.707	7.5
H5_O14	4	-8.778	-9.489	-0.711	8.1
N3_O14	4	-4.725	-5.443	-0.718	15.2
H25_O30	4	-9.880	-10.648	-0.768	7.8
C18_H33	4	-1.247	-2.017	-0.770	61.8
C2_C6	4	-1.077	-1.862	-0.785	72.9
H4_O14	4	-3.271	-4.088	-0.816	25.0
C6_H23	4	-1.995	-2.902	-0.907	45.5
H23_O30	4	-2.346	-3.261	-0.915	39.0
H26_O31	4	-2.454	-3.410	-0.956	39.0
C18_H19	4	-2.076	-3.049	-0.972	46.8
O14_H19	4	-2.439	-3.433	-0.994	40.7
O30_H32	4	-3.659	-4.656	-0.997	27.3
H24_O30	4	-3.728	-4.735	-1.007	27.0
O8_H10	4	-2.458	-3.487	-1.029	41.9
C1_O8	4	-9.589	-11.859	-2.270	23.7
C21_O31	4	-10.704	-13.563	-2.859	26.7
O14_C15	4	-11.420	-14.488	-3.068	26.9
C7_O30	4	-11.438	-14.580	-3.142	27.5

Table S5: Exchange energies, calculated at HF and B3LYP level, for all 1,2; 1,3 and 1,4 interatomic bonded interactions within the alloxan dimer. Shading highlights the interactions for which $|V_{X,HF}{}^{AB}| > |V_{X,B3LYP}{}^{AB}|$. The absolute percentage difference in the right column is defined as $100x|V_{X,B3LYP}{}^{AB} - V_{X,HF}{}^{AB}|/ V_{X,HF}{}^{AB}$. The energy in green refers to the intramolecular O…O interactions in the alloxan monomers

Interaction	1,*n*	$V_{X,HF}$ / kJ/mol	$V_{X,B3LYP}$ / kJ/mol	ΔV_X (B3LYP-HF)	|%|
C14_C22	2	-714.979	-712.994	1.985	0.3
C14_C15	2	-715.573	-713.815	1.758	0.2
C2_C3	2	-717.598	-716.504	1.094	0.2
C2_C10	2	-717.301	-716.286	1.015	0.1
N21_H23	2	-578.468	-602.430	-23.962	4.1
H16_N17	2	-570.838	-595.129	-24.291	4.3
H4_N5	2	-567.464	-593.320	-25.856	4.6
N9_H11	2	-565.124	-591.553	-26.428	4.7
C2_O7	2	-1051.237	-1118.391	-67.153	6.4
C14_O19	2	-1044.126	-1115.165	-71.039	6.8
C15_N17	2	-681.594	-780.008	-98.414	14.4
C3_N5	2	-689.331	-787.782	-98.451	14.3
N21_C22	2	-681.729	-781.409	-99.680	14.6
N9_C10	2	-698.908	-798.786	-99.878	14.3
C3_O8	2	-934.665	-1035.430	-100.764	10.8
C15_O20	2	-926.025	-1029.098	-103.073	11.1
C22_O24	2	-924.183	-1027.307	-103.124	11.2
C10_O12	2	-908.238	-1014.537	-106.298	11.7
C13_N21	2	-645.059	-765.743	-120.684	18.7
C13_N17	2	-656.126	-777.349	-121.223	18.5
C1_N5	2	-648.080	-769.414	-121.334	18.7
C1_N9	2	-636.802	-758.355	-121.552	19.1
C1_O6	2	-861.856	-994.267	-132.412	15.4
C13_O18	2	-848.753	-983.124	-134.371	15.8
N5_N9	3	-69.981	-55.151	14.830	21.2
N17_N21	3	-69.090	-54.455	14.635	21.2
O6_N9	3	-111.647	-97.643	14.004	12.5
N5_O6	3	-111.538	-97.908	13.630	12.2
O18_N21	3	-111.896	-98.286	13.610	12.2
N17_O18	3	-110.779	-97.608	13.170	11.9
N17_O20	3	-121.832	-112.892	8.940	7.3
N21_O24	3	-123.099	-114.488	8.611	7.0
N5_O8	3	-123.340	-114.894	8.446	6.8
N9_O12	3	-122.139	-113.893	8.246	6.8
C2_N5	3	-32.591	-31.369	1.222	3.7
C14_N17	3	-31.516	-30.393	1.123	3.6
C14_N21	3	-30.849	-29.883	0.967	3.1
C2_N9	3	-31.651	-30.845	0.806	2.5
O7_C10	3	-47.464	-48.026	-0.562	1.2
O19_C22	3	-46.553	-47.454	-0.901	1.9
C15_H16	3	-2.874	-3.845	-0.971	33.8
C3_H4	3	-2.887	-3.880	-0.993	34.4
C10_H11	3	-2.933	-3.955	-1.022	34.8
C22_H23	3	-2.979	-4.004	-1.026	34.4
C15_O19	3	-45.475	-46.550	-1.075	2.4
C1_H11	3	-2.612	-3.801	-1.189	45.5
C1_H4	3	-2.694	-3.897	-1.203	44.6

13_H16	3	-2.787	-4.014	-1.226	44.(
3_O7	3	-44.946	-46.299	-1.353	3.0
15_C22	3	-12.881	-14.322	-1.441	11.:
2_O8	3	-41.630	-43.113	-1.482	3.6
3_C10	3	-12.833	-14.392	-1.559	12.:
14_O20	3	-40.270	-41.849	-1.579	3.9
13_C22	3	-3.082	-4.787	-1.706	55.:
13_C15	3	-3.161	-4.909	-1.748	55.:
1_C3	3	-3.132	-4.883	-1.751	55.!
1_C10	3	-3.129	-4.881	-1.752	56.(
14_O24	3	-39.413	-41.183	-1.771	4.5
2_O12	3	-39.660	-41.447	-1.787	4.5
16_N21	4	-1.305	-1.558	-0.253	19.4
4_N9	4	-1.292	-1.553	-0.261	20.:
5_H11	4	-1.283	-1.547	-0.264	20.(
17_H23	4	-1.338	-1.612	-0.274	20.!
2_H11	4	-1.014	-1.333	-0.320	31.(
14_H16	4	-1.042	-1.365	-0.322	30.!
2_H4	4	-1.047	-1.371	-0.324	30.!
14_H23	4	-1.061	-1.399	-0.338	31.!
17_C22	4	-2.567	-2.985	-0.418	16.:
C15_N21	4	-2.485	-2.928	-0.443	17.8
C3_N9	4	-2.572	-3.024	-0.452	17.6
C3_O12	4	-1.890	-2.383	-0.493	26.1
C13_C14	4	-0.770	-1.267	-0.498	64.7
C15_O24	4	-1.906	-2.408	-0.501	26.3
O20_C22	4	-1.927	-2.442	-0.515	26.7
C1_C2	4	-0.773	-1.291	-0.517	66.9
O8_C10	4	-1.954	-2.486	-0.532	27.2
H16_O18	4	-2.372	-3.253	-0.881	37.2
H4_O6	4	-2.330	-3.213	-0.883	37.9
O6_H11	4	-2.318	-3.207	-0.889	38.3
O18_H23	4	-2.484	-3.432	-0.948	38.2
N17_O19	4	-3.743	-4.806	-1.063	28.4
O19_N21	4	-3.750	-4.839	-1.088	29.0
N5_O7	4	-3.886	-5.012	-1.126	29.0
O7_N9	4	-4.093	-5.331	-1.238	30.3
H11_O12	4	-3.196	-4.576	-1.380	43.2
H16_O20	4	-3.394	-4.828	-1.434	42.2
H4_O8	4	-3.341	-4.811	-1.471	44.0
H23_O24	4	-3.578	-5.139	-1.561	43.6
C15_O18	4	-2.423	-4.141	-1.718	70.9
C1_O12	4	-2.371	-4.115	-1.744	73.6
O18_C22	4	-2.441	-4.207	-1.766	72.3
C3_O6	4	-2.543	-4.387	-1.844	72.5
C13_O24	4	-2.538	-4.438	-1.900	74.8
C1_O8	4	-2.591	-4.540	-1.949	75.2
O6_C10	4	-2.723	-4.728	-2.005	73.6
C13_O20	4	-2.642	-4.658	-2.017	76.3
O7_O12	**4**	**-25.716**	**-29.513**	**-3.797**	**14.8**
O19_O24	**4**	**-26.805**	**-31.146**	**-4.340**	**16.2**
O19_O20	**4**	**-26.866**	**-31.221**	**-4.356**	**16.2**
O7_O8	**4**	**-26.997**	**-31.427**	**-4.431**	**16.4**

Table S6: Average internuclear distances (in Å) for all type of interactions in the alloxan dimer

1,n	Average Internuclear Distance
1,2	1.28
1,3	2.30
1,4	3.13
1,5	4.20
1,6	5.11
Inter (all)	5.49
Inter $\lvert V_{X,HF}^{AB}\rvert > \lvert V_{X,B3LYP}^{AB}\rvert$	6.51
Inter $\lvert V_{X,B3LYP}^{AB}\rvert > \lvert V_{X,HF}^{AB}\rvert$	5.25

Table S7: Net atomic charges q (QTAIM or QCT) for all 12 atoms in NMA calculated at MP2, HF and B3LYP level

	q		
Atom	MP2	HF	B3LYP
c2	1.5427	1.7760	1.4678
c4	-0.0169	0.1147	0.0041
c9	0.4052	0.6050	0.3824
h10	0.0055	-0.0324	0.0030
h11	0.0468	0.0216	0.0462
h12	0.0084	-0.0266	0.0079
h5	-0.0135	-0.0471	-0.0167
h6	0.0260	-0.0002	0.0228
h7	0.0352	0.0111	0.0321
h8	0.3890	0.4020	0.3688
n1	-1.2362	-1.4658	-1.1376
o3	-1.1923	-1.3581	-1.1808

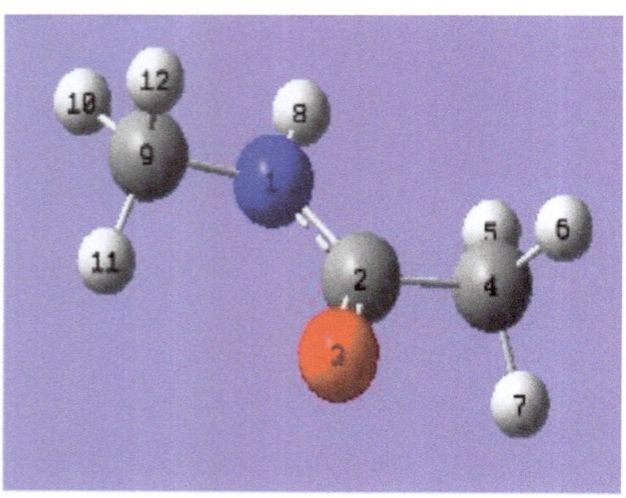

	q		
Atom	MP2	HF	B3LYP
c1	0.4926	0.6833	0.4659
c15	0.5008	0.6907	0.4726
c18	1.4854	1.7338	1.4435
c2	1.4912	1.7356	1.4519
c21	0.5060	0.6922	0.4652
c22	0.0739	0.2082	0.0643
c29	1.4946	1.7283	1.4528
c6	1.4826	1.7255	1.4427
c7	0.5027	0.7001	0.4741
h10	0.4323	0.4487	0.4193
h11	0.0580	0.0372	0.0635
h12	-0.0019	-0.0323	0.0039
h16	0.0582	0.0398	0.0637
h17	-0.0082	-0.0395	-0.0031
h19	0.4297	0.4448	0.4169
h23	0.4430	0.4601	0.4309
h24	-0.0091	-0.0381	-0.0033
h25	0.0387	0.0170	0.0427
h26	0.4362	0.4501	0.4246
h27	0.0413	0.0136	0.0474
h28	-0.0025	-0.0409	0.0021
h32	0.0053	-0.0211	0.0105
h33	0.0001	-0.0373	0.0062
h4	-0.0051	-0.0392	-0.0017
h5	0.0820	0.0679	0.0866
n13	-1.3462	-1.5652	-1.2709
n20	-1.3475	-1.5691	-1.2754
n3	-1.3473	-1.5700	-1.2740
n9	-1.3477	-1.5702	-1.2724
o14	-1.1450	-1.3262	-1.1488
o30	-1.1820	-1.3564	-1.1827
o31	-1.1696	-1.3485	-1.1719
o8	-1.1419	-1.3237	-1.1467

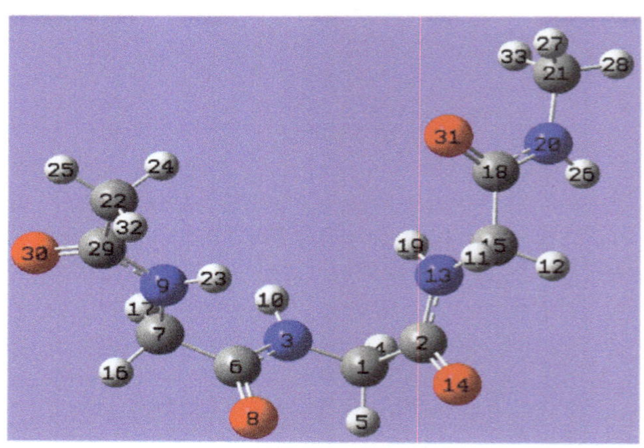

Table S8: Net atomic charges q (QTAIM or QCT) for all 33 atoms in TriGly calculated at MP2, HF and B3LYP level

Table S9: Net atomic charges q (QTAIM or QCT) for all 24 atoms in the alloxan dimer calculated at MP2, HF and B3LYP level

	q		
Atom	MP2	HF	B3LYP
c1	2.0009	2.4046	1.9147
c10	1.5421	1.8125	1.5051
c13	1.9939	2.4011	1.9057
c14	1.1283	1.2638	1.1152
c15	1.5589	1.8271	1.5247
c2	1.1092	1.2362	1.0943
c22	1.5496	1.8154	1.5157
c3	1.5487	1.8223	1.5145
h11	0.5122	0.5467	0.5029
h16	0.5063	0.5390	0.4967
h23	0.4979	0.5303	0.4873
h4	0.5097	0.5439	0.5003
n17	-1.3523	-1.6104	-1.2605
n21	-1.3709	-1.6245	-1.2786
n5	-1.3641	-1.6222	-1.2719
n9	-1.3656	-1.6246	-1.2728
o12	-1.1558	-1.3286	-1.1561
o18	-1.1857	-1.3559	-1.1794
o19	-1.0679	-1.1924	-1.0645
o20	-1.1246	-1.2927	-1.1275
o24	-1.1320	-1.2991	-1.1356
o6	-1.1575	-1.3315	-1.1526
o7	-1.0679	-1.1845	-1.0623
o8	-1.1124	-1.2790	-1.1150

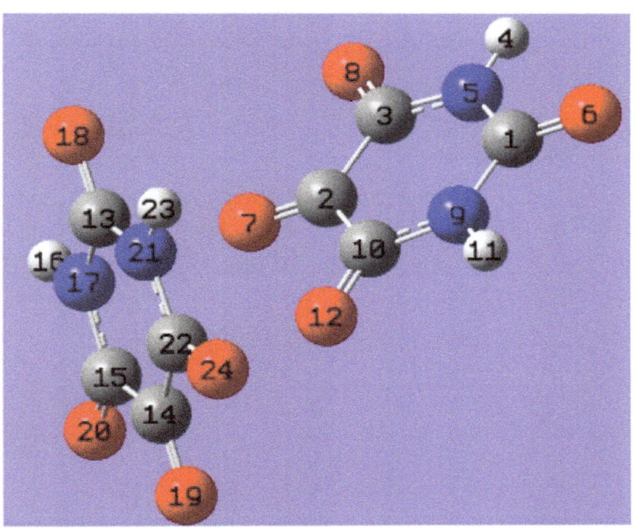

ACKNOWLEDGEMENTS

PM thanks the BBSRC for a studentship and PP the EPSRC for an Established Career Fellowship. Gratitude is due to Dr Todd Keith for closely working with him towards an AIMAll code that handles B3LYP functions for the first time, now some time ago. His private communications are much appreciated. It should be clear that Dr Keith has singled-handedly implemented the B3LYP functional in AIMAll, the source code of which he holds. AMP acknowledges funding from MINECO project CTQ2012-31174.

REFERENCES

1. R. F. W. Bader and D. Bayles, J. Phys. Chem. A, 2000, 104, 5579–5589 .
2. P. L. A. Popelier, M. Devereux and M. Rafat, Acta Crystallogr., Sect. A: Found. Crystallogr., 2004, A60, 427–433 .
3. A. Martín Pendás, E. Francisco and M. A. Blanco, Faraday Discuss., 2007, 135, 423–438.
4. K. Kitaura and K. Morokuma, Int. J. Quantum Chem., 1976, 10, 325–331 .
5. F. M. Bickelhaupt and E. J. Baerends, Rev. Comput. Chem., 2000, 15, 1.
6. E. F. Glendening and A. Streitwieser, J. Chem. Phys., 1994, 100, 2900 .
7. F. Weinhold and C. Landis, Valency and Bonding. A Natural Bond Orbital Donor-Acceptor Perspective, Cambridge University Press Cambridge, Great Britain, 2005.
8. P. L. A. Popelier, in Structure and Bonding. Intermolecular Forces and Clusters, ed. D. J. Wales, Springer, Heidelberg, Germany, 2005, vol. 115, pp. 1–56.
9. R. F. W. Bader, Atoms in Molecules. A Quantum Theory, Oxford Univ. Press, Oxford, Great Britain, 1990.
10. M. A. Blanco, A. Martín Pendás and E. Francisco, J. Chem. Theory Comput., 2005, 1, 1096–1109 .
11. T. A. Keith, AIMAll (Version 14.04.17), (http://aim.tkgristmill.com) and T. G. S. Todd A. Keith, Overland Park KS, USA, (aim.tkgristmill.com), 2014.
12. O. A. Syzgantseva, V. Tognetti and L. Joubert, J. Phys. Chem. A, 2013, 117, 8969–8980 .
13. A. Martín Pendás, M. A. Blanco and E. Francisco, J. Chem. Phys., 2006, 125, 184112 .
14. J. M. Guevara-Vela, R. Chavez-Calvillo, M. Garcia-Revilla, J. Hernandez-Trujillo, O. Christiansen, E. Francisco, A. Martín Pendás and T. Rocha-Rinza, Chem. – Eur. J., 2013, 19, 14304–14315 .
15. V. Tognetti and L. Joubert, J. Chem. Phys., 2013, 138, 024102 .
16. I. Cukrowski, J. H. de Lange and M. Mitoraj, J. Phys. Chem. A, 2014, 118, 623–637 .
17. T. I. Madzhidov, G. A. Chmutova and A. Martín Pendás, J. Phys. Chem. A, 2011, 115, 10069–10077 .
18. T. M. Klapoetke, B. Krumm, R. Moll, S. F. Rest, Y. V. Vishnevskiy, C. Reuter, H.-G. Stammler and N. W. Mitzel, Chem. – Eur. J., 2014, 20, 1–13 .
19. K. Eskandari and C. Van Alsenoy, J. Comput. Chem., 2014, 35, 1883–1889 .
20. A. Martín Pendás, M. A. Blanco and E. Francisco, J. Comput. Chem., 2009, 30, 98–109 .
21. E. Francisco, A. Martín Pendás and M. A. Blanco, J. Chem. Theory Comput., 2006, 2, 90–102 .
22. J. Dillen, Int. J. Quantum Chem., 2013, 113, 2143–2153 .
23. A. Martín Pendás, E. Francisco and M. A. Blanco, J. Phys. Chem. A, 2006, 110, 12864–12869 .

CHAPTER 5

FIRST-PRINCIPLES CALCULATIONS, EXPERIMENTAL STUDY, AND THERMODYNAMIC MODELING OF THE AL-CO-CR SYSTEM

Xuan L. Liu[1], Thomas Gheno[2], Bonnie B. Lindahl[3], Greta Lindwall[1], Brian Gleeson[2] and ZiKui Liu[1]

[1]Department of Materials Science and Engineering, The Pennsylvania State University, University Park, Pennsylvania, 16802, United States of America

[2]Department of Mechanical Engineering and Materials Science, University of Pittsburgh, Pittsburgh, Pennsylvania, 15261, United States of America

[3]Department of Materials Science and Engineering, KTH Royal Institute of Technology, SE-100 44, Stockholm, Sweden

ABSTRACT

The phase relations and thermodynamic properties of the condensed Al-Co-Cr ternary alloy system are investigated using first-principles calculations based on density functional theory (DFT) and phase-equilibria experiments that led to X-ray diffraction (XRD) and electron probe micro-analysis (EPMA) measurements. A thermodynamic description is developed by means of the calculations of phase diagrams (CALPHAD) method using experimental and computational data from the present work and the literature. Emphasis is placed on modeling the bcc-A2, B2, fcc-γ, and tetragonal-σ phases in the temperature range of 1173 to 1623 K. Liquid, bcc-A2 and fcc-γ phases are modeled using substitutional solution descriptions. First-principles special quasirandom structures (SQS) calculations predict a large bcc-A2 (disordered)/B2 (ordered) miscibility gap, in agreement with experiments. A partitioning model is then used for the A2/B2 phase to effectively describe the order-disorder transitions. The critically assessed thermodynamic description describes all phase equilibria data well. A2/B2 transitions are also shown to agree well with previous experimental findings.

INTRODUCTION

Nickel-base superalloys used in the hot sections of gas turbines require corrosion-resistant coatings to withstand the harsh thermo-chemical conditions prevailing in these combustion environments [1,2]. Overlay MCrAlY (M = Ni, Co or NiCo) coatings are widely used for this purpose, either solely or in conjunction with a thermally insulating ceramic topcoat, most commonly yttria-stabilized zirconia, to comprise a thermal barrier coating (TBC) system [3–6]. During service, microstructural evolution of the MCrAlY layer arises as a result of temperature variations, reactions with the environment such as selective oxidation removing Al from the subsurface, or interdiffusion with the substrate material, driven by differences in chemical potentials. A good knowledge of the alloy thermodynamics is critical to understanding and predicting these evolutionary processes, which in turn affect the corrosion resistance and mechanical properties of the system.

Computational design of effective MCrAl-based coating compositions necessitates a thermodynamic description methodology which can predict phase compositions and fractions at the temperatures of interest. One such method is CALPHAD, which has been used extensively and successfully in the thermodynamic modeling of many multi-component systems including Ni-base superalloys [7,8]. The CALPHAD approach parameterizes the Gibbs energy functions of all phases in unary, binary and ternary systems using computational and

Citation: Liu XL, Gheno T, Lindahl BB, Lindwall G, Gleeson B, Liu Z-K (2015) First-Principles Calculations, Experimental Study, and Thermodynamic Modeling of the Al-Co-Cr System. PLoS ONE 10(4): e0121386. doi:10.1371/journal.pone.0121386.

experimental data on phase equilibria and thermochemistry. The parameters are entered in a database and extrapolations to multi-component systems can be made. Limitations are known to exist in the current description of the quaternary Al-Co-Cr-Ni system for Co contents in excess of 20 wt. % [9,10] which are common in practice [2,11,12]. While assessment of the Al-Cr-Ni system has been well documented [7], the thermodynamic properties of the Al-Co-Cr system have not been adequately assessed.

Phases important to coating systems are B2 (β, $Pm\bar{3}m$, simple cubic type) and fcc-A1 (γ, $Fm\bar{3}m$, disordered f.c.c.). However, bcc-A2 (α, $Im\bar{3}m$, disordered b.c.c.), hcp-A3 (ε, $P6_3$/mmc, disordered h.c.p.) and sigma (σ, $P4_2$/mnm, Frank-Kasper) must also be modeled to allow for a complete description. One important feature of the Al-Co-Cr system is the A2/B2 second-order transformation at temperatures above 1373 K [13]. The present study aims to construct a thermodynamic model accounting for the A2/B2 ordering phenomenon, where atoms take on distinct lattice sites in B2. The two phases have vastly different compositions at low temperatures, but merge to similar compositions at higher temperatures which has to be accounted for in the thermodynamic model.

As phase equilibrium data are scarce, especially in the important B2+γ+σ region, experiments are conducted in this study at 1173, 1273, and 1373 K to aid the thermodynamic modeling. Additionally, first-principles calculations based on density functional (DFT) theory are incorporated to supplement the lack of experimentally measured thermochemical data in the ternary system [14] as well as to study the A2+B2 miscibility gap that is prevalent over the entire phase diagram. Ultimately, this system will be used in the construction of a multicomponent Al-Co-Cr-Ni-Y database. As a result, this database needs to be compatible with previously modeled Al-Co-Ni [15] and Al-Cr-Ni [7] systems.

The Al-Co thermodynamic model assessed by Dupin and Ansara [16] is adopted in the present work. This will allow compatibility with Dupin's [15] model of the Al-Co-Ni system. The Co-Cr system modeled by Oikawa et al. [17] is found to successfully capture experimental activity [18], enthalpy [18] and phase boundary data [17], which are crucial toward the extension to the Al-Co-Cr ternary, and is therefore adopted here. It also allows compatibility with the modeling of the Co-Cr-Ni system by Yang et al. [19], which would enable modeling of the quaternary Al-Co-Cr-Ni system. In Oikawa et al.'s [17] treatment of the Co-Cr system, no indication of B2 ordering was reported, hence the B2 phase will be modeled as a metastable phase with data from first-principles calculations. Saunders' [20] assessment of the Al-Cr system, used in COST507, is used here. Its description of the composition range of some intermetallics is not complete, but this will be shown to have no impact on the Al-Co-Cr system at the temperatures of interest in this study. Although XRD data from Helander and Tolochko [21] indicate a possible B2 ordering, no A2 to B2 second-order transition has been conclusively established in the Al-Cr system. Following Dupin et al. [7] who noted that a A2/B2 transition would modify the A2-$AlCr_2$ phase boundary and require additional experimental data for proper assessment, the B2-AlCr phase is not included here.

Ishikawa et al. [13] investigated the Al-Co-Cr system in the temperature range 1273–1623 K using multi-phase alloys and diffusion couples. Isotherms at 1573 and 1623 K showed complete dissolution of σ and a particularly large composition range for the B2 phase. The experimental work also provided compositions for the A2/B2 order-disorder transition at 1473 K and above, which are important to the present modeling. Previous models are combined with ternary data from the literature and this work to produce the overall thermodynamic description.

MATERIALS AND METHODS

Experimental Procedures

Ingots of nominal compositions (at. %) Co-20Cr-20Al-0.4Y (# A4), Co-26Cr-11Al-0.1Y (A5), Co-35Cr-6Al-0.1Y (A6) and Co-35Cr-11Al-0.1Y (A7) were prepared by arc-melting and drop-cast in a chamber evacuated and back-filled with 0.5 atm of argon. Specimens were cut to approximate dimensions of 10 x 10 x 1 mm, vacuum-encapsulated in quartz capsules, homogenized for 48 h at 1423 K in a tube furnace, and slowly brought to equilibration temperature. Equilibration treatments were conducted at 1173, 1273 and 1373 K for 500, 200 and 100 h, respectively, followed by water quenching to retain the equilibrated microstructures. Phase constitutions were studied by X-ray diffraction (XRD) with a PANalytical Empyrean instrument, using a Co radiation source ($K_{\alpha 1}$ = 1.789 Å).

Polished sections of the heat-treated alloys were prepared by standard metallographic procedures. Phase compositions were determined by electron probe micro-analysis (EPMA) using a JEOL JXA-8530F field emission gun instrument. For each element, measured intensities were converted to concentrations by interpolation via a calibration curve built using a series of standards of known compositions (chemical analysis by inductively coupled plasma mass spectrometry). The probe size used during measurements was about 1 µm, and the alloy microstructures were sufficiently coarse for each phase to be analyzed individually.

CALPHAD THERMODYNAMIC MODELS

Solution phases: bcc, fcc (γ), hcp (ε), and liquid.

In the present work, the Gibbs energies of γ, ε, and liquid solution phases are modeled using the following equation based on a single sublattice,

$$G_m^{\Phi} = \sum x_i {}^{o}G_i^{\Phi} + RT\sum x_i ln x_i + {}^{XS}G_m^{\Phi} + RT\,{}^{Magnetic}G_m^{\Phi} \quad (1)$$

where ${}^{o}G_i^{\Phi}$ represents the molar Gibbs energy of the pure elements, i = Al, Co, Cr, with structure Φ taken from the SGTE Unary PURE4 database [22], and the other symbols have their usual meanings. In Eq. 1, the first term represents the physical mixing of the elements and the second term is the contribution from the ideal entropy of mixing. The excess term, ${}^{XS}G_m^{\Phi}$, is modeled using a Redlich-Kister polynomial [23] to represent the non-ideal interactions between Al, Co and Cr, as shown below:

$$^{XS}G_m^{\Phi} = \sum_i \sum_{j>i} x_i x_j \sum_{v=0} {}^{v}L_{i,j}^{\Phi}(x_i - x_j)$$

$$+\sum_i \sum_{j>i} \sum_{k>j} x_i x_j x_k [{}^{1}L_{i,j,k}^{\Phi}(x_i + \delta_{i,j,k}) + {}^{2}L_{i,j,k}^{\Phi}(x_j + \delta_{i,j,k}) + {}^{3}L_{i,j,k}^{\Phi}(x_k + \delta_{i,j,k})] \quad (2)$$

Where $\delta_{i,j,k} = (1-x_i-x_j-x_k) / 3$. The binary and ternary interaction terms, ${}^{v}L_{i,j}^{\Phi}$ and ${}^{v}L_{i,j,k}^{\Phi}$, are expressed in the form A + B×T where A and B are the model parameters to be evaluated. Due to the ferromagnetic nature of Co and the antiferromagnetic nature of Cr, ${}^{Magnetic}G_m^{\Phi}$ is included to capture magnetic contributions to the Gibbs energy [24,25]. Hillert and Jarl [24] expressed the magnetic contribution as:

$$^{Magnetic}G_m^{\Phi} = RTln(\beta + 1)f(\tau) \quad (3)$$

where β represents the average magnetic moment and the function f(τ) is related to the Curie temperature (T_C):

$$f(\tau) = 1 - \frac{1}{A}\left[\frac{79\tau^{-1}}{140p} + \frac{474}{497}\left(\frac{1}{p} - 1\right)\left(\frac{\tau^3}{6} + \frac{\tau^9}{135} + \frac{\tau^{15}}{600}\right)\right], if\ \tau = \frac{T}{T_C} \leq 1$$

$$f(\tau) = -\frac{1}{A}\left(\frac{\tau^{-5}}{10} + \frac{\tau^{-15}}{315} + \frac{\tau^{-25}}{1500}\right), if\ \tau = \frac{T}{T_C} \geq 1$$

$$A = \frac{518}{1125} + \frac{11,692}{15,975}\left(\frac{1}{p} - 1\right) \quad (4)$$

The constant p takes on values of 0.28 for fcc and hcp metals and 0.4 for bcc metals.

To account for the A2/B2 order-disorder transition, the bcc phase is modeled using the partitioning model, which treats the ordered and disordered components separately [26]. The disordered component of A2 is described using Eq. 1, and the model for the ordered component is described in the next section. Additionally, triple-defect mechanisms are especially important when considering site ordering in B2-aluminide-containing systems such as Al-Co [16], Al-Fe [27] or Al-Ni [7]. To account for this phenomenon [29] when the A2 phase is combined with the B2 phase, vacancies have been introduced in the substitutional sublattice of the A2 phase model as in Ref. [26]. Ordering of the bcc phase will be described in the following section.

Intermetallic Phases: B2 and σ

The two ordered phases, B2 and σ, are described using sublattice models based on the compound energy formalism (CEF) [28]. In order to describe the A2/B2 ordering for the bcc phase, both A2 and B2 are modeled using a single Gibbs energy function where the ordered part (B2) is described by a sublattice formula $(Al,Co,Cr,Va)_1(Al,Co,Cr,Va)_1$ and the disordered part (A2) by $(Al,Co,Cr,Va)_1$. Specifically, B2 is appended to the already existing A2 model. Then, as implemented in Thermo-Calc, free energy minimization determines whether A2 and/or B2 are stable depending on the input conditions. In addition, a set of parameterization constraints derived by Dupin and Ansara [26] are used to partition the ordered and disordered parts of the A2/B2 Gibbs energy to allow independent evaluations. The B2 phase contains six stoichiometric compounds, or "end-members" (excluding vacancies) which represent the reference states generated when pure components fully occupy a sublattice; i.e. B2-(Al)(Al), B2-(Al)(Co), B2-(Al)(Cr), B2-(Co)(Cr), B2-(Co)(Co), and B2-(Cr)(Cr). The end-members of the B2 phase in the present system all come from binary systems and as a result, their formation energies are fixed by the previous Al-Co and Al-Cr binary assessments in this model; B2 is not included in the Co-Cr assessment by Oikawa et al. [17] and is therefore modeled in the present work.

The partitioning model describes the A2 and B2 with one single Gibbs energy function and the same sublattice model, $(Al,Co,Cr,Va)_1(Al,Co,Cr,Va)_1$. As a result, the Gibbs energy of A2 combined with B2 is described as:

$$G_m^{\alpha/\beta} = G_m^{\beta}(x_i) + \Delta G_m^{order}(y_i^{(s)}) \quad (5)$$

to include contributions from the disordered solution, $G_m^{\beta}(x_i)$, in A2 as well as from the ordered B2 itself, $\Delta G_m^{order}(y_i^{(s)})$. Here, $\Delta G_m^{order}(y_i^{(s)})$ can be separated into two terms, and the total Gibbs energy of A2 and B2 becomes:

$$G_m^{\alpha/\beta} = G_m^{\alpha}(x_i) + G_m^{\beta(O)}(y_i^{(s)}) - G_m^{\beta(O)}(y_i^{(s)} = x_i) \quad (6)$$

where $y_i^{(s)}$ denotes the site fraction of element i on sublattice s. It can be seen that the ordering independent term, $G_m^{\alpha}(x_i)$, and the ordering dependent terms $G_m^{\beta(O)}(y_i^{(s)})$ and $G_m^{\beta(O)}(y_i^{(s)} = x_i)$ are separated in a way to allow each to be modeled independently. The term $G_m^{\beta(O)}(y_i^{(s)})$ takes on the partitioned form, as described by Dupin and Ansara [26]:

$$G_m^{\beta(O)}(y_i^{(s)}) = \sum_i\sum_j (y_i' y_j'' \, {}^oG_{i:j}^{\beta(O)} + y_j' y_i'' \, {}^oG_{j:i}^{\beta(O)}) + RT[\sum_i y_i' ln(y_i') + \sum_j y_j'' ln(y_j'')]$$
$$+\sum_i\sum_j\sum_k \{y_i' y_j' y_k'' [\sum_k {}^vL_{i,j:k}^{\beta(O)}(y_i' - y_j')^v]\} + \sum_i\sum_j\sum_k \{y_k' y_i'' y_j'' [\sum_k {}^vL_{k:i,j}^{\beta(O)}(y_i'' - y_j'')^v]\}$$
$$+\sum_i\sum_j\sum_k\sum_l y_i' y_j' y_k'' y_l'' \, {}^vL_{i,j:k,l}^{\beta(O)} + \sum_i\sum_j\sum_k\sum_l (y_i' y_j' y_k' y_l'' \, {}^vL_{i,j,k:l}^{\beta(O)} + y_i'' y_j'' y_k'' y_l' \, {}^vL_{l:i,j,k}^{\beta(O)}) \tag{7}$$

When the phase becomes disordered, i.e. $y_i' = y_i'' = x_i$, $G_m^{\beta(O)}(y_i^{(s)}) - G_m^{\beta(O)}(y_i^{(s)} = x_i)$ equals zero and eliminates the ordering energy contribution. The equivalence of the CEF and partitioned models is obtained using $2G_m^{Partition}(y_i^{(s)}) = G_m^{CEF}(y_i^{(s)})$. In Eq. 7, ${}^oG_{i:j}^{\beta(O)}$ represents the Gibbs energy of a B2 end-member expressed by Eq. 10 with i and j in the first and second sublattices, respectively. $L_{i:j,k}^{\beta(O)}$ is the interaction term between end-members of i:j and i:k that takes the form of A + B×T where A and B are the model parameters to be evaluated in modeling process. Relationships between the interaction parameters used in the CEF and partitioned models exist and are derived by Dupin and Ansara [26]. B2 end-members have both sites equivalent to one another so A2 disordering is possible when those sites have the same disordered occupancy. This crystallographic site equivalency is taken into account with the following expressions [26]:

$${}^oG_{i:j}^{\beta} = {}^oG_{j:i}^{\beta},\ {}^vL_{i,j:k}^{\beta} = {}^vL_{k:i,j}^{\beta},\ {}^vL_{i,j:k,l}^{\beta} = {}^vL_{k,l:i,j}^{\beta},\ \text{and}\ {}^vL_{i,j,k:l}^{\beta} = {}^vL_{l:i,j,k}^{\beta} \tag{8}$$

As adopted from the Co-Cr binary [17], σ is modeled with the sublattice formulation $(Al, Co)_8 (Al, Co, Cr)_{18}(Cr)_4$. The phase σ contains σ-$(Al)_8(Al)_{18}(Cr)_4$, σ-$(Al)_8(Cr)_{18}(Cr)_4$, σ-$(Co)_8(Co)_{18}(Cr)_4$, σ-$(Co)_8(Cr)_{18}(Cr)_4$ binary and σ-$(Al)_8(Co)_{18}(Cr)_4$, σ-$(Co)_8(Al)_{18}(Cr)_4$ ternary end-members. To extend the solubility of σ from the binary Co-Cr system, Al must be introduced while keeping consistency with the original model for σ. Joubert [29] suggested that Al should only be allowed to mix in the first and second sublattices given its size and electronic characteristics. In accordance with this, a model described as $(Al, Co)_8(Al, Co, Cr)_{18}(Cr)_4$ is established. The Gibbs energy of σ in per mole of formula has the form,

$$G_m^{\sigma} = \sum_i\sum_j y_i' y_j'' \, {}^oG_{i:j:Cr}^{\sigma} + RT[8\sum_i y_i' ln(y_i') + 18\sum_j y_j'' ln(y_j'')]$$
$$+\sum_{i_1}\sum_{i_2}\sum_j \{y_{i_1}' y_{i_2}' y_j'' [\sum_k {}^kL_{i_1,i_2:j:Cr}^{\sigma}(y_{i_1}' - y_{i_2}')^k]\}$$
$$+\sum_i\sum_{j_1}\sum_{j_2} \{y_i' y_{j_1}'' y_{j_2}'' [\sum_k {}^kL_{i:j_1,j_2:Cr}^{\sigma}(y_{j_1}'' - y_{j_2}'')^k]\} \tag{9}$$

where y_i' and y_i'' represent the site fractions of i in the first and second sublattices, ${}^oG_{i:j:Cr}^{\sigma}$ represents the Gibbs energy of particular σ end-members as shown by Eq. 10, and $L_{i:j,k:Cr}^{\sigma}$ is the interaction term between end-members of i:j:Cr and i:k:Cr.

The described binary and ternary end-members for B2 and σ are modeled as follows,

$$G_m^{Al_xCo_yCr_z} = x^oG_{Al}^{fcc} + y^oG_{Co}^{hcp} + z^oG_{Cr}^{bcc} + \Delta_f G^{Al_xCo_yCr_z} \tag{10}$$

where $\Delta_f G^{Al_xCo_yCr_z} = \Delta_f H^{Al_xCo_yCr_z} - T\Delta_f S^{Al_xCo_yCr_z}$ and represents the Gibbs energy of formation of a particular B2 or σ end-member with the composition $Al_xCo_yCr_z$. The Debye-Grüneisen model is used to predict the enthalpy and entropy of the compounds as a function of temperature from 0 K properties obtained by the DFT calculations. Details for these calculations are presented in the following section. The PARROT module within Thermo-Calc [30] is used to assess all model parameters for each phase in the system using the phase equilibrium data from our experimental measurements and the results by Ishikawa et al. [13] as well as first-principles thermochemical data calculated in the present work. PARROT is a thermodynamic data assessment module which has been developed to fit model parameters to experimental data by a least mean square method [31].

First-Principles Methodologies

The Helmholtz energy, F(V,T), of a condensed phase, in terms of the quasiharmonic approach, from first-principles calculations based on DFT is expressed as follows [32,33]:

$$F(V,T) = E_{0K}(V) + F_{Vib}(V,T) + F_{T-el}(V,T) \tag{11}$$

In the above expression, $F_{Vib}(V,T)$ and $F_{T-el}(V,T)$ represent the temperature-dependent vibrational and thermal-electronic contributions, respectively. In the present work, the Helmholtz energy is taken approximately as the Gibbs energy due to the negligible ambient pressure used in the modeling. The thermal electronic contribution to the Helmholtz energy is estimated based on the electronic density of states and calculated using Fermi-Dirac statistics for metallic systems [32]. $E_{0K}(V)$ is the static contribution at 0 K without the zero-point vibrational energy. It is obtained using a four-parameter Birch-Murnaghan (BM4) equation of state (EOS) [32,34]:

$$E_{0K}(V) = a + bV^{-2/3} + cV^{-4/3} + dV^{-2} \quad (12)$$

where a, b, c, and d are fitting parameters. Energy versus volume (E-V) data used in the fitting are relaxed with respect to ionic positions and cell shape at the given volumes.

Lattice vibrations are modeled with the Debye-Grüneisen model with the benefit of both accuracy and efficiency; the relevant equations have been described in detail in previous publications [35,36]. The scaling factor is implemented to scale the Debye temperature as a consequence of the differences in transverse and longitudinal phonon modes [37]. It has been shown that the scaling factor is highly dependent on the crystal structure [38,39]. This factor can be estimated from elastic constant calculations using DFT and then averaged as an isotropic medium. Elastic constant calculations are implemented in this work for fcc-Al, hcp-Co, bcc-Co, B2, and σ using the method proposed by Shang et al. [40].

In order to predict A2 solution mixing, calculations of the enthalpy of mixing based on special quasirandom structures calculations (SQS) [41] are performed using the 16-atom binary A2 model developed by Jiang et al. [42] and the 32- or 36-atom ternary A2 model developed by Jiang [43]. Additionally, ternary B2 solution mixing calculations are also performed using 8- and 16-atom supercells generated for the isostructural B2-AlNi system by Jiang et al. [44]. SQS calculation procedures for the present work are performed with the method recently implemented by Lieser et al. [45]. The SQS supercells are first relaxed with respect to cell volume only, and then to only cell volume and shape, and finally to cell volume, shape as well as ion positions simultaneously. Radial distribution functions (RDF) of relaxed supercells are compared with the ideal bcc structure after each relaxation step [45]. The structures with the lowest energy that retain the required structural symmetry are used in the present work.

The Vienna ab-initio Simulation Package (VASP) [46] is used for spin-polarized DFT calculations due to the ferro- and antiferromagnetic natures of Co and Cr, respectively. Electron-ion interactions are described by the accurate projector augmented-wave (PAW) method [47,48]. The generalized gradient approximation (GGA) as implemented by Perdew, Burke, and Ernzerhof (PBE) [49] is used to describe the electron exchange and correlation. A plane-wave cutoff energy of 400 eV is consistently used to ensure enough basis sets are included, as recommended by the VASP manual [50]. Reciprocal k-meshes used for fcc-Al/Co, hcp-Co, bcc-Al/Co/Cr, B2 and σ are 21×21×21, 23×23×12, 17×17×17, 15×15×15, and 6×6×11, respectively. The structures are relaxed by implementing the Methfessel-Paxton method [51] to minimize the forces acting on the atoms. After relaxations, a final calculation using the tetrahedron with Blöchl corrections [52] is applied to ensure an accurate total energy calculation.

RESULTS AND DISCUSSION

Experimental Results

Alloy phase constitutions were determined through a combination of XRD (selected spectra shown in Fig 1) and phase composition analyses (EPMA results given in Table 1). At 1173 and 1273 K, the alloys A4, A5 and A7 are within the B2-γ-σ three-phase triangle, which is an invariant equilibrium in a three-component system. Small differences measured between the phase compositions are larger than the experimental standard deviation, and may reflect the slight influence of Y on phase equilibria. Selected microstructures observed after equilibration at 1273 and 1373 K are shown in Fig 2. In Fig 2A and 2B, the dark matrix is B2, the two light phases are γ and σ, and the bright precipitates are Y-containing intermetallics (denoted MY). These phases are identified based on their measured compositions and associated XRD results. This phase constitution is typical for equilibrium at 1173 and 1273 K. At 1373 K, the B2 phase exhibits compositions that vary widely between the four alloys studied. These are substantially different from those measured at 1173 and 1273 K, with Al contents as low as 16.2 at.% and Cr contents as high as 36.9 at.%. Nevertheless, XRD confirmed that the ordered B2 structure is maintained at 1373 K. In the absence of σ reflections, it is concluded that σ has been replaced by A2, which dissolves significantly more Al (Table 1). The A2 and B2 phases each have a cubic structure, and their peaks cannot be resolved with the instrument available. Analysis by EPMA (not shown here) indicated that two types of Co-rich yttrides were present, containing ~ 7 at.% and ~ 9 at.% Y, with 18–25 at.% Cr and 9–14 at.% Al. Due to the low volume fraction of these phases, their structures could not be determined by XRD.

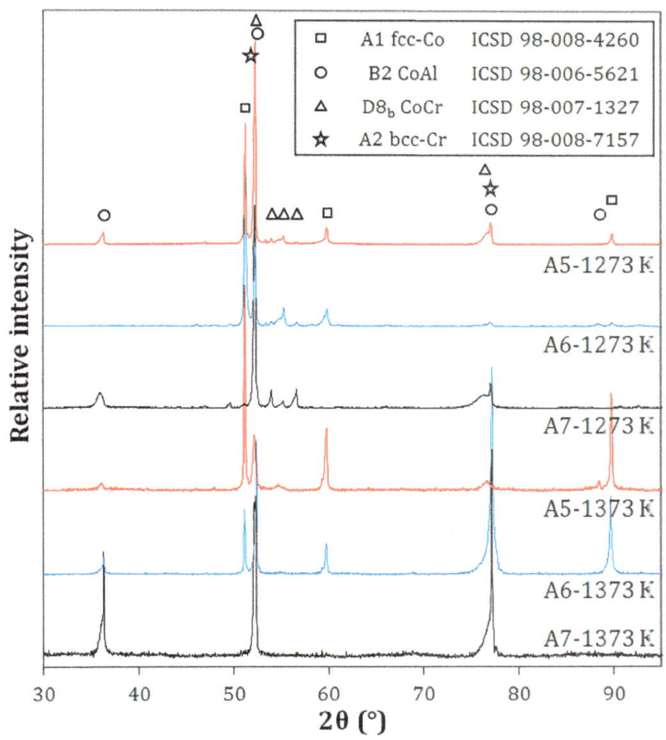

Figure 1. XRD analysis of selected alloy phase constitutions at 1273 K and 1373 K. Note that preferential orientations inherent to cast microstructures were still present after annealing. Specimens were rotated in-plane to ensure that all phases were detected.

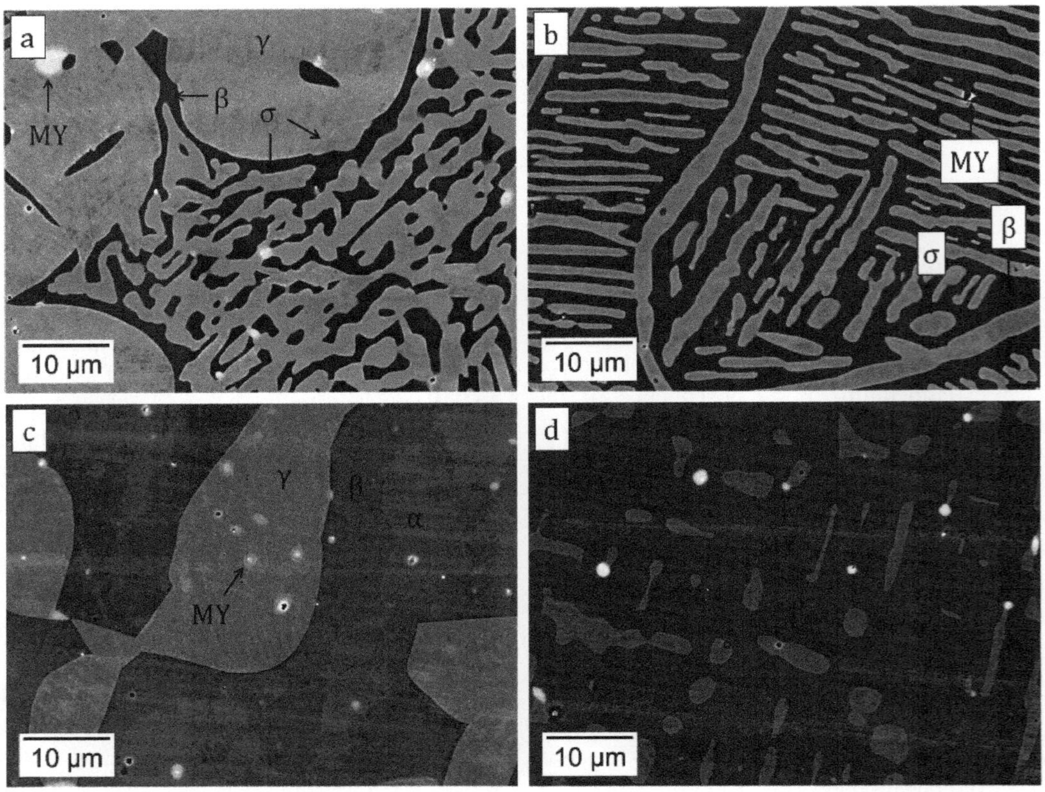

Figure 2. Microstructure of alloys (a,c) A6 and (b,d) A7 equilibrated at (a,b) 1273 K and (c,d) 1373 K. The bright precipitates are Y-containing intermetallics (noted MY).

Table 1. Phase compositions of the CoCrAlY alloys measured by EPMA (at.%)

T (K)	Ref.	B2			γ			σ			A2		
		Al	Cr	Co	Al	Cr	Co	Al	Cr	Co	Al	Cr	Co
1173	A4	36.6	13.2	50.3	5.6	34.3	60.1	2.9	54.0	43.1			
	A5	36.3	*13.6*	50.1	5.5	34.8	59.7	2.9	54.5	42.6			
	A6	36.3	14.2	49.5	5.5	35.3	59.2	2.9	54.8	42.3			
	A7	36.8	14.6	48.7				2.9	57.4	39.7			
1273	A4	32.0	18.1	49.9	6.6	35.8	57.6	4.0	53.5	42.5			
	A5	31.2	19.2	49.7	6.8	36.2	57.0	4.0	53.8	42.2			
	A6	31.7	18.9	49.5	6.7	36.2	57.1	3.9	54.1	42.0			
	A7	33.1	18.5	48.3				3.8	56.8	39.4			
1373	A4	28.2	20.7	51.1	8.4	33.7	57.9						
	A5	26.0	23.6	50.3	8.4	34.5	57.2						
	A6	16.2	36.9	46.8	8.0	36.5	55.5				12.4	41.6	46.0
	A7	22.3	33.4	44.2							11.7	48.4	39.9

The present experimental results confirm and refine the phase relationships originally documented by Ishikawa et al. [13]. Between 1273 and 1373 K, the B2+γ+σ three-phase region is replaced by an A2+B2+γ triangle, and the solubility of Cr in B2 drastically increases. Both are due to the shrinkage of the A2/B2 miscibility gap as the temperature increases, and to the associated dissolution of σ into A2. Ishikawa et al. [13] observed the A2/B2 order-disorder transition at 1473 K. The present measurements show very close compositions for B2 and A2 in equilibrium with γ at 1373 K and therefore, indicate that the miscibility gap is closed slightly above 1373 K.

First-Principles Results

To provide a benchmark for the first-principles methodology used in the present work, fundamental properties of Al, Co, Cr and relevant phases, shown in Table 2, are compared with previous first-principles calculations [53,54] and data from various experiments [55–59]. Good agreement is seen for these properties for all three elements in their standard element reference (SER) state (pure, most stable at 298 K and 1 bar) [22]. It should be noted that Cr takes on an anti-ferromagnetic state at 0 K. The results agree with previous calculations and experiments [56,59,60] for anti-ferromagnetic Cr. The lattice parameters and bulk modulus of B2-(Al)(Co) agree with previous results [61,62] and hence, the predicted values for B2-(Co)(Cr) and B2-(Cr)(Al) can be adopted with confidence. No measured bulk moduli for σ-CoCr could be found in the literature and hence, only experimental lattice parameters are compared. The calculated values for the end-members σ-$(Co)_8(Co)_{18}(Cr)_4$ and σ-$(Co)_8(Cr)_{18}(Cr)_4$ are compared to the lattice parameter of a σ-$Co_{13}Cr_{17}$ alloy and are in good agreement with measurements by Dickins et al. [55].

Table 2. Various properties of the Al, Co, Cr, and the end-members of B2 and σ

Phase	V_0 (Å^3/atom)	B_0 (GPa)	Scaling factor	Source
Al (fcc)	16.50	78	0.63	Present work
	17.08	65		DFT[54]
		72		Experiment[56]
		79 (0 K)		Experiment[57]
Co (hcp)	10.88	210	0.78	Present work
	11.07	204		DFT[54]
		191		Experiment[56]
		196 (0 K)		Experiment[58]
Cr (bcc)	11.59	176	0.88	Present work
	11.56	189		DFT[60]
	11.58	258		DFT[54]
		190		Experiment[56]
		192 (0 K)		Experiment [59]
B2 Phase				
(Al)(Co)	11.61	178	0.87	Present work
		157		DFT[61]
		162±5		Experiment[61]
(Co)(Cr)	11.60	214	0.77	Present work
(Cr)(Al)	14.03	125	0.82	Present work
σ Phase				
$(Al)_8(Al)_{18}(Cr)_4$	15.70	86	0.78	Present work
$(Al)_8(Cr)_{18}(Cr)_4$	12.36	199	0.82	Present work
$(Co)_8(Al)_{18}(Cr)_4$	13.07	141	0.78	Present work
$(Al)_8(Co)_{18}(Cr)_4$	11.65	174	0.78	Present work
$(Co)_8(Co)_{18}(Cr)_4$	11.12	197	0.69	Present work
$(Co)_8(Cr)_{18}(Cr)_4$	11.33	249	0.75	Present work

These properties are derived from the energy vs. volume curves using the 4-parameter Birch-Murnaghan EOS. B_0 denotes the bulk modulus. The bulk modulus at room temperature of Al, Co, and Cr are also presented, as reported by Kittel [56]. Other experimental temperatures are shown if known; reported 0 K values are extrapolated from low temperature data.

doi:10.1371/journal.pone.0121386.t002

Results from our recent study on the Debye-Grüneisen model [63] indicate that using a calculated scaling factor for the Debye temperature is an accurate and efficient method to predict thermodynamic properties of pure elements and intermetallic phases in comparison with the more computationally demanding phonon supercell approach. Table 2 shows that the pure-element properties derived from the EOS fitting match experiments [56–58] and the DFT predictions [54,60]. Therefore, the Debye approximation is used to estimate the thermodynamic properties for all end-members of B2 and σ. Predicted enthalpies of formation ($\Delta_f H$) for each end-member of B2 and σ are calculated at 298 K and are shown in Table 3. Note that the non-SER reference B2 formation energies are calculated with respect to the bcc phase of the pure elements and that σ energies are taken with respect to the fcc phase in the first sublattice, bcc in the second and bcc in the third following Ref. [7,17]. Experimental thermochemical data for B2 and σ end-members are unavailable at 298 K and consequently, only previous CALPHAD assessments and DFT results are compared in Table 3. Stein et al. [64] determined values of$\Delta_f H_{298}$ = -64.45 kJ/mol-atom and $\Delta_f S_{298}$ = -11.43 J/mol-atom of B2 based on modeling of experimental enthalpy data at 1100 K. The present DFT predictions using the scaling factor Debye-Grüneisen model are in excellent agreement with these values.

Table 3. Predicted enthalpies and entropies of formation of B2 and **σ** end-members at 298 K

Phase	End-member	$\Delta_f H$(kJ/mol-atom)	$\Delta_f H$ *(SER)*(kJ/mol-atom)	$\Delta_f S$ (J/mol-a)	$\Delta_f S$ *(SER)* (J/mol-a)	Source
(B2)	**(Al)(Co)**	**-66.89**	**-57.67**	-13.73	**-8.23**	Present work
Binary		-64.45		-11.43		CALPHAD[64]
	(Co)(Cr)	11.98	16.60	-1.73	1.03	Present work
	(Cr)(Al)	**-7.37**	**-2.78**	**-9.18**	**-6.45**	Present work
(σ)	**$(Co)_8(Al)_{18}(Cr)_4$**	**-30.56**	**-24.58**	**-7.33**	**-4.05**	Present work
Ternary	**$(Al)_8(Co)_{18}(Cr)_4$**	**-18.93**	**-13.39**	**-4.44**	**-1.13**	Present work
(σ)	$(Co)_8(Co)_{18}(Cr)_4$	3.60	9.62	0.28	3.60	Present work
Binary	0 K		9.84			Present work
	0 K		11.65			DFT[65]
	$(Co)_8(Cr)_{18}(Cr)_4$	5.72	6.19	0.16	0.16	Present work
	0 K		6.20			Present work
	0 K		8.39			DFT[65]
	$(Al)_8(Al)_{18}(Cr)_4$	5.73	11.24	-6.02	-2.74	Present work
	$(Al)_8(Cr)_{18}(Cr)_4$	1.84	1.84	-3.32	-3.32	Present work

Energies are shown in units of J/mol-formula and atom with the most stable end-members shown in bold text. Also, energies taken with respect to standard states are denoted with SER. Energies used for CALPHAD modeling are taken with different reference states depending on the sublattice models used. B2 formation energies are calculated with respect to the bcc phase of the pure elements. For σ, energies are taken with respect to the fcc phase in the first sublattice, bcc in the second and bcc in the third; same as the sublattice model implemented in the current work.

doi:10.1371/journal.pone.0121386.t003

Unfortunately, calorimetric measurements for σ-CoCr have proven unreliable, as reported measured and calculated enthalpies of formation range from -3 to +10 kJ/mol-atom, shown in Ref. [65–67]. Also, the stability of σ in the Co-Cr system was suggested to range from 30 to 45 at.% Co, which is quite different from the calculated end-member compositions of 26.6 and 86.6% at.% Co. However, Downie and Arslan [67] measured the formation enthalpy of one σ–CoCr sample close to the composition of the end-member σ-$(Co)_8(Cr)_{18}(Cr)_4$ at 473 K and found $\Delta_f H$ = 6.77 kJ/mol-atom. While no direct comparison can be made due to the composition and site fractions, the predicted value of $\Delta_f H_{298}$ = 6.19 kJ/mol-atom for σ-$(Co)_8(Cr)_{18}(Cr)_4$ is in good agreement with the measured value. Table 3 shows the final values (non-SER columns) for B2 and σ that are used as input for the thermodynamic modeling in the present work.

For all binary and ternary bcc SQS calculations, RDFs indicate that the simultaneous relaxation of cell volume, shape and ion positions reduce the symmetry of the SQS cell environment. Additionally, ternary SQS calculations show symmetry reduction when only cell volume and shape are allowed to relax. Therefore, SQS supercells with the lowest energy and a coordination environment sufficiently close to the ideal bcc structure are used in the analysis. Calculated binary and ternary enthalpies of mixing are shown in Table 4 and Fig 3 for A2 and B2. Along with the symmetry, the magnetism of the SQS supercells is checked. Cobalt is ferromagnetic at 0 K with an average μB/atom of 2.2 while Al is non-magnetic. At 0 K, Cr is antiferromagnetic with no net average magnetic moment. Our calculations show that Al and Cr additions, both with no net magnetic moment, decrease the μB/atom of Co as a function of composition.

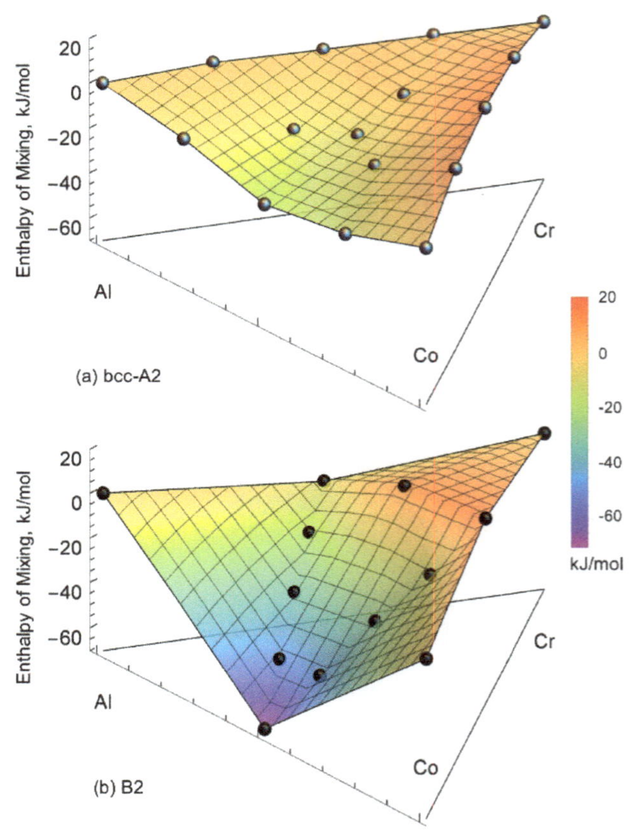

Figure 3. Predicted A2 (a) and B2 (b) enthalpies of mixing. These calculations are based on 16-atom binary and 32/36-atom ternary SQS supercells at 0 K. Grey points in (a) represent distinct A2 SQS compositions and black points in (b) represent B2 SQS compositions. A color map is added to guide the reader in viewing the energy surface.

Table 4. Calculated enthalpies of mixing for solution A2 and B2

Al	Co	Cr	SQS prototype	ΔH_{mix} (kJ/mol-atom)
		A2		
		Al-Co		
0.75	0.25		16	-8.71
0.5	0.5		16	-21.69
0.25	0.75		16	-18.57
		Co-Cr		
	0.75	0.25	16	9.99
	0.5	0.5	16	11.98
	0.25	0.75	16	9.59
		Al-Cr		
0.75		0.25	16	2.45
0.5		0.5	16	1.44
0.25		0.75	16	1.01
		Al-Co-Cr		
0.5	0.25	0.25	32	-9.36
0.25	0.5	0.25	32	-6.86
0.25	0.25	0.5	32	-1.17
0.33	0.33	0.33	36	-7.86
Al	Co	Cr	SQS prototype	ΔH_{mix} (kJ/mol-atom)
		B2		
		(Al)(Co)-(Co)(Cr) section		
0.5	0.5	0		-67.75
0.375	0.5	0.125	32	-47.66
0.25	0.5	0.25	8	-27.20
0.125	0.5	0.375	32	-9.96
0	0.5	0.5		11.99
		(Co)(Cr)-(Cr)(Al) section		
0	0.5	0.5		11.99
0.25	0.25	0.5	8	8.22
0.5	0	0.5		-7.96
		(Cr)(Al)-(Al)(Co) section		
0.5	0	0.5		-7.96
0.5	0.125	0.375	32	-18.49
0.5	0.25	0.25	8	-32.77
0.5	0.375	0.125	32	-49.54
0.5	0.5	0		-67.75

These calculations are based on binary and ternary SQS calculations at 0 K with references taken as bcc-A2 for Al, Co, and Cr.

doi:10.1371/journal.pone.0121386.t004

The SQS calculations, shown in Fig 3 and Table 4, predict the formation of a low-temperature miscibility gap in both Co-Cr and Al-Cr while strong negative mixing is seen in Al-Co. These predictions agree well with experimental observations as well as previous models of the Al-Co [16,64] and Co-Cr [17] binary systems currently adopted for the thermodynamic modeling. In the Co-Cr binary modeled by Oikawa et al. [17], a low temperature A2 miscibility gap is seen at 298 K. Furthermore, the predicted A2 $Al_{0.5}Co_{0.5}$ mixing enthalpy (Table 3) is in good agreement with the Al-Co assessment [16]. However, the Al-Cr model [20] does not produce a low-temperature A2 miscibility gap and hence, disagrees with our current predictions. This is due to a lack of low-temperature thermochemical data available in the literature at the time when Al-Cr was originally modeled. Ternary A2 miscibility gaps are also predicted (Fig 3A), especially along the $Al_{0.5}Co_{0.5}$-Cr cross-section where the composition $Al_{0.5}Co_{0.5}$ shows the lowest mixing enthalpy on the whole energy surface. A convex energy surface would produce a tangent from $Al_{0.5}Co_{0.5}$ to pure Cr that generates a large miscibility gap across the entire $Al_{0.5}Co_{0.5}$-Cr cross-section. When B2 mixing is also taken into account, as shown in Fig 3B, this A2 and B2 miscibility gap becomes more prominent across the $Al_{0.5}Co_{0.5}$-Cr cross-section in accordance with the experiments by Ishikawa et al. [13]. The extremely high formation energy of the end-member (Al)(Co) is responsible for this behavior. Additionally, the SQS results show that A2 is more stable than B2 near Cr because ΔH_{mix} = -1.17 kJ/atom for A2 at the composition $Al_{0.25}Co_{0.25}Cr_{0.5}$ while ΔH_{mix} = 8.22 kJ/atom for B2 at the same composition. These SQS results are in good agreement with measured phase compositions by Ishikawa et al. [13] as well as the present experimental findings. As shown later in Fig 4, this A2/B2 miscibility gap is still observed at 1173 K.

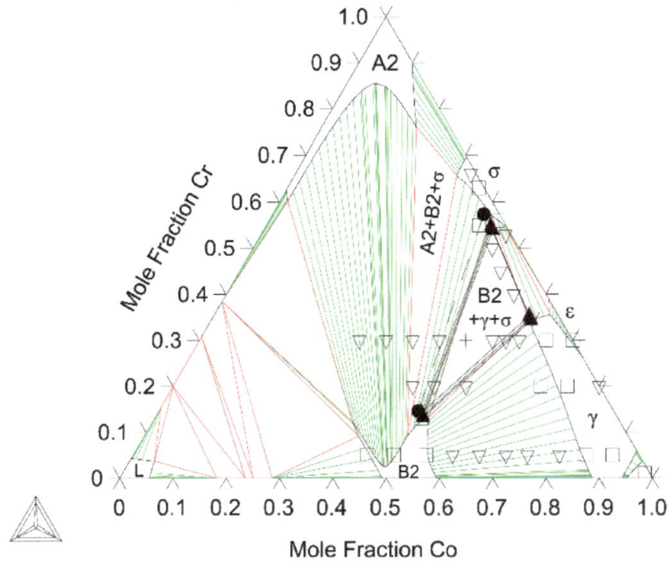

Figure 4. Al-Co-Cr Isothermal section at 1173 K. Shown with phase equilibria data from Moskvitina et al. [68]: single phase (□), 2-phase (▿), and 3-phase (+). Phase equilibria data from the present work: 2-phase (●), 3-phase (▲).

CALPHAD Modeling Results

The three binary models, Al-Co [16], Co-Cr [17], and Al-Cr [20], are extrapolated to the ternary system using the partitioning model [26]. In addition, thermodynamic parameters from DFT, listed in Table 3, are used as input. The binary B2-(Al)(Co), B2-(Cr)(Al), σ-$(Co)_8(Co)_{18}(Cr)_4$and σ-$(Co)_8(Cr)_{18}(Cr)_4$ end-members are fixed by previous models and cannot be modified without reassessment of those systems. The initial ternary extrapolation produces a phase diagram with satisfactory phase boundaries for fcc, A2, and B2 at 1173 K compared to experiments, but is at variance with the experimental isothermal sections at all other temperatures [13,68]. A2/B2 regions do not separate at the order/disorder compositions even at temperatures as high as 1623 K, which is at significant variance with experiments. The poor extrapolation is a result of fixing the B2-(Cr)(Al) end-member to the value proposed by Dupin et al. [7] in their modeling of the Al-Cr-Ni ternary system. Dupin et al. [7] assessed $\Delta_f H$ of B2-(Cr)(Al) to be -13.719 kJ/mol-atom, which is much more stable than the value of -7.37 kJ/mol-atom predicted by DFT in the present work. However, modifying this energy will require significant reassessment of Al-Cr-Ni ternary description and is outside the scope of this work. As a consequence of this fixed B2 end-member, additional interaction parameters for the B2 phase are needed to correctly reproduce its miscibility gap.

The binary B2-(Co)(Cr) end-member and the Co-Cr interaction parameters are evaluated by fitting to the present EPMA results at 1173, 1273, and 1373 K (Table 1) data from Ishikawa et al. [13] as well as first-principles thermochemical data. Given the complexity of the A2/B2 miscibility gap region, as seen in Figs 4–8, ternary, temperature dependent interaction parameters are needed to describe the B2-(Al)(Co) extension into the ternary as well as the A2/B2 transition temperatures. Interaction parameters for the A2 are also needed to capture the Al and Co solubility. The evaluated parameters are shown in Table 5 and built into a thermodynamic database.

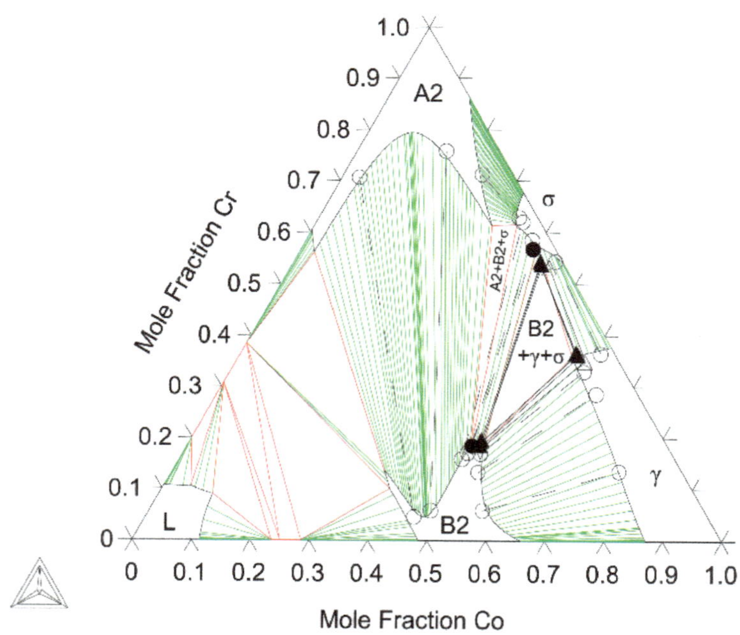

Figure 5. Al-Co-Cr Isothermal section at 1273 K. Shown with phase equilibria data from Ishikawa et al. [13]: 2-phase (옃), and 3-phase (▿). Experimental phase equilibria data from the present work: 2-phase (●), 3-phase (▲).

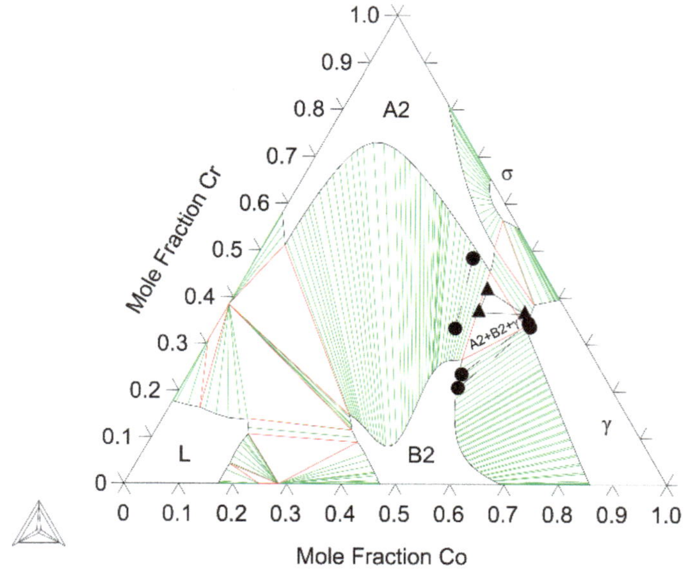

Figure 6. Al-Co-Cr Isothermal section at 1373 K. Shown with phase equilibria data from the present work: 2-phase (●), 3-phase (▲).

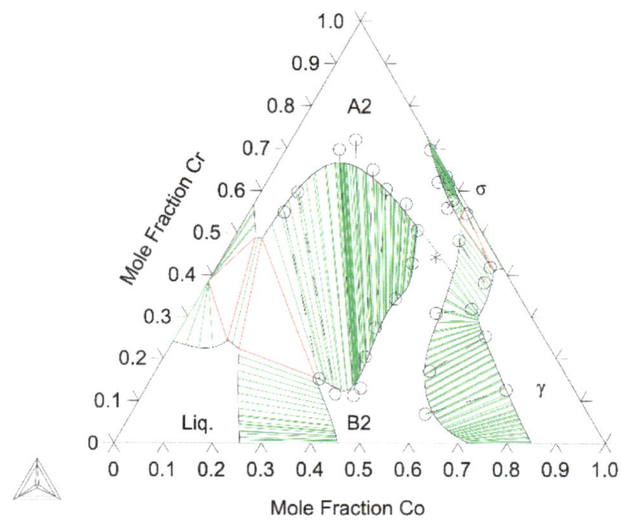

Figure 7. Al-Co-Cr isothermal section at 1473 K. Phase equilibria data from Ishikawa et al. [13]: 2-phase (였), order-disorder transition (*). The calculated order-disorder transition is shown with (· · ·).

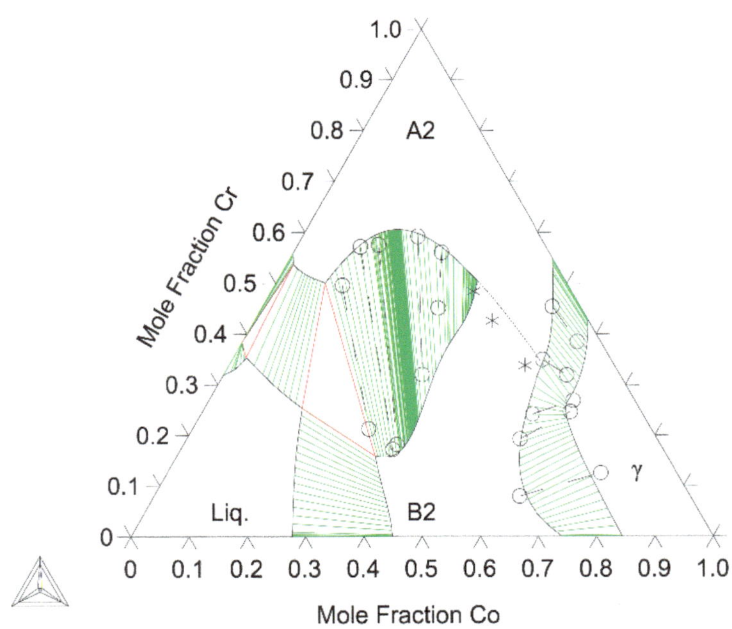

Figure 8. Al-Co-Cr Isothermal section at 1573 K. Phase equilibria data from Ishikawa et al. [13]: 2-phase (였), order-disorder transition (*). The calculated order-disorder transition is shown with (· · ·).

Table 5. Model parameters and functions for the ternary Al-Co-Cr system

Phase	Parameters	Values	References
liquid (L) phase: (Al,Co,Cr)$_1$	${}^{0}L^{Liq}_{Al,Co,Cr}$	+30000	Present work
fcc (γ) phase: (Al,Co,Cr)$_1$	${}^{0}L^{fcc}_{Al,Co,Cr}$	+17295	Present work
hcp (ε) phase: (Al,Co,Cr)$_1$	${}^{0}L^{hcp}_{Al,Co,Cr}$	+25000	Present work
bcc (A2) phase: (Al,Co,Cr,Va)$_1$	${}^{0}L^{A2}_{Al,Co}$	*+GB2ALCO–LB2ALCO*	Present work
	${}^{0}L^{A2}_{Al,Cr}$	–54900 + 10 ×*T*	*COST 507*[20]
	${}^{0}L^{A2}_{Co,Cr}$	+1033–1.481 × *T*	*Oikawa et al.*[17]
	${}^{1}L^{A2}_{Co,Cr}$	+11972–13.374 × *T*	*Oikawa et al.*[17]
	${}^{0}L^{A2}_{Al,Va}$	*+GB2ALVA–LB2ALVA*	*Dupin et al.*[7]
	${}^{0}L^{A2}_{Co,Va}$	*+GB2COVA–LB2COVA*	Present work
	${}^{0}L^{A2}_{Cr,Va}$	+100000	*Dupin et al.*[7]
	${}^{0}L^{A2}_{Al,Co,Cr}$	*–LB2ALCO + LBALCOMCR +LBCOCRMAL +L1BCRALMCO*	Present work
	${}^{1}L^{A2}_{Al,Co,Cr}$	*+LALCO2CR–LB2ALCO +LBALCOMCR + LBCOCRMAL*	Present work
	${}^{2}L^{A2}_{Al,Co,Cr}$	*–LB2ALCO + LBALCOMCR +LBCOCRMAL –L1BCRALMCO*	Present work
beta (B2) phase: (Al,Co,Cr,Va)$_{0.5}$ (Al,Co,Cr, Va)$_{0.5}$	${}^{0}G^{B2}_{Al:Co} = {}^{0}G^{B2}_{Co:Al}$	+0.5 × *GB2ALCO*–0.5 × *LB2ALCO*	Present work
	$TC^{B2}_{Al:Co} = TC^{B2}_{Co:Al}$	-1450	Present work
	$BMAG^{B2}_{Al:Co} = BMAG^{B2}_{Co:Al}$	-1.35	Present work
	${}^{0}G^{B2}_{Co:Cr} = {}^{0}G^{B2}_{Cr:Co}$	+0.5 × *GB2COCR*	Present work
	${}^{0}G^{B2}_{Cr:Al} = {}^{0}G^{B2}_{Al:Cr}$	+0.5 × *GB2CRAL*	Present work
	${}^{0}G^{B2}_{Al:Va} = {}^{0}G^{B2}_{Va:Al}$	+0.5 × *GB2ALVA*–0.5 × *LB2ALVA*	Present work
	${}^{0}G^{B2}_{Co:Va} = {}^{0}G^{B2}_{Va:Co}$	+0.5 × *GB2COVA*–0.5 × *LB2COVA*	Present work
	${}^{0}G^{B2}_{Cr:Va} = {}^{0}G^{B2}_{Va:Cr}$	0	Present work
	${}^{0}L^{B2}_{Co,Cr:Al} = {}^{0}L^{B2}_{Al:Co,Cr}$	+0.5 × *LBCOCRMAL*	Present work
	${}^{0}L^{B2}_{Al,Cr:Co} = {}^{0}L^{B2}_{Co:Al,Cr}$	+0.5 × *LBCRALMCO*	Present work
	${}^{0}L^{B2}_{Al,Co:Cr} = {}^{0}L^{B2}_{Cr:Al,Co}$	+0.5 × *LBALCOMCR* + *BU1ALCO*	Present work
	${}^{1}L^{B2}_{Co,Cr:Al} = {}^{1}L^{B2}_{Al:Co,Cr}$	+0.5 × *L1BCOCRMAL*	Present work
	${}^{1}L^{B2}_{Al,Cr:Co} = {}^{1}L^{B2}_{Co:Al,Cr}$	+0.5 × *L1BCRALMCO*	Present work
	${}^{1}L^{B2}_{Al,Co:Cr} = {}^{1}L^{B2}_{Cr:Al,Co}$	+0.5 × *L1BALCOMCR*	Present work
	${}^{0}L^{B2}_{Al,Co,Cr:Al} = {}^{0}L^{B2}_{Al:Al,Co,Cr}$	+0.5 × *LAL2COCR*	Present work
	${}^{0}L^{B2}_{Al,Co,Cr:Co} = {}^{0}L^{B2}_{Co:Al,Co,Cr}$	+0.5 × *LALCO2CR*	Present work
	${}^{0}L^{B2}_{Al,Co,Cr:Cr} = {}^{0}L^{B2}_{Cr:Al,Co,Cr}$	+0.5 × *LALCOCR2*	Present work
sigma (σ) phase: (Al,Co)$_8$ (Al,Co,Cr)$_{18}$ (Cr)$_4$	${}^{0}G^{\sigma}_{Al:Al:Cr}$	+161148	Present work
	${}^{0}G^{\sigma}_{Al:Cr:Cr}$	+47886	Present work
	${}^{0}G^{\sigma}_{Co:Co:Cr}$	–16899–29.814 × *T*	*Oikawa et al.*[17]
	${}^{0}G^{\sigma}_{Co:Cr:Cr}$	–259935 + 85.097 × *T*	*Oikawa et al.*[17]
	${}^{0}G^{\sigma}_{Al:Co:Cr}$	–617537	Present work
	${}^{0}G^{\sigma}_{Co:Al:Cr}$	–931862	Present work
	${}^{0}L^{\sigma}_{Co:Al,Cr:Cr}$	–195992	Present work
Function	**Value**		
LAL2COCR	0		Present work
LALCO2CR	–78970 + 89.123 × *T*		Present work
LALCOCR2	0		Present work

Phase	Parameters	Values	References
LBCOCRMAL	28320–16.474 × *T*		Present work
LBCRALMCO	0		Present work
LBALCOMCR	–46432		Present work
L1BCOCRMAL	0		Present work
L1BCRALMCO	13276		Present work
L1BALCOMCR	0		Present work
GB2ALCO	–138500 + 34.620 × *T*		Dupin and Ansara [16]
LB2ALCO	54531–37.04 × *T*		Present work
GB2COCR	35909–16.474 × *T*		Present work
GB2CRAL	–4000		*Dupin et al.*[7]
BU1XY (X,Y = Al,Co,Cr)	–0.5 × *LB2XY*		Present work

Only A2, B2, and σ binary parameters are listed in full for their importance, all other binary parameters can be found in the respective binary Al-Co[16], Co-Cr[17], and Al-Cr[20] references as well as the attached database file. Parameters are in units of J/mol-formula.

doi:10.1371/journal.pone.0121386.t005

Fig 4 shows the calculated isothermal section at 1173 K compared with the present EPMA results as well as experimental data from Moskvitina et al. [68]. A good agreement between calculations and experiments is reached. Fig 5 shows the calculated isothermal section at 1273 K, which agrees very well with the present experiments as well as measurements by Ishikawa et al. [13]. Fig 6 shows a calculated isothermal section at 1373 K, which agrees well with our experiments, except for the B2 composition in the A2+B2+γ three phase tie-triangle and the A2-B2 tie-line. This is attributed to the fact that the A2/B2 miscibility gap disappears very rapidly with increasing temperature, and as a consequence the morphology of this region of the diagram is extremely sensitive to temperature variations around 1373 K. The A2+B2+γ three phase triangle is predicted to disappear at 1423 K measurements by Ishikawa et al. [13] at 1473 K. Fig 7 plots the 1473 K isothermal section, which fits experimental measurements from Ishikawa et al. [13] quite well. At this temperature, the A2/B2 miscibility gap is replaced by a second-order transition represented by a dotted line. The calculation near the transition region shows that the B2 ordering gravitates towards the disordered composition with more Cr additions on the Co-Cr rich side of the phase diagram. This is in good agreement with the experimental results on the order/disorder alloys found by Ishikawa et al. [13]. Fig 8 shows the 1573 K isothermal sections. Furthermore, the A2/B2 and B2/γ tie-lines as well as A2/B2 transition points show good agreement at 1573 and 1623 K compared with data from Ishikawa et al. [13] (not included here). An overall good thermodynamic description of Al-Co-Cr, which takes into account predicted first-principles thermochemical data, is produced in this work. We have demonstrated a model that is able to reproduce the rapid replacement of one three-phase σ+B2+γ tie-triangle by another and the sharp increase in the solubility of Cr in B2 from 1273 to 1373 K, as observed experimentally (see Section 6.1).

CONCLUSIONS

The Al-Co-Cr system is investigated thoroughly using first-principles calculations, XRD, and EPMA measurements to produce a complete CALPHAD thermodynamic description. First-principles DFT calculations predict a large A2/B2 miscibility gap in the ternary system, which is demonstrated experimentally using phase composition measurements from 1173 to 1373 K. In addition, the experimentally-measured phase compositions for A2, B2, γ and σ are in good agreement with previous experimental results [13]. To aid in the modeling of the ternary system, the Debye-Grüneisen model is used to predict finite-temperature data of B2 and σ such as heat capacities, entropies, and enthalpies. It is found that the complex A2/B2 phase region, which includes an order-disorder transition, can be accurately described with a partitioned bcc model. The calculated B2 ordering compositions are also shown to be in agreement with previous as well as present experimental studies. Overall, a consistent thermodynamic description of the Al-Co-Cr system is produced and its accuracy for predicting thermodynamic properties of all phases relevant for MCrAl-base coatings is confirmed.

ACKNOWLEDGMENTS

Computing clusters LION and Cyberstar are provided by the Materials Simulation Center and the Research Computing and Cyberinfrastructure Group at the Pennsylvania State University. Additionally, this work used the Extreme Science and Engineering Discovery Environment (XSEDE), which is supported by National Science Foundation grant number ACI-1053575. We would like to also thank Dr. Yi Wang and Richard Otis for stimulating discussions about the computational aspects of this work.

AUTHOR CONTRIBUTIONS

Conceived and designed the experiments: XLL TG BG ZKL. Performed the experiments: XLL TG. Analyzed the data: XLL TG BL GL BG ZKL. Contributed reagents/materials/analysis tools: TG BG ZKL. Wrote the paper: XLL TG BL GL BG ZKL. Implementing the thermodynamic model: BL.

REFERENCES

1. Wood JH, Goldman E, in Sims CT, Stoloff NSWCH. Superalloys II. 2nd ed. New York, NY: Wiley-Interscience; 1987.
2. Reed RC. The Superalloys. 1st ed. Cambridge, UK: Cambridge University Press; 2006.
3. Goward GW. Progress in coatings for gas turbine airfoils. Surf Coatings Technol. 1998 Oct;108–109:73–9. doi: 10.1016/s0257-8972(98)00667-7
4. Clarke DR, Levi CG. Materials Design for the Next Generation Thermal Barrier Coatings. Annu Rev Mater Res. 2003;33:383–417.
5. Darolia R. Thermal barrier coatings technology: critical review, progress update, remaining challenges and prospects. Int Mater Rev. 2013 Aug 18;58(6):315–48. doi: 10.1179/1743280413y.0000000019
6. Evans AG, Mumm DR, Hutchinson JW, Meier GH, Pettit FS. Mechanisms controlling the durability of thermal barrier coatings. Prog Mater Sci. 2001 Jan;46(5):505–53. doi: 10.1016/s0079-6425(00)00020-7
7. Dupin N, Ansara I, Sundman B. Thermodynamic Re-Assessment of the Ternary System Al-Cr-Ni. Calphad. 2001;25(2):279–98. doi: 10.1016/s0364-5916(01)00049-9
8. Kattner UR. Construction of a Thermodynamic Database for Ni-Base Superalloys: A Case Study. TMS: CALPHAD and Alloy Thermodynamics. 2002.
9. Achar DRG, Munoz-Arroyo R, Singheiser L, Quadakkers WJ. Modelling of phase equilibria in MCrAlY coating systems. Surf Coatings Technol. 2004 Oct;187(2–3):272–83. doi: 10.1016/j.surfcoat.2004.02.018
10. Toscano J, Gil A, Hüttel T, Wessel E, Naumenko D, Singheiser L, et al. Temperature dependence of phase relationships in different types of MCrAlY-coatings. Surf Coatings Technol. 2007 Dec;202(4–7):603–7. doi: 10.1016/j.surfcoat.2007.06.044
11. Nicholls JR. Designing oxidation-resistant coatings. JOM. 2000 Jan;52(1):28–35. doi: 10.1007/s11837-000-0112-2

12. Wright IG, Gibbons TB. Recent developments in gas turbine materials and technology and their implications for syngas firing. Int J Hydrogen Energy. 2007 Nov;32(16):3610–21. doi: 10.1016/j.ijhydene.2006.08.049
13. Ishikawa K, Ise M, Ohnuma I, Kainuma R, Ishida K. Phase Equilibria and Stability of the BCC Aluminide in the Co-Cr-Al System. Ber Bunsenges Phys Chem. 1998;102:1206–10. doi: 10.1002/bbpc.19981020924
14. Liu Z-K. First-Principles Calculations and CALPHAD Modeling of Thermodynamics. J Phase Equilibria Diffus. 2009 Sep 3;30(5):517–34. doi: 10.1007/s11669-009-9570-6
15. Dupin N. Contribution à l'évaluation thermodynamique des alliages polyconstitués à base de nickel (Ph.D. Thesis). INP Grenoble; 1995.
16. Dupin N, Ansara I. Thermodynamic Assessment of the System Al-Co. Rev Métallurgie. 1998;95(9):1121–9. pmid:21267112
17. Oikawa K, Qin G-W, Ikeshoji T, Kainuma R, Ishida K. Direct evidence of magnetically induced phase separation in the fcc phase and thermodynamic calculations of phase equilibria of the Co–Cr system. Acta Mater. 2002 May;50(9):2223–32. doi: 10.1016/s1359-6454(01)00433-5
18. Havrankova J, Vrestal J, Tomiska J. Mass-spectrometric determination of thermodynamic properties of liquid Co-Cr alloys. Kov Mater. 1999;37:34–41.
19. Yang S, Jiang M, Li H, Liu Y, Wang L. Assessment of Co-Cr-Ni ternary system by CALPHAD technique. Rare Met. 2012 Jan 26;31(1):75–80. doi: 10.1007/s12598-012-0466-y
20. Ansara I, Dinsdale AT, Rand MH. COST 507: Definition of Thermochemical and Thermophysical Properties to Provide a Database for the Development of New Light Alloys—Thermochemical Database for Light Metal Alloys—Volume 2. Luxembourg: Office for Official Publications of the European Communities; 1998.
21. Helander T, Tolochko O. An Experimental Investigation of Possible B 2-Ordering in the Al-Cr System. J Phase Equilibria. 1999;20(March 1998):57–60. doi: 10.1361/105497199770335947
22. Dinsdale AT. SGTE data for pure elements. Calphad. 1991 Oct;15(4):317–425. doi: 10.1016/0364-5916(91)90030-n
23. Redlich O, Kister AT. Algebraic Representation of Thermodynamic Properties and the Classification of Solutions. Ind Eng Chem. American Chemical Society; 1948 Feb 1;40(2):345–8. doi: 10.1021/ie50458a036
24. Hillert M, Jarl M. A Model for Alloying Effects in Ferromagnetic Metals. Calphad. 1978 Jan;2(3):227–38. doi: 10.1016/0364-5916(78)90011-1
25. Inden G. The role of magnetism in the calculation of phase diagrams. Phys B+C. 1981 Jan;103(1):82–100. doi: 10.1016/0378-4363(81)91004-4
26. Dupin N, Ansara I. On the sublattice formalism applied to the B2 phase. Zeitschrift für Met. 1999;90:76–85. doi: 10.1107/S0108767309007235. pmid:19349661
27. Sundman B, Ohnuma I, Dupin N, Kattner UR, Fries SG. An assessment of the entire Al–Fe system including D03 ordering. Acta Mater. 2009 Jun;57(10):2896–908. doi: 10.1016/j.actamat.2009.02.046
28. Hillert M. The compound energy formalism. J Alloys Compd. 2001 May;320(2):161–76. doi: 10.1016/s0925-8388(00)01481-x
29. Joubert J-M. Crystal chemistry and Calphad modeling of the [sigma] phase. Prog Mater Sci. 2008 Mar;53(3):528–83. doi: 10.1016/j.pmatsci.2007.04.001
30. Andersson J-O, Helander T, Höglund L, Shi P, Sundman B. Thermo-Calc & DICTRA, computational tools for materials science. Calphad. 2002 Jun;26(2):273–312. doi: 10.1016/s0364-5916(02)00037-8
31. Jansson B. Evaluation of parameters in thermochemical models using different types of experimental data simultaneously. Technical Report TRITAMAC-0234. Stockholm, Sweden; 1984.
32. Shang SL, Wang Y, Kim D, Liu Z-K. First-principles thermodynamics from phonon and Debye model: Application to Ni and Ni3Al. Comput Mater Sci. 2010 Feb;47(4):1040–8. doi: 10.1016/j.commatsci.2009.12.006
33. Wang Y, Liu Z-K, Chen L-Q. Thermodynamic properties of Al, Ni, NiAl, and Ni3Al from first-principles calculations. Acta Mater. 2004 May;52(9):2665–71. doi: 10.1016/j.actamat.2004.02.014
34. Birch F. Finite Elastic Strain of Cubic Crystals. Phys Rev. American Physical Society; 1947 Jun 1;71(11):809–24. doi: 10.1103/physrev.71.809
35. Zacherl CL, Shang S-L, Saengdeejing A, Liu Z-K. Phase stability and thermodynamic modeling of the Re-Ti system supplemented by first-principles calculations. Calphad. 2012 Sep;38:71–80. doi: 10.1016/j.calphad.2012.05.001
36. Liu XL, Hargather CZ, Liu Z-K. First-principles aided thermodynamic modeling of the Nb–Re system. Calphad. Elsevier; 2013 Jun;41:119–27. doi: 10.1016/j.calphad.2013.02.006
37. Alers GA. Use of Sound Velocity Measurements in Determining the Debye Temperature of Solids. Physical Acoustics: Volume 3 Part B. 1965. p. 1–42. doi: 10.1016/b978-0-12-395669-9.50008-4
38. Chen Q, Sundman B. Calculation of Debye Temperature for Crystalline Structures—A Case Study on Ti, Zr, and Hf. Acta Mater. 2001;49:947–61. doi: 10.1016/s1359-6454(01)00002-7
39. Lu X-G, Selleby M, Sundman B. Calculations of thermophysical properties of cubic carbides and nitrides using the Debye–Grüneisen model. Acta Mater. 2007 Feb;55(4):1215–26. doi: 10.1016/j.actamat.2006.05.054
40. Shang SL, Wang Y, Liu Z-K. First-principles elastic constants of α- and θ-Al2O3. Appl Phys Lett. 2007 Mar 7;90(10):101909. doi: 10.1063/1.2711762
41. Zunger A, Wei S-H, Ferreira LG, Bernard JE. Special quasirandom structures. Phys Rev Lett. 1990 Jul 16;65(3):353–6. pmid:10042897 doi: 10.1103/physrevlett.65.353

42. Jiang C, Wolverton C, Sofo J, Chen L-Q, Liu Z-K. First-principles study of binary bcc alloys using special quasirandom structures. Phys Rev B. 2004 Jun 28;69(21):214202. doi: 10.1103/physrevb.69.214202
43. Jiang C. First-principles study of ternary bcc alloys using special quasi-random structures. Acta Mater. 2009 Sep;57(16):4716–26. doi: 10.1016/j.actamat.2009.06.026
44. Jiang C, Chen L-Q, Liu Z-K. First-principles study of constitutional point defects in B2 NiAl using special quasirandom structures. Acta Mater. 2005 May;53(9):2643–52. doi: 10.1016/j.actamat.2005.02.026
45. Lieser AC, Zacherl CL, Saengdeejing A, Liu Z-K, Kecskes LJ. First-principles calculations and thermodynamic re-modeling of the Hf-W system. Calphad. 2012 Sep;38:92–9. doi: 10.1016/j.calphad.2012.04.005
46. Kresse G, Furthmüller J. Efficient iterative schemes for ab initio total-energy calculations using a plane-wave basis set. Phys Rev B. 1996 Oct 15;54(16):11169–86. pmid:9984901 doi: 10.1103/physrevb.54.11169
47. Blöchl PE. Projector augmented-wave method. Phys Rev B. 1994 Dec 15;50(24):17953–79. pmid:9976227 doi: 10.1103/physrevb.50.17953
48. Kresse G, Joubert D. From ultrasoft pseudopotentials to the projector augmented-wave method. Phys Rev B. 1999 Jan 15;59(3):1758–75. doi: 10.1103/physrevb.59.1758
49. Perdew J, Burke K, Ernzerhof M. Generalized Gradient Approximation Made Simple. Phys Rev Lett. 1996 Oct 28;77(18):3865–8. pmid:10062328 doi: 10.1103/physrevlett.77.3865
50. Kresse G, Marsman M, Furthmüller J. VASP: The Guide. Wein, Austria; 2013.
51. Methfessel M, Paxton AT. High-precision sampling for Brillouin-zone integration in metals. Phys Rev B. American Physical Society; 1989 Aug 15;40(6):3616–21. pmid:9992329 doi: 10.1103/physrevb.40.3616
52. Blöchl PE, Jepsen O, Andersen OK. Improved tetrahedron method for Brillouin-zone integrations. Phys Rev B. 1994 Jun 15;49(23):16223–33. pmid:10010769 doi: 10.1103/physrevb.49.16223
53. Wang Y, Curtarolo S, Jiang C, Arroyave R, Wang T, Ceder G, Chen LQ, Liu ZK. Ab initio lattice stability in comparison with CALPHAD lattice stability. Calphad. 2004 Mar;28(1):79–90. doi: 10.1016/j.calphad.2004.05.002
54. Shang SL, Saengdeejing A, Mei ZG, Kim DE, Zhang H, Ganeshan S, et al. First-principles calculations of pure elements: Equations of state and elastic stiffness constants. Comput Mater Sci. 2010 Jun;48(4):813–26. doi: 10.1016/j.commatsci.2010.03.041
55. Dickins GJ, Douglas a. MB, Taylor WH. The crystal structure of the Co–Cr σ phase. Acta Crystallogr. 1956 Mar 1;9(3):297–303. doi: 10.1107/s0365110x56000826
56. Kittel C. Introduction to Solid State Physics. Wiley; 1995.
57. Kamm GN, Alers G a. Low-Temperature Elastic Moduli of Aluminum. J Appl Phys. 1964;35(2):327. doi: 10.1063/1.1713309
58. Mori N. Magnetic Contribution to the Bulk Modulus of 3d Transition Metals and Alloys. Phys B. 1988;149:226–31. doi: 10.1016/0378-4363(88)90246-x
59. Moruzzi VL, Marcus PM. Antiferromagnetism in 3d Transition Metals. Phys Rev B. 1990;42(13):8361–6. pmid:9995009 doi: 10.1103/physrevb.42.8361
60. Hafner R, Spišák D, Lorenz R, Hafner J. Magnetic ground state of Cr in density-functional theory. Phys Rev B. 2002 May;65(18):184432. doi: 10.1103/physrevb.65.184432
61. Mehl MJ, Osburn JE, Papaconstantopoulos DA, Klein BM. Structural properties of ordered high-melting-temperature intermetallic alloys from first-principles total-energy calculations. Phys Rev B. 1990 May;41(15):10311–23. pmid:9993438 doi: 10.1103/physrevb.41.10311
62. Hammer L, Blum V, Schmidt C, Wieckhorst O, Meier W, Müller S, et al. Role of Co antisite segregation in the CoAl(111) surface. Phys Rev B. 2005 Feb;71(7):075413. doi: 10.1103/physrevb.71.075413
63. Liu XL, VanLeeuwen BK, Shang SS-L, Du Y, Liu Z-K. On the scaling factor in Debye-Grüneisen model: A case study of the Mg-Zn binary system. Comput Mater Sci. 2015 Feb;98:34–41. doi: 10.1016/j.commatsci.2014.10.056
64. Stein F, He C, Dupin N. Melting behaviour and homogeneity range of B2 CoAl and updated thermodynamic description of the Al–Co system. Intermetallics. 2013 Aug;39:58–68. doi: 10.1016/j.intermet.2013.03.011
65. Pavlů J, Vřešťál J, Šob M. Ab initio study of formation energy and magnetism of sigma phase in Cr–Fe and Cr–Co systems. Intermetallics. 2010 Feb;18(2):212–20. doi: 10.1016/j.intermet.2009.07.018
66. Bell HB, Hajra JP, Putland FH, Spencer PJ. The Determination of the Thermodynamic Properties of Cobalt–Chromium Alloys Using Solid-Electrolyte EMF and High-Temperature Calorimetric Techniques. Met Sci. 1973 Jan;7(1):185–90. doi: 10.1179/030634573790445442
67. Downie DB, Arslan F. Enthalpies of formation of (cobalt + chromium) alloys at 473 K. J Chem Thermodyn. 1983 Jul;15(7):645–9. doi: 10.1016/0021-9614(83)90078-2
68. Moskvitina ES, Kuznetsov VN, Guzei LS. Refinement of the Co-Cr-Al Phase-Diagram. Vestn Mosk Univ Seriya 2 Khimiya (Moscow Univ Chem Bull). 1992;33(4):373–4.

24. D. Ferro-Costas and R. A. Mosquera, Phys. Chem. Chem. Phys., 2015, 17, 7424–7434.
25. M. Garcia-Revilla, E. Francisco, P. L. A. Popelier and A. Martín Pendás, ChemPhysChem, 2013, 14, 1211–1218 .
26. A. Martín Pendás, M. A. Blanco and E. Francisco, J. Comput. Chem., 2007, 28, 161–184 .
27. I. Cukrowski, J. H. de Lange, A. S. Adeyinka and P. Mangondo, Comput. Theor. Chem., 2015, 1053, 60–76.
28. K. Eskandari and M. Lesani, Chem. – Eur. J., 2015, 21, 4739–4746 .
29. P. Matczak, M. Domagała and S. Domagała, Struct. Chem., 2015 DOI:10.1007/s11224-015-0643-3.
30. I. Cukrowski, Comput. Theor. Chem., 2016, 1066, 62–75 .
31. A. D. Becke, J. Chem. Phys., 1993, 98, 5648–5652 .
32. P. L. A. Popelier, Int. J. Quantum Chem., 2015, 115, 1005–1011 .
33. V. Tognetti and L. Joubert, Phys. Chem. Chem. Phys., 2014, 16, 14539–14550.
34. M. Rafat and P. L. A. Popelier, in Quantum Theory of Atoms in Molecules, ed. C. F. Matta and R. J. Boyd, Wiley-VCH, Weinheim, Germany, 2007, vol. 5, pp. 121–140.
35. P. L. A. Popelier, in The Chemical Bond – 100 years old and getting stronger, ed. M. Mingos, Springer, Germany, 2016.
36. J. Poater, M. Sola, M. Duran and X. Fradera, Theor. Chem. Acc., 2002, 107, 362–371 .
37. X. Fradera, M. A. Austen and R. F. W. Bader, J. Phys. Chem. A, 1999, 103, 304–314 .
38. R. F. W. Bader and C. F. Matta, Inorg. Chem., 2001, 40, 5603–5611 .
39. P. Maxwell and P. L. A. Popelier, Mol. Phys., 2016.
40. K. Mueller, F. Diederich and R. Paulini, Angew. Chem., Int. Ed., 2005, 44, 1788–1805 .
41. A. Goerling and M. Levy, Phys. Rev. A: At., Mol., Opt. Phys., 1994, 50, 196–204 .
42. W. A. Kohn and L. J. Sham, Phys. Rev. A: At., Mol., Opt. Phys., 1965, 140, 1133–1138 .
43. R. Chávez-Calvillo, M. García-Revilla, E. Francisco, A. Martín Pendás and T. Rocha-Rinza, Comput. Theor. Chem., 2015, 1053, 90–95 .
44. T. A. Keith, AIMAll (Version 13.10.19), (http://aim.tkgristmill.com) and T. G. S. Todd A. Keith, Overland Park KS, USA, (aim.tkgristmill.com), 2013.
45. P. J. Stephens, F. J. Devlin, C. F. Chabalowski and M. J. Frisch, J. Phys. Chem., 1994, 98, 11623–11627 .
46. M. J. Frisch, G. W. Trucks, H. B. Schlegel, G. E. Scuseria, M. A. Robb, J. R. Cheeseman, G. Scalmani, V. Barone, B. Mennucci, G. A. Petersson, H. Nakatsuji, M. Caricato, X. Li, H. P. Hratchian, A. F. Izmaylov, J. Bloino, G. Zheng, J. L. Sonnenberg, M. Hada, M. Ehara, K. Toyota, R. Fukuda, J. Hasegawa, M. Ishida, T. Nakajima, Y. Honda, O. Kitao, H. Nakai, T. Vreven, J. A. Montgomery, J. E. Peralta, F. Ogliaro, M. Bearpark, J. J. Heyd, E. Brothers, K. N. Kudin, V. N. Staroverov, R. Kobayashi, J. Normand, K. Raghavachari, A. Rendell, J. C. Burant, S. S. Iyengar, J. Tomasi, M. Cossi, N. Rega, N. J. Millam, M. Klene, J. E. Knox, J. B. Cross, V. Bakken, C. Adamo, J. Jaramillo, R. Gomperts, R. E. Stratmann, O. Yazyev, A. J. Austin, R. Cammi, C. Pomelli, J. W. Ochterski, R. L. Martin, K. Morokuma, V. G. Zakrzewski, G. A. Voth, P. Salvador, J. J. Dannenberg, S. Dapprich, A. D. Daniels, Ö. Farkas, J. B. Foresman, J. V. Ortiz, J. Cioslowski and D. J. Fox, 2009.
47. Y. Zhao, J. Pu, B. J. Lynch and D. G. Truhlar, Phys. Chem. Chem. Phys., 2004, 6, 673–676.
48. M. Rafat, M. Devereux and P. L. A. Popelier, J. Mol. Graphics Modell., 2005, 24, 111–120 .
49. M. Rafat and P. L. A. Popelier, J. Comput. Chem., 2007, 28, 2602–2617 .
50. F. M. Aicken and P. L. A. Popelier, Can. J. Chem., 2000, 78, 415–426 .
51. G. J. Bartlett, A. Choudhary, R. T. Raines and D. N. Woolfson, Nat. Chem. Biol., 2010, 6, 615–620 .
52. W. Bolton, Acta Crystallogr., 1963, 17, 147–152 .
53. J. D. Dunitz and W. B. Schweizer, CrystEngComm, 2007, 9, 266.
54. H. L. Williams and C. F. Chabalowski, J. Phys. Chem. A, 2001, 105, 646 .
55. A. Hesselmann, G. Jansen and M. Schütz, J. Chem. Phys., 2005, 122, 014103

CHAPTER 6

THEORETICAL INVESTIGATION OF THE MOLECULAR STRUCTURE, VIBRATIONAL SPECTRA, NMR, UV, NBO ANALYSIS, HOMO AND LUMO ANALYSIS OF 2-(1-PIPERAZINYL) ETHANOL

Mekala R, Mathammal R and Sangeetha M

Department of Physics, Sri Sarada College for Women, Salem, Tamil Nadu, India

ABSTRACT

Quantum chemical calculation of geometries and vibrational wavenumbers of 2-(1-piperazinyl) ethanol in a ground state level are carried out by using density functional theory (DFT/B3LYP) method with 6-31+G(d,p)basis set. The harmonic vibrational frequencies are calculated and scaled values have been compared with experimental FTIR, FT-RAMAN spectra. The stability of the molecule arising from hyper conjugative interaction and the charge delocalization has been analyzed using natural bond orbital (NBO) analysis. A study on the electronic properties such as HOMO and LUMO energies are performed by time-dependent DFT (TD-DFT). NLO property of the title compound is calculated. The 1H and 13C NMR chemical shifts of the molecule are calculated by the gauge independent atomic orbitals method. The mapping of electrostatic potential energy surface (MEP) is performed for the title compound.

INTRODUCTION

2-(1-Piperazinyl) ethanol has a general formula $C_6H_{14}N_2O$ which belongs to the family of **piperazine**. Molecular weight of 2-(1-piperazinyl) ethanol 130.18 g/mol. Piperazine derivatives are widely used in pharmacological and electro optic applications. The piperazine template forms the molecular backbone; possess versatile binding properties with a frequency occurring binding motif which provides potent and selective ligands for a range of different biological targets in medicinal chemistry. Piperazines are currently the most important building blocks in drug discovery, with a high number of positive hits encountered in biological screens of this heterocyclic and its congeners across a number of different therapeutic areas [1,2]; anticancer [3-5], antifungal [6], antibacterial, antimalarial and antipsychotic agents [7], as well as HIV protease inhibitors [8,9]. Piperazines kill tumor cells directly through the induction of apoptosis. Their anti-tumour modes of action are quite distinct and are significantly more potent. They are active against a variety of different tumor types and orally bioavailable [10] piperazine derivatives can act as stimulants at low doses, while causing **hallucinations** at higher doses [11] and are designed as designer drugs.

To the best of our knowledge, neither quantum chemical calculations, nor vibrational analysis study of 2-(1-piperazinyl) ethanol has not been reported yet. DFT technique is employed to study the complete vibrational spectra of the title compound and to identify the various normal modes with greater wave number accuracy. Therefore, in the present study, FTIR and FT-RAMAN spectral analysis of 2-(1-piperazinyl) ethanol, have been recorded and verified by using density functional theory (DFT). UV and NMR studies are also carried out by B3LYP /6-31+G(d,p) basis set. The redistribution of electron

Citation: Mekala R, Mathammal R, Sangeetha M (2015) Theoretical Investigation of the Molecular Structure, Vibrational Spectra, NMR, UV, NBO Analysis, Homo and Lumo Analysis of 2-(1-Piperazinyl) Ethanol J Theor Comput Sci 2: 127. doi: 10.4172/2376-130X.1000127

density (ED) in various bonding, **anti-bonding** and E2 energies had been calculated by natural bonding orbital (NBO) analysis to give clear evidence of stabilization originating from the hyper conjugation of various intra molecular interactions. The HOMO and LUMO analysis have been used to elucidate information regarding charge transfer within the molecule.

Experimental Details

The sample of the present compound 2-(1-piperazinyl) ethanol was purchased from Merck chemical company, with spectroscopic grade and it is used as such without further purification. The FT-IR spectrum of the compound has been recorded in Perkin-Elmer spectrometer between 4000 and 400 cm^{-1}. The spectral resolution is ± 1 cm^{-1}. The FT-Raman spectrum of the compound is also recorded in the same instrument with FRA 106 Raman module equipped with Nd: YAG laser source operating at 1.064 µm line widths with 200 mW powers. The spectra are recorded with scanning speed of 30 cm^{-1}. The frequencies of all sharp bands are accurate to ± 1 cm^{-1}. ^{1}H and ^{13}C nuclear magnetic resonance (**NMR**)(400 MHz; $CDCl_3$) spectra were recorded on a Bruker HC instrument. Chemical shifts for protons are reported in parts per million scales (δ scale) downfield from tetramethylsilane (TMS).

Quantum Chemical Calculations

Calculations of the title compound were carried out with Gaussian 09 software program [12] using the B3LYP/6-31G+(d,p) to predict the molecular structure and vibrational wave numbers. Calculations were carried out with Beck's three parameter hybrid model using the Lee-Yang-Parr correlation functional (B3LYP) method. The structure optimization was performed to confirm the structure and hence to find the optimized geometry of the examined species, no imaginary wave number modes were obtained by B3LYP/6-31+G(d,p). The wavenumbers values computed contain known systematic errors and therefore, have used the constant scaling factor value of 0.956 for DFT method [13]. The DFT hybrid B3LYP functional method tends to overestimate the wavenumbers of functional modes; therefore scaling factors have to be used for obtaining a considerably better agreement with experimental data [13]. The obtained geometrical parameters (B3LYP) are given in Table 1. The assignment of the calculated wave numbers are aided by the animation option of Gauss view program, which gives a visual presentation of the vibrational modes [14].

RESULTS AND DISCUSSION

Molecular Geometry

The stability of the molecule was studied with the help of conformational analysis. Two conformations gauche and anti with respect to the C16-C19 has been carried out. The global minimum energy was found to be for the most stable conformer -421.76758126 Hartee (anti). The optimized geometrical parameters namely bond lengths, bond angles and dihedral angles are calculated by B3LYP/6- 31+G(d,p) which are listed in Table 1. The molecular structure of 2-(1-piperazinyl) ethanol with the atoms numbering are shown in the Figure 1.

Table 1: Optimized geometrical parameters of 2-(1-piperazinyl) ethanol by B3LYP/6-31G+(d,p)

Parameters	Theoritical bond length (Å)
C1-H2	1.0957
C1-H3	1.0973
C1-C4	1.5362
C1-N13	1.466
C4-H5	1.0971
C4-H6	1.1104
C4-N15	1.4614
C7-H8	1.1104
C7-H9	1.097
C7-C10	1.5361
C7-N15	1.4614
C10-H11	1.0973
C10-H12	1.0957
C10-N13	1.466
N13-H14	1.0184
N15-C16	1.459
C16-H17	1.0967
C16-H18	1.0967
C16-C19	1.5364
C19-H20	1.0994
C19-H21	1.0994
C19-O22	1.4304
O22-H23	0.9651
Bond angle (Å)	
H2-C1-H3	107.7412
H2-C1-C4	110.1545
H2-C1-N13	108.7599
H3-C1-C4	108.7978
H3-C1-N13	107.8154
C4-C1-N13	113.397
C1-C4-H5	110.0597
C1-C4-H6	108.6202
C1-C4-N15	109.8313
H5-C4-H6	107.343
H5-C4-N15	108.6565
H6-C4-N15	112.2923
H8-C7-H9	107.3428
H8-C7-C10	108.6196
H8-C7-N15	112.2912
H9-C7-C10	110.0631
H9-C7-N15	108.6577
C10-C7-N15	109.8287
C7-C10-H11	108.7871
C7-C10-H12	110.1583
C7-C10-N13	113.3937

H11-C10-H12	107.742
H11-C10-N13	107.8138
H12-C10-N13	108.7611
C1-N13-C10	111.3619
C1-N13-H14	109.781
C10-N13-H14	109.7811
C4-N15-C7	111.3619
C4-N15-C16	115.437
C7-N15-C16	115.4382
`N15-C16-H17	108.3226
N15-C16-H18	108.3198
N15-C16-C19	116.9573
H17-C16-H18	106.8048
H17-C16-C19	108.0033
H18-C16-C19	107.9999
C16-C19-H20	110.5693
C16-C19-H21	110.5696
C16-C19-O22	106.3501
H20-C19-H21	108.134
H20-C19-O22	110.6175
H21-C19-O22	110.6175
C19-O22-H23	109.3781
H2-C1-C4-H5	64.1556
H2-C1-C4-H6	-53.0962
H2-C1-C4-N15	-176.2596
H3-C1-C4-H5	-53.7368
H3-C1-C4-H6	-170.9887
H3-C1-C4-N15	65.8479
N13-C1-C4-H5	-173.6985
N13-C1-C4-H6	69.0497
N13-C1-C4-N15	-54.1138
H2-C1-N13-C10	173.5921
H2-C1-N13-H14	51.7992
H3-C1-N13-C10	-69.8454
H3-C1-N13-H14	168.3617
C4-C1-N13-C10	50.6738
C4-C1-N13-H14	-71.1191
C1-C4-N15-C7	57.74
C1-C4-N15-C16	-167.5245
H5-C4-N15-C7	178.1783
H5-C4-N15-C16	-47.0862
H6-C4-N15-C7	-63.2369
H6-C4-N15-C16	71.4987
H8-C7-C10-H11	171.0103
H8-C7-C10-H12	53.1152
H8-C7-C10-N13	-69.0327
H9-C7-C10-H11	53.7572
H9-C7-C10-H12	-64.1379
H9-C7-C10-N13	173.7142

N15-C7-C10-H11	-65.8296
N15-C7-C10-H12	176.2753
N15-C7-C10-N13	54.1275
H8-C7-N15-C4	63.2266
H8-C7-N15-C16	-71.5084
H9-C7-N15-C4	-178.1888
H9-C7-N15-C16	47.0763
C10-C7-N15-C4	-57.7471
C10-C7-N15-C16	167.5179
C7-C10-N13-C1	-50.6807
C7-C10-N13-H14	71.1122
H11-C10-N13-C1	69.8345
H11-C10-N13-H14	-168.3727
H12-C10-N13-C1	-173.6024
H12-C10-N13-H14	-51.8096
C4-N15-C16-H17	55.6992
C4-N15-C16-H18	171.196
C4-N15-C16-C19	-66.5558
C7-N15-C16-H17	-171.1758
C7-N15-C16-H18	-55.679
C7-N15-C16-C19	66.5692
N15-C16-C19-H20	-59.8388
N15-C16-C19-H21	59.89
N15-C16-C19-O22	-179.9744
H17-C16-C19- H20	177.7407
H17-C16-C19-H21	-62.5305
H17-C16-C19-O22	57.6051
H18-C16-C19-H20	62.5752
H18-C16-C19-H21	-177.696
H18-C16-C19-O22	-57.5604
C16-C19-O22-H23	-179.9632
H20-C19-O22-H23	59.9325
H21-C19-O22-H23	-59.8585

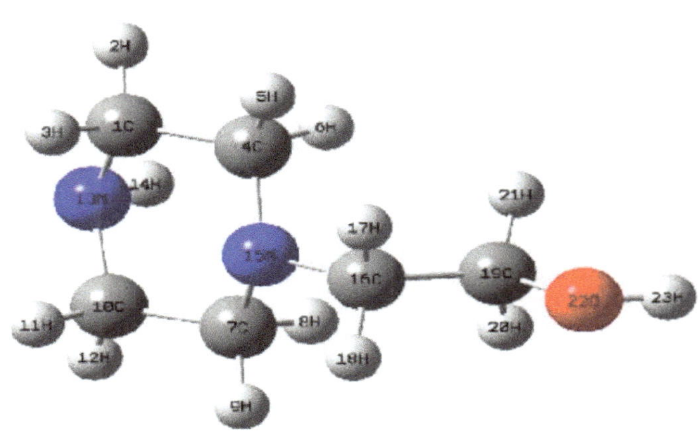

Figure 1: Molecular Structure of 2-(1-piperazinyl) ethanol with atom numbering scheme.

From the table, C-H bonds are having the bond lengths in the range of 1.0957 Å to 1.1104 Å. The presence of the ethanol group attached to the N15 atom has shifted the bond distances of C4-H6 , C7-H8 to higher value (1.1104 Å) The bond lengths of C1-H3 , C7-H_9, C10-H_9 and C10-H11 are greater than the other C-H bonds lengths, shows that the electronegative atoms N_{13} is nearly to those bonds. The C–C bond values are ranging from 1.5361 Å to 1.5364 Å. The highly electro negative oxygen atom distorted the angle C16- C19-H20 and C16-C19-H21 to higher angle (110.56°).

The steric hindrance on N atom and column interaction between the H atoms of CH_2 and CH3 groups give rise to the lone pair to be oriented in axial position while an ethanol group stays at equatorial position. The N-C bond lengths are found in the range of 1.45 Å to 1.465 Å which is comparing shorter than the experimental N-C bond length. This is due to the conjugation of its electro negativity character of the atom O_{22}which is present in the ethanol group. The N-H bond length takes the value 1.0184 Å. The other O-H and C-O bond lengths are found at 0.965 Å and 1.4304 Å respectively.

Piperazine can have different conformations which are chair, halfchair, boat, twist boat and envelope forms. Chair conformation was found to be the most stable conformer and Hendrickson proposed that for chair–chair inter conversion the most stable transition state would be one of the possible chair forms [15-17]. The dihedral angle calculated for C1-C4-N15-C16 is found to be -167.5245. Thus, the obtained optimized geometrical parameters from the Table 1 confirm the chair conformation of 2-(1-piperazinyl) ethanol.

Vibrational Analysis

The aim of the vibrational analysis is to find the vibrational modes connected with calculated specific molecular structure of the compound. The molecule consists of 23 atoms which undergo 63 normal modes of vibration. The title molecule belongs to C1 point group symmetry. Vibrational band assignments of 2-(1-piperazinyl) ethanol have been made by using Gauss view molecular visualization program [18]. After applying the scale factors, the theoretical calculations reproduce the experimental data which are good agreement with literature value. The observed and scaled theoretical frequencies, IR intensities, Raman activities and normal modes of vibrations are listed in Table 2. The theoretical and experimental FTIR and **FT-RAMAN** spectrum are shown in the Figure 2.

Table 2: Observed and calculated vibrational frequencies of 2-(1-piperazinyl) ethanol by B3LYP/631+G (d,P)

S.No	Observed frequencies		Calculated frequencies(*)		IR intensity	Raman	PED (%)
	IR	Raman	Unscaled IR	Scaled IR			
1			70	67	10.91	0.04	τOCHH(63), τHOCC(11)
2			95	91	1.72	0.33	τCCNC(62), τHCNC(18)
3			102	97	0.42	0.32	τOCCN(58), τHOCC(13)
4			228	218	2.43	0.26	τCCNH(33), τHCNC(18)
5			254	243	8.25	0.29	δCCNC(43), τHCNC(14)
6			258	247	114.91	1.68	τHOCC(87), τOCHH(10)
7			312	298.	6.12	2.10	τOCCN(50), τHOCC(30)
8		329	336	321	15.51	0.67	τ HCNC(44), τHHCN(13)
9			371	355	0.22	0.18	τ HHCN(58), τHCNC(28)
10			400	383	11.91	1.48	τHCNC(47), τHCCH(10)
11	460	483	492	471	3.02	2.38	τHCCN(54), τHCCH(23)
12			524	500	0.90	0.59	τCCNH(20), τHCCH(12)
13	585		641	613	35.85	1.45	δCCNC(27), τHCNC(9)
14			773	739	40.85	11.61	τCNHC(44), τHCNC(14)
15			790	755	100.70	4.54	τCNCH(42), τHCNC(20)
16		769	799	764	0.16	0.52	τHCCH(54), τCCNC(17)
17			847	810	0.19	1.18	τ CNCC(17), τHHCN(13)
18		852	896	857	11.81	4.47	τHNCC(13), τHHCN(11)
19	879		911	871	2.78	0.10	βCCN(38), νCC(19) , βCCH(10)
20	936		985	942	44.13	8.78	βCCN(65), νCC(30)
21			1011	967	28.66	7.06	νCC(61), νCN(22), βCH(10)
22			1038	993	2.84	0.34	τHCNC(20), τHHCN(10)
23			1041	995	10.93	1.87	τHCNC(22), τHHCN(13)

24	1011		1061	1014	68.32	10.20	νCO(80), νCN(15)
25	1057	1035	1091	1043	11.94	2.20	νCN(73), βCH(22)
26	1065	1066	1144	1094	12.10	2.55	νCN(70), βCH(13)
27			1144	1094	30.12	0.58	νCN(60), νCC(17), βCH(13)
28	1120		1182	1130	80.05	6.40	νCN(38) , βHCH(18), βCOH(11)
29	1142		1193	1140	9.98	1.36	νCN(41), βHCN(13)
30	1182		1212	1158	6.48	5.30	βHCN(32), νCN(14)
31	1196		1224	1170	21.09	7.23	βCOH(60), βHCH(28), νCN(09)
32	1202	1199	1246	1191	0.68	0.98	νCC(53) , βCCN(18), βHCH(12)
33			1303	1246	5.06	8.74	βCCH(48), νCN(22)
34	1268		1327	1269	7.36	3.83	βCCN(59), νCC(18)
35	1286		1350	1291	1.72	10.9	βHCN(69), βCNC(28)
36			1352	1290	19.58	4.14	βCCN(28), βCNC(10)
37	1297	1299	1356	1296	7.28	0.46	βHCC(58), βHOH(22)
38	1306		1365	1305	4.88	1.88	βHOC(50)
39	1321		1389	1328	1.40	1.40	βHCC(61)
40	1339		1396	1334	8.31	3.58	βHCC(40), νCC(20) , νCN(13)
41	1364		1416	1353	7.07	2.22	νCC(63), βCCN(20)
42			1450	1386	0.81	1.47	βCOH(51), βHCH(13)
43	1407		1477	1412	0.25	20.23	βHCH(79)
44			1480	1415	8.89	3.89	βHNC(47), βHCH(38)
45			1484	1418	8.37	3.79	βHCH RING +ETHANOL(80)
46			1488	1422	4.25	11.60	β HCH RING +ethanol(78)
47			1497	1431	2.72	3.58	βHCH(57), βNHC(21)
48	1445	1447	1503	1437	12.00	2.25	βHCH ring+ethanol(79)
49	1466	1466	1529	1467	3.27	3.71	βHCH ethanol(80)
50	2746		2882	2756	33.66	22.59	νCH(85)
51			2892	2765	145.72	153.17	νCH(83)
52	2879		3002	2870	49.82	112.65	νCH(78)
53			3035	29O2	56.89	88.98	νCH (88)
54			3036	2903	5.04	32.75	νCH(90)
55			3039	2905	36.3817	170.3565	νCH(95)
56			3041	2907	43.7761	48.6382	νCH(92)
57	2917		3059	2924	74.6573	50.1508	νCH(93)
58			3061	2924	32.5528	250.5377	νCH(96)
59	2946	2947	3089	2953	7.3905	122.6584	νCH(98)
60			3089	2953	46.2799	112.7917	νCH(98)
61			3088	2954	66.5198	10.9260	νCH(99)
62	3382		3504	3350	0.6450	73.0198	νNH(100)
63			3837	366	30.0620	164.2640	νOH(100)

(*) –The frequencies calculated for the most stable conformer at the B3LYP/6-31+(d,p).

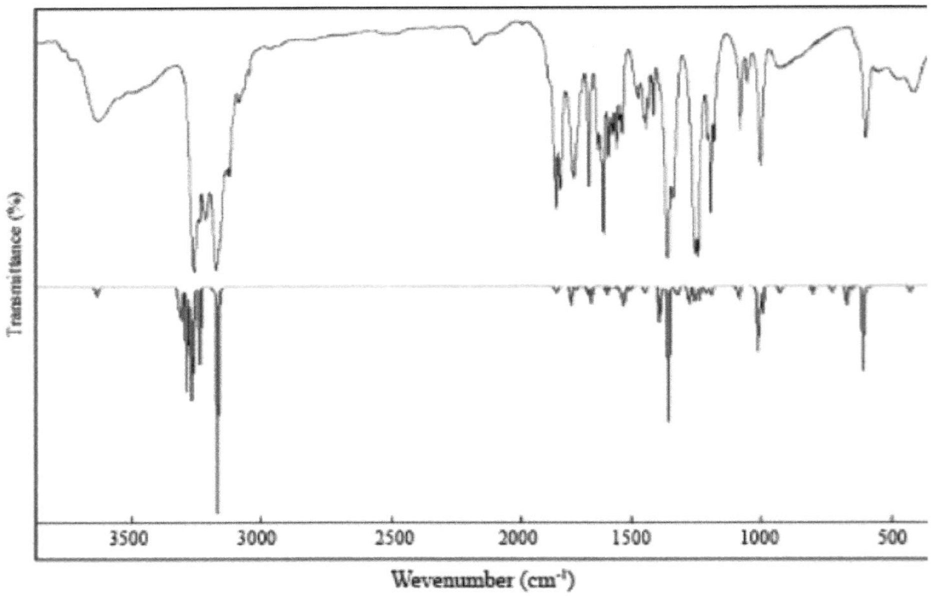

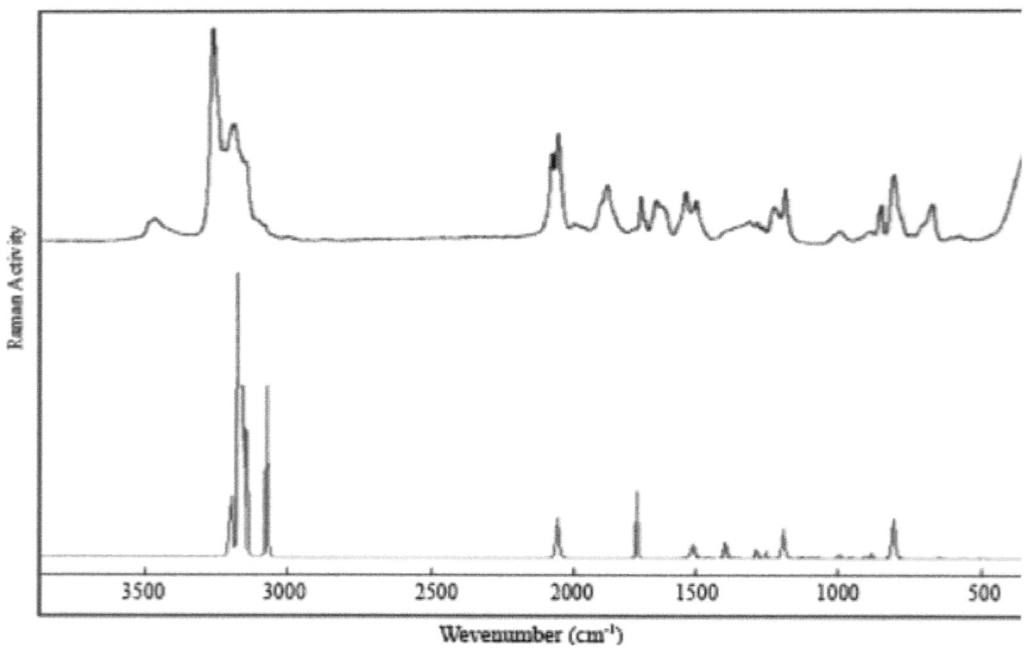

Figure 2: Comparison of experimental and calculated FTIR and FTraman spectrum of 2-(1-piperazinyl) ethanol.

CH_2 Vibrations

For the assignments of CH_2 group frequencies, basically six fundamentals can be associated with each CH_2 group namely, CH_2 symmetric stretch; CH_2 asymmetric stretch; CH_2 scissoring and CH_2 rocking, which belongs to in-plane vibrations and two out-of plane vibrations, viz., CH_2 wagging and CH_2 twisting modes, which are expected to be polarized. The asymmetric CH_2 stretching vibrations are generally observed above 3000 cm^{-1}, while the symmetric stretch will appear between 3000 and 2900 cm^{-1} [19].

The hetero aromatic compounds and its derivatives are structurally very close to benzene. The C-H stretching vibrations of the aromatic and hetero aromatic structures occur in the region 2800-3100 cm^{-1}. This permits the ready identification of the structure. The C-H stretching vibrations are identified at 2954, 2953, 2924, 2907, 2905, 2903, 2902, 2870, 2765, 2756 cm^{-1}. The CH_2 bending vibrations are presented at 1461, 1436, 1431, 1422, 1418, 1412 cm^{-1}. The out of-plane vibrations are also identified and the values are presented in the Table 2. The values of CH_2 vibrations are good agreement with the experimental data and literature [15].

N-H Vibrations

Hetero cyclic compounds containing an N-H group exhibit N-H stretching absorption in the region from 3500-3200 cm^{-1} [20]. Hence, the N-H stretching vibrations of the title compound identified at 3350 cm^{-1}. The in-plane bending vibrations of HCN, HNC, are observed at 1291, 1415 cm^{-1}. The out-of-plane bending vibrations are also calculated and tabulated in Table 2.

Ring Vibrations

With the help of theoretical calculations, C-N, C-C vibrations are identified and assigned in this study. The heterocyclic compound Pyrimidine's absorb strongly at 1600-1500 cm^{-1} due to the C=C and C=N ring stretching vibrations. Absorptions are also observed at 1640-1620 cm^{-1}, 1580-1520 cm^{-1} 1000-960 cm^{-1} and 875-775 cm^{-1} [21]. Hence, the title compound have the C-N stretching modes are calculated at 1043, 1094, 1094, 1130, 1140 cm^{-1}. The calculated C-C stretching vibrations are predicted at 967, 1191, 1335 cm^{-1}. Due to the electron with drawing effect of nitrogen in the ring and oxygen in ethanol, the force constant of C-C bond is decreased and so the stretching frequency is lowered. The in-plane vibrations of HCC, CCN, HCN, CNC, HNC are identified in the range 1158-1415 cm^{-1} and the other values are tabulated in Table 2.

C-O and O-H Vibrations

The C-O stretching vibrations in alcohols and phenols produce a strong band near 1260-1000 cm^{-1} and sensitive to the nature of the substituent's bonded to carbonyl carbon. For ethyl alcohol, the intense broad band near 3360 cm^{-1} represents the hydrogen bonded O-H stretching vibration and the weak intensity bands at 1330 and 1270 cm^{-1} exhibit O-H in plane bending and C-O stretching coupled vibrations. The broad absorption near 650 cm^{-1} is consistent with O-H out-of-plane bending vibration [22,23].

Hence the title compound, observed the calculated value of C-O stretching at 1014 cm^{-1}. The O-H stretching frequency is predicted at 3368 cm^{-1} and in plane bending vibrations of COH, HOC, COH predicted at 1170,1305, 1386 cm^{-1} respectively. The C-O and O-H frequencies are good agreement with available data.

^{13}C NMR and ^{1}H NMR Analysis

The isotropic chemical shifts are frequently used as an aid in identification of reactive organic as well as ionic species. It is recognized that accurate predictions of molecular geometries are essential for reliable calculations of magnetic properties. Gauge-independent atomic orbital (GIAO) ^{1}H and ^{13}C chemical shift values (with respect to TMS) were calculated using the DFT (B3LYP) method with 6-31+G(d,p) basis set and compared with experimental ^{1}H and ^{13}C chemical shift values. The results of this calculation are shown in Table 3 together with the experimental values. The result shows that the range of ^{1}H and ^{13}C NMR chemical shift of the typical organic molecule is usually >100 ppm [24,25] the accuracy ensures reliable interpretation of spectroscopic parameters.

Table 3: Experimental and calculated ^{13}C NMR and ^{1}H NMR chemical shift (ppm) 2-(1-piperzinyl) ethanol

Atoms	Experimental	B3LYP/6-31+G(d,p)
C16	58.04	51.00
C19	60.91	47.1
C7	54.54	45.24
C4	54.54	45.24
C10	45.83	39.87
C1	45.83	39.87
H20	3.63	4.007
H21	3.63	4.007
H17	2.52	3.203
H18	2.52	3.203
H3	2.89	3.108
H11	2.89	3.108
H12	2.89	2.87
H2	2.89	2.87
H5	2.48	2.74
H9	2.48	2.74
H8	2.48	2.48
H6	2.48	2.48
H14	-	0.554
H23	-	0.47

The important aspect is that, hydrogen attached or nearby electron withdrawing atom or group can decrease the shielding and move the resonance of attached proton towards to a higher chemical shift. By contrast electron donating atom or group increases the shielding and moves the resonance towards to a lower chemical shift. In the present investigation, the chemical shift values calculated for hydrogen atom (with respect to TMS) are 4.007 to 0.47 ppm. Due to the electron withdrawing atom of N_{13} and O_{22} which decreases the chemical shift value for the hydrogen atom H14 and H23. The calculated ^{13}C chemical shift values are ranging from (with respect to TMS) 51 to 39.87 ppm. Most of the ^{13}C and ^{1}H chemical shift results match with the experimental data.

Frontier Molecular Orbitals

The highest occupied molecular orbital (HOMO) and lowest unoccupied molecular (LUMO) are named as frontier molecular orbitals (FMO). The FMOs plays an important role in the optical and electrical properties, as well as in quantum chemistry [26]. The HOMO represents the ability to donate an electron, LUMO as an electron acceptor represents the ability to obtain an electron. The energy gap between HOMO and LUMO determines the kinetic stability, chemical reactivity and optical polarizability and chemical hardness-softness of the molecule [27-31]. In order to evaluate the energetic behavior of the title compound, the HOMO and LUMO energy gap was calculated at the B3LYP/6-31+G(d,p) level, which reveals that the energy gap reflects the chemical activity of the molecule. The energies of four important molecular orbitals of the title compound in gas: the second highest and highest occupied MO's (HOMO and HOMO-1), the lowest and the second lowest unoccupied MO's (LUMO and LUMO+1) were calculated. The calculated energy value of HOMO is -5.7044 eV and LUMO is -0.2019 eV in gaseous phase. The value of energy separation between the HOMO and LUMO is 5.5022 eV explains the eventual charge transfer interaction within the molecule, which influences the chemical reactivity of the molecule. The calculated HOMO and LUMO energy values are presented in the Table 4. Consequently, the lowering of the HOMO- LUMO band gap is essentially a consequence of the large stabilization of the LUMO due to strong electron-acceptor ability of the electron group. Other quantum descriptors like electronegativity (χ), chemical hardness (η), electrophilicity (ψ) and softness (ζ) of 2-(1-piperazinyl)ethanol are 5.6035, -5.8053, -0.9652, 0.8612 eV respectively, for the title molecule. The HOMO and LUMO energy diagram is shown in the Figure 3.

Table 4: HOMO-LUMO energy calculated by B3LYP/6-31+G (d,p) for 2-(1-piperzinyl)ethanol

Parameters	B3LYP/6-31+G(d,p)
HOMO	-5.7044
LUMO	-0.2019
HOMO-1	-6.3703
LUMO+1	-0.1801
HOMO-LUMO	5.5022

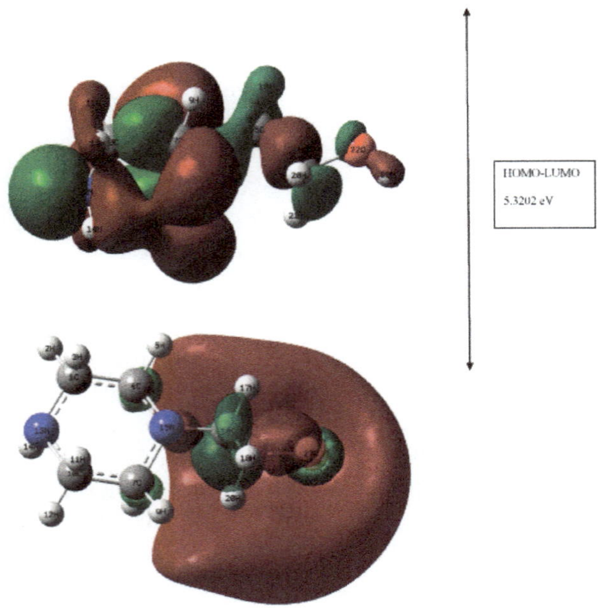

Figure 3: HOMO and LUMO plot of 2-(1-piperazinyl) ethanol.

Analysis of Molecular Electrostatic Potential (MESP), Mulliken Atomic Charges

The molecular electrostatic potential surface (MESP) is a method of mapping electrostatic potential onto the iso-electron density surface simultaneously displays electrostatic potential (electron+nuclei) distribution, molecular shape, size and dipole moments of the molecule and it provides a visual method to understand the relative polarity. Electrostatic potential maps illustrate the charge distributions of molecules three dimensionally. These maps allow us to visualize variably charged regions of a molecule. Knowledge of the charge distributions can be used to determine how molecules interact with one another. One of the purposes of finding the electrostatic potential is to find the reactive site of a molecule. In the electrostatic potential map, the semi-spherical blue shapes that emerge from the edges of the above electrostatic potential map are hydrogen atoms [31]. The molecular electrostatic potential (MEP) at a point r in the space around a molecule (in atomic units) can be expressed as

$$v(r) = \sum Z_A \,|\, R_A - r \,| - \int \rho(r') / \,| r'- \qquad (1)$$

Where, ZA is the charge on nucleus A, is the electronic density function for the molecule. The first and second terms represent the contributions to the potential due to nuclei and electrons, respectively. V(r) is the resultant at each point r, which is the net electrostatic effect produced at the point r by both the electrons and nuclei of the molecule. The total electron density and MESP surfaces of the molecules under investigation are constructed by using B3LYP/6- 31+G(d,p) method. The electrostatic potential increases in the order red<orange<yellow<green<blue. The color code in the range between -5.337 (red) to +5.337 (blue) in the title compound, where the color scheme of MESP in the negative electrostatic potentials are shown in red, the intensity of which is proportional to the absolute value of the potential energy, and positive electrostatic potentials are shown in blue while Green indicates surface areas where the potentials are close to zero. Figure 4 shows the molecular electrostatic potential surface of (2-(1-piperazinyl) ethanol. Local negative electrostatic potential (red) signal indicates in the O_{22} and N_{13} atoms whereas local positive electrostatic potentials (blue) signal indicates in the H23 in ring. Green areas cover parts of the molecule where electrostatic potentials are close to zero (C–C bonds). GAUSSVIEW 5.0.8 visualization program has been utilized to construct the MESP surface.

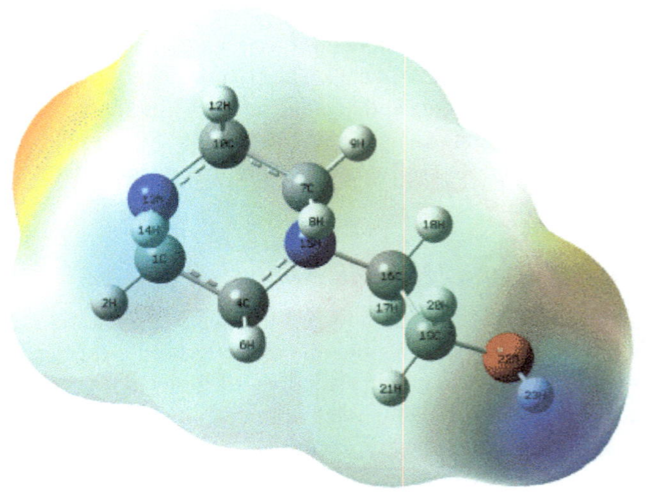

Figure 4: Molecular electrostatic potential of 2-(1-piperazinyl) ethanol.

The Mulliken atomic charge calculation has an impotent role in the application of quantum chemical calculations to molecular system [28,29] because of atomic charges effect dipole moment, molecular polarizability, electronic structure and more a lot of properties of molecular system. The charge distribution over the atoms suggests the formation of donor and acceptor pairs involving the charge transfer in the molecule. The Mulliken population analysis in 2-(1-piperaziniyl) ethanol was calculated using B3LYP level6-31+G (d,p) basis set and are listed in Table 5. All the carbon atoms are having the negative value due to the attachment of the nitrogen and oxygen atoms (N_{13}, N15 and O_{22}). The hydrogen atom H14 and H23 near to N_{13} and O_{22} atoms accommodate higher positive charge than the other hydrogen atoms. This is due to presence of electronegative nitrogen and oxygen atoms, the hydrogen atom attract the positive charge from the nitrogen and oxygen atoms.

Table 5: Mulliken atomic charge and natural atomic charges of 2-(1-piperazinyl) ethanol

Atoms	Mulliken charges	Atomic charges
C1	-0.289	-0.29055
H2	0.132	0.24053
H3	0.152	0.23699
C4	-0.162	-0.27939
H5	0.138	0.24457
H6	0.120	0.19444
C7	-0.162	-0.27939
H8	0.120	0.19444
H9	0.130	0.24457
C10	-0.289	-0.29055
H11	0.152	0.23700
H12	0.132	0.24053
N13	-0.426	-0.71303
H14	0.290	0.38740
N15	-0.124	-0.55738
C16	-0.251	-0.28235
H17	0.147	0.24603
H18	0.148	0.24603
C19	-0.017	-0.13711
H20	0.124	0.20036
H21	0.124	0.20036
O22	-0.528	-0.78573
H23	0.346	0.50218

NBO Analysis

The natural bond orbital (NBO) calculations were performed using NBO 3.1 program [23] as implemented in the Gaussian 09 package at the DFT/B3LYP level in order to understand various second-order interactions between the filled orbital of one subsystem and vacant orbital of another subsystem, which is a measure of the intermolecular delocalization or hyperconjucation. A useful aspect of the NBO method is that it gives information about interactions of the both filled and virtual orbital spaces that could enhance the analysis of intra molecular interactions.

The second order Fock-matrix was carried out to evaluate the donor –acceptor interactions in the NBO basis. The interactions result in a loss of occupancy from the localized NBO of the idealized Lewis structure into an empty non-Lewis orbital. For each donor (i) and acceptor (j) the stabilization energy (E2) associated with the delocalization i→j is determined as qi donor orbital occupancy, E_i, E_j, diagonal elements, F_{ij}, the off diagonal NBO Fock Matrix element.

$$E(2) = \Delta E_{i,j} = q_1 \frac{(F_{i,j})^2}{(E_j - E_i)} \qquad (2)$$

In NBO analysis large E (2) value shows the intensive interaction between electron donor and electron acceptors and greater the extent of conjugation of the whole system, the possible intensive interactions are given in the Table 3. The second order perturbation theory analysis of Fock matrix in NBO basis shows strong intra-molecular hyper conjugative interactions of σ electrons.

The strong intra-molecular hyper conjugative interaction of C1-C4 and C7-C10 from of N_{13}, C4-H6, C7-H8 and C16-C19 from of N15 and also C19-H20 C19-H21 from of O_{22} of which increases ED that weakens the respective bonds leading to the stabilization of ~7.81, 9.28 and 5.74 kcal mol^{-1} respectively.

UV-Spectral Analysis

All the structures allows strong π→π* or σ→σ* transition in the UV visible region with high extinction coefficients. TD-DFT/B3LYP/6- 31+G(d,p) calculations have been used to determine the low-lying excited states of 2-(1-piperazinyl) ethanol. The calculated results involving the vertical excitation energies, oscillator strength (f) and wavelength are carried out and it can be seen in Table 6. The calculated absorption maxima values for 2-(1-piperazinylethanol) have been found to be 255.84, 242.88 and 232.16 nm. The oscillator strength for 232.16 is higher in magnitude. These three absorption bands are mainly derived from the contribution of bands due to π→π* transitions. The theoretical spectrum of UV-VIS is shown in Figure 5.

Table 6: Theoretical electronic absorption spectra value of 2-(1-piperazinyl) ethanol

TD DFT/B3LYP/6-31+G(d,p)		
Wave length λ (nm)	**Excitation energy (eV)**	**Oscillator strength (f)**
232.16	5.3406	0.0225
242.88	5.1048	0.0094
255.84	4.8461	0.0164

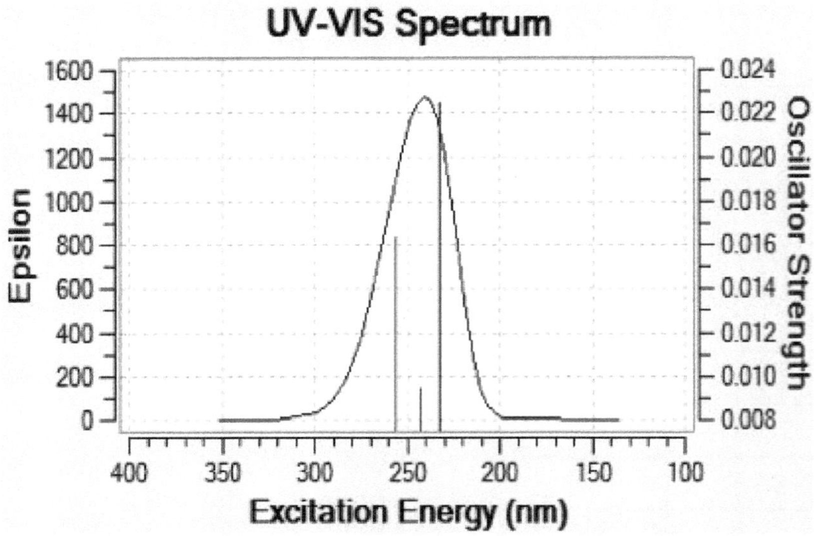

Figure 5: Theoretically calculated UV-vis spectrum of 2-(1-piperazinyl) ethanol.

Polarizability and First Order Hyperpolarizability

In order to investigate the relationships among molecular structures and nonlinear optical properties (NLO), the polarizibilities and first order hyperpolarizibilities of the 2-(1-piperazinylethanol) compound was calculated using DFT/B3LYP method with 6-31+G (d,p) basis set, based on the finite field approach.

The polarizability and hyper polarizability tensors (α_{xx}, α_{xy}, α_{yy}, α_{xz}, α_{yz}, α_{zz} and β_{xxx}, β_{xxy}, β_{xyy}, β_{yyy}, β_{xxz}, β_{xyz}, β_{yyz}, β_{xzz}, β_{yzz}, β_{zzz}) can be obtained by a frequency job output file of Gaussian. The mean polarizability (α_{tot}), anisotropy of polarizability ($\Delta\alpha$) and the average value of the first order hyperpolarizibilities (β_{tot}) can be calculated using the equations.

$$\alpha_{tot} = \alpha_{xx} + \alpha_{yy} + \alpha_{zz} / 3 \qquad (3)$$

$$\Delta\alpha = \frac{1}{\sqrt{2}}[(\alpha_{xx} - \alpha_{yy})^2 + (\alpha_{yy} - \alpha_{zz})^2 + (\alpha_{zz} - \alpha_{xx})^2 + 6\alpha^2_{xx}]^{1/2} \qquad (4)$$

$$\beta = [(\beta_{xxx} + \beta_{xyy} + \beta_{xzz})^2 + (\beta_{yyy} + \beta_{yzz} + \beta_{yxx})^2 + (\beta_{zzz} + \beta_{xxz} + \beta_{yyz})^2]^{1/2} \qquad (5)$$

In Table 7, the calculated parameters described above and electronic dipole moment $\mu i(i=x,y,z)$ and the total dipole moment μ for the title compound are listed. The total dipole moment can be calculated using the following equation.

$$\mu = \sqrt{\mu_x^2 + \mu_y^2 + \mu_z^2} \qquad (6)$$

It is well known that the higher values of dipole moment, molecular polarizibility and first order hyper polarizability are important for more active NLO properties. The first order hyper polarizability (β_{tot}) and the component of hyper polarizability β_x, β_y, β_z of 2-(1-piperazinylethanol) along with the related properties (μ, α_{tot}, $\Delta\alpha$) are reported in Table 7.

Table 7: Electric dipole moment μ (debye), mean polarizibilityα_{tot} ($\times 10^{24}$ e.s.u), anisotropy polarizibility β_{tot}($\times$ 1030 e.s.u) for 2-(1-piperazinyl) ethanol

Parameters	Values
μ_x	0.5692
μ_y	-0.3034
μ_z	-0.7781
μ_{tot}	1.3549
α_{xx}	101.2163
α_{xy}	-1.4535
α_{yy}	88.8266
α_{xz}	8.2137
α_{yz}	-2.6602
α_{zz}	86.2593
α_{tot}	13.49×10^{-24}e.s.u
$\Delta\alpha$	3240.33
β_{xxx}	42.1762
β_{xxy}	-9.8546
β_{xyy}	4.8639
β_{yyy}	-31.615
β_{xxz}	-12.9023
β_{xyz}	-8.7636
β_{yyz}	-41.5162
β_{xzz}	4.4543
β_{yzz}	-4.1116
β_{zzz}	-105.0195
β_{tot}	259.79×10^{-30} e.s.u

The calculated value of dipole moment was found to be 1.3549 Debye. The highest value of dipole moment is observed at 0.5692 Debye for component μ_x and the lowest value of the dipole moment is observed at -0.7781 Debye for component μ_z. The calculated polarizibility and anisotropy of polarizability 2-(1-piperazinylethanol) are 13.6492 × 10-24 e.s.u and 26.056 × 10^{-24} e.s.u., respectively. The magnitude of the molecular hyperpolarizibility (β) is one of the key factors in a NLO system. The B3LYP/6-31+G(d,p) calculated first order hyper polarizability (β_{tot}) value of 2-(1-piperazinylethanol) is equal to 1.499 × 10^{-30}e.s.u, which is 11.5 times that of urea (0.13 × 10-30 e.s.u) [32]. This result clearly indicates that the title compound is a strong candidate to develop a nonlinear optical material.

CONCLUSION

The geometrical parameters of the optimized structure were studied in detail and the influence of the substituted groups is explained. The vibrational frequency analysis by B3LYP method agrees satisfactorily with experimental results. On the basis of calculated potential energy distribution result, assignments of the fundamental vibrational frequencies have been made unambiguously. The HOMOLUMO energy gap and other related molecular properties were discussed and reported. The value of energy gap indicates that it is a good chemical reactive molecule. NBO study reveals that the lone pair orbital participates in electron donation to stabilize the compound. The MEP study shows that the electrophilic attack takes place at the N_{13} and O_{22} position of the title compound. The Mulliken atomic charges and natural atomic charges obtained are tabulated that gives the proper understanding of the atomic theory. Thus the present investigation provides complete vibrational assignments, structural information properties of the compound, ^{1}H and ^{13}C NMR chemical shifts values satisfactorily coincide with experimental results. The UV data indicated that the electronic transition in the compound has π-π* transitions. The values of dipole moment (μ_{tot}), linear polarizability (α_{tot}) and first-order hyperpolarizability (β_{tot}) of the molecule were calculated. It has been found that the value of first-order hyperpolarizability is 11.5 times greater than that of urea, which shows that the molecule is a good NLO material.

REFERENCES

1. Berkheij M, van der Sluis L, Sewing CJ, den Boer D, Terpstra JW, et al.(2005) Tetrahedron Lett 46:2369-2371.
2. Guo CC1, Tong RB, Li KL (2004) Chloroalkylpiperazine and nitrogen mustard porphyrins: synthesis and anticancer activity.Bioorg Med Chem 12: 2469-2475.
3. Rosen K, Weissman SA, Sager J, Reamer RA, Askin D, et al.(1995) Asymmetric hydrogenation of tetrahydropyrazines: Synthesis of (S)-piperazine-2-tert-butylcarboxamide, an intermediate in the preparation of the HIV protease inhibitor indinavir. Tetrahedron Lett 36: 6419-6422.
4. Gillet R1, Jeannesson P, Sefraoui H, Arnould-Guérin ML, Kirkiacharian S, et al. (1998) Piperazine derivatives of butyric acid as differentiating agents in human leukemic cells. Cancer ChemotherPharmacol 41: 252-255.
5. Gabriel FE, Gu J,Slater LM, Hara K, Jacobs JW (2000) Tumor apoptosis induced by epoxide-containing piperazines, a new class of anti-cancer agents. Cancer ChemotherPharmacol 45:183-191.
6. Hulme C, Cherrier MP (1999) Novel applications of ethyl glyoxalate with the Ugi MCR. Tetrahedron Lett 40: 5295-5299.
7. Upadhayaya RS, Sinha N, Jain S, Kishore N, Chandra R, et al. (2004) Optically active antifungal azoles: synthesis and antifungal activity of (2R,3S)-2-(2,4-difluorophenyl)-3-(5-[2-[4-aryl-piperazin-1-yl]-ethyl]-tetrazol-2-yl/1-yl)-1-[1,2,4]-triazol-1-yl-butan-2-ol. Bioorg Med Chem 12: 2225-2238.
8. Chaudhary P, Kumar R, Verma AK, Singh D, Yadav V, et al. (2006) Synthesis and antimicrobial activity of N-alkyl and N-aryl piperazine derivatives. Bioorg Med Chem 14: 1819-1826.
9. Dorsey BD, Levin RB, McDaniel SL, Vacca JP, Guare JP, et al. (1994) L-735,524: the design of a potent and orally bioavailable HIV protease inhibitor. J Med Chem 37: 3443-3451.
10. Askin D, Eng KK, Rossen K, Purick RM, Wells KM, et al. (1994)Highly diastereoselective reaction of a chiral, non-racemic amide enolate with (S)-glycidyltosylate. Synthesis of the orally active HIV-1 protease inhibitor L-735,524. Tetrahedron Lett 35: 673-676.
11. Yarim M1, Koksal M, Durmaz I, Atalay R (2012) Cancer Cell Cytotoxicities of 1-(4-Substitutedbenzoyl)-4-(4-chlorobenzhydryl)piperazine Derivatives.Int J MolSci 13: 8071-8085.
12. Frisch MJ, Trucks GW, Schlegel HB, Scuseria GE, Robb MA, et al. (2009) Gaussian Inc, Wallingford CT.
13. ForesmanJB, Frisch E (1996) Exploring chemistry with Electronic structure method:A Guide to Using Gaussian,Pittsburg,PA.
14. Dennington R, Keith T, Millam J (2009) Gauss view,Version 5,Semichem Inc,Shawnee Mission,KS,2009.
15. Gunasekaran S, Anita B (2008) Spectral investigation and normal coordinate analysis of piperazine. Indian J Pure &ApplPhys 46: 833-838.
16. Hendrickson JB (1961) Molecular Geometry. I. Machine Computation of the Common Rings. J Am ChemSoc 83: 4537- 4547.
17. Lett RG, Petrakis L, Ellis AF, Jensen RK (1970)Nuclear magnetic resonance spectral parameters and ring interconversion of a series of piperazines. J PhysChem 74: 2816-2822.
18. Dennington R,Keith T,Millam J (2007) Gaussview,version4.12 semichemIncShawnee missionK.S.
19. Matulkova I, Nemec I, Teubner K, Nemec P, Micker Z (2008)Novel compounds of 4-amino-1,2,4-triazole with dicarboxylic acids – crystal structures, vibrational spectra and non-linear optical properties. JMolStruct 873: 46-60.

20. Socrates G (2001) Infrared and Raman characteristic group frequencies. Tables and charts. (3rdedn) John Wiley and Sons, Chichester,USA.

21. Socrates G (1987) Infrared characteristic group Frequencies, John Wiley and Sons,Newyork, USA.

22. Mohan J (2001) Organic Spectroscopy principles and applications.(2nd edn) New Age International (P) Limited Publishers,New Delhi.

23. Glendening ED, Reed AE, Carpenter JE, WeinholdF, NBO Version 3.1 Gaussian IncPittsburghPA.

24. Srivastava A, Tandon P, Jain S, Asthana BP(2011) Antagonistic properties of a natural product-Bicuculline with the gamma-aminobutyric acid receptor: studied through electrostatic potential mapping, electronic and vibrational spectra using ab initio and density functional theory. SpectrochimActaA 84: 144-155.

25. Karthick T, Balachandran V, Perumal S, Nataraj A (2013) Vibrational (FT-IR and FT-Raman) spectra and quantum chemical studies on the molecular orbital calculations, chemical reactivity and thermodynamic parameters of 2-chloro-5-(trifluoromethyl) aniline.SpectrochimActa A MolBiomolSpectrosc 107: 72-81.

26. Fleming L (1976) Frontier Orbitals and Organic Chemical reactions, John Wiley & Sons, London.

27. Kosar B1, Albayrak C (2011) Spectroscopic investigations and quantum chemical computational study of (E)-4-methoxy-2-[(p-tolylimino) methyl] phenol.SpectrochimActa A MolBiomolSpectrosc 78: 160-167.

28. Sun YX,Hao QL,Yu ZX,Wei WX,Lu LD, et al. (2009) Experimental and density functional studies on 4-(4-cyanobenzylideneamino)antipyrine. MolPhys107: 223-235.

29. Reed AE, Weinstock RB, Weinhold F (1985) Natural population analysis. J ChemPhys83: 735-746.

30. Adant C,Dupuis M,Bredas JL (2004)Ab initio study of the nonlinear optical properties of urea: Electron correlation and dispersion effectsInt.JQuantum Chem56: 497-507.

31. Arivazhagan M1, Meenakshi R (2011) Quantum chemical studies on structure of 1-3-dibromo-5-chlorobenzene. SpectrochimActa A MolBiomolSpectrosc 82: 316-326.

32. Bevan OJ,Boerio-Goates J (2000) Calculations from statistical Thermodynamics, Acadamic Press.

CHAPTER 7 VIBRATIONAL SPECTRA AND STRUCTURES OF SI_nC CLUSTERS (n = 3–8)

Nguyen Xuan Truong[a], Marco Savoca[a], Dan J. Harding[b,c], André Fielicke[a] and Otto Dopfer[a]

[a]Institut für Optik und Atomare Physik, Technische Universität Berlin, Hardenbergstraße 36, D-10623 Berlin, Germany

[b]Institut für Physikalische Chemie, Georg-August-Universität Göttingen, Tammannstraße 6, D-37077 Göttingen, Germany

[c]Department of Dynamics at Surfaces, Max-Planck-Institut für Biophysikalische Chemie, Am Faßberg 11, D-37077 Göttingen, Germany

ABSTRACT

The effects of doping bare silicon clusters with carbon on their physical properties are of fundamental interest for the chemistry of the interstellar medium and the development of novel nanostructures in materials science. Carbon-doped silicon clusters (Si_nC, n = 3–8) are characterized in the gas phase with infrared-ultraviolet two-color ionization (IR-UV2CI) spectroscopy, mass spectrometry, and quantum chemical calculations. Structural identification is achieved by comparing the measured and calculated vibrational absorption spectra of the low-energy Si_nC isomers identified by global optimization algorithms. Except for planar Si_3C, the most stable Si_nC clusters have three-dimensional configurations. While the Si_3C and Si_6C structures are uniquely assigned, several stable isomers of Si_4C, Si_5C, Si_7C, and Si_8C may co-exist under the present experimental conditions. Interestingly, some of the structures observed here are different from the ground state structures predicted previously. For the small neutral clusters ($n \leq 5$), structures similar to those reported previously for the anions are observed. The highly stable Si_3C unit with a nearly linear Si–C–Si motif is identified as characteristic building block in several of the most stable Si_nC structures. In all identified structures, a large negative charge of almost −2$e$is located on the C atom, indicating its role as electron donor in the Si_n host moiety. The B3LYP/cc-pVTZ level proves reliable in finding the experimentally observed isomers.

INTRODUCTION

Silicon–carbon based nanotubes, nanowires, and clusters are of great interest as promising building blocks for the synthesis of novel materials at the nanoscale.[1] In addition, small silicon–carbon clusters, *e.g.*, SiC_m with m = 1–4, were found in interstellar space.[2–7] Hence understanding the nature of Si–C bonding might, on the one hand, open a way toward engineering of such nanostructures at the atomic level. On the other hand, it is crucial for several chemical processes and observations in the interstellar medium.[8–11] Although located in the same main group of the periodic table, silicon and carbon exhibit very different physical and chemical properties. Silicon prefers multidirectional single bonds, while carbon can easily form single, double, and triple bonds. Generally, substituting a single Si atom with a C atom in a pure Si_n cluster induces strong distortions, but the geometric topology maintains.[12,13] When doped into silicon clusters, the C atom favors locations with more than threefold coordinated bonding in the ground state structures, making most Si_nC_m clusters more stable than their pure silicon counterparts.[13]

There have been numerous theoretical studies on silicon–carbon clusters.[13–33] For small-sized clusters, for example, the

Citation: Truong NX, Savoca M, Harding DJ, Fielicke A, Dopfer O. Vibrational spectra and structures of SinC clusters (n = 3–8). Phys. Chem. Chem. Phys. 2015, 17, 18961-18970, DOI: 10.1039/C5CP02588E.

ground state structures and vibrational spectra of Si_3C and Si_2C_3were determined using Hartree–Fock (HF) and second-order many body perturbation (MP2) theories.[14,15] A clear trend of favorable geometries in the transition from pure carbon to pure silicon clusters was observed, *i.e.*, the chain-like carbon-rich and three-dimensional silicon-rich clusters. This has also been observed for Si_nC_m ($n + m \leq 8$) cluster anions in both theory[32]and experiments.[12] Most close to our work, two lowest-energy Si_4C structures were found at the CCSD (T) level, *i.e.*, the minimum trigonal C_{3v} pyramid and a distorted edge capped C_{2v}pyramid (about 21 kJ mol^{-1} higher).[33] At the HF level, however, the latter structure lies about 8 kJ mol^{-1} lower than the former. Additional calculations of infrared (IR) spectra, isotopic shifts, and dipole moments indeed hinted at their coexistence.[33]

In contrast to theory, spectroscopic characterization of silicon–carbon clusters is largely lacking. Most previous studies have concentrated on charged clusters isolated in the gas phase or neutrals in an inert matrix. For instance, photoelectron (PE) spectra of $Si_nC_m^-$ ($1 \leq n \leq 7$, $1 \leq m \leq 5$) cluster anions found that Si_nC^- ($3 \leq n \leq 7$) and their Si_{n+1}^- counterparts have a similar topology.[12] The PE spectrum of SiC_3^- shows electronic transitions, corresponding to both linear and cyclic forms.[34] However, no conclusion about the true global minimum on the SiC_3potential energy surface could be drawn. With Fourier-transform infrared spectroscopy in Ar matrices combined with quantum chemical calculations, the molecular structures of small Si_nC_m ($n + m \leq 5$) clusters were determined, *e.g.*, a linear centrosymmetric geometry for Si_2C_3 and a rhomboidal C_{2v} structure for Si_3C.[35–38] Furthermore, Si_3C_2 is a pentagon, and Si_2C_4 has a linear symmetric geometry in the ground state.[39] Similarly, C-rich clusters such as SiC_4, SiC_5, SiC_7, and SiC_9, were found to exhibit chain-like structures.[38,40–43] For small Si_nC_m clusters with $n + m \leq 4$, the electronic structure has been studied by resonant photoionization.[11,44,45] In general, little spectroscopic information is available for somewhat larger Si-rich Si_nC_m clusters, which is the topic of the present work.

Through recent advances in the generation of intense IR laser radiation, novel spectroscopic methods for neutral clusters in the gas phase have been developed, such as IRMPD (infrared multiple photon dissociation),[46,47] IR-REMPI (infrared resonance enhanced multiple photon ionization),[48] and IR-UV2CI (infrared-ultraviolet two-color ionization).[49] While IR-REMPI spectra are often broadened due to the absorption of several hundred IR photons, the IR-UV2CI spectra can be better resolved and more closely related to the linear IR absorption spectra. Employing the IR-UV2CI technique for neutral Si_nC_m clusters, we recently observed the evolution from 3D to 1D chain-like configurations in Si_nC_{6-n} clusters by sequentially substituting Si with C atoms in pure Si_6.[50] In addition, the influence of first-row dopant atoms on the properties of bare silicon clusters (Si_6X with X = Be, B, C, N, O) has been determined.[51] As expected, the geometric and electronic properties of silicon containing clusters can substantially be modified by changing the dopant atom.

Here we identify the geometric and electronic structures of Si_nC clusters with n = 3–8 *via* their IR-UV2CI spectra. Combining quantum chemical calculations with global optimization,[52,53] the low-energy Si_nC isomers are determined. Structural identification is achieved through comparison of the measured IR-UV2CI spectra with the IR spectra predicted for the corresponding low-energy isomers. Results for Si_5C and Si_6C have been reported elsewhere,[50,51] but are included here with additional insights for completeness. Electronic properties and charge distributions of selected low-energy isomers will also be discussed.

EXPERIMENTAL AND COMPUTATIONAL METHODS

The IR-UV2CI setup has been described elsewhere.[47,49,50] Briefly, carbon-doped silicon clusters are produced by laser ablation of a pure silicon rod within a pulsed flow of He gas containing 1% CH_4 and thermalized to about 100 K in a liquid-nitrogen cooled expansion channel. After passing through a skimmer, the neutral Si_nC_m clusters are overlapped with counter-propagating IR radiation from the 'Free Electron Laser for Infrared experiments' (FELIX)[54] and then post-ionized by an unfocused F_2 laser (E_{F2} = 7.87 eV) in the extraction zone of a reflectron time-of-flight mass spectrometer. For clusters of specific sizes, with an ionization energy (IE) close to E_{F2}, prior resonant excitation with IR photons from a FELIX pulse raises the internal energy and thereby enhances the efficiency for ionization by the UV laser pulse that is about 30 μs delayed. Resonant excitation of a vibrational mode is followed by rapid internal vibrational energy redistribution, leading to complete thermalization on this experimental timescale. The ionization efficiency usually follows a *S*-curve behavior as a function of excitation energy, with a slope depending on the Franck–Condon factor for ionization. An increase of the internal energy of the cluster upon IR absorption therefore results in an enhancement of the ionization yield and its IR wavelength dependence closely reflects the linear vibrational absorption spectrum of the neutral cluster. The IR-UV2CI spectra in Fig. 1 are obtained from the relative ionization enhancement normalized by the IR photon flux.[55] The observed widths of the bands of 15–45 cm^{-1} arise from a combination of unresolved rotational structure, sequence hot band transitions involving low-frequency modes, the FELIX bandwidth (*ca.* 0.5–1% full width at half maximum (FWHM) of the central wavelength), and possibly the multiple photon absorption process.

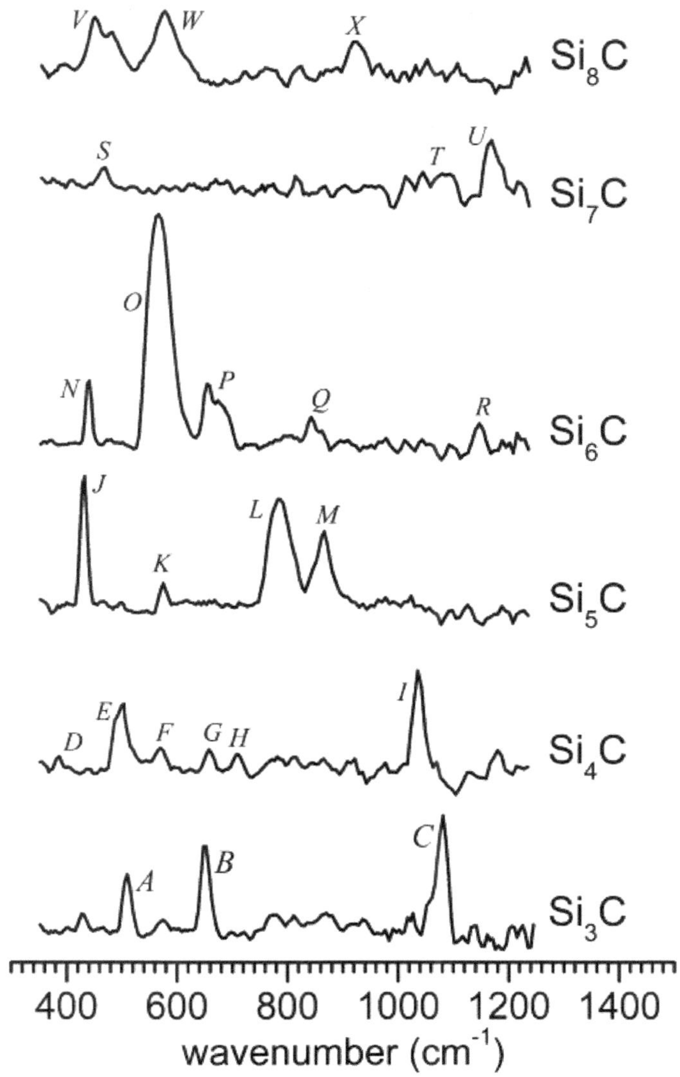

Figure 1: Measured IR-UV2CI spectra of Si_nC (with n = 3–8). All spectra are drawn to the same scale, according to the relative IR enhancement normalized by laser fluence.

Quantum chemical calculations are performed for both singlet and triplet states of each cluster. In an effort to thoroughly explore the complex potential energy surface of the cluster, we employ global optimization (GO) techniques such as the genetic (GA) and basin hopping (BH) algorithms.[56,57] Details of our BH implementation have been given elsewhere.[51,52] Basically, there are two major steps. First, up to five thousand candidate structures are evaluated in terms of the total energy by the GO coupled with DFT calculations at the RI-BP86/def-SVP level as implemented in TURBOMOLE V6.3.1.[58] Second, the first twenty nonequivalent lowest-energy isomers are tightly re-optimized at the B3LYP/cc-pVTZ level including the Grimme dispersion correction (D3)[59] with the resolution-of-the-identity approximation as implemented in GAUSSIAN09.[60] At this level, the linear IR absorption spectra, ionization and binding energies, and natural bond orbital (NBO) populations are obtained. To facilitate convenient comparison with the experimental spectra, the theoretical IR absorption stick spectra are convoluted with a 20 cm^{-1} FWHM Gaussian profile. The reported vibrational frequencies are unscaled. Similar calculations for selected clusters have been performed at the TPSS/cc-pVTZ and MP2/cc-pVTZ levels, which have also proven reliable for doped Si clusters.[47,49,50] The vertical electronic transitions of selected Si_nC clusters are calculated at the second-order approximate coupled cluster RI-CC2/def2-TZVP and TD-DFT/cc-pVTZ levels using TURBOMOLE V6.3.1.[58] As the excited-state calculation for high-symmetry structures (*e.g.*, C_{3v}, C_{4v}, and C_{5v}) is not supported at the CC2 level, the C_1 point group is used for these clusters.

RESULTS AND DISCUSSION

The IR-UV2CI spectra measured for Si_nC (n = 3–8), the linear IR absorption spectra and the geometries of the respective lowest-energy isomers calculated at the B3LYP/cc-pTVZ level are shown in Fig. 1–7. The computational spectra are all drawn to the same linear intensity scale for direct comparison. The relative energies, ionization and binding energies are listed in Tables 1 and 2, respectively. Vertical excitations (Table 3) of selected Si_nC isomers are also provided. Natural bond orbital charges, vibrational and geometric data of all structures considered are available. In the following discussion, the optimized structures obtained at the B3LYP level are used if not stated otherwise.

Table 1: Relative energies (in kJ mol^{-1}) of the most stable Si_nC (n = 3–8) isomers calculated at the B3LYP, TPSS, and MP2 levels using the cc-pVTZ basis set

Cluster	Isomer	Symmetry (State)	Relative energy			Cluster	Isomer	Symmetry (State)	Relative energy		
			B3LYP	TPSS	MP2				B3LYP	TPSS	MP2
a **Relaxing to a (C_{3v}, $^3A_1$) structure.** ***b*** **C_{2v} ($^1A_1$).**											
Si_3C	**3a**	C_{2v} ($^1A_1$)	**0.0**	**0.0**	**0.0**	Si_6C	**6a**	C_{5v} ($^1A_1$)	**0.0**	**0.0**	**0.0**
	3b	C_{2v} ($^3B_1$)	116.6	117.1	187.0		**6b**	C_s ($^1A'$)	12.2	36.8	86.5
	3c	C_s ($^3A''$)	183.1	157.8^a	199.0		**6c**	C_1 (^1A)	29.3	56.1	105.5
							6d	C_s ($^1A'$)	47.2	59.9	86.5
Si_4C	**4a**	C_2 (^1A)	**0.0**	19.9	24.2	Si_7C	**7a**	C_1 (^1A)	**0.0**	7.2	21.0
	4b	C_{3v} ($^1A_1$)	1.1	**0.0**	**0.0**		**7b**	C_1 (^1A)	0.2	6.0	29.2
	4c	C_s ($^1A'$)	18.7	38.3	28.9^b		**7c**	C_{3v} ($^1A_1$)	9.9	8.8	31.2
	4d	C_s ($^1A'$)	27.9	52.9	55.0		**7d**	C_s ($^1A'$)	14.4	**0.0**	**0.0**
						Si_8C	**8a**	C_s ($^1A'$)	**0.0**	**0.0**	64.6
Si_5C	**5a**	C_{2v} ($^1A_1$)	**0.0**	**→5b**	**→5b**		**8b**	C_1 (^1A)	3.1	7.1	**0.0**
	5b	C_{4v} ($^1A_1$)	0.1	**0.0**	**0.0**		**8c**	C_1 (^1A)	17.6	13.8	54.4
	5c	C_s ($^1A'$)	33.2	44.1	76.5		**8d**	C_s ($^1A'$)	22.1	10.6	**0.0**
	5d	C_s ($^3A''$)	123.4	133.9	197.2		**8e**	C_s ($^1A'$)	39.8	18.0	49.7

Table 2: Vertical ionization energy (VIE), adiabatic ionization energy (AIE), and averaged binding energy (BE) of the low-energy Si_nC isomers calculated at the B3LYP, TPSS, and MP2 levels using the cc-pVTZ basis set (separated by semicolons, left to right, respectively)

Cluster	Isomer	VIE [eV]	AIE [eV]	BE [eV]
# calculations failed to converge.				
Si_3C	**3a**	8.13; 8.21; 8.43	7.80; 8.00; 8.12	4.90; 5.14; 5.25
	3b	6.71; 6.89; 6.42	6.64; 6.83; 6.18	4.60; 4.83; 4.76
	3c	7.16; 7.48; 7.10	5.91; 6.43; 6.06	4.43; 4.73; 4.73
Si_4C	**4a**	7.90; 8.02; 8.54	7.30; 7.46; 7.44	4.75; 5.06; 5.15
	4b	8.15; 8.25; 8.72	7.94; 8.11; 8.37	4.75; 5.10; 5.20
	4c	7.46; 7.60; 7.77	7.10; 7.27; 7.39	4.71; 5.02; 5.14
	4d	7.32; 7.45; 7.64	7.17; 7.31; 7.50	4.69; 4.99; 5.08
Si_5C	**5a**	8.19; 8.27; 8.72	7.83; 7.93; 8.61	4.86; 5.26; 5.48
	5b	8.19; 8.21; 8.72	7.95; 7.98; 8.61	4.86; 5.26; 5.48
	5c	8.00; 8.03; 8.74	7.62; 7.67; #	4.81; 5.19; 5.35
	5d	7.51; 7.48; 8.69	6.62; 6.54; #	4.65; 5.03; 5.14
Si_6C	**6a**	7.85; 7.85; 8.16	7.56; 7.69; 8.16	4.73; 5.17; 5.40
	6b	8.00; 7.78; 8.00	7.70; 7.33; 7.36	4.72; 5.12; 5.28
	6c	7.18; 7.58; 7.80	7.04; 7.13; 7.17	4.69; 5.09; 5.25
	6d	7.42; 7.44; 7.68	7.05; 7.10; 7.41	4.66; 5.08; 5.28
Si_7C	**7a**	7.59; 7.66; 8.18	7.32; 7.38; #	4.69; 5.09; 5.27
	7b	7.60; 7.65; 8.17	7.31; 7.39; 7.91	4.69; 5.09; 5.26
	7c	7.81; 7.84; 8.40	7.41; 7.43; 7.76	4.68; 5.09; 5.25
	7d	7.52; 7.53; 8.06	7.07; 7.13; 7.45	4.67; 5.10; 5.30
Si_8C	**8a**	7.52; 7.63; 8.24	7.12; 7.33; 7.73	4.69; 5.12; 5.29
	8b	7.75; 7.73; 8.23	7.48; 7.49; #	4.69; 5.11; 5.37
	8c	7.47; 7.44; 8.49	7.24; 7.25; #	4.67; 5.10; 5.30
	8d	7.47; 7.48; 8.23	7.24; 7.28; 8.08	4.66; 5.11; 5.37
	8e	7.41; 7.42; 7.67	7.18; 7.24; 7.37	4.64; 5.10; 5.31

Table 3: Vertical excitations of selected Si_nC (*n* = 3–8) isomers calculated at the RI-CC2/def2-TZVP and B3LYP/cc-pVTZ levels[a]

Cluster	Isomer (point group, state)	Transition energy [eV]	
		RI-CC2/def2-TZVP	B3LYP/cc-pVTZ
a For structures 4b (C_{3v}), 5b (C_{4v}), and 6a (C_{5v}), calculations with C_1 symmetry are performed at the RI-CC2 level. Numbers in parentheses are the oscillator strengths in 10^{-3}.			
Si_3C	**3a** (C_{2v}, $^1A_1$)	2.08 ($^1B_2$, 3.21)	1.87 ($^1B_2$, 1.54)
		2.83 ($^1A_2$, 0.00)	3.08 ($^1B_2$, 27.82)
		3.09 ($^1A_2$, 0.00)	3.77 ($^1B_2$, 17.18)
Si_4C	**4a** (C_2, ^1A)	2.25 (^1B, 17.7)	2.20 (^1B, 10.39)
		2.49 (^1A, 8.65)	2.39 (^1B, 8.03)
		2.63 (^1B, 4.19)	2.40 (^1A, 5.54)
	4b (C_{3v}, $^1A_1$)	2.77 (^1A, 10.6)	2.57 ($^1A_1$, 2.62)
		2.79 (^1A, 0.00)	2.63 (^1E, 38.28)
		2.83 (^1A, 1.56)	3.19 (^1E, 12.43)
Si_5C	**5a** (C_{2v}, $^1A_1$)	3.40 ($^1A_2$, 0.00)	3.17 ($^1A_1$, 0.15)
		3.51 ($^1A_2$, 0.00)	3.28 ($^1A_1$, 0.18)
		3.57 ($^1A_1$, 0.00)	3.42 ($^1B_2$, 0.03)
	5b (C_{4v}, $^1A_1$)	3.38 (^1A, 0.00)	3.18 ($^1A_1$, 0.24)
		3.49 (^1A, 0.00)	3.49 (^1E, 2.65)
		3.55 (^1A, 0.00)	3.77 (^1E, 26.42)
Si_6C	**6a** (C_{5v}, $^1A_1$)	2.59 (^1A, 0.00)	3.26 ($^1E_1$, 23.87)
		3.47 (^1A, 0.00)	3.44 ($^1A_1$, 12.62)
		3.51 (^1A, 10.17)	3.56 ($^1E_1$, 2.80)
Si_7C	**7a** (C_1, ^1A)	2.03 (^1A, 2.01)	2.22 (^1A, 0.77)
		2.24 (^1A, 0.34)	2.39 (^1A, 0.67)
		2.43 (^1A, 0.39)	2.49 (^1A, 1.36)
	7b (C_1, ^1A)	2.38 (^1A, 0.81)	2.42 (^1A, 0.66)
		2.52 (^1A, 0.35)	2.59 (^1A, 0.41)
		2.68 (^1A, 1.91)	2.68 (^1A, 0.71)
	7c (C_{3v}, $^1A_1$)	2.09 (^1A, 0.36)	1.94 (^1E, 0.05)
		2.48 (^1A, 0.00)	2.43 (^1E, 0.56)
		2.50 (^1A, 0.07)	2.78 (^1E, 1.08)
Si_8C	**8a** (C_s, $^1A'$)	2.19 ($^1A'$, 1.38)	1.99 ($^1A'$, 0.37)
		2.28 ($^1A''$, 14.69)	2.17 ($^1A''$, 0.10)
		2.55 ($^1A'$, 11.90)	2.34 ($^1A''$, 2.40)
	8b (C_1, ^1A)	2.15 (^1A, 0.00)	2.12 (^1A, 0.53)
		2.32 (^1A, 2.29)	2.24 (^1A, 2.86)
		2.51 (^1A, 8.64)	2.47 (^1A, 1.71)
	8c (C_1, ^1A)	1.65 (^1A, 0.54)	2.07 (^1A, 2.34)
		2.08 (^1A, 2.91)	2.29 (^1A, 4.29)
		2.14 (^1A, 2.83)	2.42 (^1A, 1.22)

Vibrational Spectra and Geometries

Si_3C. The experimental IR-UV2CI spectrum of Si_3C is compared in Fig. 2 with the linear absorption spectra calculated for the first three lowest-energy isomers 3a–3c. Three bands are observed at 509 (A), 651 (B), and 1080 cm^{-1} (C), which can readily be explained with structure 3a. In the investigated spectral range (350–1250 cm^{-1}), 3a exhibits significant absorptions at 509 (a_1), 652 (a_1), and 1109 cm^{-1} (b_2), arising from the Si–C–Si symmetric stretching mode, the symmetric breathing mode, and the Si–C–Si asymmetric stretching mode, respectively. Our results are consistent with previous work based on Ar-matrix Fourier-transform IR spectroscopy combined with *ab initio* calculations[37] and excited-state measurements.[44,45] In the former, five of the six vibrational fundamental frequencies (in cm^{-1}) were determined as 309.5 (a_1), 357.6 (b_2), 511.8 (a_1), 658.2 (a_1), and 1101.4 (b_2). The shifts from the IR-UC2CI data are rather small and may be related to matrix effects and/or the multiphotonic IR excitation in the gas-phase experiments.

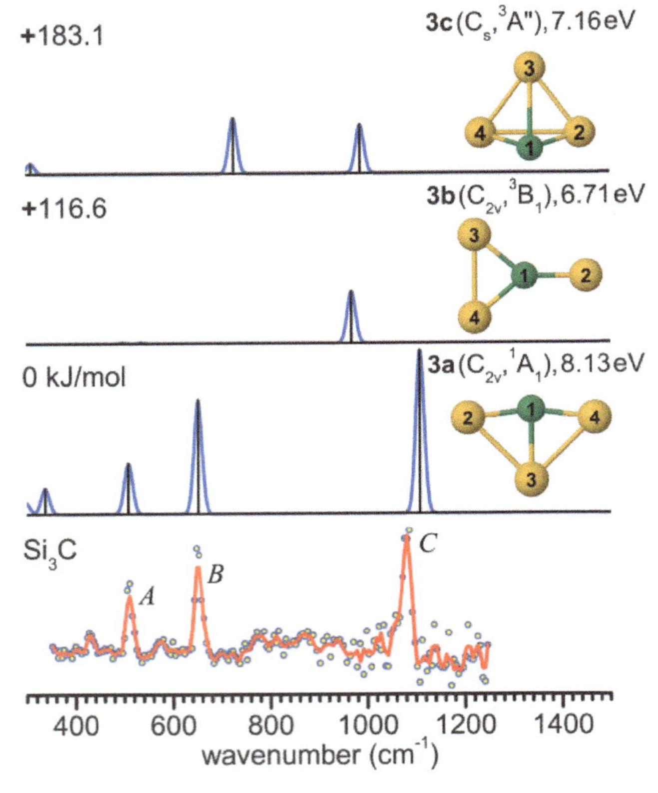

Figure 2: Comparison of the experimental IR-UV2CI spectrum of Si_3C (bottom) with linear IR spectra of the three lowest-energy isomers (3a–3c) calculated at the B3LYP/cc-pVTZ level. Molecular symmetries and electronic states are given in parentheses, while vertical ionization energies are given in eV. Relative energies are given in kJ mol^{-1}. The red line is the three-point adjacent average of the original data (circles). The experimental peak positions (in cm^{-1}) are 509 (A), 651 (B) and 1080 (C).

Structure 3a has been predicted as the ground state of Si_3C.[13,14,22,61,62] While the ground state of neutral Si_4 has a D_{2h} (^1A_g) rhomboidal geometry,[63–65] 3a is a distorted rhombus (C_{2v}, $^1A_1$) with shorter and stronger C–Si bonds and a nearly linear Si–C–Si bridge (∠4Si–1C–2Si = 164.9°). The calculated C–Si bond lengths of 1.764 and 1.944 Å are close to the previous results.[44] The anion counterpart Si_3C^- has been concluded to be a distorted planar rhombus (C_{2v}, $^2A_2$), due to the similarity between the PE spectra of Si_3C^- and Si_4^-.[11,12,61] The good agreement observed between the measured and predicted frequencies of Si_3C in Fig. 2 supports the use of the B3LYP/cc-pVTZ level for the structural and vibrational assignment of the larger carbon-doped silicon clusters described below. The next two stable isomers 3b (C_{2v}, $^3B_1$) and 3c (C_s, $^3A''$) are triplet states with relative energies well above 100 kJ mol^{-1} (Fig. 2). Isomer 3bexhibits a planar Y-shaped geometry, while 3c is a distorted triangular pyramid. At the MP2 level, isomer 3c relaxes to a C_{3v} ($^3A_1$) distorted tetrahedron.

Si_4C. Fig. 3 compares the measured IR-UV2CI spectrum of Si_4C with linear IR spectra predicted for the four lowest-energy isomers 4a–4d. Six distinct bands are observed at 388 (D), 499 (E), 571 (F), 659 (G), 710 (H), and 1037 cm^{-1} (I). Except for band F, all features can be explained with the fundamental modes of the two most stable isomers **4a** and 4b. Specifically, bands D and E are assigned to Si–C–Si bending modes of 4a predicted at 378 (a) and 490 cm^{-1} (a), while band I is attributed to the asymmetric stretching of the nearly linear 4Si–1C–5Si moiety (176.8°) at 1047 cm^{-1} (b). Bands G and H with similar peak intensities are assigned to the breathing mode at 660 cm^{-1} (a_1) and the asymmetric Si–C stretching mode at 721 cm^{-1} (e) of isomer **4b**. In addition, the breathing mode at 395 cm^{-1} (a_1) of 4b fits well to band D with a small redshift of about 10 cm^{-1}.

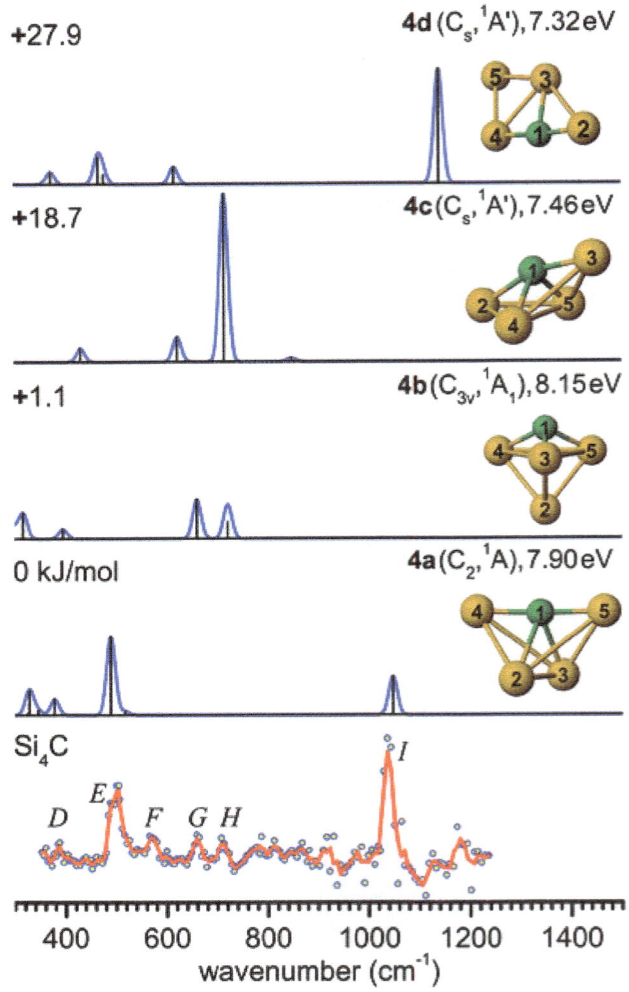

Figure 3: Comparison of the experimental IR-UV2CI spectrum of Si_4C (bottom) with linear vibrational spectra of the four lowest-energy isomers (**4a–4d**) calculated at the B3LYP/cc-pVTZ level. Molecular symmetries and electronic states are given in parentheses. Relative energies are given in kJ mol^{-1}, while vertical ionization energies are given in eV. The red line is the three-point adjacent average of the original data (circles). The experimental peak positions (in cm^{-1}) are 388 (D), 499 (E), 571 (F), 659 (G), 710 (H), and 1037 (I).

Interestingly, isomer 4b (C_{3v}, $^1A_1$) is often predicted as the ground state structure of Si_4C,[12,13,61,64] as also obtained here with the TPSS and MP2 methods (Table 1). This structure is formed by replacing the Si atom at one of the apexes of the trigonal bipyramidal Si_5 (D_{3h}, $^1A_1'$) cluster.[66] However, the energy difference between 4a and 4b is known to be strongly dependent on the theoretical level.[33] For example, the present B3LYP calculations predict a lower-symmetry ground state structure 4a (C_2, ^{1}A), and the next stable structures are 1.1 (4b), 18.7 (4c, C_s, ^{1}A′), and 27.9 kJ mol^{-1} (4d, C_s, ^{1}A′) higher in energy. Structure 4a with an estimated IE of about 7.9 eV (Table 2) is expected to yield a somewhat different IR-UVC2I enhancement than 4b with a slightly higher IE (8.15 eV).

Nonetheless, the observed relative band intensities in Fig. 3 suggest that the population of both isomers are of similar order of magnitude. The ground state structure of Si_4C (4a) is similar to the most stable structure of its anion counterpart Si_4C^- (C_{2v}, $^2A_1$) predicted at the MP2/6-31G* level.[12,61] A significant contribution of 4c and 4d to the IR-UV2CI spectrum is ruled out, because of their substantially higher energies and the mismatch of their predicted IR spectrum with the measured one. Finally, we note that the Si_4C structures assigned from the IR-UV2CI spectrum are different from those postulated from the analysis of the photodetachment experiments of the corresponding anions.[61] This example illustrates the importance of the direct spectroscopic study of the neutral clusters for the reliable experimental determination of the true ground state.

Si_5C. The experimental IR-UV2CI spectrum of Si_5C[50] is compared in Fig. 4 to the linear IR spectra of the four lowest-energy isomers (5a–5d). Four bands are observed at 427 (J), 573 (K), 785 (L), and 868 cm^{-1} (M), which are mostly related to structures 5a/b. The ground state structure of Si_5C has widely been predicted as a distorted tetragonal bipyramid 5b (C_{4v}, $^1A_1$), which has a similar geometry as the corresponding Si_6 (D_{4h}, $^1A_{1g}$) cluster.[65–67] This scenario is also supported by our TPSS and MP2 calculations (Table 1), and the IR-UV2CI spectrum has indeed been assigned to **5b** in our previous study by the comparison with the calculated TPSS spectrum.[50] The bands J, K, and L, agree well with the predicted fundamentals, while band M has been explained with the overtone of J,[50] although its intensity is surprisingly high for an overtone.

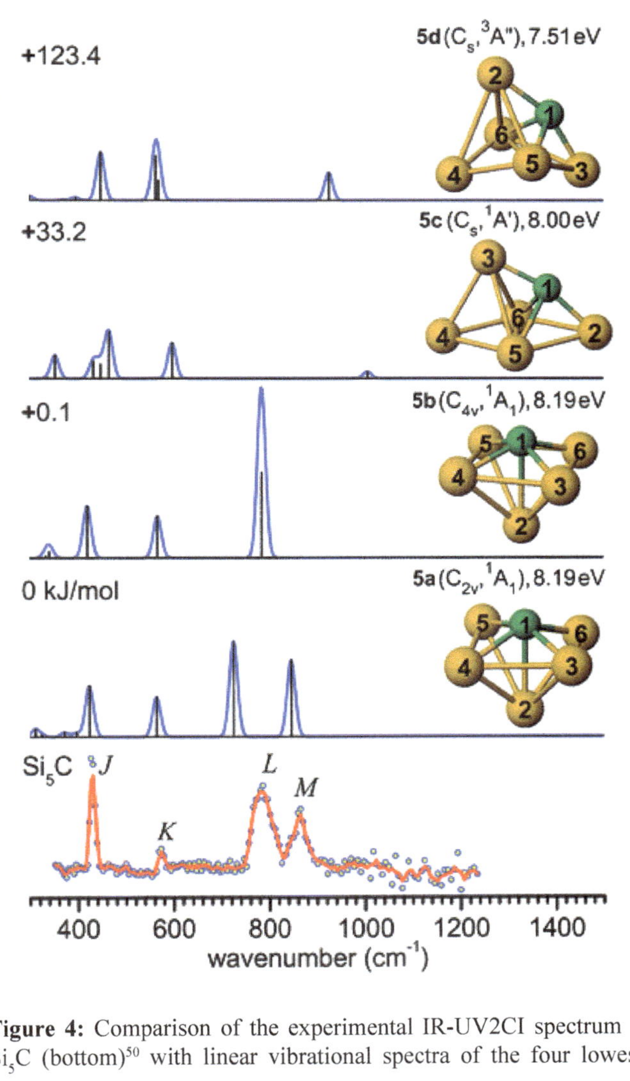

Figure 4: Comparison of the experimental IR-UV2CI spectrum of Si_5C (bottom)[50] with linear vibrational spectra of the four lowest-energy isomers (**5a–5d**) calculated at the B3LYP/cc-pVTZ level. Molecular symmetries and electronic states are given in parentheses, while vertical ionization energies are given in eV. Relative energies are given in kJ mol^{-1}. The red line is the three-point adjacent average of the original data (circles). The experimental peak positions (in cm^{-1}) are 430 (J), 574 (K), 786 (L), and 863 (M).

The B3LYP/cc-pVTZ level suggests an alternative scenario. In contrast to the MP2 and TPSS levels, B3LYP yields a ground state structure 5a (C_{2v}, $^1A_1$) resulting from a distortion of the C_{4v} cage 5b, along with an energy lowering of only 0.1 kJ mol^{-1}, while 5b is a low-lying transition state, leading to a fluxional structure. Similar to 5a, the low frequency bands J and K can again be assigned to deformation and breathing modes at 424 (a_1) and 565 cm^{-1} (a_1). While in the higher frequency range, the 5b spectrum shows a single intense degenerate peak at 785 cm^{-1} (e) in excellent agreement with band L, 5a exhibits two split vibrational modes at 726 cm^{-1} (b_1, asymmetric Si–C stretching of 4Si–1C–6Si) and 847 cm^{-1} (b_2, asymmetric Si–C stretching of 3Si–1C–5Si). These may be assigned to bands L and M, however with unusually large redshifts of up to 50 cm^{-1} from experiment, which possibly arise from unusual anharmonicity along the distortion coordinate.

Similar findings were noted for isovalent Si_6.[47,49] Two possible ground state structures have also been proposed for Si_6 showing the best agreement with the measured IR-UV2CI spectra, *i.e.*, a C_{2v} ($^1A_1$) structure found by DFT and a D_{4h} ($^1A_{1g}$) structure predicted with MP2. This difference has been reasoned as distortions in Si_6 induced by a pseudo Jahn–Teller effect.[67,68] It was impossible to draw a firm conclusion on the true ground state structure of Si_6, as a potential band splitting for the C_{2v} structure was unresolved experimentally. For Si_5C, the splitting is larger (~120 cm^{-1}) and hence observable. More theoretical efforts need to be devoted to the pseudo Jahn–Teller effects in Si_6 and Si_5C to fully understand the measured IR spectra. Such efforts are, however, beyond the scope of the present work.

For comparison, quantum chemical calculations for Si_5C^- have indicated a C_{2v} ($^2A_1$) ground state for the anion, but the PE spectrum has been interpreted to mainly originate from a C_s($^2A'$) anion into a C_s ($^1A'$) neutral state (corresponding to 5c).[12,61] Similar to Si_4C, the structures of the neutral Si_5C clusters derived from the IR and PE spectra are different, due to the vertical transitions favored by the Franck–Condon principle.

Si_6C. The results for Si_6C published recently[51] are included here for completeness. Fig. 5 compares the IR-UV2CI spectrum of Si_6C with linear IR spectra of the four lowest-energy isomers (6a–6d). Five bands are detected at 438 (N), 566 (O), 664 (P), 842 (Q) and 1137 cm^{-1} (R). All these features can readily be explained by structure 6a (C_{5v}, $^1A_1$) with the C atom at the apex of a pentagonal bipyramid, which has widely been accepted as the ground state structure of Si_6C.[13,22,32,51] Its shape is similar to the D_{5h} ($^1A_1'$) ground state of Si_7 determined in experiments[49,66,69] and theory.[63,66,70] Band N is assigned to the breathing mode at 425 cm^{-1} (a_1), while band O is assigned to the asymmetric and symmetric Si–C stretching modes at 544 cm^{-1} (e_1) and 550 cm^{-1} (a_1), respectively. A fundamental mode near 660 cm^{-1} (P) is not predicted for any of the low-energy isomers. Therefore, we tentatively assign it to a combination band of the a_1 mode predicted at 425 cm^{-1} with the low frequency modes at 226 (e_1) or 257 cm^{-1} (a_1). Similarly, the weaker band Q might be an overtone of the mode at 425 cm^{-1} (a_1), and band *R* may be related to an overtone and/or combination of the modes at 544 cm^{-1} (e_1) and 550 cm^{-1} (a_1).

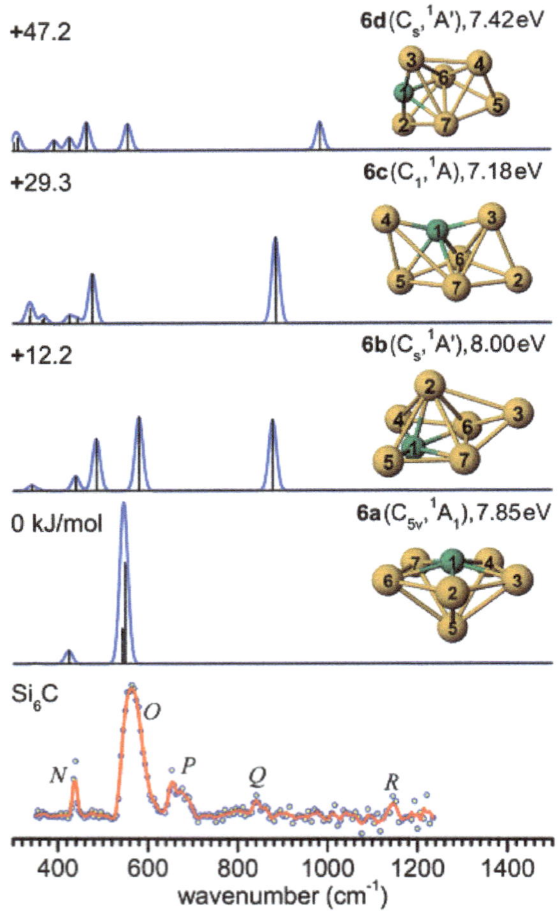

Figure 5: Comparison of the experimental IR-UV2CI spectrum of Si_6C (bottom) with linear vibrational spectra of the four lowest-energy isomers (**6a–6d**) calculated at the B3LYP/cc-pVTZ level. Molecular symmetries and electronic states are given in parentheses. Relative energies are given in kJ mol^{-1}, while vertical ionization energies are given in eV. The red line is the three-point adjacent average of the original data (circles). The experimental peak positions (in cm^{-1}) are 438 (N), 566 (O), 664 (P), 842 (Q), and 1137 (R).

Si_7C. The IR-UV2CI spectrum of Si_7C and the linear absorption spectra and geometries of the four lowest-energy isomers (7a–7d) are displayed in Fig. 6. Three bands are observed at 464 (S), 1075 (T), and 1163 cm^{-1} (U). The two intense bands S and U can be explained with structure 7b (C_1, ^{1}A), which at the B3LYP level is slightly higher in energy (0.2 kJ mol^{-1}) than the most stable one, 7a (C_1, ^{1}A). Band S is assigned to the Si–C–Si bending mode at 461 cm^{-1} (a), while band U is attributed to the asymmetric Si–C–Si stretching mode at 1195 cm^{-1} (a). Although structures 7a and 7c (C_{3v}, $^1A_1$) also exhibit IR active modes close to band S, neither of their intense bands at 680 (7a) and 768 cm^{-1} (7c) is observed in the experiment. We therefore exclude their substantial contribution to the observed spectrum. Actually, closer inspection of Fig. 6 reveals that the small signal at 678 cm^{-1} marked by an arrow may be attributed to 7a. Indeed, band T fits well to the asymmetric Si–C–Si stretching mode of **7a** calculated at 1103 cm^{-1} (a). Similar to 3a and 4a, the Si–C stretching modes at high frequency are characterized by the nearly linear Si–C–Si motifs present in 7a (∠5Si–1C–3Si = 175.8°) and 7b (∠4Si–1C–3Si = 168.4°). Interestingly, while 7d (C_s, ^{1}A′) has been predicted as the ground state structure of Si_7C (see ref. 22 and 32 and our TPSS and MP2 calculations), no signal of its most intense stretching mode at 526 cm^{-1} (a″) is observed in the IR-UV2CI spectrum. This structure is obtained by substituting a Si atom by a C atom in a deformed bicapped Si_8 octahedron (C_{2h}, ^1A_g).[47,70,71] We note that the relative energies of the four considered isomers 7a–7dare quite small (<15 kJ mol^{-1} at the B3LYP and TPSS levels, Table 1). Their calculated IE values (7.5–7.8 eV, Table 2) are relatively low compared to the photon energy of the F_2 laser. This may strongly affect the possibilities of detecting substantial enhancements in the ionization yield upon resonant IR absorption.[47]

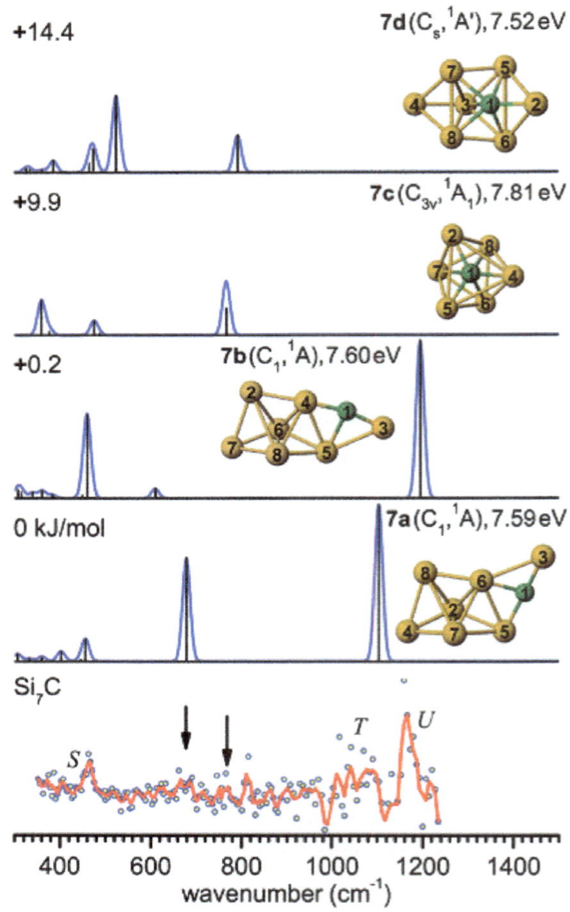

Figure 6: Comparison of the experimental IR-UV2CI spectrum of Si_7C (bottom) with linear vibrational spectra of the four lowest-energy isomers (**7a–7d**) calculated at the B3LYP/cc-pVTZ level. Molecular symmetries and electronic states are given in parentheses. Relative energies are given in kJ mol^{-1}, while vertical ionization energies are given in eV. The red line is the three-point adjacent average of the original data (circles). The experimental peak positions (in cm^{-1}) are 464 (S), 1075 (T) and 1163 (U). Arrows indicate weak bands at 678 and 769 cm^{-1}, which might be attributed to the modes at 679 (**7a**) and 768 cm^{-1} (**7c**), respectively.

Si_8C. The experimental IR-UV2CI spectrum of Si_8C and the calculated linear IR spectra of the five lowest-energy isomers 8a–8e are compared in Fig. 7. The present B3LYP and TPSS calculations predict 8a (C_s, ^{1}A′) as the most stable structure, which is in agreement with previous work.[13] Structure 8a is analogous to its Si_9 counterpart (C_s, ^{1}A′).[47,70] Other stable isomers (8b–8e) higher in energy have similar topological geometries. The IR-UV2CI spectrum of Si_8C shows three major bands at 449 (V), 574 (W), and 922 cm^{-1} (X). These features can readily be explained with isomer 8b, which is only 1.3 kJ mol^{-1} above the putative global minimum 8a. Specifically, band V and W are assigned to Si–C–Si bending modes at 452 cm^{-1} (a) and 563 cm^{-1} (a), respectively. Band X is attributed to the asymmetric Si–C–Si stretching mode at 962 cm^{-1} (a) of the almost linear Si–C–Si unit (∠7Si–1C–6Si = 160°).

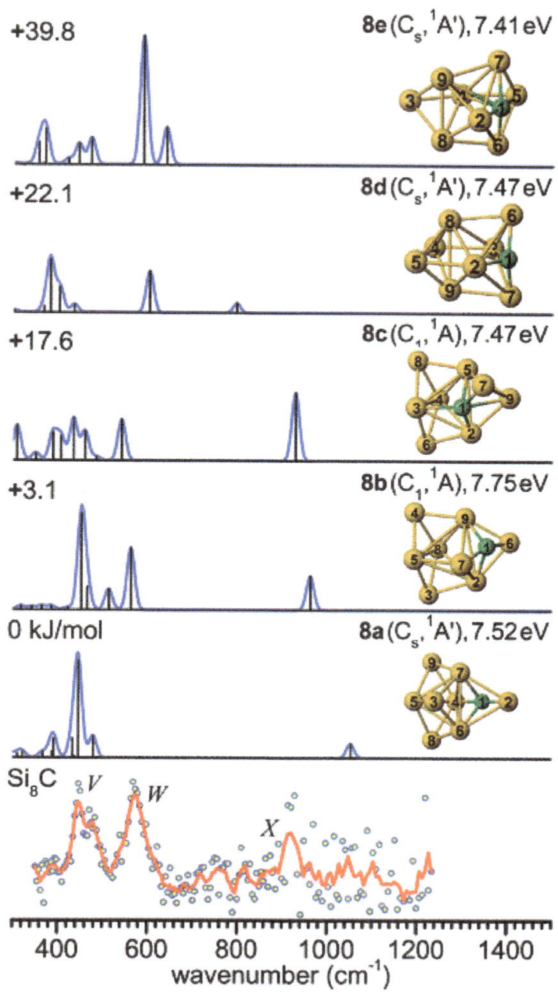

Figure 7: Comparison of the experimental IR-UV2CI spectrum of Si_8C (bottom) with linear vibrational spectra of the five lowest-energy isomers (8a–8e) calculated at the B3LYP/cc-pVTZ level. Molecular symmetries and electronic states are given in parentheses. Relative energies are given in kJ mol^{-1}, while vertical ionization energies are given in eV. The red line is the three-point adjacent average of the original data (circles). The experimental peak positions (in cm^{-1}) are 449 (V), 574 (W), and 922 (X).

The main feature in the IR spectrum of isomer 8a at 446 (a′) coincides with band V. Interestingly, structure 8a also contains the linear Si–C–Si block with ∠4Si–1C–2Si = 179.5°. The corresponding asymmetric Si–C–Si stretching mode at 1051.7 cm^{-1} (a′) is, however, weak. Therefore, a contribution of this isomer to the measured spectrum cannot be ruled out from the calculated IR spectrum. However, its calculated IE value is rather low (7.52 eV), so that the IR-UV2CI enhancement using 7.87 eV photons is probably weak. Similar conclusions apply to all higher energy isomers (7.41–7.47 eV for 8c–8e) with the notable exception of the identified isomer 8b (7.75 eV). Hence, our favored assignment of the measured spectrum is mainly to isomer 8b, while the presence of further isomers in the beam cannot be excluded. No other experimental information about Si_8C is available for comparison.

Structural and Electronic Properties

The evolution of structures with increasing size for Si_nC (n = 3–8) is summarized in the left part of Fig. 8 showing the isomers that explain the major features of the experimental spectra. Most of these structures contain a Si_3C substructure with a (nearly) linear Si–C–Si unit, except Si_5C and Si_6C. Interestingly, the Si_3C cluster (along with Si_2C) has been identified as a particular stable size in fragmentation experiments.[72,73] This stability, thermodynamically and as structural element, may be explained in terms of its valence structure as shown in Fig. 8(right). Boys localization of the eight valence orbitals reveals the presence of three lone pairs, each at one of the Si atoms, two Si–C σ-bonds in the Si–C–Si unit, as well as three delocalized bonds, one of them being a π-bond delocalized over the entire cluster. The resulting formal bond orders range from 0.53 between neighboring Si atoms to 1.53 within the linear Si–C–Si unit. The high bond order in the Si–C–Si unit goes along with relatively short Si–C bond lengths of 176 pm as compared to the third Si–C bond, which is 194 pm long and has a formal bond order of only 0.87. The larger clusters also have similar bond lengths within the Si–C–Si bridge unit. The average of the corresponding Si–C distances in this unit is a measure of its strength and can be directly related with the frequency of the asymmetric Si–C–Si stretching mode observed experimentally. For the structures in which we identify this element (3a, 4a, 7b, 8b) this particular mode is found at 1080, 1037, 1163 and 922 cm^{-1}, respectively, and the average Si–C distances are 176, 178, 174 and 183 pm, *i.e.* there is a clear anti-correlation between bond length and stretching frequency.

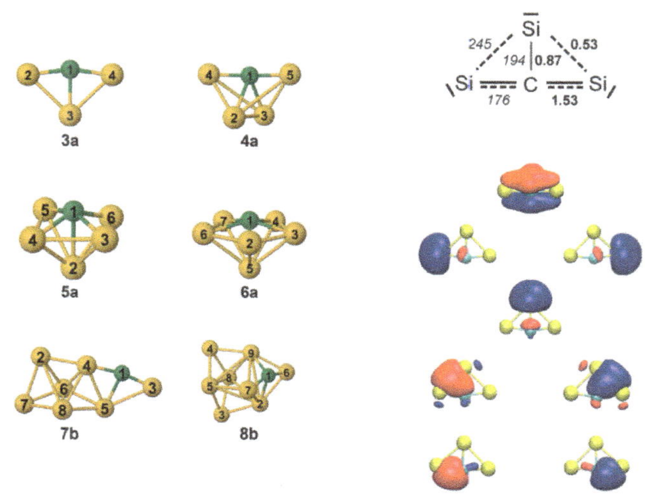

Figure 8: (left) Overview of the structures observed for Si_nC (n = 3–8). Most of these structures contain a Si_3C substructure with a (nearly) linear Si–C–Si unit, except Si_5C and Si_6C. (right) Valence structure and localized molecular orbitals for Si_3C as obtained from a Boys localization. Formal bond orders (bold) are compared to the bond lengths in pm (italics).

The NBO charge analysis is increasingly used for a view of local charge instead of the unreliable Mulliken population analysis. As expected for Si_nC clusters, a large negative charge is located at the C atom, which is more electronegative than Si. The positive part is mainly shared by the Si atoms directly bound to the C atom. This result suggests that these Si atoms are very good electron donors, as observed previously for the first-row element doped silicon clusters.[51]

Vertical excitations of selected stable clusters calculated with the second-order approximate coupled cluster and TD-DFT methods are listed in Table 3 for the first three transitions. The first electronic transition energies for the considered clusters are relatively high (>1.65 eV).

CONCLUSIONS

Carbon-doped silicon clusters are characterized with IR-UV2CI spectroscopy combined with quantum chemical simulations. The structures of the Si_nC (n = 3–8) clusters are determined by comparison of the measured IR-UV2CI spectra with linear IR absorption spectra of the low-energy isomers calculated at the B3LYP/cc-pVTZ level. Optimized stable structures are found employing global optimization algorithms coupled with electronic structure (DFT) methods. This technique is robust and reliable in searching for the ground state structures of such type of clusters, in which electronic effects might play an important role. For Si_3C and Si_6C, only the lowest-energy isomers are present in the prepared cluster sample, while for Si_4C, Si_5C, Si_7C, and Si_8C multiple stable structures may co-exist. The nearly linear motif (Si–C–Si) is identified as characteristic building block in several of the most stable Si_nC structures (n = 3–5, 7, 8). The NBO analysis shows that most charge is located on the C atom and its nearest-neighbor Si atoms. Although the theoretical approaches used in this work agree well on predicting low-energy isomers, the B3LYP method proves to be the most reliable one in finding the ground state structures.

ACKNOWLEDGEMENTS

This work was supported by the Deutsche Forschungsgemeinschaft within the research unit FOR 1282 (DO 729/5, FI 893/4). We gratefully acknowledge the support of the Stichting voor Fundamenteel Onderzoek der Materie (FOM) in providing beam time of FELIX and the FELIX staff for their skillful assistance, in particular, B. Redlich and A. F. G. van der Meer. The project has received funding from the European Community's Seventh Framework Programme (FP7/2007–2013) under Grant Agreement No. 226716. We thank J. Langer and A. Lagutschenkov for their initial contribution.

REFERENCES

1. P. Melinon, B. Masenelli, F. Tournus and A. Perez, *Nat. Mater.*, 2007, 6, 479–490.
2. I. Cherchneff, Y. H. Le Teuff, P. M. Williams and A. G. G. M. Tielens, *Astron. Astrophys.*, 2000, 357, 572–580.
3. K. P. Lim and F. W. Lampe, *Int. J. Mass Spectrom.*, 1990, 101, 245–256.
4. S. K. Mandal and H. W. Roesky, *Chem. Commun.*, 2010, 46, 6016–6041.
5. D. S. N. Parker, A. V. Wilson, R. I. Kaiser, N. J. Mayhall, M. Head-Gordon and A. G. G. M. Tielens, *Astrophys. J*, 2013, 770, 33.
6. P. Thaddeus, S. E. Cummins and R. A. Linke, *Astrophys. J.*, 1984, 283, L45–L48.
7. M. Frenklach, C. S. Carmer and E. D. Feigelson, *Nature*, 1989, 339, 196–198.
8. F. Combes, M. Boquien, C. Kramer, E. M. Xilouris, F. Bertoldi, J. Braine, C. Buchbender, D. Calzetti, P. Gratier, F. Israel, B. Koribalski, S. Lord, G. Quintana-Lacaci, M. Relano, M. Rollig, G. Stacey, F. S. Tabatabaei, R. P. J. Tilanus, F. van der Tak, P. van der Werf and S. Verley, *Astron. Astrophys.*, 2012, 539, A67.
9. K. Furuya, Y. Aikawa, K. Tomida, T. Matsumoto, K. Saigo, K. Tomisaka, F. Hersant and V. Wakelam, *Astrophys. J.*, 2012, 758, 86.
10. M. Satta, T. Grassi and F. A. Gianturco, *Mon. Not. R. Astron. Soc.*, 2013, 429, 269–274.
11. M. Steglich and J. P. Maier, *Astrophys. J.*, 2015, 801, 119.
12. A. Nakajima, T. Taguwa, K. Nakao, M. Gomei, R. Kishi, S. Iwata and K. Kaya, *J. Chem. Phys.*, 1995, 103, 2050.
13. Q. Y. Chu, B. X. Li and J. Yu, *J. Mol. Struct.*, 2007, 806, 67–76.
14. C. M. L. Rittby, *J. Chem. Phys.*, 1992, 96, 6768–6772.
15. C. M. L. Rittby, *J. Chem. Phys.*, 1994, 100, 175–180.
16. R. S. Grev and H. F. Schaefer, *J. Chem. Phys.*, 1984, 80, 3552–3555.
17. R. S. Grev and H. F. Schaefer, *J. Chem. Phys.*, 1985, 82, 4126–4130.
18. R. S. Grev and H. F. Schaefer, *Chem. Phys. Lett.*, 1985, 119, 111–118.
19. I. L. Alberts, R. S. Grev and H. F. Schaefer, *J. Chem. Phys.*, 1990, 93, 5046–5052.
20. Z. Y. Jiang, X. H. Xu, H. S. Wu, F. Q. Zhang and Z. H. Jin, *J. Mol. Struct.*, 2002, 589, 103–109.
21. Z. Y. Jiang, X. H. Xu, H. S. Wu, F. Q. Zhang and Z. H. Jin, *Chin. J. Struct. Chem.*, 2003, 22, 459–463.
22. Z. Y. Jiang, X. H. Xu, H. S. Wu and Z. H. Jin, *J. Phys. Chem. A*, 2003, 107, 10126–10131.
23. Z. Y. Jiang, X. H. Xu, H. S. Wu, F. Q. Zhang and Z. H. Jin, *Chem. Phys.*, 2003, 290, 223–231.
24. Z. Y. Jiang, X. H. Xu, H. S. Wu, F. Q. Zhang and Z. H. Jin, *J. Mol. Struct.*, 2003, 624, 61–67.
25. Z. Y. Jiang, X. H. Xu, H. S. Wu, F. Q. Zhang and Z. H. Jin, *J. Mol. Struct.*, 2003, 621, 279–284.
26. J. Hou and B. Song, *J. Chem. Phys.*, 2008, 128, 154304.
27. B. Song, Y. Yong, J. Hou and P. He, *Eur. Phys. J. D*, 2010, 59, 399–406.
28. J. Zhang, W. C. Lu, Q. J. Zang, L. Z. Zhao, C. Z. Wang and K. M. Ho, *J. Phys.: Condens. Matter*, 2011, 23, 205305.
29. S. Erkoc and L. Turker, *Physica E*, 2000, 8, 50–56.
30. M. Cui, M. Ge, J. Feng, H. Zhang, W. Tian and C. Sun, *J. Mol. Struct.*, 1999, 492, 241–256.
31. Q. X. Li, W. C. Lu, Q. J. Zang, L. Z. Zhao, C. Z. Wang and K. M. Ho, *Comput. Theor. Chem.*, 2011, 963, 439–447.
32. S. Hunsicker and R. O. Jones, *J. Chem. Phys.*, 1996, 105, 5048–5060.
33. A. D. Zdetsis, G. Froudakis, M. Muhlhauser and H. Thumnel, *J. Chem. Phys.*, 1996, 104, 2566–2573.

34. G. E. Davico, R. L. Schwartz and W. C. Lineberger, *J. Chem. Phys.*, 2001, 115, 1789–1794.

35. J. D. Presilla-Marquez, S. C. Gay, C. M. L. Rittby and W. R. M. Graham, *J. Chem. Phys.*, 1995, 102, 6354–6361.

36. J. D. Presilla-Marquez and W. R. M. Graham, *J. Chem. Phys.*, 1994, 100, 181–185.

37. J. D. Presilla-Marquez and W. R. M. Graham, *J. Chem. Phys.*, 1992, 96, 6509–6514.

38. P. A. Withey and W. R. M. Graham, *J. Chem. Phys.*, 1992, 96, 4068–4072.

39. J. D. Presilla-Marquez, C. M. L. Rittby and W. R. M. Graham, *J. Chem. Phys.*, 1996, 104, 2818–2824.

40. T. H. Le, C. M. L. Rittby and W. R. M. Graham, *J. Chem. Phys.*, 2014, 140, 064314.

41. X. D. Ding, S. L. Wang, C. M. L. Rittby and W. R. M. Graham, *J. Phys. Chem. A*, 2000, 104, 3712–3717.

42. X. D. Ding, S. L. Wang, C. M. L. Rittby and W. R. M. Graham, *J. Chem. Phys.*, 1999, 110, 11214–11220.

43. J. D. Presilla-Marquez, C. M. L. Rittby and W. R. M. Graham, *J. Chem. Phys.*, 1997, 106, 8367–8373.

44. N. J. Reilly, M. Steglich, D. L. Kokkin, J. P. Maier, J. F. Stanton and M. C. McCarthy, *J. Mol. Spectrosc.*, 2015, 310, 135–140.

45. J. F. Stanton, J. Dudek, P. Theule, H. Gupta, M. C. McCarthy and P. Thaddeus, *J. Chem. Phys.*, 2005, 122, 124314.

46. P. Gruene, D. M. Rayner, B. Redlich, A. F. G. van der Meer, J. T. Lyon, G. Meijer and A. Fielicke, *Science*, 2008, 321, 674–676.

47. M. Haertelt, J. T. Lyon, P. Claes, J. de Haeck, P. Lievens and A. Fielicke, *J. Chem. Phys.*, 2012, 136, 064301.

48. G. von Helden, D. van Heijnsbergen and G. Meijer, *J. Phys. Chem. A*, 2003, 107, 1671–1688.

49. A. Fielicke, J. T. Lyon, M. Haertelt, G. Meijer, P. Claes, J. de Haeck and P. Lievens, *J. Chem. Phys.*, 2009, 131, 171105.

50. M. Savoca, A. Lagutschenkov, J. Langer, D. J. Harding, A. Fielicke and O. Dopfer, *J. Phys. Chem. A*, 2013, 117, 1158–1163.

51. N. X. Truong, M. Savoca, D. J. Harding, A. Fielicke and O. Dopfer, *Phys. Chem. Chem. Phys.*, 2014, 16, 22364–22372.

52. D. J. Harding, C. Kerpal, G. Meijer and A. Fielicke, *J. Phys. Chem. Lett.*, 2013, 4, 892–896.

53. R. Gehrke and K. Reuter, *Phys. Rev. B: Condens. Matter Mater. Phys.*, 2009, 79, 085412.

54. D. Oepts, A. F. G. van der Meer and P. W. van Amersfoort, *Infrared Phys. Technol.*, 1995, 36, 297–308.

55. M. Haertelt, A. Fielicke, G. Meijer, K. Kwapien, M. Sierka and J. Sauer, *Phys. Chem. Chem. Phys.*, 2012, 14, 2849–2856.

56. S. Heiles and R. L. Johnston, *Int. J. Quantum Chem.*, 2013, 113, 2091–2109.

57. D. J. Wales, *Energy landscapes*, Cambridge University Press, Cambridge, 2003.

58. E. M. Fernández, J. M. Soler, I. L. Garzón and L. C. Balbás, *Phys. Rev. B: Condens. Matter Mater. Phys.*, 2004, 70, 165403.

59. S. Grimme, J. Antony, S. Ehrlich and H. Krieg, *J. Chem. Phys.*, 2010, 132, 154104.

60. M. J. Frisch, G. W. Trucks, H. B. Schlegel, G. E. Scuseria, M. A. Robb, J. R. Cheeseman, G. Scalmani, V. Barone, B. Mennucci, G. A. Petersson, H. Nakatsuji, M. Caricato, X. Li, H. P. Hratchian, A. F. Izmaylov, J. Bloino, G. Zheng, J. L. Sonnenberg, M. Hada, M. Ehara, K. Toyota, R. Fukuda, J. Hasegawa, M. Ishida, T. Nakajima, Y. Honda, O. Kitao, H. Nakai, T. Vreven, J. A. Montgomery Jr, J. E. Peralta, F. Ogliaro, M. Bearpark, J. J. Heyd, E. Brothers, K. N. Kudin, V. N. Staroverov, R. Kobayashi, J. Normand, K. Raghavachari, A. Rendell, J. C. Burant, S. S. Iyengar, J. Tomasi, M. Cossi, N. Rega, N. J. Millam, M. Klene, J. E. Knox, J. B. Cross, V. Bakken, C. Adamo, J. Jaramillo, R. Gomperts, R. E. Stratmann, O. Yazyev, A. J. Austin, R. Cammi, C. Pomelli, J. W. Ochterski, R. L. Martin, K. Morokuma, V. G. Zakrzewski, G. A. Voth, P. Salvador, J. J. Dannenberg, S. Dapprich, A. D. Daniels, Ö. Farkas, J. B. Foresman, J. V. Ortiz, J. Cioslowski and D. J. Fox, *GAUSSIAN09, Rev.D.01*, Gaussian Inc., Wallingford CT, 2009.

61. R. Kishi, M. Gomei, A. Nakajima, S. Iwata and K. Kaya, *J. Chem. Phys.*, 1996, 104, 8593–8604.

62. P. S. Yadav, R. K. Yadav, S. Agrawal and B. K. Agrawal, *Physica E*, 2006, 33, 249–262.

63. K. Raghavachari and V. Logovinsky, *Phys. Rev. Lett.*, 1985, 55, 2853–2856.

64. K. Raghavachari, *Abstr. Pap. Am. Chem. Soc.*, 1986, 192, 96.

65. K. Raghavachari, *J. Chem. Phys.*, 1986, 84, 5672–5686.

66. S. Li, R. J. Vanzee, W. Weltner and K. Raghavachari, *Chem. Phys. Lett.*, 1995, 243, 275–280.

67. A. D. Zdetsis, *J. Chem. Phys.*, 2007, 127, 014314.

68. P. Karamanis, D. Zhang-Negrerie and C. Pouchan, *Chem. Phys.*, 2007, 331, 417–426.

69. E. C. Honea, A. Ogura, C. A. Murray, K. Raghavachari, W. O. Sprenger, M. F. Jarrold and W. L. Brown, *Nature*, 1993, 366, 42–44.

70. K. Raghavachari and C. M. Rohlfing, *J. Chem. Phys.*, 1988, 89, 2219–2234.

71. G. Meloni and K. A. Gingerich, *J. Chem. Phys.*, 2001, 115, 5470.

72. B. W. Ticknor and M. A. Duncan, *Chem. Phys. Lett.*, 2005, 405, 214–219.

73. M. Pellarin, C. Ray, P. Melinon, J. Lerme, J. L. Vialle, P. Keghelian, A. Perez and M. Broyer, *Chem. Phys. Lett.*, 1997, 277, 96–104

CHAPTER 8

ATOMIC SPECTROMETRY UPDATES: REVIEW OF ADVANCES IN ATOMIC SPECTROMETRY AND RELATED TECHNIQUES

E. Hywel Evans[a], Jorge Pisonero[b], Clare M. M. Smith[c] and Rex N. Taylor[d]

[a]School of Geography, Earth, and Environmental Sciences, University of Plymouth, Drake Circus, Plymouth, UK PL4 8AA

[b]University of Oviedo, Faculty of Science, Department of Physics, c/ Calvo Sotelo s/n, 33006 Oviedo, Spain

[c]St. Ambrose High School, Blair Road, Coatbridge, Lanarkshire, UK ML5 2EW

[d]Ocean and Earth Science, University of Southampton, NOC, Southampton, UK SO14 3ZH

ABSTRACT

This review covers developments in 'Atomic Spectrometry'. It covers atomic emission, absorption, fluorescence and mass spectrometry, but excludes material on speciation and coupled techniques which is included in a separate review. It should be read in conjunction with the other related reviews in the series.[1–6] A critical approach to the selection of material has been adopted, with only novel developments in instrumentation, techniques and methodology being included. A number of themes have emerged from this year›s review. Novel developments in sample preparation include the use of magnetic nanoparticles for solid phase extraction and an increase in the number of viable methods for use in quantitation of peptides, proteins and DNA by elemental tagging, particularly the development of multiplexed techniques. There were also some advances in electrochemical vapour generation, such as the use of reducing reagents immobilized on electrodes and development of electrolytic membrane diffusion cells. New instrumentation has been to the fore, with research into some novel and interesting miniaturized plasma sources being of particular interest, plus the development of a new distance of flight (DOF) mass spectrometer. There have been some insightful fundamental studies of particle atomization and ionization in ICP-MS, and subsequent space-charge effects after ion sampling, which should enable practical steps to solving some of these problems. One of the main drivers of research in isotope ratio analysis has been to increase detection power to enable smaller and smaller quantities of analyte to be measured, which yields advantages in geochronology and nuclear forensics. Laser ablation has played a large part in many types of analysis where high resolution is required, including analysis of single particles, small inclusions or elemental mapping of tissues, so the development of LA methods with extremely small spot sizes and efficient sample transport to the source has been key. In addition, several novel methods have been developed, such as laser ablation molecular isotopic spectrometry (LAMIS) for isotope ratio analysis, and stand-off LIBS techniques such as 'underwater LIBS'. In contrast, very little research into the use of chemometrics has been reported, and fundamental studies are generally confined to characterization of new plasma sources.

Citation: E. Hywel Evans, Jorge Pisonero, Clare M. M. Smith and Rex N. Taylor, Atomic spectrometry updates: Review of advances in atomic spectrometry and related techniques, J. Anal. At. Spectrom. 2014, 29, 773-794, DOI: 10.1039/C4JA90019G.

SAMPLE INTRODUCTION

Liquids

Liquid sample introduction relates to methods wherein the sample is introduced into the instrument in the form of a liquid, such as through nebulization or into a thermal vapourizer; whereas in vapour generation the sample is initially in the form of a liquid but is converted to a vapour prior to introduction into the instrument.

Sample Pre-Treatment

Solid Phase Extraction

A predominant theme in this review has been the use of magnetic nano-particles for solid phase extraction (SPE). The use of MNPs is advantageous because they can be easily separated from the sample after preconcentration by immobilisation in a magnetic field, then the analyte is eluted into a much smaller volume. This property of immobilisation also makes them ideal for preconcentration from a flowing stream. The field has been reviewed (74 references) by Giakisikli and Anthemidis[7] in which they gave a clear overview of the literature, separating applications into those utilising 'inorganic' and 'organic' coated NPs used for metal pre-concentration, with comprehensive application tables arranged by analyte. They focused on the benefits of MNP-SPE for automated methods, particularly the convenience of incorporating them into flowing systems. The same authors[8] reported the use of octadecylsilane functionalised maghemite MNPs as a SPE medium. They used two NdFeB magnets to immobilise the MNPs in a FI system, thereby avoiding the use of frits, and used this for online formation and preconcentration of a DDTC–Cd complex from waters. Analysis by ETAAS yielded an LOD of 3 ng L^{-1} and RSD of 3.9% at the 50 ng L^{-1} level, with stable column operation for 600 cycles. In a similar approach, Chen et al.[9] utilized 'on-chip' SPE for speciation of Se yeast cells. For this, they prepared a column on a microfluidic chip packed with sulfonated, polystyrene-coated MNPs (Fe_3O_4@PSS). The chip had seven separate inlets to allow introduction of sample solutions, reagents and standards at flow rates of approximately 1 μL min^{-1}, thereby allowing on line enzyme digestion of selenium-enriched yeast cells followed by SPE, subsequent elution with 0.5 mol L^{-1} Na_2CO_3 and analysis using HPLC-ICP-MS. Close reading of the paper seems to suggest that the principal advantage of using MNPs in this application was the ease with which the microfluidic column could be packed in the presence of a magnetic field, rather than to hold the MNPs in place. The authors compared the performance of on-line and off-line MSPE for speciation of $SeCys_2$, MeSeCys, SeMet, GluMeSeCys, and SeEt, and achieved LODs in the range from 0.022 to 0.149 μg L^{-1}, with the latter method yielding LODs at the lower end of the range. They validated the method by analysis of selenium-enriched yeast cells (S. cerevisiae BY4742), obtaining recoveries of between 85.3 and 106% for a 50 μg g^{-1} spike.

Cui et al.[10] have coated Fe_3O_4@Si–OH MNPs with the ionic liquid methyltrioctylammonium chloride ($MTOA^+$ Cl^-) for use in extracting Cd, Cu and Zn from biological samples. The target analytes were determined in the final eluate solution of 0.8 mL of 1.0 mol L^{-1} HNO_3, with LODs of 0.33, 0.54 and 0.56 μg L^{-1}, respectively. The method was found to be tolerant of concentrations of other transition and heavy elements such as Co, Cr, Fe, Mn, Ni and Pb up to between 1 and 4 mg L^{-1}, but also much higher concentrations (ca. 1000×) of polyanions, alkaline and alkaline earth elements. The method was validated by analysis of the CRMs GBW09103 human urine and GBW07601 human hair, and spiked samples, with recoveries between 89 and 120%.

Selectivity in SPE has long been a goal of the analyst, and ion-imprinted polymers (IIPs) are a class of adsorbent that holds some promise. Ebrahimzadeh et al.[11] have developed a novel IIP for Au preconcentration, by incorporating a dipyridyl ligand in ethylene glycol dimethacrylate, then coating it onto porous silica particles. They achieved an LOD of <0.2 ng mL^{-1} using FAAS and selectivity of preconcentration in the presence of palladium and platinum. Another novel IIP has been reported by de Lima et al.[12] who formed the polymer by reduction of $Se^{IV}$to Se^0 prior to reaction with 4-vinylpyridine and vinylimidazole, hence they designated the product as an atom imprinted polymer (AIP). Preconcentration of Se was achieved in FI mode by first reducing the analyte to Se^0, using 0.5% (w/v) citric acid, and then preconcentration on the AIP which was immobilised in a mini-column. The Se^0 was then eluted by oxidation to Se^{IV}using hypochlorite, subsequently reduced to SeH_2 using $NaBH_4$ and analysed by HG-FAAS. The method was optimized using a Doehlert design to give an LOD of 53 ng L^{-1} and preconcentration factor of 232. Validation was by the determination of Se in DORM-2 dogfish muscle, with 98% recovery. Selenium was also successfully determined in Brazil nuts, apricot, white beans, rice flour and milk powder with recoveries in the range 100 to 109%. It is worth noting that significant matrix interferences (>10%) were observed in the presence of less than 10-fold excess of Cr^{3+} and Cu^{2+}, so the authors recommended that the method of standard additions be used.

Bauer et al.[13] reported a novel SPE method whereby trace metals were preconcentrated from environmental waters using silica NPs. The NPs were functionalized with strong cation exchange ligands based on 3-mercaptopropyltriethoxysilane modified with acetonitrile, trifluoroacetic anhydride and urea hydrogen peroxide. These particles had diameters in the range from 300 to 700 nm so could be suspended as a slurry and nebulized directly into ICP-OES without the need for elution of the analytes. The authors found that the analytical signal initially decreased by 10% 2.5 min after initial ultrasonic suspension of the slurry, but stayed constant thereafter up to 12.5 min. They attributed the initial decrease to settling out of larger particles. An average preconcentration factor of 10.4 was achieved, yielding LODs between 0.03 μg L^{-1} for Be to 0.48 μg L^{-1} for Fe, and the method was validated by the determination of Al, Be, Cd, Co, Cr, Cu, Fe, Mn, Ni and Pb in NIST1643e fresh water.

Ozdemir et al.[14] have reviewed (160 references) the use of microbacteria as SPE media, with particular attention paid to preconcentration of metals using bacteria immobilised on resins.

Liquid Phase Extraction

The method of dispersive liquid–liquid microextraction (DLLME) has been reviewed (126 references) by Andruch et al.,[15] who outlined the basic principles and produced application tables ordered by analyte for quick reference. Another review (99 references) by Miro and Hansen[16] covered liquid phase microextraction (LPME) and solid phase microextraction (SPME), taking a selective approach to illustrate key developments in on-line methods and covering the various manifestations of these techniques. The attraction of LPME methods is that they minimise the amount of organic solvent used for the extraction step by using small amounts of highly dispersed droplets of the extraction solvent (usually an organic solvent), in either batch or sequential injection mode. An alternative approach was adopted by Boucherit et al.[17] whereby they directly nebulized an aqueous sample into an organic extracting solvent. In this way the principle of fine dispersion was retained but the advantage of reducing the amount of organic solvent was lost. Nevertheless, they found an improvement in the speed of extraction (compared with traditional solvent extraction) of Mo from aqueous HCl (into tributyl phosphate in n-hexane) and H_3PO_4 (into bis(2-ethylhexyl) phosphoric acid in n-hexane) solutions. Equilibration was achieved almost instantaneously between the aerosol and the extractant, but took up to 180 s and 900 s for the respective traditional extraction methods. The analysis was performed using ICP-OES with an LOD of 2 μg L^{-1} and a 10-fold enrichment.

Two of the key requirements for DLLME are the initial dispersion of the extracting solvent and subsequent centrifugation of the mixed phases in order to separate the highly dispersed organic droplets from the aqueous phase. This can also be achieved using an ionic liquid (IL) in place of an organic solvent. The role of ILs in extraction methods has been reviewed (113 references) by Han et al.,[18] covering liquid and solid phase extraction utilising hollow fibres, single drop and dispersive methods for both organic and inorganic analytes.

There have been some novel variations on the standard procedure. Naeemullah et al.[19] extracted a vanadium-8-HQ complex (0.1–1.0 mL of a 0.1% w/v solution of 8-HQ) from waters using 1-butyl-3-methylimidazolium hexafluorophosphate (20–70 mL of a 1 : 1 mixture of the IL and chloroform). In this case dispersion of the IL was aided by using chloroform in the mixture and high pressure argon (15 kPa for 5 min). The sample was centrifuged for 10 min at 3000 rpm and dissolved in 200 mL of 0.1 mol L^{-1} HNO_3 in ethanol–water (1 : 1 v/v) before analysis using ETAAS, to yield an LOD and enrichment factor of 20 ng L^{-1} and 100, respectively. The use of a hollow fibre, in conjunction with room temperature ILs, has been reported by Zeng and Tang.[20] They used butyl- octyl- or hexyl- 3-methylimidazoliu mhexafluorophosphate (or tetraborate) ILs to aid complexation of Au with dithizone, with subsequent extraction into 9 μL of 1-octanol trapped inside a hollow fibre. This solution was then flushed out with a microsyringe, mixed with 65 μL of 95% (v/v) alcoholic solution with 1% (v/v) HNO_3 and siphoned into a FAA spectrometer for analysis. They found that the use of the ILs improved extraction efficiency by five-fold to give an enrichment factor of 130 and LOD of 0.9 ng $mL^{-1}$for Au. Shah et al.[21] performed a microextraction of Pb using an IL inside a 10 mL syringe. The pH buffered sample was placed in the syringe with 0.01% (w/v) dithizone, then 500 μL of 0.075% (v/v) Triton X-114 and 100 μL of 1-Butyl-3-methyl-imidazolium hexafluorophosphate added to extract a Pb–dithizone complex. Separation of the IL phase was effected by decanting from the syringe and diluting in 0.5 mL of methanol, before injecting 100 μL aliquots into the sample uptake tube of FAAS. The method yielded an enrichment factor of 82.6 and LOD of 0.281 μg L^{-1}.

Elemental Tagging

Quantitation of proteins by elemental tagging and ICP-MS detection is a method which has threatened to become mainstream for a number of years, but has yet to be adopted on a routine basis. Several reviews have been published over the past year which highlighted the current status and possible future applications. Yan et al.[22] emphasised protein-labelling and element-tagging strategies based on biospecific interactions, ranging from protein to whole cell quantification. Sanz-Medel et al.[23] reviewed (83 references) the last five years of research in this area and identified application niches for elemental tagging such as quantification of metalloproteins in biological systems, improved sensitivity, imaging of biological tissues and certification of peptides. Kretschy et al.[24] made a similar overview (140 references) with particular emphasis on the use of HPLC coupled with ICP-MS, including an application table of frequently used metal chelating molecules and conjugation reagents that have been used for elemental tagging. Giesen et al.[25] reviewed (125 references) the use and potential of ICP-MS as a detector for immunoassays. They pointed out that future developments, which capitalize on the high sensitivity and multiplexing advantage, rely on the development of high-throughput methods, and predicted that signal amplification and spatial analysis using LA-ICP-MS offers some potential for single cell analysis.

Normally, absolute quantitation of peptides and proteins by IDA requires the use of an isotopically enriched isotopomer of the target peptide, which will likely have to be synthesized for the purpose. This goes some way to explaining why the technique has not been widely adopted. Esteban-Fernandez et al.[26] have used the metal coded affinity tag (MeCAT) approach for absolute quantitation of proteins using ID-ICP-MS. The MeCAt reagent was loaded with (separately) natural and isotopically enriched Yb then tagged to various proteins via an iodoacetamide (IA) cysteine-reactive group on the reagent to form a MeCAT(Yb)-IA-Protein complex. Proteins were separated by gel electrophoresis (GE) and analyzed either directly by LA-ICP-MS or after excision and digestion of the gel spots. Absolute quantitation was performed by IDA after spiking with standard proteins labelled with enriched Yb. Crucially, both natural and enriched MeCAT(Yb)-IA-Protein complexes migrated equally during the GE separation, and a stable ^{172}Yb:^{171}Yb ratio was observed along the spot when using LA-ICP-MS. The same group[27] used the same labelling strategy and compared this method with LC-ESI-MS, LC-ESI-MS-MS and HPLC-ICP-MS for the relative quantitation of alpha-lactalbumin, using Eu, Ho, Lu, Tb and Tm in a tetraplex labelling approach, and SDS-PAGE separation prior to analysis. They found that all three methods yielded quantitative results.

Liu et al.[28] reported a non-species-specific IDA method, using LC-ICP-MS, for absolute quantitation of bradykinin. They used commercially available 2′,2″-(10-(2-((2,5-dioxopyrrolidin-1-yl)oxy)-2-oxoethyl)-1,4,7,10-tetraazacyclododecane-1,4,7-triyl) tri-acetic acid (DOTA–NHS–ester) to react with bradykinin, then labelled the resultant conjugate by forming a bradykinin-DOTA-Eu chelate at pH6, by using a two-fold molar excess of Eu to DOTA. Non-species-specific IDA was achieved by post-column perfusion of a ^{151}Eu spike solution (4.27 ng g^{-1}) via a three-way connector. The method was validated by comparison with a species-specific GC-MS method, in which the bradykinin sample was spiked with ^{13}C-labelled phenylalanine, then hydrolysed to yield free phenylalanine before derivatization with methyl chloroformate. The two methods agreed to within 4%, with 5% RSD at concentrations in the range from 20 to 500 pmol. An LOD of 7.3 fmol was estimated for the species-non-specific LC-ICP-MS method. Wang et al.[29] developed a different approach based on labelling standard peptides with Tb to act as an internal standard. They also used a DOTA–NHS–ester to form peptide -DOTA conjugates with synthesized standard peptides and the equivalent endogenous target peptide, which were then chelated with either Tb or Tm, respectively. Analysis was performed by LC-MRM-MS using external calibration, but also by using the Tb-DOTA-peptide:Tm-DOTA-peptide ratios ascertained from the widely separated mass fragments in co-eluting peaks as an internal standardization method. They achieved an LOD of 0.8 fmol μL^{-1}, so the technique might well benefit from using ICP-MS as the detection method.

Yet another solution to this problem has been reported by Konz et al.[30] who utilised the intrinsic ability of the hormone hepcidin-25 to complex with Cu^{II} ions. They used two approaches, both involving IDA and LC-ICP-MS: post-column addition of a non-species-specific ^{65}Cu spike; and synthesis of a labelled ^{65}Cu-hepcidin-25 complex to use as a species-specific spike. The stability of the Cu-hepcidin-25 complex was checked using ESI-MS, and found to be stable under physiological conditions and to exist as an equimolar complex, however, they did note that spiking with the ^{65}Cu-hepcidin-25 complex should occur directly prior to species specific IDA because Cu-exchange reactions were possible; a problem that did not occur with post-column addition. They used both methods to determine hepcidin-25 in human serum samples and found an average concentration of 21.6 ng mL^{-1}, which agreed well with those reported in the literature of 19.2 ng mL^{-1}. In an unrelated but relevant paper Arnquist and Holcombe[31] studied metal-protein binding equilibria between Cu^{II} ions and bovine serum albumin (BSA) using preparative ultracentrifugation to separate bound and unbound Cu, with ICP-MS detection. In this way they determined the apparent (K_{app}) and intrinsic (K_{int}) binding affinities under different conditions of pH; information which could be used to predict the type of exchange reactions reported by Konz et al.[30]

Absolute DNA quantitation by elemental tagging poses a similar challenge to protein quantitation. Han et al.[32] have described a novel and elegant multiplexing approach to this problem. They utilized 15 DNA hybridization probes for separate DNA targets, of 25–30 bases length, related to clinical diseases. The probes were 3′ end-functionalized with thiol, derivatized with maleimide-DOTA (MMA-DOTA), then individually labelled by complexation with a specific REE. A mixture of these DNA-MMA-DOTA-REE complexes was then added to the target DNA sequences to allow hybridisation to occur, and these were then captured by further hybridization with capture probes which had been immobilised on MNPs. The MNPs with the attached probe-target-capture sequences were separated from the mixture containing the excess REE labelled probes. Finally, the REE labelled probes were released from the target sequences by raising the temperature to 95 °C and analyzed by ICP-MS using external calibration with standard REE labelled probes. They further developed the method to allow an IDA approach to absolute quantification by using a spike-DNA-MMA-DOTA-REE probe complexed with an enriched isotope; in this case separate ^{161}Dy-enriched and ^{168}Er-enriched probes were tested. The method was as before, except that the MNP-immobilized capture probes contained an extra sequence which was complementary to the spike-DNA-MMA-DOTA-REE probe, and the quantitation step was performed by ID-ICP-MS. The method was applied to the analysis of B. anthracis and severe acute respiratory syndrome (SARS) targets and good agreement with a UV-vis method was obtained.

An alternative approach to DNA quantitation was reported by Leclerc et al.[33] who used ICP-MS (MC and Q) to quantify the total amount of DNA extracted from cells of Legionella pneumophila by determining the mass fraction of intrinsic P. They compared the results of the ICP-MS analysis with UV-vis and quantitative polymerase chain reaction (qPCR) methods, and found good agreement between the three. The ICP-MS method could be used with about 2 mL of DNA solution at about 30 ng g^{-1} P per measurement and had 11× lower uncertainty compared with qPCR because of the uncertainty in the qPCR standards. Swart[34] has addressed the important metrological difficulties of protein quantitation. The main problem seems to be that there are few SI-traceable reference materials for metallo-proteins, which is a serious lack when comparable results for samples of human origin are required. The author discussed the problems inherent in quantitative analysis and the production of traceable standards for this class of compounds using the examples of hemoglobin, transferrin, superoxide dismutase, ceruloplasmin, and C-reactive protein.

Nebulization

Miniaturization and automation in analytical sciences have evolved tremendously over the past decade, particularly within the framework of "green" analytical chemistry. Miro and Hansen described the merits of advanced flow methodology using mesofluidic platforms to simplify and speed up the overall analytical process.[35] Introduced in 2000, lab-on-a-valve (LOV), coupled with sequential injection analysis, has been developed as a downscaled analytical tool for pressure-driven sampling at the low μL level. Uses of this technique include on-chip sample processing involving renewable micro-solid phase extraction, non-chromatographic speciation and chemical vapour generation using programmable flow. These can be coupled with on-chip optical and electrochemical detection, or used as a front end to modern detection equipment or column separation systems. The authors provided a critical view of the future of LOV through a survey of currently used mesofluidic systems and identified what they consider to be the major challenges involved in the development of LOV for applications that are unsuitable for lab-on-a-chip microfluidic approaches.

Determination of metals in volume limited biological samples is critical to metallomics research and microfluidic chips provide an excellent platform for the miniaturization of extraction. In this context, Wang et al.[36] designed and fabricated a chip-based LPME device to combine with ETV-ICP-MS for the determination of trace Bi, Cd, Cu, Hg, Pb and Zn in cell and human serum samples. Using sodium DDTC as the chemical modifier for low-temperature ETV-ICP-MS and the extracting reagent for the chip-based LPME, the chip parameters influencing the extraction efficiency were optimized by orthogonal design experiments. Under optimized conditions, LODs were in the range from 6.6 to 89.3 pg mL^{-1} with the RSDs ranging from 4.23 to 8.15% ($n = 7$, $c = 1$ ng mL^{-1}) for the target metal ions. The linear range covered three orders of magnitude with correlation coefficients higher than 0.99. The method was validated by analysis of CRMs GBW07605 human hair and GBW09152 human serum. Metal determination in HepG2 cells, Jurkat T cells and a human serum sample was achieved with the recoveries between 86.6 and 119%. This rapid and sensitive method has low sample and reagent consumption and offers potential for the analysis of quantity-limited biological samples such as cells.

The analysis of engineered nanoparticles continues to be a subject of some interest, requiring the development of new methods for characterization of size, mass, elemental composition, and number concentration. Elzey et al.[37] developed a sample introduction system to allow the separation and analysis of gold nanoparticles. An ES-differential mobility analyzer (ES-DMA) was used to achieve real-time upstream size discrimination. A condensation particle counter and ICP-MS were employed as downstream detectors, providing information on number density and elemental composition, respectively, of aerosolized nanoparticles versus the upstream size selected by ES-DMA. A gas-exchange device was designed and optimized to improve the conversion of air flow (from the electrospray) to argon flow required to sustain the ICP-MS plasma, the key compatibility issue for instrumental hyphenation. This initial work demonstrates successful measurement of the key analytical factors for nanoparticle characterization.

A similar approach, to determine the size and mass of metallic engineered nanoparticles, was reported by Gschind et al.[38] using a monodisperse microdroplet generator (MDG) for sample introduction into ICP-MS. Vertical and horizontal sample introduction configurations were evaluated. The horizontal introduction system resulted in shorter measurement time by reducing temporal jitter. Quantitation of Ag, Au and CeO_2 nanoparticles of different sizes and polydispersities was achieved. Results were compared to those obtained using TEM and asymmetric flow FFF, and the advantages and limitations of the proposed technique were discussed.

Finally, Pereira et al.[39] have developed a multiple nebulizer based on flow blurring hydrodynamics and used it to reduce matrix effects on As and Se determination by ICP-OES. This so-called flow blurring multiple nebulizer (FBMN) is essentially an array of three pneumatic nebulizers arranged in close proximity to each other which produces droplets <53 µm diameter (Sauter mean diameter = 16.3 µm) when operated at gas and liquid flow rates of 0.6 L min^{-1} and 0.3 mL, respectively. They studied matrix effects caused by Ca, K, Mg and Na when using the FBMN (with and without internal standardization using In, Ge or Y) and compared the results with a standard pneumatic nebulizer. It is not clear what advantage the FBMN nebulizer affords, given that internal standardization is an established approach to matrix correction and the propensity for the authors to present key comparative findings in the electronic supplementary information rather than the main body of the paper.

Thermal Vaporization

Badliei et al. described electrothermal, near-torch vaporization (NTV) sample introduction system for micro- or nano-volume samples.[40] Samples were pipetted onto coiled-filament assemblies that were purposely developed to be removable and interchangeable. These were inserted into a small-volume vaporization chamber, that could be clipped onto any ICP torch with a ball joint, dried and then vaporized in situ. Interchangeable assemblies of various volume capacities were constructed to be small and lightweight so that they could be portable. This horizontally-operated, NTV sample introduction system was interfaced to an axially-viewed ICP-OES and optimized for 8 elements (Ba, Be, Ca, Cd, Mg, Pb, V and Zn). Precision and LODs obtained were comparable with existing ETV systems. Analytical capability and utility was demonstrated using the determination of Pb at pg mL^{-1} levels in a diluted natural water CRM, and the determination of Zn in 80 nL volumes of the liquid extracted from an individual vesicle. In a further paper, this system was applied to Cr speciation in seawater.[41] Samples were electrodeposited in the field on portable coiled-filament assemblies made from Re. Assemblies with dried residues of Cr^{III} and Cr^{III} + Cr^{VI} were transported to the lab for determination by NTV-ICP-OES. Electrodeposition allowed selective species deposition, preconcentration and matrix clean up from seawater samples. Free Cr^{VI} was electrodeposited at −0.3 V and Cr^{III} + Cr^{VI} at −1.6 V (both vs. Ag/AgCl). Interestingly, at 0 V and in the absence of an electrodeposition potential only Cr^{VI} was spontaneously and selectively adsorbed onto the coil. Due to preconcentration afforded by electrodeposition, the detection limits obtained after 60 s at the voltages stated, using buffered (pH = 4.7) artificial seawater spiked with either Cr^{III} or Cr^{VI}, were 20 and 10 pg mL^{-1}, respectively. In comparison, the LOD for Cr obtained by pipetting 5 µL of diluted standard solution directly onto the coil was 500 pg mL^{-1}, demonstrating that electrodeposition offered between 40 and 60 fold improvement. The method was validated in the lab using buffered artificial seawater and it was used in the field for the first time by sampling seawater, buffering it and immediately electrodepositing Cr species on portable assemblies on-site. Electrodeposition in the field helped avoid species transformation during sample pre-treatment, such as acidification during transport and storage. An additional benefit was the ability to ship the portable assemblies with Cr species on them rather than shipping large volumes of sample to the lab.

Vapor Generation

Chemical Vapor Generation

Gao et al.[42] have reviewed (101 references) the application of VG in ICP-MS, covering most of the common methods developed in the last twelve years, with particular reference to CVG-ID-ICP-MS. The review contains a number of tables arranged alphabetically by analyte for ease of reference.

Yilmaz et al.[43] have investigated the efficacy of using $K_3[Mn(CN)_6]$ as an additive for HG of Pb reduced to plumbane using $NaBH_4$. The point being that this reagent should enhance the reaction in the same way that $K_3[Fe(CN)_6]$ has been shown to do by other workers. They did indeed find that a 1% v/v solution improved sensitivity, when coupled with ICP-MS, by as much as 48-fold. The method did suffer from severe signal suppression in the presence of Cu at a concentration of 0.1 µg mL^{-1}. This was eliminated by increasing the $K_3[Mn(CN)_6]$ concentration to 1%, but does not bode well for the analysis of real samples. Kratzer et al.[44] used radioactive ^{212}Pb to investigate the HG and trapping efficiency of plumbane using a similar method which also utilised $NaBH_4/K_3[Fe(CN)_6]$ as an enhancing reagent. They prepared the ^{212}Pb tracer by precipitation of PbS from $Th(NO_3)_4$ with subsequent purification by ion exchange separation to yield $^{212}PbCl_2$. The efficiency of plumbane generation, trapping and subsequent volatilization was determined by measuring the activity of gamma radiation from the ^{212}Pb tracer with a scintillation detector. The HG apparatus was made in-house from a reagent mixing manifold and gas–liquid separator, which was interfaced with an electrically heated quartz trap or charcoal column depending which part of the process was to be investigated. Results showed that the efficiency of HG was ca. 96%; trapping efficiency was 88% at 300 °C; and volatilization efficiency was 85% at 800 °C. They also found that the presence of Bi at a concentration of 400 ng mL^{-1} reduced the volatilization efficiency to 13%.

A novel method of CVG has been developed by Zhu et al.,[45] who used a cylindrical dielectric barrier discharge (DBD) plasma reactor made from two concentric glass tubes with a copper wire in the center, and the outer tube wrapped with copper foil to act as the other electrode. The sample solution was run down the outside surface of the inner tube to form a thin film, from which hydrides of Cd were generated by reaction with hydrogen, produced after instigation of an 18 W DBD between the electrodes. Detection was performed using FI-QF-AFS to yield LODs of 0.03 ng mL^{-1} and 0.008 ng mL^{-1}, using DBDs generated with argon and helium, respectively. No interference from first row TMs at concentrations of 1 mg L^{-1} was observed, but signal suppression down to as low as 40% (for Ni) occurred at 10 mg L^{-1}.

Ai et al.[46] used Fe_3O_4 MNPs for the oxidation of Hg species prior to CVG. The system involved mixing an HPLC eluent with H_2O_2 prior to a column containing the MNPs, in which organo-Hg species were oxidized by $OH^{\cdot}$ free radicals formed as a result of the catalytic breakdown of H_2O_2. Subsequent CVG using KBH_4 and AFS detection yielded LODs of 0.7, 1.1, 0.8 and 0.9 mg L^{-1} (as Hg) for Hg^{2+}, MeHg, EtHg and PhHg, respectively. The method was validated by determination of MeHg in CRM DORM-2 dogfish muscle, with 95% recovery.

Electrochemical Vapor Generation

Zhang et al.[47] have developed a novel method for speciation of Hg^{2+} and $MeHg^+$ using an L-cysteine modified graphite electrode. The cleaned electrode surface was conditioned by applying a +1.8 V, 6 min potential step in 0.1 mol L^{-1} H_2SO_4 to create carboxylic acid surface groups. The electrode was then immersed in 15 g L^{-1} of L-cysteine for 8 h at room temperature to ensure peptide bond formation. They found that only Hg^{2+} was reduced at a low current of 0.2 A, whereas both Hg^{2+} and $MeHg^+$ were reduced at 2.2 A, thereby allowing electrochemical speciation to be performed. The LODs were 0.098 and 0.073 µg L^{-1}, respectively, using AFS detection. The system was relatively free from interference due to Ni, Fe and Co up to 100 mg L^{-1}, but Cu and Pb caused signal suppression at a concentration above 0.5 mg L^{-1}. Validation was performed by analysis of the CRM DORM-2 dogfish muscle, with excellent recovery, however, the authors also reported that the electrode required renewal after ca. 20 cycles.

Hranicek et al.[48] constructed and evaluated fifteen different electrolytic cells of varying types (non-membrane, thin-layer and tubular) for electrochemical vapor generation (EVG) of H_2Se and detection by QF-AAS. In all of the cells lead and platinum were used as cathode and anode, respectively, and a Nafion membrane was used to separate the chambers. After optimization, the EVG efficiency, and spatial distribution of the analyte, for the five best-performing cells was evaluated using ^{75}Se as a radioactive tracer. In this way it was possible to use autoradiography to determine the distribution of ^{75}Se-containing compounds in the various parts of the cells by means of an activity budget. The LODs were sub-ng mL^{-1} for all five cells, approaching the performance of CHG (using $NaBH_4$) which was used as a benchmark. The tubular cell, with an LOD of 0.4 ng L^{-1}, was deemed the best-performing cell for FI because it had the highest efficiency (78%), lowest cathode volume (43 mm^3) and fastest response time. The thin layer electrolytic cell was performed similarly, with cathode volume, LOD and efficiency of 225 mm^3, 0.31 ng mL^{-1} and 75%, respectively, but did not have such a rapid response.

An EVG cell comprising a Ti anode, separated by a cation exchange membrane between two Pb cathodes, was designed and tested by Shen-Tu et al.[49] Using an electrolysis current of 3.5 A, and with anolyte (0.4 mol L^{-1} H_2SO_4) and catholyte (0.3 mol L^{-1} HCl) flows of 4.0 and 1.5 mL min^{-1}, respectively, they were able to generate and determine AsH_3 and SbH_3 with LODs of 0.65 and 0.9 µg L^{-1}, respectively, using AFS. Various types of cation exchange membrane were tested, including polyethylene and Nafion 324. The authors found that longevity was greatly increased to several months continuous use by including porous epoxy resin spacers either side of the membrane. The method was validated by analysis of the CRMs GBW07405 and GBW07406 soil, with recoveries of between 96 and 107%.

Photochemical Vapor Generation

Suzuki et al.[50] utilized a tandem UV photochemical vapor generation (PVG) system for the quantitation of Se species using QF-AAS. The dual lamp system consisted of a 40 W lamp for oxidation followed by an 8 W, 254 nm germicidal UV-lamp. The 8 W lamp was arranged in parallel with 3 mm i.d. × 4 mm o.d. quartz tubing, coated inside with TiO_2 to act as a catalytic reactor. Selenite, selenomethionine, L-selenocystine, γ-glutamyl-Se-methylselenocysteine and selenocystamine were directly photolysed to volatile Se species (presumed to be SeCO and SeH_2) using only TiO_2 catlaysed photolysis with the 8 W lamp (i.e., without the 40 W lamp), with efficiencies of between 73 and 100%, in the presence of 5% acetic acid. However, selenobetaine and selenate required pre-oxidation with the 40 W lamp to obtain Se^{VI} first. The system was optimized for post-HPLC detection of Se species.

Sturgeon and Luong[51] developed two PVG reactors constructed from arrays of UV LEDs with either 0.3 mW and 245–260 nm output, or 0.3 mW and 360 nm output. They compared the results of PVG of inorganic and methylmercury using formic acid as the reducing medium, with those obtained using a thermal reactor at 85 °C. No reduction of the species was observed using the 360 nm LEDS but the 245–260 LEDS were effective. An LOD of 0.68 ng was obtained using the 245–260 nm LEDs using a 9 min irradiation time. Analysis of PACS-2 marine sediment and DOLT-4 fish liver tissue using the method of standard additions resulted in recoveries of almost 100%.

Matusiewicz and Slachinski[52] combined an ultrasonic nebulizer and UV photolysis spray chamber by inserting an 8 W, UV pen lamp directly into a cyclonic spray chamber. This enhanced the production of As, Bi, Sb, Se and Sn hydrides, Hg vapour and volatile Cd species by PVG to yield LODs of 11, 22, 16, 19, 20, 15 and 11 μg L^{-1}, respectively. Acetic acid (40% v/v) was used as the generation medium and detection was performed by MIP-AES. The main advantages were the ability to use a very small (<30 μL) sample size by ultrasonic nebulization and the relative lack of reducing agents.

Solids

Methods of solids analysis are either direct, where vaporization, atomization and excitation of the sample occurs together, such as arc/spark ablation, glow discharge analysis and secondary ion mass spectrometry; or indirect, where vaporization is separate to atomization and/or excitation, such as laser ablation or thermal vaporization coupled with a separate atomization/excitation source.

Direct Methods

The processes occurring during slurry sampling of Al_2O_3 for ICP-OES were investigated by Wang et al.[53] The conditions for sampling were optimized with relation to aerosol mass transport, with a cut-off point for aerosol particle size being set at between 7 and 10 μm for the sample introduction system. By using a customized computer model and experimental results, the maximum particle size for complete vaporization was found to be approximately 7 μm. For a gas flow of 0.8 L min^{-1}, particles with a diameter of up to 8 μm could be evaporated with an efficiency of 68% and particles as large as 5 μm could be evaporated completely within the nebulization gas flow region. In order to produce slurries for analysis an Al_2O_3 sintering block was ground using a mortar combined with a mixer mill device for particle reduction. The sample slurry was prepared by directly dispersing powdered Al_2O_3 in an aqueous solution with an addition of 0.5 wt% poly(acrylate amine) as the dispersant. The accuracy of the results was compared with the data obtained through high-pressure digestion with acid. The NIST SRM 699 alumina was analyzed and good agreement was found with the certified values.

Glow Discharge

In a review by Ganeev et al.,[54] various types of GDMS systems and their applications were considered. Special attention was paid to RF and pulsed GD combined with TOF-MS. Applications of GDMS were reviewed, with focus on analysis of semiconductors, polymers, and dielectrics, as well as isotope analysis and depth profiling. Methods of standard-free MS analysis of solids were considered and some interesting prospects for combining GDMS with preliminary chromatographic separation were also discussed.

Depth profiling with pulsed GD is a promising technique that has been the subject of investigation for a number of years. The use of a pulsed voltage for sputtering reduces the sputtering rate and thermal stress, and thereby improves the analysis of thin layered and thermally fragile samples. Pulsed glow discharge has not been widely studied, however, and this has limited its practical use. Efimova et al.[55] investigated the factors that must be optimized when pulsed mode is used, i.e., duty cycle, frequency and pulse length, in order to obtain the best sputtering and crater shape. Pulse length was found to be a crucial parameter for the crater shape and thermal effects. They compared pulsed DC and RF modes and found that they produced similar sputtering effects.

Combinatorial and high-throughput techniques have become well-established in materials science to aid in the accelerated discovery of novel functional materials. Thin film composition spreads are frequently used for this purpose by performing localized measurements and exploring different properties throughout the phase diagram. The thin film composition map has to be determined in order to understand the behavior of the compound or alloy of interest. There are several established techniques that can give elemental composition maps of large areas but the total mapping time can range from several hours to tens of hours or more. Thus, faster elemental mapping techniques are needed. In a study by Gamez et al.,[56] the applicability of GD-AES elemental mapping for combinatorial and high-throughput screening samples was explored by characterizing a CuNi thin film composition spread. Qualitative analysis images showed that Cu and Ni composition gradients can be obtained in a very short time period. In addition, the use of reference materials allowed quantitative elemental analysis maps to be extracted from the emission intensity images. The images of emission intensities and curve fitting parameters are discussed. Finally, the GD-AES results are compared to profiles obtained by EDXRF. The GD-AES figures-of-merit relevant to elemental mapping were compared with selected alternative techniques from the literature to put their potential impact into perspective.

An approach to the analysis of non-conducting materials using GD-MS was suggested by Qian et al.[57] A thin coating of molten indium was placed on the sample surface allowing the sample to be sputtered and ionized. After optimization of the plasma conditions, the prepared samples were analyzed directly by DC-GD-MS with satisfactory detection limits, stability and reproducibility. Two NIST SRMs were used to validate the precision, accuracy and reproducibility of this method. The results showed that this method could avoid grinding, sustain the relatively stable discharge and enhance the stability of signals compared to the method of using a cathode made from an In pin rolled in the sample powder.

Indirect Methods

Laser Ablation

LA for direct solid analysis has advanced over the past 50 years and has applications in many disciplines. The choice of laser is normally dependent on the application requirements, however, there are fundamental effects attributed to the laser pulse duration that drive the ablation sampling process. A comparison of nanosecond and femtosecond LA by Russo et al.32 (ref. 58) provides a useful discussion with respect to analysis based on the optically induced plasma at the sample surface (such as LIBS or LA molecular isotopic spectrometry), and transport of the ablated mass aerosol to ICP-MS. The same authors provided a useful review of LA[59] covering fundamental research, applications based on detecting photons at the ablation site (LIBS and LAMIS), collection of particles for excitation in a secondary source (ICP), and future directions for the technology.

A method has been developed for the determination of Ba, Cu, Ni, Pb, Sr and Zn in solid geological and environmental samples by LA-ICP-MS using a spinning sample platform.[60] The platform, containing a sample and a standard, is spun during the ablation, allowing the quasi-simultaneous ablation of both materials. The aerosols resulting from the ablation of sample and standard were mixed in the ablation cell allowing quantitation of the analytes by standard additions. The proportion of standard versus sample could be increased by performing the ablation further from the axis of rotation. The ablated masses were determined using a strategy based on IDA. This spinning LA method has been applied to the Allende meteorite and four powdered CRMs fused in lithium borate glasses. SRM 612 trace elements in glass was also analyzed. Deviations from the certified values were found to be less than 15% for most of the elements and samples studied, with an average precision of 10%. The results demonstrated the validity of the method for the direct analysis of solid samples using a single standard.

The advantages and possibilities of using LA-ICP-MS for IR measurements were discussed by Urgast et al.[61] A comparison of quadrupole, sector-field and multi-collector systems revealed the limitations of the respective set-ups. The report focused on the possibility of using LA-ICP-MS to image isotope ratios for biological tracer studies. The authors concluded that this application has significant potential in this important field.

Isotope ratio measurements using LA-ICP-MS in air under atmospheric pressure has been performed.[62] Samples of brass and two minerals (galena and zircon) were ablated in air and the laser-generated aerosol aspirated into a gas exchange device, where the air was replaced with argon and transported into an MC-ICP-MS instrument. Results were compared to those obtained using conventional LA-MC-ICP-MS in helium in a sealed cell. Comparable in-run (0.03 to 1.61%) and external (0.002 to 0.254%) precision was obtained, and the accuracy of both methods was comparable. However, the atmospheric sampling method gave up to 5-fold lower intensities and visualization of the aerosol extraction indicated that some material was lost prior to gas exchange. Nevertheless, this method is suitable for isotopic determination of bulk materials for samples which are too large to fit into an ablation cell or too valuable to be cut into smaller pieces.

The quantitation capabilities of LA-ICP-MS are known to be restricted by elemental fractionation as a result of ablation, transport, and atomization/ionization effects, which may result in inaccuracies whenever non-matrix matched standard materials are used for calibration. Flamigni et al.[63] studied the vaporization processes of laser-produced aerosols and subsequent diffusion losses occurring inside the ICP. Both ICP-OES and ICP-MS detection was used. Na- and Ca-specific AES of aerosols produced by LA of silicate glasses or metals revealed variations in the onset and maximum position of atomic emission in the axial channel, which were in the range of a few mm. These shifts were found to arise from composition-dependent particle/aerosol penetration depths which determined the starting point of axial vaporization in the ICP. This was also found to influence ion extraction efficiency through the ICP-MS sampling orifice, with differences observed for oxide vs. metallic aerosols caused by differences in the rate of diffusion-driven expansion from the central channel. Analysis of individual particles by ICP-MS resulted in 1/e half-value signal durations of approximately 100 μs, which agree with modelled values if AES maxima are assumed to coincide with positions of instantaneous vaporization and starting points for atomic diffusion. Results suggested that differences of up to 15% in the relative ion extraction efficiency could be explained depending on whether analytes were metals or oxides. Implications of these findings on the accuracy achievable by state-of-the-art LA-ICP-MS systems were discussed.

Thermal Vaporization

Two different approaches to the direct determination of Cl using ETV have been investigated. Antes et al.[64] developed a system for the analysis of coke by ICP-MS. The ETV system consisted of a high power lamp enclosed in a glass chamber equipped with solenoid valves. The system was controlled by a printed circuit board and software developed in Lab VIEW. Temperature and heating time were optimized and samples of up to 5 mg of petroleum coke were analyzed. Calibration was performed by standard additions using the CRM SARM 19 coal. The RSD was < 9% (n = 4) and the LOQ (10σ) of Cl was 3.5 μg g^{-1}. Results obtained for the determination of Cl in petroleum coke were in agreement with those obtained using other validated methods. The benefits of the proposed method were the reductions in: analysis time (approximately 2 min per sample); risk of contamination; analyte losses; and waste generation in comparison with techniques that require sample digestion. Hashimoto et al.[65] determined Cl in ceramics using a tungsten boat vaporizer with ICP-OES. The powdered sample was vaporized in the tungsten boat and subsequently introduced into the plasma. Calibration was by standard additions with an LOD and RSD of 0.71 ng (59 ng g^{-1}) and 3.2%, respectively, for Cl in solid ceramic materials. A number of ceramic materials were analyzed and accurate determination of Cl was achieved.

INSTRUMENTATION, FUNDAMENTALS AND CHEMOMETRICS

Instrumentation

This section covers the development of novel atom cells, excitation sources, monochromators, mass spectrometers and detectors.

Plasma sources for MS and OES continue to be the focus of much development. Jankowski and Reszke[66] have published a tutorial review (133 references) of microwave plasma sources for OES and MS. They cover the generation and classification of the different types of microwave plasmas, recent advances in construction and their applications in selected fields, particularly as OES detectors for GC. They also focus on the development of mini-discharges for miniaturized spectroscopic systems and developments in elemental and molecular ionization. The review is not comprehensive and tends to concentrate on recent developments in selected laboratories.

Miniaturized plasma sources have attracted much research interest because of the desire to place as much analytical capability as possible onto a 'chip'. Luo and Duan[67] have reviewed this area (70 references), focusing on the micro-fabrication of various types of microplasma discharges used for atomic and molecular analysis by OES and MS. Most of the applications are for gaseous analysis but they also cover solids and liquids to a lesser extent, with the attendant problems of sample introduction to the fore. Frentiu et al.[68] have developed a miniaturized system comprised of a Hg CVG device with a gold filament preconcentrator, and a CCP microtorch (20 W power and 200 mL min^{-1} argon gas flow rate), interfaced with a microspectrometer. The preconcentrator was made from an 80 μm i.d. gold filament wound into a 43-turn spiral placed inside a quartz capillary, such that preconcentrated Hg could be rapidly desorbed to achieve maximum signal after 3 s. The system was used to determine Hg in waters and polymer leachates in the concentration range 0.06 to 57.4 ng L^{-1}, with an LOD of 0.02 ng L^{-1} (0.5 pg absolute). Liu et al.[69] constructed a low power (2 W) microplasma made from two platinum electrodes inside an alumina chamber of 150 nL volume, for determination of P and S in gas by AES. The argon plasma was operated at atmospheric pressure, 280 V (DC) and 8 mA, and excited emission at 253.6 and 469.4 nm to yield LODs of 68 ppb and 1.0 ppm for P and S, respectively, when using a palm-sized spectrometer for detection. A liquid electrode plasma (LEP) has been characterized for use in AES by Kohara et al.[70] The plasma was generated directly in solution between two platinum electrodes, by applying a pulsed (3 ms) voltage of up to 2500 V in a narrow channel (100 μm) etched in fused silica. They observed the plasma and Pb emission at 405.78 nm using an array of detection equipment, including a high speed camera, CCD interface with a microscope, and a CCD spectrometer with a fibre optic interface. They estimated $T_e \sim 10\,000$ K (using Wien's displacement), $T_{exc} = 8150$ K (using a Boltzmann plot of Cu emission lines) and $n_e = 1.4 \times 10^{15}$ cm^{-3} (using Stark broadening of the H_β line). Time resolved monitoring of the emission revealed a double peak over a period of several ms. Spatial imaging showed that maximum emission occurred at the centre of the expanding plasma bubble in the confines of the 100 μm channel, but also at the anode side bubble–liquid interface where the background was much lower. An LOD of 4.0 μg L^{-1} was obtained for Pb. The same group[71] then went on to integrate the LEP onto a PDMS chip and obtained an LOD of 0.27 ppm for Pb. The PDMS chip could be used for 'tens' of measurements before degradation reduced performance.

New sources for MS and OES are constantly under development. Pfeifer et al.[72] developed a low consumption argon ICP-MS. The design included an air-cooled, bulb-shaped quartz torch which fitted inside a standard rf coil, and up against a modified sampling cone with an extended tip. This arrangement allowed the torch to be operated at an argon flow of 1.5 L $min^{-1}$and ions to be sampled from inside the torch, but prevented entrainment of the cooling air. The modular design allowed a variety of sample introduction systems to be interfaced, including solution nebulization, HPLC and GC, with LODs between 0.002 and 0.046 μg L^{-1} for Al, As, Be, Cd, Co, Cr, Hg, Mn, Ni, Pb, V and Zn for solution nebulization, compared to LODs which were between 2 and 10 times lower in conventional ICP-MS mode.

A dielectric barrier discharge (DBD) was investigated by Muller et al.[73] as a possible soft ionization source. The DBD was generated in a 30 mm long glass capillary at a helium gas flow of 600 mL min^{-1}, and from 3.5 to 6.5 kV at 21.5 kHz. The discharge was examined spectroscopically in various discharge modes to determine the populations of N_2^+ and N_2 as indicators of the suitability of the plasma for soft ionization. They concluded that soft ionization should be favored by operation in homogeneous mode (below 4 kV) which had the highest production of N_2 metastable atoms. Another type of DBD, developed by Liu et al.,[74] was used as a desorption/atomization source for AFS. The argon plasma (10 kHz, 2.5 kV, 100 mL min^{-1}) was generated between two copper rings in a quartz capillary, and used to desorb and atomize Hg species from a TLC plate. The analytes were subsequently determined by QF-AFS with LODs of 0.51, 0.29 and 0.34 pg for Hg^{2+}, MeHg and PhHg, respectively.

The solution-cathode GD (SC-GD) source is attractive because it can be used for the direct analysis of a liquid. Progress in this area has been reviewed (55 references) by Jamroz et al.,[75]with emphasis on the design and analytical figures of merit of different types of system over the last 20 years. The review contains a table of LODs arranged alphabetically by analyte for those wishing for a quick lookup guide. The mechanism of solution-to-plasma transport and excitation in this type of source was investigated by Schwartz et al.[76] by high speed imaging and laser scattering. They found that the surface of the solution deformed on plasma ignition and a number of high powered plasma nodes formed across it at timescales of 100 ns to 100 μs, which were the source of most of the analyte emission. They also concluded that these nodes only occurred at low pH which also corresponded with an increase in sample transport into the plasma. Zhu et al.[77] have used a SC-GD in a novel way, as a method of generating OsO_4 vapor for introduction into an ICP-OES instrument. An LOD of 0.51 ng mL^{-1} was obtained without the need for oxidizing reagents, though there was the requirement to prepare the samples in approx 0.1 M acid solution (i.e., pH = 1). Xiao et al.[78] improved the performance of an a.c. SC-GD by adding organic compounds to the electrolyte. They found that adding 3% (v/v) formic acid enhanced emission by 13-, 17- and 78-

fold for Ag, Cd and Pb, respectively, to yield LODs of 1, 17 and 45 µg L^{-1}.

There have also been new developments in spectrometers. Gundlach-Graham et al.[79] coupled an ICP with a distance of flight (DOF) MS. As the name suggests, this is similar to TOF-MS in that ions must be pulsed into the spectrometer however, m/z separation was effected by using a spatially selective, commercially available microchannel plate (MCP)/phosphor detector. The MCP detector elements were stacked in a chevron arrangement and ions were selectively deflected into the plane of the detector. The instrumental setup allowed for a high-to-low mass range of 1.11 : 1 with m/Δm resolution (FWHM) of between 440 and 670 with an ICP source. LODs of between 2 and 30 ng L^{-1}, LDRs of 10^5 and IR precision of 0.1% RSD were obtained. One of the main advantages of DOF-MS is that low-cost, commercial detectors are available which are sufficiently large (100 mm) to allow a high-to-low mass range of 1.45 : 1, sufficient for simultaneous detection of the lanthanide elements.

Bhatia et al.[80] have developed novel zoom optics which allowed Faraday detectors with wide apertures to be fixed in place and dispersion varied using a DC quadrupole and deflection optics. This arrangement resulted in flat-topped peaks for ^{87}Sr:^{86}Sr IR measurement with precision of 8 ppm. A new type of ion funnel has been designed by Egorova et al.[81] to replace the conventional second stage, ion optics in an ICP-MS instrument. The funnel was made from 54 ring electrodes each of 0.5 mm thickness with 0.5 mm spacing between. The electrode apertures were circular with continually decreasing inner diameters, so that the stacked arrangement formed a funnel with a 35 mm i.d. at the entrance and 2 mm i.d. at the exit. Radiofrequency and DC potentials could be applied to the electrodes and adjusted to optimize the transmission of ions of different m/z. Ion transmission was found to be lower by one order of magnitude compared with conventional ion optics, but it was possible to introduce gases such as N_2 and H_2 to reduce the abundance of Ar-based polyatomic ions by one and three orders of magnitude, respectively.

Fundamentals

Coverage is confined to a consideration of fundamental studies related to instrumental methods used primarily for analytical atomic spectrometry, and excludes studies related to e.g. astronomy and sputtering sources, except where particularly relevant.

Fundamental Constants

New measurements of fundamental constants often crop up, generally in the context of physics or astronomy, but some have direct relevance to atomic spectrometry. Ruffoni[82] developed software which can be used to simplify atomic emission spectra obtained from FT spectrometers. This so-called F(TS) atomic spectrum tool (FAST) is intended for calculating atomic branching functions, transition probabilities and oscillator strengths (and uncertainties) for use in astrophysics, but there are obvious uses in analytical atomic spectrometry. The programme is written in C++, runs on Linux/Unix/Windows and has a graphical user interface. Doidge[83] has observed forbidden transitions in the ICP-OES spectra of singly ionized Ba, Ca and Sr, with n = 6, 4, or 5, respectively, of the type $ns^2S_{1/2}$-$nd\ ^2D_{3/2}$ and $ns^2S_{1/2}$-$nd\ ^2D_{5/2}$. Transitions had the following wavelengths: 216.5943 and 217.5622 nm for Ba; 175.8755 and 175.9348 nm for Ca; and 187.3609 and 187.6660 nm for Sr. Manrique et al.[84] have identified transition probabilities for 48 lines of the transition array $3d^84s$–$3d^84p$ of Ni(II) using LIBS. They found good agreement between their measurements and other workers, including the NIST Atomic Spectra Database, and also reported 17 new measurements.

Diagnostics

Konjevic et al.[85] reported a procedure for measurement of low n_e densities using the Stark broadening of hydrogen Balmer lines. In particular, they address the problem of analyzing the line profile to determine half-width and presented several formulae to correct for inaccuracies in the Voigt function, and to correct for van der Waals broadening which is significant at low n_e density close to the fine structure limit (FSL). They recommended that careful measurements of atomic hydrogen temperature and determination of the reduced mass emitter–perturber parameter (µ) was necessary, and that data tables for Stark broadening should only be used for n_e above the FSL ($6 \times 10^{20}\ m^{-3}$ for H_α, $4 \times 10^{19}\ m^{-3}$ for H_β and $10^{19} m^{-3}$ for H_γ). Rincon et al.[86] characterized an atmospheric argon MIP known as the "Torche à Injection Axiale sur Guide d›Ondes" (TIAGO), which is the latest iteration of an axial MIP in which the torch is a nozzle forming part of the waveguide. The T_{rot} was measured using a Boltzmann plot of the P branch of N_2 ($B^2\Sigma_u^+ \rightarrow X^2\Sigma_g^+$) to yield temperatures in the range 6000 to 2000 K, which decreased markedly with gas flow rate (0.5 to 5.0 L min^{-1}). The effect of power (250, 450 and 650 W) depended on gas flow rate, with temperature increasing with power when a gas flow of 1.5 L min^{-1} was used, but the trend reversing when the gas flow was increased above this. They also measured n_e using Stark broadening of H_α by using Levenberg–Marquardt nonlinear fitting algorithm to divide the Voigt width into its Gaussian and Lorentzian contributions, and taking into account both Doppler and van der Waals broadening. They assumed a T_e of 15000 ± 5000 K in order to calculate n_e, which showed an increase from ca. $0.5 \times 10^{15}\ cm^{-3}$ to $3.0 \times 10^{15}\ cm^{-3}$ as gas flow rate was increased from 0.5 to 1.5 L min^{-1}, but remained relatively constant with power. Orejas et al.[87] have taken 2D images of an atmospheric pressure, helium GD source using zoom lenses coupled to an intensified CCD detector. Wavelength regions were isolated using bandpass filters to isolate emission produced by He*, N_2^+ and O* revealing that He*, N_2^+ were concentrated in regions close to the electrodes whereas O* emission appeared in most of the plasma volume. They also observed that higher helium flow rate and greater interelectrode distances lead to higher emission signals for all species.

The use of mixed gas plasmas, or gases other than argon, in ICP-OES has been around since the first reports of the technique in the 1960s, whether for economic reasons or to modify the temperature of the plasma in some way. Scheffler and Pozebon[88] found that addition of nitrogen (20 mL min^{-1}) to the nebulizer gas flow of an axially viewed argon ICP improved its robustness by a factor of two, as evaluated by the Mg(II) 280.270 nm : Mg(I) 285.213 nm ratio. They studied matrix effects, caused by 5 g L^{-1} of Na or K, on ten spectral lines with excitation energies between 11.5 and 15.5 eV and found that they were generally reduced by the addition of nitrogen without degrading the LOD or linear range. In some cases (for high excitation energy) an enhancement was observed, and the matrix effect of Ca was more variable. Any number of theories can be proposed for these observations but it is generally the case that changes to the size and shape of the central channel must be considered in addition to atomization and excitation mechanisms. This influence of plasma shape was well illustrated by Miyahara et al.[89] who studied the effect of nitrous oxide addition to an air ICP. Addition of 40% nitrous oxide increased the plasma volume by two-fold, but atom and ion line intensities increased, and they observed a decrease in T_{rot} from 3200 to 2700 K when nitrous oxide was increased from 0 to 100%.

Similar in some ways to the use of mixed gases, Grindlay et al.[90] reported on the effect of carbon (in the form of glycerol, citric acid, potassium citrate and ammonium carbonate) on the signal for hard-to-ionize elements: As, Au, B, Be, Cd, Hg, I, Ir, Os, P, Pd, Pt, S, Sb, Se, Te and Zn in ICP-MS. They found that some elements showed an enhancement while others were unaffected, and put this down to a combination of changes in the ion distribution in the plasma and charge transfer reactions between some of the enhanced elements and carbon-containing species in the plasma.

Single particle analysis has attracted much interest in recent years. Chan et al.[91] have studied the effect of single aerosol droplets on plasma impedance by using two methods: an RF-probe coil placed inside the torch; and by tapping the phase-detector signal of the impedance matching network. They found that there was a direct correlation between temporal spikes in the impedance and H_α atomic emission, which was linearly proportional to the volume of the introduced droplet. The measurements were made over short (ca. 10 ms per pulse) timescales and the method was proposed as a trigger for single droplet introduction into the ICP with appropriate tuning of the threshold value. In addition, the authors estimated that the energy change coupled to the plasma caused by a 50 μm droplet was between 8 and 11 W (0.03 to 0.04 J).

Niemax[92] published an overview of the single droplet approach to diagnostic studies in ICP-MS, with recommendations on how to improve detection efficiency. He emphasized the spatial nature of analyte atomization, ionization and diffusion processes in the plasma, and how this interacts with the position and size of the sampling orifice. He recommended the use of a narrow injector and high gas flow rate to shift the point of atomization downstream so as to optimize the position where temperature is high and analyte ions have not had much time to diffuse away from the particle origin. However, he pointed out that optimization is a complex business that also depends on analyte mass (heavier elements diffuse more slowly) and other operating conditions. In a related report, Flamigni et al.[63] studied aerosols and single particles produced by LA in the context of atomization, ionization and diffusion effects on ion sampling in ICP-MS. They used LA to generate aerosols containing Na_2O and CaO from NIST610 silicate glass, and elemental Na and Ca from their metals, and observed that maximum emission for the oxides occurred further along the central channel than the elemental forms, ca. +3 mm for Ca and +1.5 mm for Na. This spatial shift in vaporization resulted in a discrepancy between the relative extraction efficiencies of metallic vs. oxide aerosols of 25% for Ca/CaO to 97% in case of Na/Na_2O. Modelling of penetration depth revealed that this was more pronounced for particles larger than 1 μm. The main conclusion of this interesting study was that the inherent difficulty of calibrating LA-ICP-MS without matrix matched standards will require an approach which addresses this spatial difference in vaporization of particles. Some insight into these effects might be gained by studies such as that of Aghaei et al.,[93] who developed a computational fluid dynamics model to investigate the effect of varying the pressure behind the sampler, gas flow rates and applied power in an ICP-MS instrument. This allowed them to model gas pathways through the sampler cone, concluding that it is possible to control whether only the injector gas flow passes through the orifice, or if it is accompanied by some auxiliary gas flow. In a further paper[94] they used this to predict the effect of sampling cone position and orifice diameter on plasma temperature, electron number density, gas density and gas flow velocity; and the implications for extraction of ions formed from vaporizing particles travelling through the plasma. They conclude that a large sampler orifice favors extraction of both injector and auxiliary gas, and consequently large ion clouds, while a smaller orifice favors extraction of small ion clouds.

As if spatial heterogeneity in the ICP itself was not enough to contend with, there is also the time-worn problem of space charge effects in the post-sampler ion beam. Bergeson et al.[95]used LIF to study space charge effects on Ca, caused by the matrix species Cs, Mg, and Pb, in the second vacuum stage of a commercially available ICP-MS instrument with right-angle ion-reflecting geometry. They used a CW-Ti:Sapphire laser, frequency doubled to 393.4 nm, to excite the Ca(II) $4s^2S_{1/2} \rightarrow 4p^2P_{3/2}$ transition, and monitored the subsequent fluorescence at 854.2 nm from the ion beam at the entrance aperture to the third vacuum stage using a CCD detector. Interestingly, they found that the Pb matrix (1.5 mM) caused an off-center deflection in the Ca ion beam rather than the more commonly assumed defocusing effect. They concluded that this could be corrected for by on-the-fly adjustment of the ion optics, with the caveat that this effect may well be specific to the type of ion optics studied.

Houk›s group[96] have continued their studies of metal argide (MAr^+) polyatomic ions of Co, Cr, Cu, Fe, Mn, Ni, Ti, V and Zn in the ion extraction interface of an ICP-MS instrument. They compared experimental measurements of MAr^+/M^+ ratios with theoretical ratios calculated from the expected thermodynamic equilibrium at T_{gas} = 6000 K, and found that the experimental ratios were far lower than the calculated ratios, with the weakest bound species exhibiting the largest discrepancy. Increasing the ion kinetic energies by increasing the plasma potential, via a homemade guard electrode between the RF coil and torch, was observed to further reduce the experimental ratios, and adjusting the skimmer cone geometry also had an effect. Taking this evidence in total they postulated that the low experimental ratios were caused by collision induced dissociation (CID) due to a shock wave at the tip of, or in the throat of, the skimmer cone, thereby yielding far fewer MAr^+ ions than expected. Bouchoux et al.[97] published a theoretical investigation into interferences on Se in ICP-Ms. They estimated heat of formation, ionization energy (IE), and proton affinity (PA) for Se, SeH, SeH_2, ArH, ArH_2, Ar_2, Ar_2H, Ar_2H_2, ArCl and ArHCl using quantum mechanical methods. Three composite methods (G3, G4, and W1) were used with accuracy better than 5 kJ mol^{-1} for ΔH°_{f}, PA, and IE. They also used CCSD(T) calculations using the aug-cc-pVTZ basis set for the neutral non-covalent complexes which were thought to mainly involve dispersive forces. Hence, they were able to make predictions about the thermochemical reactions between the ions $Se^{\cdot+}$, SeH^+, $SeH_2^{\cdot+}$, SeH_3^+, $Ar_2^{\cdot+}$, Ar_2H^+, $Ar_2H_2^{\cdot+}$, $Ar_2H_3^+$, $ArCl^+$, $ArClH^{\cdot+}$, and $ArClH_2^+$ with neutral gases H_2, CH_4, NH_3, O_2, CO, CO_2, NO, and N_2O used in DRC-ICP-MS. Results were presented in tables of data and as thermochemical ladders, including previously unreported values for Ar and Ar_2 hydrides and ArCl-containing neutral and ionized species. A full discussion of these results is beyond the scope of this review so readers interested in optimizing the performance of DRC-ICP-MS for Se determinations are advised to refer to the full paper.

Interferences

Dettman and Olesik[98] have used a pLTE model applied to ICP-OES to correct line-dependent changes in analyte sensitivity caused by matrix-induced changes in plasma temperature. Coupled with an Mn internal standard to correct for sample transport efficiency, they reduced matrix effects caused by 1.0 mol L^{-1} Ba and Na and 1.5 mol L^{-1} Li and Ca to less than 20% for a range of analytes. Errors caused by 1.5 mol L^{-1} La were likewise reduced to less than 40%. In a two part study, Chan and Hieftje[99,100] used spatial measurements of analyte emission across an axially-viewed ICP to indicate the presence of spectroscopic interferences, or those related to either the plasma or sample introduction system. They measured the relative analyte intensity (defined as the signal from the analyte in the presence of the matrix divided by the signal without the matrix present) over the cross-section of the axially viewed ICP to give a spatial pattern equivalent to the analyte concentration in the presence of the interferent. This was done using Ba to induce plasma-related matrix interferences, mineral acids to cause sample introduction effects, and two spectrally interfering line pairs (As and Cd at 228.81 nm, and Er and Co at 239.73 nm). In an elegantly simple approach, they interpreted a flat profile to mean absence of interference and a curved profile to mean that an interference was present. In the second study, they developed a statistical protocol to decide whether the profile was flat or curved and hence flag the presence of an interference.

Chemometrics

For the purposes of this review, chemometrics relates to a range of mathematical techniques and algorithms used for optimization, interpretation of analytical data, correction of interferences, and instrument calibration. Very little novel work has been published during the review period, except in the area of metrology wherein Mermet and co-workers have published two related papers[101,102] on accuracy and precision. In the first paper they computed an accuracy profile, based on the ISO 5725-1, 5725-4 and 21748 standards which are normally used to estimate the standard deviation of interlaboratory experiments. They used these methods to determine tolerance intervals as a function of concentration by computing the intermediate precisions for a range of analytes, in simple solutions, measured by ICP-OES and ICP-MS. This is rather time consuming so they conceded that it is best suited to method validation when a range of possible sample concentrations is likely. In the second paper[102] they applied the method to determine the LOQs for Pd (ICP-MS), U and B (ICP-OES). They concluded that the LOQ determined by the accuracy profile method was the most pessimistic (and hence most realistic), and that the weighting of the calibration regression was important, with a 1/y weighted regression being a good compromise.

LASER-BASED ATOMIC SPECTROMETRY

This section highlights the most recent developments in laser-based atomic spectrometry. In particular, it relates to the main atomic spectrometry techniques where the laser is used as either an intense energy source or a source of precise wavelength, including laser induced breakdown spectroscopy (LIBS), laser atomic absorption spectroscopy (LAAS), cavity ringdown spectroscopy (CRS), laser induced fluorescence (LIF), and other tandem techniques. The use of lasers for fundamental studies of the properties of atoms and for thin film deposition will not be covered here. Studies related to laser ablation (LA) as a method of sample introduction are covered in Section 1.3.2.1.

Laser Induced Breakdown Spectroscopy (LIBS)

Laser induced breakdown spectroscopy is one of the most important techniques where the laser is used as an energy source to induce an optical plasma. This section describes the latest instrumental developments and fundamental studies related to LIBS, but not applications.

Fundamental Studies

Determination of plasma temperature and electron density from Mg(II) emission spectral line broadening was investigated by Choi et al.[103] using double-pulse LIBS. This configuration was employed because it provided enough energy to sustain the plasma emission during the time period when the line width was at a minimum. Cvejić et al.[104] used single-pulse LIBS and high speed photography to calculate Stark broadening parameters for Mg(I) and Mg(II) lines. The high speed photography was used to determine the time of plasma decay which provided the highest plasma homogeneity. Aguilera et al.[105] reported the measurement of Stark widths of 53 Ni(II) lines in the LIBS spectrum using an experimental procedure based on accurate measurement of the electron density through the Stark broadening of the H_α line. In a similar way, the Stark broadening of two Ti(III) lines were measured by Chen et al.[106] during the early phase (i.e., between 60 and 200 ns) of the ablation process. At this point, it should be noted that Parigger[107] reviewed the limitations on the use of hydrogen Balmer series for the calculation of Stark broadening, and thus for the determination of electron number density under certain conditions. An alternative method, based on matching synthetic spectra (iteratively calculated from an equilibrium model of plasma radiation) and experimental emissivity spectra (obtained by Abel inversion of intensity spectra) has been developed by Merk et al.[108] This was used to make spatially resolved measurements of plasma temperature and number densities of plasma species in an argon LIBS plasma at atmospheric pressure. They concluded that the spectral fit method provided more accurate plasma diagnostics than the Boltzmann plot and Stark broadening methods.

The structure of the LIBS plasma was studied by Bai et al.[109] using dual-wavelength differential spectroscopic imaging. This involved the use of two filters, alternatively placed in front of the CCD detector, to eliminate the background continuum signal from a specific emission line. A pure aluminum target in an argon atmosphere was used to observe the distribution of ambient gas and target evaporated species inside the ablation plume using different ablation regimes (e.g., different laser fluences and pulse durations). This study demonstrated the key role that laser fluence has on the internal structure of the LIBS plasma. Similarly, Eschlböck-Fuchs et al.[110] investigated the expansion dynamics and the optical emission of LIBS as a function of the temperature of sample materials such as aluminum alloys, silicon wafers and metallurgical slag. They observed that higher temperatures lead to larger plume volumes and brighter broadband plasma emission in air, though other plasma parameters were found to be independent of temperature, at least for the slag samples.

Freeman et al.[111] studied ns and fs LIBS operated in vacuum and at atmospheric pressure conditions. They investigated the dynamics of plasma evolution and characteristics of the emission spectra and observed that the fs-LIBS plasma plume expanded in a narrower angular distribution, probably due to a strong, self-generated magnetic field created at the onset of plasma expansion. They also noticed that ns-LIBS spectra showed transition lines from lower-charged ions, while fs-LIBS spectra were mainly dominated by neutral species emission. The fs-LIBS technique was employed by Banerjee et al.[112] for the detection of submicron buried layers inside a semiconductor material (e.g., Si) by making use of the precise hole drilling capabilities of fs-LA. A correlation between depth resolution and processing speed was observed when carrying out these analyses. Ahmido et al.[113] used fs-LIBS to detect visible to NIR signatures from molecular fragments of the nitro (NOx) group after the ablation of potassium nitrate (KNO_3) and sodium nitrate ($NaNO_3$) coatings. A potential application of this method could be the detection of nitro-group energetic materials at standoff distances.

Identification of Spectral Lines

Barefield et al.[114] reported the identification and assignments of complex atomic emission spectra of mixed actinide oxides (e.g., UO_2, PuO_2, AmO_2, and NpO_2 simulated fuel pellets) using LIBS. They concluded that a resolving power of approximately 20,000 was adequate for analyzing complex mixtures of actinide elements within the same sample. Gaft et al.[115] studied the Zr LIBS spectra in the UV region (190–240 nm) at the onset (150–200 ns) of plasma formation in air. This allowed the identification of emission and absorption lines from highly ionized Zr ions (e.g., Zr(III) and Zr(IV)). New emission lines from Zr(I) and Zr(II) were observed in this spectral region.

Spectra in LIBS are sometimes complex and difficult to interpret, thus Labutin et al.[116] have developed an algorithm to automatically identify emission lines. This simulates the spectra corresponding to different temperatures and electron densities, then searches for the best correlated pair of experimental and model spectra constructed using information about atomic and ionic lines, broadening mechanisms and the parameters of the spectrograph. The model then attributes lines to the peaks. The algorithm was successfully applied to interrogate the spectral range from 393.34 to 413.04 nm, identifying 41 peaks.

Calibration Methods

A novel, optimized calibration procedure has been developed by Frydenvang et al.[117] The method was based on a combination of empirical and theoretical methods to determine elemental ratios. It was demonstrated to have potential for fast, real-time, in situ analyses due to the fact that sample preparation is minimized and processing time is negligible. Self-absorption is an important parameter that might affect calibration. Usually, self-absorbed emission lines are disregarded or corrected because they would otherwise introduce non-linear effects into the calibration. However, Cristoforetti et al.[118] have demonstrated that self-absorption might also provide useful information when characterizing the composition of laser-induced plasmas. The poor reproducibility of LIBS also affects the calibration, and efforts have been made to improve it, including experimental optimization, chemometric techniques or the use of internal standards. Criteria used for selection of the most appropriate internal standard have included using the most intense spectral line of a matrix element, the closest matrix line to the analytical one, or the use of a spectral line with a close upper energy level. Labutin et al.[119] have investigated an alternative criterion, based on using the best correlated pair of lines observed for variations in laser energy, and successfully used it for the analysis of high alloy steels and soils. The approach was shown to work even under non-LTE conditions.

Instrumentation

A historical review of the first five years of laser based micro-spectrochemical analysis was recently published by Baudelet et al.,[120] covering the first measurement systems after the invention of the laser in 1960 to its evolution to conventional LIBS. Nowadays, very compact (palm-top size) and powerful sub-nanosecond pulse width lasers are being investigated. For instance, Bhandari et al.[121] have developed a megawatt peak power, 266 nm UV microlaser based on a passively Q-switched Nd:YAG/Cr^{4+}:YAG microchip laser, which might be of great interest for the design of portable-LIBS systems. Scharun et al.[122] have designed a LIBS setup with a multi-kHz fibre laser which can be used for handheld quantitative analysis of metallic samples without mechanical sample preparation.

Time-resolved, thermodynamic characterization of small, high-jitter powerchip LIBS plasmas was performed by Merten et al.[123] under a variety of atmospheres (e.g., air, Ar and He). This fast-evolving plasma was evaluated for LTE in terms of both the McWhirter and temporal criteria. The former considers the electron/heavy particle collision rate required to guarantee that the total effective population and depopulation rates are dominated by electron collisions, and the latter examines the evolution of temperature and electron number density and compares their rate of change to the rate at which electron collisions can thermalize the change. These considerations were used to conclude that LTE might be present at the early stage of the powerchip LIBS plasma generation, especially under an argon atmosphere, making possible calibration-free LIBS measurements.

New Developments

Laser ablation molecular isotopic spectrometry (LAMIS) is an interesting approach for real-time isotopic analysis of samples. It is known that isotope shifts in atomic spectra are small (e.g., a few pm) because the electronic transitions are slightly affected by differences in isotopic masses. Hence, isotopic measurements in LIBS atomic spectra would require high spectral resolution and/or the use of low pressure conditions. In comparison, optical isotope shifts in molecular spectra are much higher (e.g., hundreds of pm) because the vibrational and rotational energy levels are more affected by the differences in isotopic masses. Excited molecular species are produced during LIBS plasma expansion through different chemical reactions, the precise nature of which is still relatively unknown. In LAMIS, emission spectra are collected at longer delay times after the laser–matter interaction to allow the formation of the excited molecules in the plasma. LAMIS is considered of great interest for the fundamental understanding of plasma properties and for different isotopic applications. For instance, Dong et al.[124] used LAMIS to study C_2 molecular formation, by observation of the Swan bands, in laser ablation of fabricated carbon isotopic samples containing different amounts of ^{13}C. This was performed in a neon atmosphere at 0.1 MPa. Measurement of the emission band demonstrated that C_2 was formed by a recombination process and that the $^{12}C/^{13}C$ ratio varied with time after the laser pulse and the distance above the sample surface.

Another variant of conventional LIBS, called nanoparticles enhanced LIBS (NELIBS), was investigated by Giacomo et al.[125] The method is based on the deposition of a drop of noble metal colloidal solution on the surface of a solid conducting target, in order to enhance the LIBS emission intensity. It was shown that the NELIBS enhancements were related to changes in the ablation process because a significant decrease in the ablation threshold was detected for all of the investigated metal samples. However, no significant differences were observed in the temperature and electron number density for NELIBS compared with LIBS.

LIBS is a powerful analytical technique for the analysis of samples placed in a gas environment (e.g., air, Ar, He) or in vacuum, but also for the analysis of samples immersed in liquids. Lazic et al.[126,127] studied LIBS plasma and vapour bubble formation after ablation of Al targets underwater, using dual pulse excitation, by fast photography and time resolved spectroscopy. Position-selective emission spectroscopy of a LIBS plasma produced on an Al target in water was investigated by Matsumoto et al.[128] They observed that the density at the edge of the plasma produced by long-pulse irradiation (ca. 100 ns) was low enough to achieve well-resolved, narrow spectral lines, even without time-gated detection. These results might change the current concept of the instrumentation for underwater LIBS and open new possibilities for in situ chemical analysis of solid materials in water.

The combination of LIBS with diode-laser LAAS and LIF has been investigated by Khater[129] to improve the analytical capability and figures of merit of LIBS for trace analysis of light elements (e.g., C, S, P and N) in steel samples.

Laser Atomic Absorption Spectroscopy (LAAS)

A method for time-resolved, tunable diode LAAS (DLAAS) has been developed by Adamek et al.[130] An electronic module was designed to provide two signals that were used to trigger the fine tuning of the laser wavelength and to control time-defined signal sampling from the absorption. This method was successfully used to obtain time-resolved measurement of density and temperature of Ar* in a pulsed discharge. On-line and real-time quantitation of Xe in argon was performed by Jacquet et al.[131] using tunable DLAAS in a DC glow discharge. Xe is a fission product of U that must be monitored, for safety reasons, on-line in the cover gas of a sodium cooled fast neutron reactor. In this study, continuous wave laser absorption from the lower metastable state at 823 nm was used to measure the isotopic and hyperfine structure of Xe isotopes, and to individually quantify the odd isotopes.

Cavity Ringdown Spectroscopy (CRDS)

The background and principles of CRDS and various applications of this technology were recently reviewed by Zhang et al.[132] Significant developments of plasma sources, lasers, and cavity optics were reported, highlighting the great analytical potential of plasma-CRDS for elemental detection and isotopic measurements. Nevertheless, it was pointed out that further efforts are still needed in fields such as cavity design, plasma source design, instrumental improvement and integration, and measurements of radicals and molecules.

Biscan et al.[133] investigated the production of 2^3S_1 metastable helium in a laser-induced plasma at low pressure. The absorption CRDS technique was employed to measure the density time evolution of metastable helium and ablated atoms, produced during the ns laser ablation process of various materials (e.g., graphite, aluminum and iron) in a low pressure (e.g., 0.03–2 mbar), helium atmosphere. They concluded that the width of ablated atom lines was mainly related to directional movement of atoms and that the appearance of metastable He might be attributed to excitation in collisions with ejected electrons and heavy excited particles.

Laser-Induced Fluorescence (LIF)

Frankevich et al.[134] have coupled LIF with a differential mobility analyzer in order to obtain two-dimensional information on the electrical mobility and fluorescence spectroscopy of gas-phase ions. The authors claim that this unique instrument combination could be of great interest to investigate the physical and chemical properties of fluorescent proteins.

Many light atoms (e.g., H, O, C) have LIF transitions in the VUV spectrum that are difficult to excite. Two-photon LIF has been developed to overcome this difficulty by exciting the LIF transition by absorption of two photons. One problem with this approach is that amplified spontaneous emission (ASE), which is created by the population inversion between the excited and the de-excited states, might affect the linear response of the LIF signal with respect to the number density of the target species and laser intensity. With this in mind, Li et al.[135]investigated, using a Monte Carlo model, the effects of line-narrowing of ASE, which is due to the fact that ASE photons near the central frequency are amplified more effectively. Consideration of these line-narrowing effects was necessary to avoid significant errors in the interpretation of the LIF and ASE signals.

ISOTOPE RATIO ANALYSIS (IRA)

Isotope ratio analysis is dominated by plasma source ionization, with sample introduction either by laser ablation or solution nebulization. Thermal ionization is still a valuable low-blank method which contributes to a number of applications. Precise measurements of isotopic abundance are used in a variety of disciplines including nuclear forensics, geochronology, stable isotope fractionation, isotope dilution analysis, isotope tracer studies and archaeological provenance. Instrumental developments continue to allow isotopic measurements to be made on smaller samples and more robust precision has been achieved by the application of poly-spiking techniques.

It is always appropriate to periodically re-assess the principles and knowledge-base. In this respect, a review by Baxter et al.[136] was a welcome assessment of some fundamental aspects of SF-MS. Factors which influence the IR measurement, at all stages of the process from the true ratio, through the measured ratio, to the final corrected ratio, were deliberated. The effects of Poisson counting statistics and flicker noise on ratio precision were treated mathematically, and accuracy offsets via abundance sensitivity tailing were discussed. Usefully, a thorough consideration of mass discrimination was contained in this tutorial, reviewing both the correction functions and the methodology of internal and external normalisation, as well as sample standard bracketing. Finally there was a comprehensive evaluation of dead time correction including a systematic review of the differing estimation methods which will be of great use to novice and experienced users.

Instrumental Developments

A common thread of much IRA research is a drive to lower the amount of analyte required to generate a particular, or at least useful, level of precision. Such error-reduction has often been achieved by increasing the sensitivity of the MS technique, thereby requiring less sample to achieve signals significantly above noise level. Where the amount of analyte is small, the method of sample introduction to the mass spectrometer, particularly for plasma ionization instruments, is crucial to maximising sensitivity. In such cases, the sample dilution needs to be minimised to maintain suitable signal levels of the isotopes of interest. At the same time, the volume of sample consumed during the analysis also needs to be restrained to allow acquisition of a statistically viable population of isotope ratio data. Paredes et al.[137] have furthered their previous investigations[138] into low-flow nebulization, using continuous sample introduction rates of <10 μL min^{-1}. Validation of IR determinations was completed with Sr isotopes using an internal $^{88}Sr/^{86}Sr$ mass bias correction. Best results were obtained by heating the spay chamber walls to 65 °C and using a flow rate of 9 μL min^{-1}. Sensitivity of the MC-ICP-MS still ultimately determined the amount of element consumed in the analysis and, in the optimal case, a sample solution of ~250 ng mL^{-1} was used to maintain ^{86}Sr signal levels at >0.1 V. In a companion paper[137] a torch integrated sample introduction system (TISIS) was evaluated with respect to how mass discrimination effects on Sr isotopes can be controlled by sample flow rate and spray chamber temperature. Essentially, heavy-biased mass discrimination effects increased with higher flow-rates (15 μL min^{-1}) and higher spray chamber temperatures (100 °C). The authors hypothesised that this was related to a more effective solvent evaporation process in the TISIS at higher temperatures, and a reduction in water load at low sample flow rates. The other interesting finding was that low flow rates (5 μL min^{-1}) and high temperatures resulted in a degradation of the exponential law correction model for the mass discrimination process. It was suggested that this stems from irregularities in the aerosol formation processes which, in turn, could affect mass-dependent diffusion processes in the plasma.

There have also been developments in detection systems, such as a new generation of Faraday detectors equipped with low-resistivity (10^{12} Ω) resistors reported by Koornneef et al.[139]These detectors were designed to provide lower baseline noise compared to the conventional 10^{11} Ω amplifiers, and were used to measure $^{87}Sr/^{86}Sr$ and $^{143}Nd/^{144}Nd$ for 1 ng amounts of analyte. Comparison with previous studies indicated that reproducibility was improved by factors of 2 and 1.4, respectively. Combined with increases in instrument efficiency, this has the potential to benefit applications such as radiometric dating or characterising isotopic disequilibrium in geological samples.

Hu et al.[140] assessed the effect on sensitivity of fast interface pumping and the X-skimmer and Jet-sampling cones in a commercial MC-ICP-MS instrument. Combinations of new and old cones were trialled for Hf, Lu and Yb with a view to optimizing Hf IR measurements in zircons by LA. Results indicated that the X- and Jet-cone combination improved Hf sensitivity by a factor of ~5 compared to the standard cone combination. An interesting feature of the analysis was that the Jet-sampling cone resulted in a non-linear mass-dependent fractionation that could not be corrected by standard fractionation laws. However, it was found that this fractionation effect was inhibited by the addition of small quantities of nitrogen to the central argon-helium flow from the collision cell.

The effects of mass discrimination in the sampling and post-sampling regions of ICP-MS was examined by Kivel et al.[141] Using tantalum or graphite targets positioned close behind the skimmer cone orifice, elements ranging in mass between Li and U were admitted under wet plasma conditions to the interface. Material deposited on the target was then analysed both in bulk by solution nebulization, and as individual points along a radial transect from the central axis by laser ablation. The bulk measurements revealed a heavy-isotope bias, indicating that extraction of ions from the plasma contributed in the order of 50% of the measured discrimination. Radial measurement of Cd isotope ratios demonstrated a heavy-isotope bias at the axis which progressed to a relatively light-isotope bias outwards. As the ion beam was subject to "clipping" downstream from the target position, this was taken by Kivel et al. to indicate that further light-ion removal completed the remaining fraction of the instrumental mass discrimination.

Sometimes major improvements in analytical efficiency may be too good to be true. This may well be the case with intracavity optogalvanic spectroscopy (ICOGS) for measuring radiocarbon isotope systematics. Initial reports of this technique (Murnick, D. E.; Dogru, O.; Ilkmen, E. Anal. Chem. 2008, 80, 4820–4824) suggested that it could match the LOD achieved by AMS for $^{14}C/^{12}C$. At the time, this report generated significant interest because of the potential to simplify the analysis, reduce costs and increase the size of datasets. However, Perssonet al.[142] have produced a frank assessment of the technique based on a comprehensive dataset and they have not been able to reproduce the reported optogalvanic signal dependence on ^{14}C concentration and wavelength. In addition, they have now estimated that the detection limit for ^{14}C is likely to be > 2 orders of magnitude higher than originally estimated.

Double-Spike Systematics

Conventionally, double spiking is used for in-run instrumental mass-bias corrections in ICP-MS and TIMS. De-convolution using a double spike to give bias-free isotope ratios requires an element with a minimum of four accurately measurable isotopes in the natural or the with-spike mixture. However, by using gravimetric mixing of the sample with different proportions of an isotopic spike, Chew and Walczyk[143] have developed a modification of the double spike technique capable of high-precision IR measurements using only three isotope signals. This gravimetric spiking IDA (GS-IDA) approach has the potential to increase the precision of Mg isotope ratios, benefiting studies of metabolism in humans and elemental diffusion in mineral species. Application of this technique could open other suitable isotope systems to a similar spike correction protocol.

Instigation of the double spike or GS-IDA techniques for determining isotopic composition requires calculation of the appropriate spike composition and an understanding of the optimum sample-spike mixing proportions. John[144] has developed models where the concentrations of the spike and sample are allowed to vary independently, rather than restricting spike-plus-sample to a fixed concentration mixture (within the bounds of the signal capacity of the detector). The result was a decrease of up to 30% in the theoretical error, or greater reductions where isobaric interferences were present.

Double spike systematics have also been applied to the analysis of Pt stable isotope ratios.[145] Using two separate Pt double spikes, the $^{198}Pt/^{194}Pt$ ratio could be reproduced to within 10^{-5}amu^{-1}. These findings indicated that, despite variable amounts of isobaric ^{192}Os, accuracy of the Pt isotope ratio was unaffected because ^{192}Pt was not used in the double-spike de-convolution. Application of this technique provided a redefined Pt atomic weight and indicated that Pt stable isotope fractionation is quantifiable above the reproducibility of the method.

Stable Zn isotopes have been the focus of a study by Makishima and Nakamura,[146] who have developed a low-blank method for Zn separation using TEVA resin, followed by a double spike determination of $\delta^{66}Zn$. Blank levels were reduced from greater than 1 ng reported in earlier studies to around 50 pg, enabling $\delta^{66}Zn$ to be reproduced to 0.004% RSD (2σ) for 100 ng of Zn. This method will prove valuable in high-resolution studies of Zn where material is limited and Zn concentrations are low.

Shalev et al.[147] compared MC-ICP-MS and TIMS for measuring the non-radiogenic $^{88}Sr/^{86}Sr$ ratio using a double spike. Results of the double spike bias corrections were found to be more reproducible than sample-standard bracketing for MC-ICP-MS. The two instruments produced the same $^{88}Sr/^{86}Sr$ values, with TIMS having smaller within-run errors but needing a longer integration time to achieve this.

Nuclear Forensics

Kraiem et al.[148] have introduced a method for determining the mass of uranium in single micron particles. This is driven by the need to develop particle reference materials for nuclear safeguards, which would be used to evaluate mass and isotope abundances of uranium in such materials. Their method used ID-TIMS, where individual or multiple uranium oxide particles were mounted on a filament and spiked with ^{233}U in an acid medium. A separate measurement of a particle was made without the spike to determine the U isotope ratios. The estimated error for the calculated U mass was better than 10%, thought to be related to incomplete dissolution of the particle on the filament. The precision of the $^{235}U/^{238}U$ measurement was found to be better than 1% (2σ).

The need for rapid yet accurate analysis of isotopic materials has made LA a potential sampling tool for safeguards and nuclear forensic investigations. Kappel et al.[149] used a nanosecond laser coupled with MC-ICP-MS to directly analyse U isotope ratios in single U-doped glass particles. Uncertainties were better that 1.4% and 6% (2σ) for the $^{235}U/^{238}U$ and $^{236}U/^{238}U$ ratios, respectively. A similar study of direct measurement of uranium isotopes in enriched uranium pellets and rods was undertaken by Marin et al.,[150] using a 266 nm Nd:YAG LA–SF–ICP-MS. A combined precision and accuracy of <2% RSD (σ) for the $^{235}U/^{238}U$ ratio was achieved. Each study defined a reproducibility window between 0.5% and 2% as the state of the art for $^{235}U/^{238}U$ determinations on individual particles, setting the analytical benchmark for nuclear forensic methods in the future.

Efforts to measure $^{236}U/^{238}U$ in samples with ultra-low ^{236}U have been advanced by Tanimizu et al.[151] using tandem Q-ICP-MS/MS. A key feature of this analysis was the oxidation of U in the reaction cell between the tandem quadrupoles. In this reaction only 1% of the UH^+ was converted to UOH^+. The reduction in $^{235}UOH^+$ production enabled $^{236}U^{16}O^+/^{238}U^{16}O^+$ to be measured down to 10^{-9}, which approaches the analytical LOD of TIMS.

As with many isotope systems reviewed, the common aim is to increase sensitivity, reduce background interference and improve precision of analysis, ultimately to enable IR measurement of ever smaller quantities. One requirement in the nuclear field is the monitoring of humans, particularly if exposed to anthropogenic sources of uranium in their occupation, or as a result of proximity to a nuclear installation. Arnason et al.[152] took on the challenge of measuring <40 pg mL^{-1} of uranium isotopes in urine. Solid phase extraction of the urine matrix followed by U isolation using a double-washed UTEVA resin minimized the blanks to <0.3 pg, i.e., ~1%. Based on a urine sample of as little as 1 mL and ≥50 pg U, the $^{235}U/^{238}U$ ratio was measured within 2% (2σ), which is a useful level of detection for monitoring.

Where the quantity of uranium is not restricted, $^{235}U/^{238}U$ can be determined by more conventional mass spectrometric techniques. Mathew et al.[153] used sample loads of 1 to 2 μg U to assess the total evaporation TIMS method. One finding was that the mass bias, assessed by various U standards, was consistent within individual analytical turrets (batches), but variable between turrets. Using these standard-based 'external' estimates of the mass bias resulted in precision better than 0.04% (2σ) for the $^{235}U/^{238}U$ ratio, which approaches the precision attainable by U double spike analysis. However, this external method could not detect if an individual sample had incomplete matrix separation, which may influence the degree of mass bias.

Stable Isotope Fractionation

Investigations into stable isotope fractionation of heavy elements has been the focus of a number of studies. Ohno and Hirata[154] have investigated the variation of Ce and Nd isotopes in natural geological reference materials. Interestingly, carbonate standards were found to have significantly higher Ce and Nd isotope ratios compared to igneous and other sedimentary rocks. However, no difference in the amount of isotopic fractionation was found between the two elements, implying that this fractionation is independent of changes in redox state.

Pramann et al.[155] have investigated a potential source of "broadband" interference across the Si mass spectrum in MC-ICP-MS. This took the form of a broad raised background centred at ^{29}Si–^{30}Si, present in NaOH-containing solutions but absent from KOH solutions. It appeared that this interference was detector specific because it was not observed in the central Faraday cup, only in the high- and low-mass Faraday arrays. The logical conclusion is that the interference, which significantly degrades Si isotope ratios, was caused by ion scattering in the presence of high sodium concentration. Absence of the interference from the centre Faraday was most likely related to this detector's position in a shielded location offset behind the main high-low Faraday array.

As often happens with analytical developments, sometimes two papers on essentially the same technique emerge simultaneously. Yang et al.[156] and Sanchez-Lorda et al.[157] are an example of such. Elements with relatively similar ionization efficiencies can experience similar levels of instrumental mass bias. For elements with different efficiencies, such as Hf and Yb, mass bias interrelationship can be formulated during, or prior to, the measurement. Hence, it is a logical progression to examine if elemental concentration ratios and isotope ratios can be measured within an individual multicollector analysis. In both of the aforementioned papers MC-ICP-MS was trialled for the measurement of $^{143}Nd/^{144}Nd$ in the same run as $^{147}Sm/^{144}Nd$. These ratios are of particular interest because they form the basis of Sm-Nd geochronology, and because they are preferably measured from the same dissolution and chemical separation. Conventional analysis of the Sm-Nd system is by TIMS or MC-ICP-MS and requires IDA to determine the concentration ratio and chemical separation of Sm from Nd. Both studies used Nd isotopic standards doped with variable amounts of Sm to replicate the likely natural range of Sm/Nd, and formulated a relationship between the exponential mass bias coefficients to correct the isobaric ^{144}Sm interference. Reported errors on $^{143}Nd/^{144}Nd$ and $^{147}Sm/^{144}Nd$ were similar for the two studies, and were roughly equivalent to measurements made by conventional methods. Sanchez-Lorda et al.[157] reported that $^{147}Sm/^{144}Nd$ reproducibility ranged from 0.2% to 0.7%, which is slightly higher than for IDA measurements. A significant difference in the papers is that these workers also included a new methodology to isolate the rare earths from the matrix elements prior to the combined-run $^{143}Nd/^{144}Nd$, $^{147}Sm/^{144}Nd$ MS analysis.

Recent improvements in sample introduction and plasma sampling have produced a step-change in the sensitivity of ICP systems. Using a PFA microflow nebulizer, desolvator and X-skimmer/Jet-sampler cones, Huang et al.[158] sought to precisely determine Nd isotope ratios for ng levels of analyte using MC-ICP-MS. They achieved a sensitivity of >2000 V per μg $mL^{-1}$for Nd. At this elevated level of detection, a 2.5 ng mL^{-1} solution produces a 1 V signal for ^{146}Nd ($\sim 1 \times 10^{-11}$A ion current). The analysis of small volume samples (0.5 mL) containing 0.5 ng of Nd resulted in a $^{143}Nd/^{144}Nd$ ratio for the JNdi-1 standard of 0.512101 ± 0.000029 (2σ, n = 6). With 5 ng of Nd the errors were reduced to 0.512124 ± 0.000010 which is comparable to typical long-term reproducibility statistics for NdO^+ measurements by TIMS. Huang et al.[158] have used this advance in sensitivity to evaluate the potential for this method to tackle Nd isotope determinations on seawater. Using a 5 km seawater sampling station off Bermuda as a trial for the method, the results were comparable to previously achieved Nd isotopic profiles.

Laser Ablation Isotope Analysis

In situ analysis of materials by LA continues improve with respect to the precision of IR measurements. Chen et al.[159] used a fs-laser in combination with MC-ICP-MS to examine the Pb isotope ratios in copper-based coins. Due to its high Pb content bronze is an idea material for this type of analysis. The reported precision of <60 ppm (1σ) for $^{208}Pb/^{204}Pb$ approached the level achievable by solution analysis. However, the key result of the study was the ability to rapidly discriminate between coins of different dynasties and detect heterogeneity of metal sources within an individual dynastic currency. This is quite interesting from an archaeological perspective but also has implications for the forensic analysis of ballistics and discrimination of other metal components.

The IRA of Pb by LA-MC-ICP-MS was also tackled in a study by Darling et al.[160] who compared the results from direct measurement with samples where Pb was chromatographically separated from the matrix. A suite of samples from the Sudbury impact structure, in particular the included sulfide minerals chalcophyrite, pentlandite, pyrite and pyrrhotite, were the focus of the study. Encouragingly, the Pb isotope ratios generated by the direct sampling and by chemical isolation of Pb were within analytical uncertainty, typically 0.1% (2σ) for $^{206}Pb/^{204}Pb$. This was achieved despite bias corrections being made to non-matrix matched external standard measurements. Common problems encountered with sulfide mineral analysis are the isobaric and molecular interferences from oxides, argades and sulfides of Re, Os, Pt and the rare-earth elements. However, by restricting oxide production to <0.5%, and monitoring the$^{205}Tl/^{203}Tl$ ratio, it was concluded that these isobaric interferences did not influence the measured Pb isotope ratios. An interesting facet of the results is that analyses from each sample, including the solution analyses, formed linear arrays in Pb-isotope space (e.g., $^{207}Pb/^{204}Pb$ – $^{206}Pb/^{204}Pb$). This suggests a isochronous relationship, and helps define the initial (at formation) isotope ratios across the spectrum of samples studied.

A major issue encountered in LA analysis is concentration calibration. This has often been addresses by normalising analyte signals to other elements of known concentration, or by the co-introduction of a calibrated spike solution. The best accuracy can be achieved using IDA, which Claverie et al.[161] demonstrated using a spinning platform within the ablation cell. Sample material and an isotope spike in solid form were located adjacent to each other on the sample holder, such that fast rotation (~2000 rpm) of the sample under the laser resulted in a quasi-simultaneous ablation of the spike and sample. Initial trials also included the addition of a Gd spike to both sample and reference materials to correct for any differences in the ablation rate caused by position or material properties.

In some instances, e.g., for large samples, it is not appropriate to undertake laser ablation in a standard ablation cell. Dorta et al.[62] experimented with fs-LA of samples in air, followed by conversion to an argon carrier in a gas exchange device, and introduction to a MC-ICP-MS instrument. Results for Hf, Pb, and Zn isotopes were comparable with air sampling and a conventional He atmosphere ablation cell. Single shot analyses produced an acceptable level of precision for archaeological applications.

Dating minerals by LA-MC-ICP-MS continues to evolve. Cottle et al.[162] employed a single-shot method with modified data reduction and error propagation routines for U–Pb dating of zircon and monazite. The method effectively integrated the total signal for each isotope, such that only a single isotope ratio was generated instead of a series of measured values. Key advantages of this method are that it counters any IR bias resulting from signal decay or response-time offsets caused by differences between Faraday and ion counting detectors. The method generated accurate and reproducible ages for standard zircons and monazites relatively rapidly, being around 6 times faster than conventional LA-ICP-MS. Probably the most appropriate application is where an age population is required amongst a large number of zircons. Examples of this include identification of age grouping of grains in detrital sediments and in screening zircon populations prior to more detailed age-selective analysis.

Glossary of Abbreviations

Whenever suitable, elements may be referred to by their chemical symbols and compounds by their formulae. The following abbreviations may be used without definition. Abbreviations also cover the plural form.

8-HQ	8-Hydroxyquinoline
a.c.	Alternating current
AA	Atomic absorption
AAS	Atomic absorption spectrometry
AIP	Atom-imprinted polymer
ASE	Amplified spontaneous emission
CID	Collision induced dissociation
CRDS	Cavity ringdown spectroscopy
CRM	Certified reference materials
CVG	Chemical vapour generation
DBD	Dielectric barrier discharge
DLLME	Dispersive liquid–liquid microextraction
DDTC	Diethyldithiocarbamate
DLAAS	Diode laser atomic absorption spectroscopy
DMA	Dimethylarsenic (include oxidation state if known)
DOF-MS	Distance-of-flight mass spectrometry
DRC	Dynamic reaction cell
EDXRF	Energy dispersive X-ray fluorescence
EPA	Environmental protection agency
ES	Electrospray

ESI-MS	Electrospray ionization mass spectrometry
ETV-ICP-MS	Electrothermal vapourization inductively coupled plasma mass spectrometry
FAAS	Flame atomic absorption spectrometry
FFF	Field flow fractionation
FI	Flow injection
GC-MS	Gas chromatography mass spectrometry
GD	Glow discharge
GD-MS	Glow discharge mass spectrometry
GE	Gel electrophoresis
GF	Graphite furnace
HG	Hydride generation
HPLC	High performance liquid chromatography
ICP	Inductively coupled plasma
ICP-AES	Inductively coupled plasma atomic emission spectrometry
ICP-MS	Inductively coupled plasma mass spectrometry
ICP-MS/MS	Inductively coupled plasma mass spectrometry/mass spectrometry
ICP-OES	Inductively coupled plasma optical emission spectrometry
ID	Isotope dilution
IDA	Isotope dilution analysis
IDMS	Isotope dilution mass spectrometry
IIP	Ion-imprinted polymer
IL	Ionic liquid
IR	Isotope ratio
IRA	Isotope ratio analysis
LA	Laser ablation
LA-ICP-MS	Laser ablation inductively coupled plasma mass spectrometry
LAMIS	Laser ablation molecular isotopic spectrometry
LC	Liquid chromatography
LC-MRM-MS	Liquid chromatography multiple reaction monitoring mass spectrometry
LEP	Liquid electrode plasma
LIBS	Laser induced breakdown spectroscopy
LLE	Liquid–liquid extraction
LLLME	Liquid–liquid–liquid microextraction
LOD	Limit of detection
LOV	Lab-on-a-valve

LPME	Liquid phase microextraction
MC-ICP-MS	Multicollector inductively coupled plasma mass spectrometry
MCP	Microchannel plate
MDG	Monodisperse microdroplet generator
MeCat	Metal coded affinity tag
MeSeCys	Methylselenocysteine
MIP	Microwave induced plasma
MNP	Magnetic nano-particle
MS	Mass spectrometry
NE-LIBS	Nanoparticles enhanced LIBS
NIST	National institute of standards and technology
NP	Nano-particle
NTV	Near torch vapourization
OES	Optical emission spectroscopy
PDMS	Polydimethylsiloxane
PFA	Perfluoroalkyl
QMS	Quadrupole mass spectrometry
Q-ICP-MS	Quadrupole inductively couple plasma mass spectrometry
qPCR	Quantitative polymerase chain reaction
r.f.	Radio frequency
REE	Rare earth element
RSD	Relative standard deviation
SC-GD	Solution-cathode glow discharge
SDS-PAGE	Sodium dodecyl sulfate polyacrylamide gel electrophoresis
SeCys2	Selenocystine
SeEt	Selenoethionine
SeMet	Selenomethionine
SF-ICP-MS	Sector field inductively coupled plasma mass spectrometry
SF-MS	Sector field mass spectrometry
SIMS	Secondary ion mass spectrometry
SPE	Solid phase extraction
SPME	Solid phase micro-extraction
TEM	Transmission electron microscopy
TOF-MS	Time-of-flight mass spectrometry
UV-vis	Ultraviolet-visible spectrophometry.

REFERENCES

1. O. T. Butler, W. R. L. Cairns, J. M. Cook and C. M. Davidson, J. Anal. At. Spectrom., 2013, 28(2), 177–216 .
2. S. Carter, A. S. Fisher, M. W. Hinds, S. Lancaster and J. Marshall, J. Anal. At. Spectrom., 2013, 28(12), 1814–1869 .
3. R. Clough, C. F. Harrington, S. J. Hill, Y. Madrid and J. F. Tyson, J. Anal. At. Spectrom., 2013, 28(8), 1153–1195 .
4. E. H. Evans, M. Horstwood, J. Pisonero and C. M. M. Smith, J. Anal. At. Spectrom., 2013, 28(6), 779–800 .
5. A. Taylor, M. P. Day, S. Hill, J. Marshall, M. Patriarca and M. White, J. Anal. At. Spectrom., 2013, 28(4), 425–459 .
6. M. West, A. T. Ellis, P. J. Potts, C. Streli, C. Vanhoof, D. Wegrzynek and P. Wobrauschek, J. Anal. At. Spectrom., 2013, 28(10), 1544–1590 .
7. G. Giakisikli and A. N. Anthemidis, Anal. Chim. Acta, 2013, 789, 1–16 .
8. G. Giakisikli and A. N. Anthemidis, Talanta, 2013, 110, 229–235 .
9. B. B. Chen, B. Hu, M. He, Q. Huang, Y. Zhang and X. Zhang, J. Anal. At. Spectrom., 2013, 28(3), 334–343 .
10. C. Cui, B. Hu, B. B. Chen and M. He, J. Anal. At. Spectrom., 2013, 28(7), 1110–1117 .
11. H. Ebrahimzadeh, E. Moazzen, M. M. Amini and O. Sadeghi, Microchim. Acta, 2013, 180(5–6), 445–451 .
12. G. C. de Lima, A. C. do Lago, A. A. Chaves, P. S. Fadini and P. O. Luccas, Anal. Chim. Acta, 2013, 768, 35–40 .
13. G. Bauer, M. A. Neouze and A. Limbeck, Talanta, 2013, 103, 145–152 .
14. S. Ozdemir, V. Okumus, A. Dundar and E. Kilinc, Microchim. Acta, 2013, 180(9–10), 719–739 .
15. V. Andruch, I. S. Balogh, L. Kocurova and J. Sandrejova, J. Anal. At. Spectrom., 2013, 28(1), 19–32 .
16. M. Miro and E. H. Hansen, Anal. Chim. Acta, 2013, 782, 1–11 .
17. A. Boucherit, H. Khalaf, E. Paredes and J. L. Todoli, Talanta, 2012, 99, 330–334 .
18. D. Han, B. Tang, Y. R. Lee and K. H. Row, J. Sep. Sci., 2012, 35(21), 2949–2961 .
19. M. Naeemullah, T. G. Tuzen, D. Kazi, Citak and M. Soylak, J. Anal. At. Spectrom., 2013, 28(9), 1441–1445 .
20. C. J. Zeng and L. F. Tang, Anal. Lett., 2013, 46(9), 1442–1453 .
21. C. C. Shen, C. C. Wu, H. Cheng, R. L. Edwards, Y. T. Hsieh, S. Gallet, C. C. Chang, T. Y. Li, D. D. Lam, A. Kano, M. Hori and C. Spotl, Geochim. Cosmochim. Acta, 2012, 99, 71–86 .
22. X. W. Yan, L. M. Yang and Q. Q. Wang, Anal. Bioanal. Chem., 2013, 405(17), 5663–5670 .
23. A. Sanz-Medel, M. Montes-Bayon, J. Bettmer, M. L. Fernandez-Sanchez and J. R. Encinar, Trac, Trends Anal. Chem., 2012, 40, 52–63 .
24. D. Kretschy, G. Koellensperger and S. Hann, Anal. Chim. Acta, 2012, 750, 98–110 .
25. C. Giesen, L. Waentig, U. Panne and N. Jakubowski, Spectrochim. Acta, Part B, 2012, 76, 27–39 .
26. D. Esteban-Fernandez, F. S. Bierkandt and M. W. Linscheid, J. Anal. At. Spectrom., 2012, 27(10), 1701–1708 .
27. G. Schwarz, S. Beck, D. Benda and M. W. Linscheid, Analyst, 2013, 138(8), 2449–2455 .
28. R. Liu, X. D. Hou, Y. Lv, M. McCooeye, L. Yang and Z. Mester, Anal. Chem., 2013, 85(8), 4087–4093 .
29. X. Y. Wang, X. Wang, W. J. Qin, H. J. Lin, J. F. Wang, J. Y. Wei, Y. J. Zhang and X. H. Qian, Analyst, 2013, 138(18), 5309–5317 .
30. T. Konz, M. Montes-Bayon and A. Sanz-Medel, Anal. Chem., 2012, 84(19), 8133–8139 .
31. I. J. Arnquist and J. A. Holcombe, Spectrochim. Acta, Part B, 2012, 76, 140–146 .
32. G. J. Han, S. C. Zhang, Z. Xing and X. R. Zhang, Angew. Chem. Int. Ed., 2013, 52(5), 1466–1471 .
33. O. Leclerc, P. O. Fraisse, G. Labarraque, C. Oster, J. P. Pichaut, M. Baume, S. Jarraud, P. Fisicaro and S. Vaslin-Reimann, AnBio, 2013, 435(2), 153–158 .
34. C. Swart, Anal. Bioanal. Chem., 2013, 405(17), 5697–5723 .
35. M. Miro and E. H. Hansen, Anal. Chim. Acta, 2012, 750, 3–15 .
36. H. Wang, Z. K. Wu, Y. Zhang, B. B. Chen, M. He and B. Hu, J. Anal. At. Spectrom., 2013, 28(10), 1660–1665 .
37. S. Elzey, D. H. Tsai, L. L. Yu, M. R. Winchester, M. E. Kelley and V. A. Hackley, Anal. Bioanal. Chem., 2013, 405(7), 2279–2288 .
38. S. Gschwind, H. Hagendorfer, D. A. Frick and D. Gunther, Anal. Chem., 2013, 85(12), 5875–5883 .
39. C. D. Pereira, M. A. Aguirre, J. A. Nobrega, M. Hidalgo and A. Canals, J. Anal. At. Spectrom., 2012, 27(12), 2132–2137 .
40. H. R. Badiei, B. Lai and V. Karanassios, Spectrochim. Acta, Part B, 2012, 77, 19–30 .
41. H. R. Badiei, J. McEnaney and V. Karanassios, Spectrochim. Acta, Part B, 2012, 78, 42–49 .
42. Y. Gao, R. Liu and L. Yang, ChSBu, 2013, 58(17), 1980–1991.
43. V. Yilmaz, Z. Arslan and L. Rose, Anal. Chim. Acta, 2013, 761, 18–26 .
44. J. Kratzer, S. Musil, M. Vobecky and J. Dedina, J. Anal. At. Spectrom., 2013, 28(3), 344–353 .
45. Z. L. Zhu, Q. J. Wu, Z. F. Liu, L. Liu, H. T. Zheng and S. H. Hu, Anal. Chem., 2013, 85(8), 4150–4156 .
46. X. Ai, Y. Wang, X. D. Hou, L. Yang, C. B. Zheng and L. Wu, Analyst, 2013, 138(12), 3494–3501 .
47. W. B. Zhang, X. A. Yang, Y. P. Dong and J. J. Xue, Anal. Chem., 2012, 84(21), 9199–9207 .
48. J. Hranicek, V. Cerveny, J. Kratzer, M. Vobecky and P. Rychlovsky, J. Anal. At. Spectrom., 2012, 27(10), 1761–1771 .

49. C. Shen-Tu, S. L. Zhang and C. Lei, SpecL, 2013, 46(6), 427–432 Search .

50. T. Suzuki, R. E. Sturgeon, C. B. Zheng, A. Hioki, T. Nakazato and H. Tao, Anal. Sci., 2012, 28(8), 807–811 .

51. R. E. Sturgeon and V. Luong, J. Anal. At. Spectrom., 2013, 28(10), 1610–1619 .

52. H. Matusiewicz and M. Slachcinski, SpecL, 2013, 46(5), 315–326.

53. Z. Wang, J. Y. Zhang, H. J. Zou, M. Dong, D. R. Qiu and P. Y. Yang, Talanta, 2013, 107, 338–343 .

54. A. A. Ganeev, A. R. Gubal, K. N. Uskov and S. V. Potapov, Russ. Chem. Bull., 2012, 61(4), 752–767 .

55. V. Efimova, V. Hoffmann and J. Eckert, Spectrochim. Acta, Part B, 2012, 76, 181–189 .

56. G. Gamez, G. Mohanty and J. Michler, J. Anal. At. Spectrom., 2013, 28(7), 1016–1023 .

57. R. Qian, S. J. Zhuo, Z. Wang and P. K. Robinson, J. Anal. At. Spectrom., 2013, 28(7), 1061–1067 .

58. R. E. Russo, X. L. Mao, J. L. Gonzalez and J. Yoo, Spectroscopy, 2013, 28(1), 24–39.

59. R. E. Russo, X. L. Mao, J. J. Gonzalez, V. Zorba and J. Yoo, Anal. Chem., 2013, 85(13), 6162–6177 .

60. F. Claverie, J. Malherbe, N. Bier, J. L. Molloy and S. E. Long, Anal. Chem., 2013, 85(7), 3584–3591 .

61. D. S. Urgast and J. Feldmann, J. Anal. At. Spectrom., 2013, 28(9), 1367–1371 .

62. L. Dorta, R. Kovacs, J. Koch, K. Nishiguchi, K. Utani and D. Gunther, J. Anal. At. Spectrom., 2013, 28(9), 1513–1521 .

63. L. Flamigni, J. Koch and D. Gunther, Spectrochim. Acta, Part B, 2012, 76, 70–76 .

64. F. G. Antes, E. Dullius, A. Ben da Costa, R. F. Molz, J. N. G. Paniz, E. M. M. Flores and V. L. Dressler, Microchem. J., 2013, 109, 117–121 .

65. B. Hashimoto, H. Daidoji, H. Uchihara, K. Iwasaki, Y. Okamoto and T. Fujiwara, Anal. Lett., 2013, 46(8), 1299–1305 .

66. K. Jankowski and E. Reszke, J. Anal. At. Spectrom., 2013, 28(8), 1196–1212 .

67. D. B. Luo and Y. X. Duan, Trac, Trends Anal. Chem., 2012, 39, 254–266 .

68. T. Frentiu, A. I. Mihaltan, E. Darvasi, M. Ponta, C. Roman and M. Frentiu, J. Anal. At. Spectrom., 2012, 27(10), 1753–1760 .

69. Y. Liu, W. Q. Cao, X. L. Ding, X. Yuan and Y. X. Duan, Spectrochim. Acta, Part B, 2012, 76, 152–158 .

70. Y. Kohara, Y. Terui, M. Ichikawa, T. Shirasaki, K. Yamamoto, T. Yamamoto and Y. Takamura, J. Anal. At. Spectrom., 2012, 27(9), 1457–1464 .

71. D. V. Khoai, A. Kitano, T. Yamamoto, Y. Ukita and Y. Takamura, MiEng, 2013, 111, 343–347 Search .

72. T. Pfeifer, R. Janzen, T. Steingrobe, M. Sperling, B. Franze, C. Engelhard and W. Buscher, Spectrochim. Acta, Part B, 2012, 76, 48–55 .

73. S. Muller, T. Krahling, D. Veza, V. Horvatic, C. Vadla and J. Franzke, Spectrochim. Acta, Part B, 2013, 85, 104–111 .

74. Z. F. Liu, Z. L. Zhu, H. T. Zheng and S. H. Hu, Anal. Chem., 2012, 84(23), 10170–10174 .

75. P. Jamroz, K. Greda and P. Pohl, Trac, Trends Anal. Chem., 2012, 41, 105–121 .

76. A. J. Schwartz, S. J. Ray, E. Elish, A. P. Storey, A. A. Rubinshtein, G. C. Y. Chan, K. P. Pfeuffer and G. M. Hieftje, Talanta, 2012, 102, 26–33 .

77. Z. L. Zhu, C. Y. Huang, Q. He, Q. Xiao, Z. F. Liu, S. C. Zhang and S. H. Hu, Talanta, 2013, 106, 133–136 .

78. Q. Xiao, Z. L. Zhu, H. T. Zheng, H. Y. He, C. Y. Huang and S. H. Hu, Talanta, 2013, 106, 144–149 .

79. A. Gundlach-Graham, E. A. Dennis, S. J. Ray, C. G. Enke, C. J. Barinaga, D. W. Koppenaal and G. M. Hieftje, J. Anal. At. Spectrom., 2013, 28(9), 1385–1395 .

80. R. K. Bhatia, V. K. Yadav, V. M. Mahadeshwar, M. M. Gulhane, E. Ravisankar, T. K. Saha, V. Nataraju and S. K. Gupta, Int. J. Mass spectrom., 2013, 339, 39–44 .

81. T. Egorova, R. Dietiker, B. Hattendorf and D. Gunther, Spectrochim. Acta, Part B, 2012, 76, 40–47 .

82. M. P. Ruffoni, CoPhC, 2013, 184(7), 1770–1776 .

83. P. S. Doidge, Spectrochim. Acta, Part B, 2013, 85, 100–103 .

84. J. Manrique, J. A. Aguilera and C. Aragon, J. Quant. Spectrosc. Radiat. Transf., 2013, 120, 120–124 .

85. N. Konjevic, M. Ivkovic and N. Sakan, Spectrochim. Acta, Part B, 2012, 76, 16–26 .

86. R. Rincon, J. Munoz, M. Saez and M. D. Calzada, Spectrochim. Acta, Part B, 2013, 81, 26–35 .

87. J. Orejas, J. Pisonero, N. Bordel, T. Nelis, P. Guillot and A. Sanz-Medel, Spectrochim. Acta, Part B, 2012, 76, 166–174 .

88. G. L. Scheffler and D. Pozebon, Anal. Chim. Acta, 2013, 789, 33–40 .

89. H. Miyahara, Y. Kaburaki, T. Iwai and A. Okino, Bunseki Kagaku, 2012, 61(11), 917–924 .

90. G. Grindlay, J. Mora, M. de Loos-Vollebregt and F. Vanhaecke, Spectrochim. Acta, Part B, 2013, 86, 42–49 .

91. G. C. Y. Chan, Z. L. Zhu and G. M. Hieftje, Spectrochim. Acta, Part B, 2012, 76, 87–95 .

92. K. Niemax, Spectrochim. Acta, Part B, 2012, 76, 65–69 .

93. M. Aghaei, H. Lindner and A. Bogaerts, Spectrochim. Acta, Part B, 2012, 76, 56–64 .

94. M. Aghaei, H. Lindner and A. Bogaerts, J. Anal. At. Spectrom., 2013, 28(9), 1485–1492 .

95. A. J. Edmund, S. D. Bergeson, M. Lyon, N. Taylor, I. Kalinitchenko and P. B. Farnsworth, Spectrochim. Acta, Part B, 2012, 76, 109–118 .

96. C. H. Ebert, T. M. Witte and R. S. Houk, Spectrochim. Acta, Part B, 2012, 76, 119–125 .

97. G. Bouchoux, A. M. Rashad and A. I. Helal, J. Phys. Chem. A, 2012, 116(36), 9058–9070 .

98. J. R. Dettman and J. W. Olesik, Spectrochim. Acta, Part B, 2012, 76, 96–108 .

99. G. C. Y. Chan and G. M. Hieftje, Anal. Chem., 2013, 85(1), 50–57 .

100. G. C. Y. Chan and G. M. Hieftje, Anal. Chem., 2013, 85(1), 58–65 .

101. J. M. Mermet and G. Granier, Spectrochim. Acta, Part B, 2012, 76, 214–220 .

102. J. M. Mermet, G. Granier and P. Fichet, Spectrochim. Acta, Part B, 2012, 76, 221–225 .

103. I. Choi, X. L. Mao, J. J. Gonzalez and R. E. Russo, Appl. Phys. A: Mater. Sci. Process., 2013, 110(4), 785–792 .

104. M. Cvejic, M. R. Gavrilovic, S. Jovicevic and N. Konjevic, Spectrochim. Acta, Part B, 2013, 85, 20–33 .

105. J. A. Aguilera, C. Aragon and J. Manrique, J. Quant. Spectrosc. Radiat. Transf., 2013, 114, 151–156 .

106. C. S. Chen, B. Y. Man, D. Liu, X. Song and X. J. Chen, J. Phys. B: At. Mol. Phys., 2013, 46(1), 4 Search .

107. C. G. Parigger, Spectrochim. Acta, Part B, 2013, 79–80, 4–16 .

108. S. Merk, A. Demidov, D. Shelby, I. B. Gornushkin, U. Panne, B. W. Smith and N. Omenetto, ApSpe, 2013, 67(8), 851–859 .

109. X. S. Bai, Q. L. Ma, M. Perrier, V. Motto-Ros, D. Sabourdy, L. Nguyen, A. Jalocha and J. Yu, Spectrochim. Acta, Part B, 2013, 87, 27–35 .

110. S. Eschlbock-Fuchs, M. J. Haslinger, A. Hinterreiter, P. Kolmhofer, N. Huber, R. Rossler, J. Heitz and J. D. Pedarnig, Spectrochim. Acta, Part B, 2013, 87, 36–42 .

111. J. R. Freeman, S. S. Harilal, P. M. Diwakar, B. Verhoff and A. Hassanein, Spectrochim. Acta, Part B, 2013, 87, 43–50 .

112. S. P. Banerjee, Z. J. Chen, I. Utkin and R. Fedosejevs, Appl. Phys. A-Mater. Sci. Process., 2013, 111(3), 791–798 .

113. T. Ahmido, A. Ting and P. Misra, ApOpt, 2013, 52(13), 3048–3057 .

114. J. E. Barefield, E. J. Judge, J. M. Berg, S. P. Willson, L. A. Le and L. N. Lopez, ApSpe, 2013, 67(4), 433–440 .

115. M. Gaft, L. Nagli and I. Gornushkin, Spectrochim. Acta, Part B, 2013, 85, 93–99 .

116. T. A. Labutin, S. M. Zaytsev and A. M. Popov, Anal. Chem., 2013, 85(4), 1985–1990 .

117. J. Frydenvang, K. M. Kinch, S. Husted and M. B. Madsen, Anal. Chem., 2013, 85(3), 1492–1500 .

118. G. Cristoforetti and E. Tognoni, Spectrochim. Acta, Part B, 2013, 79–80, 63–71 .

119. T. A. Labutin, S. M. Zaytsev, A. M. Popov, I. V. Seliverstova, S. E. Bozhenko and N. B. Zorov, Spectrochim. Acta, Part B, 2013, 87, 57–64 .

120. M. Baudelet and B. W. Smith, J. Anal. At. Spectrom., 2013, 28(5), 624–629 .

121. R. Bhandari and T. Taira, OptEn, 2013, 52(7), 6 Search .

122. M. Scharun, C. Fricke-Begemann and R. Noll, Spectrochim. Acta, Part B, 2013, 87, 198–207 .

123. J. A. Merten, B. W. Smith and N. Omenetto, Spectrochim. Acta, Part B, 2013, 83–84, 50–55 .

124. M. Dong, X. Mao, J. J. Gonzalez, J. Lu and R. E. Russo, Anal. Chem., 2013, 85(5), 2899–2906 .

125. A. De Giacomo, R. Gaudiuso, C. Koral, M. Dell'Aglio and O. De Pascale, Anal. Chem., 2013, 85(21), 10180–10187 .

126. V. Lazic, J. J. Laserna and S. Jovicevic, Spectrochim. Acta, Part B, 2013, 82, 42–49 .

127. V. Lazic, J. J. Laserna and S. Jovicevic, Spectrochim. Acta, Part B, 2013, 82, 50–59 .

128. A. Matsumoto, A. Tamura, K. Fukami, Y. H. Ogata and T. Sakka, Anal. Chem., 2013, 85(8), 3807–3811 .

129. M. A. Khater, Spectrochim. Acta, Part B, 2013, 81, 1–10 .

130. P. Adámek, J. Olejníček, M. Čada, Š. Kment and Z. Hubička, Opt. Lett., 2013, 38(14), 2428–2430 .

131. P. Jacquet and A. Pailloux, J. Anal. At. Spectrom., 2013, 28(8), 1298–1302 .

132. S. D. Zhang, W. Liu, X. H. Zhang and Y. X. Duan, Spectrochim. Acta, Part B, 2013, 85, 1–12 .

133. M. Biscan and S. Milosevic, Spectrochim. Acta, Part B, 2013, 81, 20–25 .

134. V. Frankevich, P. M. L. Sinues, K. Barylyuk and R. Zenobi, Anal. Chem., 2013, 85(1), 39–43 .

135. X. S. Li and L. Ma, J. Quant. Spectrosc. Radiat. Transf., 2013, 114, 157–166 .

136. D. C. Baxter, I. Rodushkin and E. Engstrom, J. Anal. At. Spectrom., 2012, 27(9), 1355–1381 .

137. E. Paredes, J. L. Todoli and C. R. Quetel, J. Anal. At. Spectrom., 2013, 28(3), 327–333 .

138. E. Paredes, D. G. Asfaha, E. Ponzevera, C. Brach-Papa, M. Van Bocxstaele, J. L. Todoli and C. R. Quetel, J. Anal. At. Spectrom., 2011, 26(7), 1372–1379 .

139. J. M. Koornneef, C. Bouman, J. B. Schwieters and G. R. Davies, J. Anal. At. Spectrom., 2013, 28(5), 749–754 .

140. Z. C. Hu, Y. S. Liu, S. Gao, W. G. Liu, W. Zhang, X. R. Tong, L. Lin, K. Q. Zong, M. Li, H. H. Chen, L. Zhou and L. Yang, J. Anal. At. Spectrom., 2012, 27(9), 1391–1399 .

141. N. Kivel, I. Gunther-Leopold, F. Vanhaecke and D. Gunther, Spectrochim. Acta, Part B, 2012, 76, 126–132 .

142. A. Persson, G. Eilers, L. Ryderfors, E. Mukhtar, G. Possnert and M. Salehpour, Anal. Chem., 2013, 85(14), 6790–6798 .

143. G. Chew and T. Walczyk, Anal. Chem., 2013, 85(7), 3667–3673 .

144. S. G. John, J. Anal. At. Spectrom., 2012, 27(12), 2123–2131 .

145. J. Creech, J. Baker, M. Handler, M. Schiller and M. Bizzarro, J. Anal. At. Spectrom., 2013, 28(6), 853–865 .

146. A. Makishima and E. Nakamura, J. Anal. At. Spectrom., 2013, 28(1), 127–133 .

147. N. Shalev, I. Segal, B. Lazar, I. Gavrieli, J. Fietzke, A. Eisenhauer and L. Halicz, J. Anal. At. Spectrom., 2013, 28(6), 940–944 .

148. M. Kraiem, S. Richter, N. Erdmann, H. Kuhn, M. Hedberg and Y. Aregbe, Anal. Chim. Acta, 2012, 748, 37–44 .

149. S. Kappel, S. F. Boulyga and T. Prohaska, J. Environ. Radioact., 2012, 113, 8–15 .

150. R. C. Marin, J. E. S. Sarkis and M. R. L. Nascimento, J. Radioanal. Nucl. Chem., 2013, 295(1), 99–104 .

151. M. Tanimizu, N. Sugiyama, E. Ponzevera and G. Bayon, J. Anal. At. Spectrom., 2013, 28(9), 1372–1376 .

152. J. G. Arnason, C. N. Pellegri and P. J. Parsons, J. Anal. At. Spectrom., 2013, 28(9), 1410–1419 .

153. K. J. Mathew, G. O'Connor, A. Hasozbek and M. Kraiem, J. Anal. At. Spectrom., 2013, 28(6), 866–876 .

154. T. Ohno and T. Hirata, Anal. Sci., 2013, 29(1), 47–53 .

155. A. Pramann, O. Rienitz and D. Schiel, Anal. Chem., 2012, 84(23), 10175–10179 .

156. Y. H. Yang, F. Y. Wu, Z. Y. Chu, L. W. Xie and J. H. Yang, Spectrochim. Acta, Part B, 2013, 79–80, 82–87 .

157. M. E. Sanchez-Lorda, S. G. de Madinabeitia, C. Pin and J. I. G. Ibarguchi, Int. J. Mass spectrom., 2013, 333, 34–43 .

158. K. F. Huang, J. Blusztajn, D. W. Oppo, W. B. Curry and B. Peucker-Ehrenbrink, J. Anal. At. Spectrom., 2012, 27(9), 1560–1567 .

159. K. Y. Chen, C. Fan, H. L. Yuan, Z. A. Bao, C. L. Zong, M. N. Dai, X. Ling and Y. Yang, Spectrosc. Spectr. Anal., 2013, 33(5), 1342–1349 .

160. J. R. Darling, C. D. Storey, C. J. Hawkesworth and P. C. Lightfoot, Geochim. Cosmochim. Acta, 2012, 99, 1–17 .

161. F. Claverie, J. Malherbe, N. Bier, J. L. Molloy and S. E. Long, Anal. Bioanal. Chem., 2013, 405(7), 2289–2299 .

162. J. M. Cottle, A. R. Kylander-Clark and J. C. Vrijmoed, ChGeo, 2012, 332, 136–147

CHAPTER 9

ELECTRONIC STRUCTURE AND SPECTROSCOPIC STUDIES OF THE MOLECULE SCS

Rawad Halabi, Mahmoud Korek

Faculty of Science, Beirut Arab University, Beirut, Lebanon

ABSTRACT

Theoretical investigation of the lowest electronic states of ScS molecule, in the representation $^{2s+1}\Lambda^{(+/-)}$, has been performed via CASSCF and MRCI (single and double excitations without Davidson correction) calculations. The calculated potential energy curves (PECs), permanent dipole moment curves (PDMCs), and spectroscopic constants are reported for the 10 lowest electronic states. The eigenvalues E_v, the rotational constants B_v, and the centrifugal distortion constants D_v have been calculated for various vibrational levels. The comparison of the present results with the available experimental data in literature shows an overall good agreement. To the best of our knowledge, 6 electronic states of the ScS molecule, between 11600 cm^{-1} and 15000 cm^{-1} are not yet observed experimentally and are investigated in the present work for the first time.

INTRODUCTION

The molecules containing transition metal are species of chemical and astrophysical importance. Scandium is the first transition metal atom which has only one d-electron in the ground state. Diatomic molecules like ScS are simple transition metal-containing systems in which d-electrons take part in bonding and provide ideal models for understanding the electronic structure and reactivity (chemiluminescent reactions).

To the best of our knowledge, the experimental work for the ScSmolecule is that of Gengler et al. [1] - [3] . They studied the $A^2\Pi$-$X^2\Sigma^+$ and $B^2\Pi$-$X^2\Sigma^+$ band system of scandium monosulfide, ScS, using Fourier transform emission spectroscopy (FT) and laser excitation spectroscopy (LIF). The only theoretical work for ScS in literature is that of Bauschlicher and Langhoff [4] for the ground $X^2\Sigma^+$ and $(1)^2\Pi$ states using ab initio method.

In the present work, ab initio investigations of the lowest lying electronic states of the ScS molecule have been performed via CAS-SCF/MRCI (Complete Active Space Self Consistent Field, Multireference Configuration Interaction) method. Multireference CI calculations (single and double excitations without Davidson corrections) in which the entire CAS-SCF configuration space is used as the reference, are performed to account the correlation effects. The potential energy curves (PECs), the transition dipole moment curves (TDMCs) together with the energy minimum for the ground state T_e, the equilibrium internuclear distance R_e, the harmonic frequency ω_e and the rotational constant B_e have been obtained

Citation: Halabi, R. and Korek, M. (2015) Electronic Structure and Spectroscopic Studies of the Molecule ScS. Journal of Modern Physics, 6, 546-552. doi: 10.4236/jmp.2015.65059.

for the 10 lowest-lying electronic states. Six electronic states have been investigated in the present work for the first time.

METHOD OF CALCULATIONS

In the present work, ab initio investigations of the lowest-lying electronic states of the ScS molecule have been performed via CASSCF method. Multireference CI calculations (single and double excitations without Davidson corrections) were performed to determine the correlation effects. The entire CASSCF configuration space was used as a reference in the MRCI calculation which have been performed via the computational chemistry program MOLPRO [5] . The scandium species is treated in an all electron scheme [6] [7] , the 21 electrons of the scandium atom are considered using the contracted Gaussian basis set used by Bauschlicher and Langhoff [4] for s, d, and f functions then we have taken p function from 6 - 31 G** basis set [5] which has 16 diffused functions contracted set of Gaussian functions. The sulfur species is treated as a system with 16 inner electrons taken into account the pseudopotential W^{ps} [8] - [10] together with the corresponding Gaussian basis set. In the range of the internuclear distance R_e around equilibrium distances of its ground state, the ScS molecule is assumed to be mainly ionic as many transition-metals Sc^+S^-. The potential energy calculations for the states $^{2s+1}\Lambda^{(+/-)}$ of the molecule ScS have been carried using a CASSCF method. Among the 37 electrons explicitly considered for ScS (21 electrons for Sc and 16 for S) 27 inner electrons were frozen in subsequent calculations so that 10 valence electrons were explicitly treated. The active space contains 4σ (Sc: 4s, $3d_0$), 1π ($3d_{\pm1}$) and 1δ (Sc: $3d_{\pm2}$), this corresponds to 6 active molecular orbitals orbitals in the C_{2v} symmetry distributed into irreducible representation in the following way: $3a_1$, $1b_1$, $1b_2$, $1a_2$, noted [3111]. The doubly occupied orbitals 2σ (S: 5s, $5p_0$), and 1π (S: $5p_{\pm1}$) of sulfur have been considered as inactive in the CASSCF calculations. Correlation effects for the ten valence electrons have been taken into account through multireference calculations MRCI (single and double excitations without Davidson correction) where the entire CASSCF configuration space has been used as reference. Calculations have been performed via the computational program MOLPRO [5] taking advantage of the graphical user interface GABEDIT [11] .

RESULTS AND DISCUSSION

Calculations have been performed for 51 internuclear distances in the range 2.00 Å - 3.02 Å for 10 states in therepresentation $^{2s+1}\Lambda^{(+/-)}$. The potential energy curves (PECs) for the states $^{2,4}\Sigma$, $^{2,4}\Pi$, $^{2,4}\Delta$ and $^{2,4}\Phi$ in the considered range of Rare drawn respectively in Figure 1, Figure 2. For each state, transition energies with respect to the minimum energy of the ground state T_e, equilibrium internuclear distance R_e, harmonic frequency ω_e and the rotational constant B_e have been calculated. These constants are displayed in Table 1 together with the literature values from both theoretical and experimental investigations.

The comparison of our results with the values for T_e calculated by different techniques in the literature for the 10 lowest-lying doublet and quartet electronic states shows that, 7 values are in good agreement where the relative difference is 1.00% [^{c3}Ref.1] $\leq \delta T_e/T_e \leq$ 4.70% [$^{c1\text{-}2}$Ref.1] for the states $X^2\Sigma^+$, $(2)^2\Sigma^+$ and $(1)^2\Pi$ and one value for the $(1)^2\Pi$ state calculated in Ref. [1] by the SDCI method shows a larger relative difference equal to 13.0%. For ω_e the comparison of our calculated values with the 16 values given in the literature for the 10 lowest electronic states showed an overall good agreement where the relative difference is 0.80% [^{c5}Ref.1 for $(1)^2\Pi$] $\leq \delta\omega_e/\omega_e \leq$ 12.3% [^{c4}Ref.1 for $(2)^2\Sigma^+$]. Two larger values are obtained for the state $(1)^2\Delta$ calculated in Ref. [1] by the CPF and SDCI + R methods. Since the value of ω_e depends on the shape of the potential energy curves near the minimum, this difference may be explained by the shape of the investigated potential curve by this method. The values of R_e and B_e obtained in literature are in excellent agreement with our calculated values with relative differences 0.3% [^{c2}Ref.1] $\leq \delta R_e/R_e \leq$ 2.1% [$^{c1\text{-}1}$Ref.1] and 1.6% [^{b3}Ref.4] $\leq \delta B_e/B_e \leq$ 4.1% [^{b1}Ref.2] respectively for the states $X^2\Sigma^+$, $(2)^2\Sigma^+$ and $(1)^2\Delta$ while the relative differences become larger for the state $(1)^2\Pi$.

By using the canonical functions approach Korek et al. [12] -[19] and the cubic spline interpolation between each two consecutive points of the potential energy curves obtained from the ab initio calculation of the ScS molecule, the eigenvalue E_v, the rotational constant B_v, the centrifugal distortion constants D_v,. The abscissas of the turning point R_{min} and R_{max} have been calculated for the considered electronic states up to the vibrational levels v = 10. These values for the states $(1)^4\Sigma$, $(1)^4\Delta$, $(1)^4\Phi$, $(2)^4\Pi$ and $(2)^4\Phi$ (as illustration) are given in Table 2.

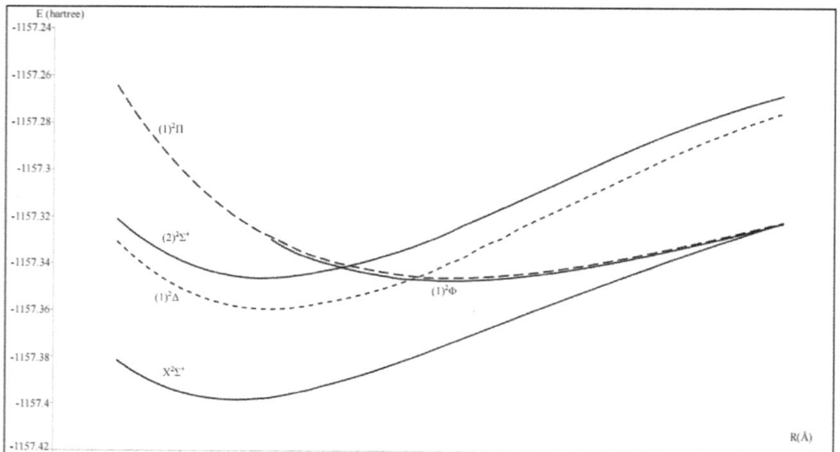

Figure 1. Potential energy curves of the electronic states $X^2\Sigma$, $(1)^2\Delta$, $(2)^2\Sigma$, $(1)^2\Pi$ and $(1)^2\Phi$ of the molecule ScS.

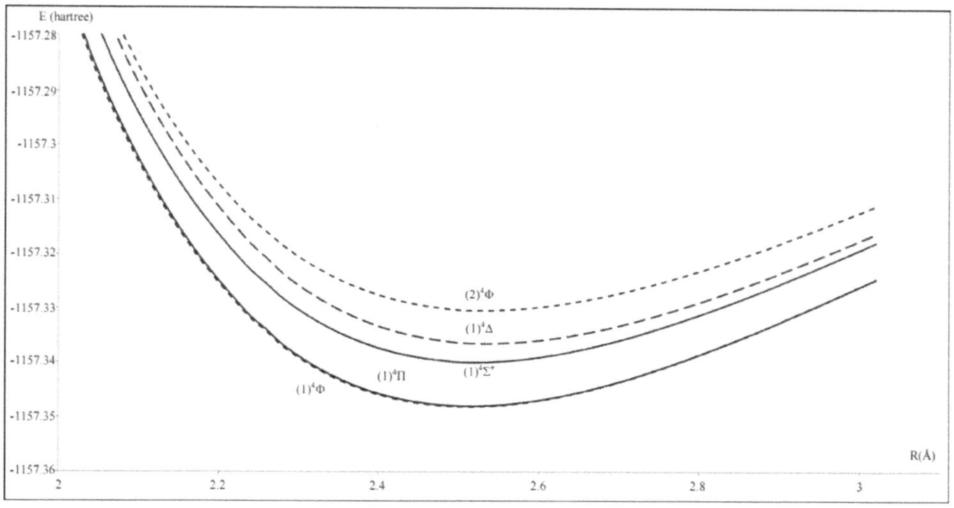

Figure 2. Potential energy curves of the electronic states $(1)^4\Sigma^+$, $(1)^4\Delta$, $(1)^4\Phi$, $(1)^4\Pi$ and $(2)^4\Phi$ of the molecule ScS.

Table 1. Calculated spectroscopic constants for the lowest 10 electronic states of ScS

States	T_e (cm^{-1})	R_e (Å)	ω_e (cm^{-1})	$B_e \times 10^2$ (cm^{-1})
$X^2\Sigma^+$	0.0[a]	2.181[a]	540.6[a]	18.929[a]
	0.0[b]	2.1375[b-1]		19.7294[b-1]
		2.144[b-2]		19.61[b-2]
		2.139[b-3]		19.71[b-3]
	0.0[c]	2.1354[c-1-1]	565.2[c-1]	
		2.167[c-2]	518[c-2]	
		2.177[c-3]	503[c-3]	
		2.126[c-4]	607[c-4]	
		2.125[c-5]	608[c-5]	
$(1)^2\Delta$	8760[a]	2.229[a]	581.6[a]	18.126[a]
		2.236[c-2]	486[c-2]	
		2.237[c-3]	488[c-3]	
		2.221[c-4]	513[c-4]	
		2.221[c-5]	515[c-5]	
$(1)^4\Phi$	11031	2.513	353.5	14.265
$(1)^4\Pi$	11058	2.515	353.1	14.247
$(1)^2\Phi$	11332	2.528	360.0	14.002
$(2)^2\Sigma^+$	11604[a]	2.2223[a]	557.4[a]	18.240[a]
	12456[b-1]	2.20391[b-1] 2.2056[b-3]	488.737[b-1]	18.5583[b-1]
	12456[b-3]	2.159[c-3]		18.53[b-3]
		2.143[c-5]	567[c-3]	
			582[c-5]	

$(1)^2\Pi$	11538.4[a]	2.497[a]	494.9[a]	14.444[a]
	11090.4[b-1]	2.2079[b-1]		18.4910[b-1]
	11000[c-1-2]			
	11691[c-2]	2.236[c-2]	468[c-2]	
	11659[c-3]	2.236[c-3]	470[c-3]	
	10698[c-4]	2.218[c-4]	488[c-4]	
	10032[c-5]	2.22[c-5]	491[c-5]	
$(1)^4\Sigma$	12828	2.522	350.6	14.158
$(1)^4\Delta$	13581	2.537	344.9	13.995
$(2)^4\Phi$	14928	2.533	333.8	14.043

[a]Present Theoretical study; [b]Ref. [1] - [3] ; [(b-1)]Ref. [1] ; [(b-2)]Ref. [2] ; [(b-3)]Ref. [3] ; [c]Ref. [4] .

Table 2. Values of the eigenvalues E_v, the abscissa of the turning points R_{min}, R_{max} the rotational constants B_v and the centrifugal distortion constants D_v for the different vibrational levels of $(1)^2\Pi$, $(2)^2\Sigma^+$, $(1)^4\Sigma^+$, $(1)^4\Delta$, $(1)^4\Phi$, $(2)^4\Pi$, and $(2)\Phi$ electronic states of the ScS molecule

	v	E_v (cm^{-1})	R_{min} (Å)	R_{max} (Å)	$B_v \times 10^2$ (cm^{-1})	$D_v \times 10^8$ (cm^{-1})
$(1)^2\Pi$	0	173.4	1.786	2.593	14.17[a] 18.45[b]	9.35[a] 10.67[b]
	1	521.2	1.791	2.649	14.15[a] 18.36[b]	8.66[a] 10.83[b]
	2	872.9	1.795	2.691	14.04	8.73
	3	1224.1	1.799	2.728	14.00	10.63
$(2)^2\Sigma^+$	0	282.9	2.171	2.277	18.16[a] 18.56[b]	7.96[a] 10.82[b]
	1	826.9	2.133	2.337	18.02[a] 18.46[b]	9.45[a] 10.88[b]
$(1)^4\Sigma^+$	0	178.1	2.453	2.596	14.13	9.33
	1	525.1	2.405	2.655	14.08	9.52
	2	868.3	2.374	2.698	14.01	9.60
	3	1208.3	2.349	2.734	13.95	9.28
	4	1547.0	2.329	2.767	13.87	9.53
	5	1883.1	2.312	2.797	13.82	9.74
	6	2216.1	2.296	2.825	13.74	8.57
	7	2549.0	2.282	2.852	13.67	
	8	2877.8	2.271	2.877		
	9	3532.0	2.247	2.926		

$(1)^4\Phi$	0	176.9	2.467	2.612	14.24	9.26
	1	529.4	2.417	2.674	14.19	9.50
	2	877.6	2.385	2.718	14.12	9.27
	3	1224.2	2.361	2.796	14.05	9.36
	4	1568.5	2.339	2.789	13.99	9.42
	5	1910.3	2.321	2.821	13.92	9.04
	6	2250.6	2.305	2.849	13.85	9.21
	7	2588.8	2.291	2.878	13.78	8.80
	8	2925.6	2.277	2.904	13.69	
	9	3261.6	2.267	2.931	11.10	
	10	3595.5	2.256	2.955	13.00	

$(2)^4\Pi$	0	176.9	2.445	2.588	14.22	9.23
	1	529.3	2.398	2.646	14.17	9.43
	2	877.8	2.367	2.688	14.10	9.28
	3	1224.4	2.343	2.724	14.04	9.25
	4	1569.0	2.323	2.756	13.97	9.45
	5	1910.70	2.305	2.786	13.90	8.98
	6	2251.10	2.289	2.814	13.83	9.19
	7	2589.4	2.276	2.841	13.76	8.92
	8	2925.8	2.265	2.866	13.68	7.11
	9	3261.4	2.254	2.891		
	10	3594.1	2.241	2.914		
$(2)^4\Phi$	0	166.1	2.463	2.609	14.01	9.97
	1	497.1	2.414	2.671	13.93	10.36
	2	822.5	2.382	2.715	13.87	10.36
	3	1144.8	2.358	2.753	13.79	9.69
	4	1466.8	2.337	2.786	13.72	10.54
	5	1785. 4	2.319	2.817	13.66	10.02
	6	2102.1	2.304	2.846	13.58	9.6501
	7	2417.5	2.289	2.874	13.21	
	8	2730.2	2.276	2.901	13.44	
	9	3041.7	2.265	2.925	13.27	
	10	3353.9	2.254	2.951		

$(1)^4\Delta$	0	171.7	2.445	2.585	13.97	9.25
	1	514.1	2.398	2.642	13.90	9.70
	2	850.4	2.367	2.684	13.85	9.82
	3	1182.6	2.344	2.718	13.78	9.01
	4	1514.9	2.324	2.749	13.71	9.99
	5	1843.3	2.307	2.779	13.65	9.44
	6	2169.8	2.291	2.806	13.57	9.05
	7	2495.1	2.278	2.831	13.50	
	8	2817.3	2.267	2.856	13.40	
	9	3138.6	2.256	2.879	13.21	

[a]Present work; [b]Ref. [3] .

The absence of the comparison with other results is because of the calculation of these values here for the first time. Then, using the canonical functions approach, the eigenvalues E_v, the rotational constants B_v and the centrifugal distortion constants D_v are calculated for different vibrational levels v for the seven molecular states of ScS ($(1)^2\Pi$, $(2)^2\Sigma$, $(1)^4\Sigma$, $(1)^4\Delta$, $(1)^4\Phi$, $(2)^4\Pi$ and $(2)^4\Phi$). These results are displayed in Table 2, respectively. The comparison of our calculated values with those obtained experimentally for the rotational constants B_v and the centrifugal distortion constants D_v showed a good agreement for the state $(2)^2\Sigma$ with averaged relative differences $\delta B_v/B_v$ = 2.1% and $\delta D_v/D_v$ = 35.8% for v = 0, and these differences are $\delta B_v/B_v$ = 2.4% and $\delta D_v/D_v$ = 15.1% for v = 1. This agreement becomes less for the state $(1)^2\Pi$, with averaged relative differences $\delta B_v/B_v$ = 23.2% and $\delta D_v/D_v$ = 14.1% for v = 0, and these differences are $\delta B_v/B_v$ = 22.9% and $\delta D_v/D_v$ = 25.1% for v = 1.

CONCLUSION

Rawad Halabi, Mahmoud Korek (12,24)">12] -[19] , the eigenvalue E_v, the rotational constant B_v, the centrifugal distortion constants D_v, and the abscissas of the turning points R_{min} and R_{max} were calculated up to the vibrational level v = 10. Six electronic states have been investigated in the present work for the first time.

REFERENCES

1. Gengler, J., Chen, J.H., Steimle, T.C., Ram, R.S. and Bernath, P.F. (2006) Journal of Molecular Spectroscopy, 237, 36-45. http://dx.doi.org/10.1016/j.jms.2006.02.006
2. Steimle, T.C., Marr, A.J. and Goodridge, D.M. (1997) Journal of Chemical Physics, 107, 10406-10414. http://dx.doi.org/10.1063/1.474204
3. Fenot, B., Femenias, J.L. and Stringat, R. (1979) Journal of Molecular Spectroscopy, 78, 40-50. http://dx.doi.org/10.1016/0022-2852(79)90034-1
4. Bauschlicher Jr., C.W. and Langhoff, S.R. (1986) Journal of Chemical Physics, 85, 5936.http://dx.doi.org/10.1063/1.451505
5. Werner, H.J., Knowles, P.J., Lindh, R., Manby, F.R., Schütz, M., Celani, P., Korona, T., Rauhut, G., Amos, R.D., Bernhardsson, A., Berning, A., Cooper, D.L., Deegan, M.J.O., Dobbyn, A.J., Eckert, F., Hampel, C., Hetzer, G., Lloyd, A.W., McNicholas, S.J., Meyer, W., Mura, M.E., Nicklass, A., Palmieri, P., Pitzer, R., Schumann, U., Stoll, H., Stone, A.J., Tarroni, R. and Thorsteinsson, T. MOLPRO Is a Package of Ab-Initio Programs.
6. Chong, D.P., Langhoff, S.R., Bauschlicher, C.W., Walch, S.P. and Partridge, H. (1986) Journal of Chemical Physics, 85, 2850-2860. http://dx.doi.org/10.1063/1.451044
7. Langhoff, S.R., Pettersson, L.G.M., Bauschlicher, C.W. and Partridge, H. (1987) Journal of Chemical Physics, 87, 481. http://dx.doi.org/10.1007/s00214-006-0168-3
8. Dunning Jr., T.H. and Hay, P.J. (1985) Journal of Chemical Physics, 82, 2664.http://dx.doi.org/10.1063/1.448263
9. Hurley, M.M., Pacios, L.F., Christiansen, P.A., Ross, R.B. and Ermler, W.C. (1986) Journal of Chemical Physics, 84, 6840. http://dx.doi.org/10.1063/1.453069
10. La John, L.A., Christiansen, P.A., Ross, R.B., Atashroo, T. and Ermler, W.C. (1987) Journal of Chemical Physics, 87, 2812. http://dx.doi.org/10.1063/1.450689
11. Allouche, A.R. (2011) Journal of Computational Chemistry, 32, 174-182.http://dx.doi.org/10.1002/jcc.21600
12. Kobeissi, H. and Korek, M. (1982) International Journal of Quantum Chemistry, 22, 23-29. http://dx.doi.org/10.1002/qua.560220104
13. Kobeissi, H. and Korek, M. (1994) Journal of Physics B: Atomic, Molecular and Optical Physics, 27, 3653. http://dx.doi.org/10.1088/0953-4075/27/16/012
14. Kobeissi, H., Korek, M., Dagher, M. and Chaalan, A. (1983) Journal of Computational Chemistry, 4, 218-225. http://dx.doi.org/10.1002/jcc.540040212

15. Kobeissi, H. and Korek, M. (1985) Journal of Physics B: Atomic, Molecular and Optical Physics, 18, 1115. http://dx.doi.org/10.1088/0022-3700/18/6/016

16. Korek, M. and Kobeissi, H. (1993) Canadian Journal of Chemistry, 71, 313-317.http://dx.doi.org/10.1139/v93-046

17. Korek, M. and Kobeissi, H. (1992) Journal of Computational Chemistry, 13, 1103-1108.http://dx.doi.org/10.1002/jcc.540130909

18. Korek, M. (1999) Computer Physics Communications, 119, 169-178.http://dx.doi.org/10.1016/S0010-4655(98)00180-5

19. Kobeissi, H., Korek, M. and Dagher, M. (1989) Journal of Molecular Spectroscopy, 138, 1-12. http://dx.doi.org/10.1016/0022-2852(89)90092-1

CHAPTER 10

MOLECULAR DYNAMICS, MONTE CARLO SIMULATIONS, AND LANGEVIN DYNAMICS: A COMPUTATIONAL REVIEW

Eric Paquet[1,2] and Herna L. Viktor[2]

[1]Vaccine Program, National Research Council, 1200 Montreal Road, Ottawa, ON, Canada K1A 0R6

[2]School of Electrical Engineering and Computer Science, University of Ottawa, 800 King Edward Road, Ottawa, ON, Canada K1N 6N5

ABSTRACT

Macromolecular structures, such as neuraminidases, hemagglutinins, and monoclonal antibodies, are not rigid entities. Rather, they are characterized by their flexibility, which is the result of the interaction and collective motion of their constituent atoms. This conformational diversity has a significant impact on their physicochemical and biological properties. Among these are their structural stability, the transport of ions through the M2 channel, drug resistance, macromolecular docking, binding energy, and rational epitope design. To assess these properties and to calculate the associated thermodynamical observables, the conformational space must be efficiently sampled and the dynamic of the constituent atoms must be simulated. This paper presents algorithms and techniques that address the abovementioned issues. To this end, a computational review of molecular dynamics, Monte Carlo simulations, Langevin dynamics, and free energy calculation is presented. The exposition is made from first principles to promote a better understanding of the potentialities, limitations, applications, and interrelations of these computational methods.

INTRODUCTION

The ability to properly sample configurational and conformational properties and to subsequently describe at the atomic level the dynamical evolution of complex macromolecular systems has wide application. This research is of paramount importance in the study of macromolecular stability of mutant proteins [1], molecular recognition, ions, and small molecule transportation of the influenza M2 channel [2, 3], protein association, the role of protein flexibility for influenza A RNA binding [4, 5], folding and hydration, influenza neuraminidase inhibitor [6–9], drug resistance [10], enzymatic reactions, folding transitions [11, 12], screening [13], accessibility assessment (see Figure 1), and hemagglutinin fusion peptide [14]. One should also mention multivalent binding mode [15], docking [16], drug (e.g., Oseltamivir and Zanamivir) efficiency against mutants [17, 18], structural biochemistry [19], biophysics, molecular biology, influenza multiple dynamics interactions [20], enzymology, pharmaceutical chemistry [21], biotechnology, rational epitope design [22], computation vaccinology [23], binding [24], and free energy [25, 26]. For instance, one may wish to calculate the free energy to assess the strength and the stability of the bond in between a monoclonal antibody (mAb)

Citation: Eric Paquet and Herna L. Viktor, "Molecular Dynamics, Monte Carlo Simulations, and Langevin Dynamics: A Computational Review," BioMed Research International, vol. 2015, Article ID 183918, 18 pages, 2015. doi:10.1155/2015/183918.

and an antigen, such as the viral hemagglutinin, to quantify the efficiency of the neutralization process.

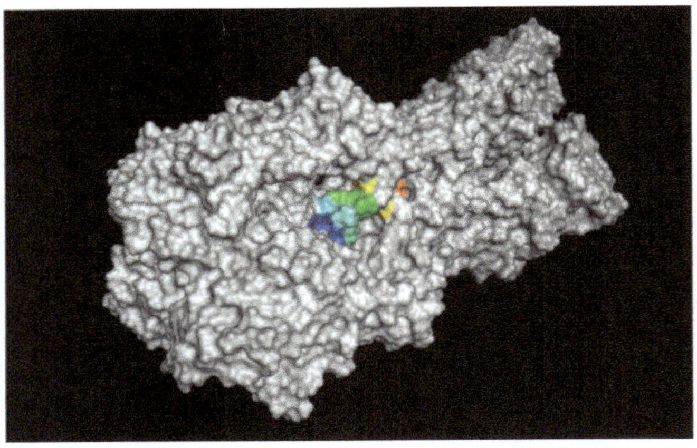

Figure 1: Accessibility assessment of a region of the influenza A virus (A/swine/Iowa/15/1930 (H1N1)).

This paper presents an algorithmic review from the first principles of Monte Carlo simulation, molecular dynamics, and Langevin dynamics (i.e., techniques that have been shown to address the abovementioned scenario). We focus our attention on the algorithmic aspect, which, within the context of a review, has not received sufficient attention. Our objective is not only to explain the algorithms but also to highlight their potential, limitations, applicability, interrelations, and generalization in the context of molecular dynamics. To this end, a number of algorithmic approaches are presented in detail, and the pros and cons of each are highlighted. The algorithms are illustrated with examples related to the influenza virus.

This paper is organized as follows. Monte Carlo simulations are reviewed in Section 2. Section 3 is concerned with molecular dynamics in the microcanonical ensemble, that is, at constant energy. Section 4 extends molecular dynamics to the canonical and the isobaric-isothermal ensemble. Constrained molecular dynamics, hybrid molecular dynamics, and steered molecular dynamics are also presented. Section 5 introduces Langevin and self-guided Langevin dynamics, and Section 6 is concerned with the calculation of the free energy. The application of molecular dynamics to macromolecular docking is addressed in Section 7. Finally, the connection in between molecular dynamics and quantum mechanics (ab initio simulations) is outlined in Section 8. This is followed by a short conclusion.

MONTE CARLO SIMULATIONS

The objective of a Monte Carlo (MC) simulation is to generate an ensemble of representative configurations under specific thermodynamics conditions for a complex macromolecular system [27]. Applying random perturbations to the system generates these configurations. To properly sample the representative space, the perturbations must be sufficiently large, energetically feasible and highly probable. Monte Carlo simulations do not provide information about time evolution. Rather, they provide an ensemble of representative configurations, and, consequently, conformations from which probabilities and relevant thermodynamic observables, such as the free energy, may be calculated. Monte Carlo simulations are not only important on their own right, but they also play a fundamental role when designing complex and hybrid molecular dynamic (MD) algorithms [28].

This section is dedicated to Monte Carlo simulations. In Section 2.1 we review some important notions about Lagrangian and Hamiltonian dynamics, which are pervasive for both Monte Carlo simulations and molecular dynamics. In Section 2.2 we introduce the partition function and the probability density function, as well as the calculation of thermodynamics observable associated with a macromolecule such as the hemagglutinin or the neuraminidase. The partition function is instrumental in computing such observables. In Section 2.3 we explain how to efficiently sample the representative space. For that, we introduce the notions of emission probability, transition probability, acceptance probability, and detailed balance.

Sampling is useful only when performed in realistic experimental conditions. For this reason we explain how to sample in the canonical ensemble (with a constant number of particles, volume, and temperature) and also in the isothermal-isobaric ensemble (with a constant number of particles, pressure, and volume) in Sections 2.4and 2.5, respectively. Finally, in Section 2.6 we address the problem of sampling in the presence of numerous minima. This is a problem particularly acute when studying influenza macromolecular structures such as the hemagglutinin and the neuraminidase.

Lagrangian and Hamiltonian Dynamics or How to Formulate Our Problem

This section presents some important notions about Lagrangian and Hamiltonian dynamics, which are pervasive and recurrent for both MC and MD. Lagrangian and Hamiltonian dynamics provide an ideal framework for the description of complex macromolecular systems, both in Cartesian and generalized coordinates [29]. The Lagrangian is defined as the difference in between the kinetic and the potential energy:

$$\mathscr{L}\left(q^{3N},\dot{q}^{3N}\right) = \mathscr{K}\left(\dot{q}^{3N}\right) - \mathscr{U}\left(q^{3N}\right), \tag{1}$$

where the kinetic energy is given by

$$\mathscr{K} = \sum_{\alpha=1}^{3N} \frac{1}{2} m_\alpha \dot{q}_\alpha^2, \tag{2}$$

and potential energy $\mathscr{U}(q^{3N})$ is a function of the positions of the constituent atoms. The q^{3N} are the generalized coordinates, and the $\dot{q}^{3N}$ are the generalized velocities. For instance, a generalized coordinate may be a bound length, a bound angle, or a dihedral angle. The space of all generalized coordinates and velocities is called the configuration space. If one introduces the generalized momentum:

$$p^\alpha \equiv \frac{\partial \mathscr{L}}{\partial \dot{q}_\alpha}, \tag{3}$$

one may define the Hamiltonian, which is the Legendre transformation of the Lagrangian:

$$\mathscr{H}\left(q^{3N},p^{3N}\right) = \sum_{\alpha=1}^{3N} p^\alpha q_\alpha - \mathscr{L}\left(q^{3N},\dot{q}^{3N}\left(q^{3N},p^{3N}\right)\right). \tag{4}$$

The Hamiltonian obeys the so-called Hamilton's equations, which is just another formulation of Newton's equations:

$$\frac{d\mathcal{H}}{dt} = 0; \qquad \dot{q}_\alpha = \frac{\partial \mathcal{H}}{\partial p^\alpha}; \qquad \dot{p}^\alpha = -\frac{\partial \mathcal{H}}{\partial q_\alpha}. \tag{5}$$

Throughout the text, we will use both generalized and Cartesian coordinates: the Cartesian coordinates of the constituent atom are noted $\mathbf{r}_i \equiv [x_i, y_i, z_i]^T$. The set of all Cartesian coordinates is noted as

$$\mathbf{r}^N \equiv \mathbf{r}_1, \ldots, \mathbf{r}_N, \tag{6}$$

and the associated differential volume element is expressed as

$$d\mathbf{r}^N \equiv d\mathbf{r}_1 \ldots d\mathbf{r}_N. \tag{7}$$

In the next section, we introduce the notions of partition function and probability density function.

Partition Functions, Probability Density Functions, and Expectation or How to Compute Observables

Partition functions are pervasive to all Monte Carlo simulations. They are required to determine the number of microstates associated with a macromolecule, the probability of occurrence of a specific conformation, and the ensemble averages of observables, such as the enthalpy or thermodynamical quantities like the free energy from which the strength of a bound between a drug, such as Oseltamivir [30], and a viral neuraminidase may be asserted.

The number of microstates may be obtained by

$$\Omega(N,V,E) \equiv E_0 C_{\{N\}} \int d\mathbf{p}^N d\mathbf{r}^N \delta\left(\mathcal{H}\left(\mathbf{r}^N, \mathbf{p}^N\right) - E\right), \tag{8}$$

where

$$C_{\{N\}} = \frac{1}{h^{3N}\left(N_A! N_B! \cdots\right)} \tag{9}$$

is a quantum factor, which accounts for the indiscernibility of the various atomic species A, B, and so forth, h is the Planck constant, and $\delta(x)$ is Dirac delta function. The function $\Omega(N, V, E)$ counts the number of states of constant energy E in a system. It is directly related to the entropy

$$S(N,V,E) = k_B \ln \Omega(N,V,E), \tag{10}$$

where k_B is the Boltzmann constant. The canonical partition function is defined as

$$Z(N,V,T) \equiv \int d\mathbf{p}^N d\mathbf{r}^N \exp\left[-\beta\mathcal{H}\left(\mathbf{r}^N, \mathbf{p}^N\right)\right], \tag{11}$$

where

$$\beta \equiv \frac{1}{k_B T} \tag{12}$$

and T is the temperature. The canonical partition function is a functional, which is uniquely determined by the Hamiltonian of the corresponding macromolecular system. If the indiscernibility factor is included in the definition, the microcanonical partition function is noted as

$$Q(N,V,T) \equiv C_{\{N\}} Z(N,V,T). \tag{13}$$

The probability that the macromolecule is in a state characterized by atomic positions r^N, and atomic momenta p^N is given by

$$\Pr\left(\mathbf{r}^N, \mathbf{p}^N\right) d\mathbf{p}^N d\mathbf{r}^N = \frac{\exp\left[-\beta\mathcal{H}\left(\mathbf{r}^N, \mathbf{p}^N\right)\right] d\mathbf{p}^N d\mathbf{r}^N}{Z(N,V,E)}. \tag{14}$$

Consequently, the ensemble average of an observable O (such as the enthalpy) is obtained by weighting the various realizations of the observable by their corresponding probability:

$$\langle O \rangle = \int d\mathbf{p}^N d\mathbf{r}^N O\left(\mathbf{r}^N, \mathbf{p}^N\right) \Pr\left(\mathbf{r}^N, \mathbf{p}^N\right). \tag{15}$$

The uncertainty (standard deviation) associated with the observable is given by

$$\sigma(O) = \sqrt{\langle O^2 \rangle - \langle O \rangle^2}. \tag{16}$$

Unfortunately, it is not possible to integrate the partition function or to compute the probability directly. This is due to the large number of degrees of freedom. For instance, the Homo sapiens influenza hemagglutinin is formed of approximately 23.000 atoms, which means that the partition function must be integrated in a 138.000-dimensional space.

In the next section, we explain how to perform such integration efficiently.

Stochastic Sampling or How to Sample Efficiently Thermodynamical Quantities

The multidimensional integrals associated with the probability and the partition function may be efficiently calculated with a procedure called Monte Carlo integration. In this approach the integration space is sampled according to a Markovian process and the integral is approximated by the average of the corresponding sampled states. Such an approach is efficient if the sampled states have a high probability of occurrence.

A sufficient, but not necessary, condition for such an efficient sampling to hold is called detailed balance:

$$\Pr(\mathcal{I})\, T(\mathcal{I} \longrightarrow \mathcal{J})\, A(\mathcal{I} \longrightarrow \mathcal{J}) = \Pr(\mathcal{J})\, T(\mathcal{J} \longrightarrow \mathcal{I})\, A(\mathcal{J} \longrightarrow \mathcal{I}), \tag{17}$$

where $\Pr(\mathcal{I})$ is the probability (emission probability) that the system is in the state $\mathcal{I} \equiv \{r^N, p^N\}$, $T(\mathcal{I} \rightarrow \mathcal{J})$ is the transition probability from state $\mathcal{I}$ to state $\mathcal{J}$, and $A(\mathcal{I} \rightarrow \mathcal{J})$ is the acceptance probability of such a transition. If we assume that the transition probability is symmetrical

$$T(\mathcal{I} \longrightarrow \mathcal{J}) = T(\mathcal{J} \longrightarrow \mathcal{I}), \tag{18}$$

then the detailed balance equation reduces to

$$\frac{A(\mathcal{I} \longrightarrow \mathcal{J})}{A(\mathcal{J} \longrightarrow \mathcal{I})} = \frac{\Pr(\mathcal{J})}{\Pr(\mathcal{I})} = \exp\left[-\beta\left(\mathcal{U}(\mathcal{J}) - \mathcal{U}(\mathcal{I})\right)\right] \tag{19}$$

A possible solution for this equation is

$$A(\mathcal{I} \longrightarrow \mathcal{J}) = \min\left\{1, \exp\left[-\beta\left(\mathcal{U}(\mathcal{J}) - \mathcal{U}(\mathcal{I})\right)\right]\right\}. \tag{20}$$

This equation is the celebrated Metropolis algorithm [31]. Consequently, each state is defined from the previous one (Markovian process). A transition to a lower energy is always accepted, while a transition to a higher energy is accepted with probability $\exp[-\beta(\mathcal{U}(\mathcal{J}) - \mathcal{U}(\mathcal{I}))]$.

Numerous variations have been designed based on this algorithm [32, 33]. For instance, the local elevation method enhances sampling by adding penalty potential to any state previously sampled (also known as a taboo search algorithm). Although useful, the microcanonical partition function is not realistic from an experimental point of view. Indeed, most observations are performed either at constant volume and temperature (canonical ensemble) or at constant pressure and temperature (isobaric-isothermal ensemble). These distributions are introduced in the next two sections.

Canonical Ensemble (NVT) Sampling or How to Sample in Realistic Experimental Conditions

The canonical ensemble is the ensemble associated with the observations made at constant volume and constant temperature. The configurational canonical partition function associated with such an ensemble is obtained by marginalizing the momenta in (11):

$$Z(N,V,T) \equiv \int d\mathbf{r}^N \exp\left[-\beta\mathcal{U}\left(\mathbf{r}^N\right)\right]. \tag{21}$$

It may also be defined as

$$Q(N,V,T) \equiv M_{\{N\}} Z(N,V,T), \tag{22}$$

where the constant

$$M_{\{N\}} = \frac{1}{\left(\sqrt{h^2\beta/2\pi m_A}\right)^{3N_A} N_A! \left(\sqrt{h^2\beta/2\pi m_B}\right)^{3N_B} N_B! \cdots} \tag{23}$$

takes into account the indiscernibility of the constituent atoms. Like in the microcanonical ensemble, the probability of occurrence of a given state $\{r^N\}$ is equal to

$$\Pr_{NVT}\left(\mathbf{r}^N\right) d\mathbf{r}^N = \frac{\exp\left[-\beta\mathcal{U}\left(\mathbf{r}^N\right)\right] d\mathbf{r}^N}{Z(N,V,T)}. \tag{24}$$

Consequently, the average value of an observable is given by

$$\langle O \rangle = \frac{1}{Z(N,V,T)} \int d\mathbf{r}^N \exp\left[-\beta\mathcal{U}\left(\mathbf{r}^N\right)\right] O\left(\mathbf{r}^N\right). \tag{25}$$

In the case of the canonical partition function, the acceptance probability associated with Monte Carlo method reduces to

$$A_{NVT}\left(\mathbf{r}^N \longrightarrow \mathbf{r}'^N\right) = \min\left\{1, \exp\left[-\beta\left(\mathcal{U}\left(\mathbf{r}'^N\right) - \mathcal{U}\left(\mathbf{r}^N\right)\right)\right]\right\}. \tag{26}$$

From the canonical partition function, it is possible to obtain various thermodynamical quantities such as the Helmholtz free energy:

$$F(N,V,T) = -k_B T \ln Q(N,V,T). \tag{27}$$

Still, most observations are performed at constant pressure and temperature. To address this limitation, the isobaric-isothermal ensemble, as presented in the next section, is introduced.

Isobaric-Isothermal Ensemble (NPT) Sampling or How to Sample in Even More Realistic Experimental Conditions

The isobaric-isothermal ensemble is representative of many experimental conditions. From the microcanonical formalism, it is possible to demonstrate that the isobaric-isothermal configurational partition function is equal to

$$Z(N,P,T) \equiv \int dV \exp\left[-\beta PV\right] \int d\mathbf{r}^N \exp\left[-\beta\mathcal{U}\left(\mathbf{r}^N\right)\right] \tag{28}$$

As usual, if the indiscernibility of the constituent atoms is taken into account, the partition function becomes

$$Q(N,P,T) = \frac{M_{\{N\}}}{V_0} Z(N,P,T). \tag{29}$$

Such a partition function assumes that the deformations of the macromolecular structure are isotropic (the same in all directions). These deformations occur to maintain the pressure constant. If the deformations are anisotropic, the partition function must be modified as follows:

$$Z(N,P,T) = \int d\mathbf{H} Z'(N,P,T,\mathbf{H})\,\delta(\det\mathbf{H} - V), \tag{30}$$

where H is the tensor associated with an elementary parallelepiped volume, which must be integrated (marginalized) over all possible variations of the elementary shape. The probability that a macromolecular system is in a state r^N is given by

$$\Pr_{NPT}\left(\mathbf{r}^N\right) d\mathbf{r}^N = \frac{\exp\left[-\beta PV\right]\exp\left[-\beta\mathcal{U}\left(\mathbf{r}^N\right)\right] d\mathbf{r}^N}{Z(N,P,T)}. \tag{31}$$

Again, various thermodynamical quantities may be defined from the partition function such as the Gibbs free energy:

$$G(N,P,T) = -k_B T \ln Q(N,P,T). \tag{32}$$

The isobaric-isothermal acceptance probability associated with the Monte Carlo method is

$$A_{NPT}\left(\mathbf{r}^N, V \longrightarrow \mathbf{r}'^N, V'\right) = \min\left\{1, \exp\left[-\beta\left(\mathcal{U}\left(\mathbf{r}'^N, V'\right) - \mathcal{U}\left(\mathbf{r}^N, V\right)\right)\right] \times \exp\left[-\beta P\left(V' - V\right) + N \ln\frac{V'}{V}\right]\right\}. \tag{33}$$

Irrelevant of the ensemble in which the calculations are performed, the Metropolis algorithm may be impaired by local minima [27]. Indeed, the acceptance probability may become trapped in a local minimum of the potential energy, which may result in an inadequate sampling of the macromolecular states, as seen in the conformational states [32, 33]. This issue is addressed in the following section.

Sampling and Local Minima or When Temperature May Help to Escape Local Minima

Many biomolecular processes associated with influenza involve activated processes in which a high-energy barrier exists in between the initial and the final state [34]. In order to efficiently sample the macromolecular states, this type of barrier must be overcome.

An efficient, although computationally expensive, approach to overcome such a barrier is called replica exchange (refer to [34] and, in the same spirit, [35]). These methods involve a certain number of non-interacting simulations, called replicas, which are performed in parallel. Each simulation is characterized by its own temperature: low temperature simulations tend to explore local minima, while high temperature simulations may overcome energy barriers and consequently move in between local minima. To favor a better exploration of the macromolecular states, the replicas are periodically exchanged (swapped) according to the following acceptance probability:

$$A_R(\mathscr{I} \longrightarrow \mathscr{J}) = \min\left\{\mathscr{I}, \exp\left[-\left(\beta_{\mathscr{J}} - \beta_{\mathscr{I}}\right) \times \left(\mathscr{U}(\mathscr{J})|_{T_{\mathscr{J}}} - \mathscr{U}(\mathscr{I})|_{T_{\mathscr{I}}}\right)\right]\right\}, \quad (34)$$

where

$$\beta_{\mathscr{I}} \equiv \frac{1}{k_B T_{\mathscr{I}}}. \quad (35)$$

This acceptance probability is similar to the ones introduced before, except for the fact that each state is characterized by its own temperature. Once the exchange is completed, the simulations resume normally until another exchange is performed. The whole procedure allows for a better sampling of the macromolecular states. For instance, this approach has been utilized recently, in conjunction with simulated annealing, for creating the infectious disease model of the H1N1 influenza pandemic [36].

Until now, we have restricted ourselves to symmetrical transition functions. Often, a better sampling may be obtained if a nonsymmetrical sampling function is employed. Let us consider the particular case in which the nonsymmetrical function depends uniquely on the final conformation:

$$T(\mathscr{I} \longrightarrow \mathscr{J}) = \pi(\mathscr{U}(\mathscr{J})). \quad (36)$$

Then, the acceptance probability becomes

$$A(\mathscr{I} \longrightarrow \mathscr{J}) = \min\left\{1, \frac{\pi(\mathscr{U}(\mathscr{I}))}{\pi(\mathscr{U}(\mathscr{J}))} \exp\left[-\beta(\mathscr{U}(\mathscr{J}) - \mathscr{U}(\mathscr{I}))\right]\right\}. \quad (37)$$

This is the so-called bias sampling algorithm [37], which considerably increases the conformational sampling efficiency of large macromolecular chains [38].

Although MC simulations allow us to sample the most probable macromolecular states, they do not provide us with their temporal evolution. The study of the temporal evolution of a macromolecular state is called molecular dynamics and is the subject of the next section.

MOLECULAR DYNAMICS OR WHEN TIME MATTERS

Molecular dynamics studies the temporal evolution of the coordinates and the momenta (the state) of a given macromolecular structure. Such an evolution is called a trajectory. A typical trajectory is obtained by solving Newton's equations. The trajectory is important in assessing numerous time-dependent observables [39] such as the accessibility of a given molecular surface [40], the interaction in between a small molecule (e.g., a drug) and the hemagglutinin or the neuraminidase of a given influenza strain, the interaction epitope-paratope in between an antigen (e.g., hemagglutinin) and an antibody (e.g., CR8020), the appearance and disappearance of a particular channel or cavity, and the fusion of the hemagglutinin with a cell membrane (fusion peptide), amongst others.

From an MD trajectory, it is possible to compute a temporal average of an observable by averaging this observable over time along the trajectory:

$$\overline{O} = \lim_{t \to \infty} \frac{1}{t} \int_0^t d\tau O(r(\tau), \dot{r}(\tau)). \quad (38)$$

Although it has never been formally proven (and that it is not always applicable: for instance, when the trajectory is periodic or when the phase space is constituted of disconnected regions), the ergodicity principle is often invoked [41]. The ergodicity principle states that the average over periods of time along a given trajectory of an observable is, at the limit, identical to the ensemble average of this observable as obtained, for instance, from Monte Carlo simulations:

$$\overline{O} \approx \langle O \rangle. \quad (39)$$

As we will see later, ergodicity is instrumental in performing MD simulations in the canonical and isobaric-isothermal ensemble. The next section is devoted to the potential or force field.

Potential or How to Approximate the Force Field

The choice of a proper potential is of the utmost importance in obtaining accurate molecular dynamics simulations [42]. The potential must be physically sound as well as computationally tractable. An approximate potential may be calculated from quantum mechanics and from the Born-Oppenheimer approximation in which only the positions of the atomic nucleus bonding are considered [42]. The potentials may be divided into bonding potentials and long-range potentials. The bonding potentials involve interaction with two atoms (bound lengths), three atoms (bound angles), and four atoms (dihedral angles). Long-range interactions are associated with the Lennard-Jones potential (van der Waal) and the Columbic potential. The harmonic approximation is utilized for the bonding potentials, which means that solely small displacements are accurately represented. The general form of the potential is

$$\mathscr{U}\left(\mathbf{r}^N\right) = \sum_d k_d (d - d_0)^2 + \sum_S k_S (S - S_0)^2 + \sum_\theta k_\theta (\theta - \theta_0)^2 + \sum_\chi k_\chi (1 + \cos(n\chi - \delta))$$

$$+\sum_{\varphi} k_{\varphi} (\varphi - \varphi_0)^2$$

$$+\sum_{i,j} \varepsilon_{ij} \left(\left(\frac{r_{ij}^0}{r_{ij}} \right)^{12} - \left(\frac{r_{ij}^0}{r_{ij}} \right)^{6} + \frac{q_i q_j}{\varepsilon_l r_{ij}} \right), \quad (40)$$

where d is the bound length, S is the Urey-Bradley bound length, θ is the bound angle, χ is the dihedral angle, φ is the improper dihedral angle, rij is the distance in between atom i and j, k_d, k_S, k_θ, k_χ, and k_φ are constants, d_0, S_0, θ_0, φ_0, and r^0_{ij} are equilibrium positions, ε_{ij} is related to the Lennard-Jones well depth, and ε_l is the effective dielectric constant. Finally, q_i is the partial atomic charge associated with atom i: the partial charge comes from the asymmetrical distribution of the electrons in the chemical bounds. The first term on the last line is the van der Waal interaction (or Lennard-Jones potential), and the last term on the last line is the Columbic interaction. The parameters of the model are determined experimentally and from quantum mechanics. Among the most popular potentials are CHARMM and AMBER [42, 43]. The two differ mostly in the manner in which the parameters are estimated. These potentials may model proteins, lipids, ethers, and carbohydrates, as well as small molecules (e.g., drugs).

The number of interactions involved in long-range interactions rapidly becomes prohibitive. For instance, for the Columbic potential, there are potentially $N!/2!(N - 2)!$ interactions, which correspond to approximately a quarter of billon interactions for an influenza hemagglutinin. To reduce the computational burden, their action range is truncated. The truncation should be performed in such a way as not to introduce artificial discontinuities, which may result in computational artefacts.

The next section is concerned with the refinement of experimentally determined macrostructures.

How to Minimize the Energy of the Conformation or How to Refine Experimentally Determined Structures

The position of the constituent atoms of a macromolecular structure is usually determined either through X-ray crystallography for the larger structure or through nuclear magnetic resonance (NMR) for the smaller molecules. If only the amino acid sequence is available, the three-dimensional structure may be inferred either from methods based on homology, such as threading, or from ab initio methods, which predict the structure from the sequence alone [44]. Among the larger structures associated with influenza are the hemagglutinin and the neuraminidase. Because a protein has to be crystallized to apply X-ray crystallography, the position of its constituent atoms may be distorted from their natural positions by the crystallization process. Consequently, bond lengths and bond angles may be distorted and steric clashes in between atoms may occur. Therefore, it is recommended to minimize the potential energy of the macromolecular structure to remediate this deficiency and to create a more realistic structure [45].

The global optimization of nonlinear functions, such as the potential, is a notoriously difficult problem because of the complexity of the energy landscape and the profusion of local minima [46]. Usually, only local optimization is performed. Such a minimization may be achieved through various algorithms [46] such as the steepest descent algorithm, the conjugate gradient algorithm, and the Newton-Raphson method. The first two are based on the gradient, while the latter is based on the Hessian. In most cases a local optimization is sufficient to refine the structure.

If a global optimization is suited or required, an approach such as simulated annealing must be utilized [47]. Simulated annealing is an MC method. The position of the atoms is subjected to small random displacements. The acceptance probability of such a displacement is given by

$$A(\mathcal{J} \longrightarrow \mathcal{J}) = \min\left\{1, \exp\left[-\beta_k \left(\mathcal{U}(\mathcal{J})|_{T_k} - \mathcal{U}(\mathcal{J})|_{T_k} \right)\right]\right\}, \quad (41)$$

where

$$T_k \in \{T_1, \ldots, T_K\} \mid T_{k+1} < T_k. \quad (42)$$

This means that the temperature acts as a control parameter. Initially, the temperature is high, which implies that transitions from lower to higher energy are allowed with nonnegligible probability in being able to escape local minima. Subsequently, the temperature is gradually reduced (cooling) to decrease the occurrence of such a transition. Transitions to lower energy are always accepted. With a proper choice of temperatures, a global optimization may be achieved. The position of the global minimum associated with the energy landscape may be further refined with local optimization.

The next section is devoted to the solvation of macromolecules.

Implicit and Explicit Solvation, Ions, and Poisson-Boltzmann Equation or How to Obtain Realistic Experimental Conditions

Macromolecules do not exist in isolation. Water molecules and ions surround them. Often, to obtain a realistic simulation, the structure of interest must be solvated. Solvation is a vast and complex subject and we refer the reader to the literature for technical details [4, 44, 45, 48]. This section is devoted to some aspects of solvation, which are particularly relevant to MD.

The solvation may be either implicit or explicit. In the case of an implicit solvation [11], the water molecules are replaced by a potential, which describe their average action while, in the case of an explicit solvation, the macromolecule is surrounded by a solvation box constituted of water molecules. It follows that computers have a limited amount of memory, and thus, the size of this box cannot be infinite. To reduce the number of water molecules, various shapes may be utilized such as cubic, rhombic, dodecahedron truncated octahedron, and sphere. The shape of the solvation box is chosen to minimize the number of molecules required for solvation while maintaining at all times a minimum buffer of solvent. Because of its finite dimensions, the solvation box presents unnatural boundary effects, which should be minimized. This may be partially achieved with a larger solvation box or with periodic boundary conditions (PBC). For a rectangular solvation box, periodic boundary conditions are defined as

$$O(x_i) = O(x_i + L_x)\big|_{i=1,\dots,N},$$
$$O(y_i) = O(y_i + L_y)\big|_{i=1,\dots,N},$$
$$O(z_i) = O(z_i + L_z)\big|_{i=1,\dots,N}. \tag{43}$$

Much care must be taken when using periodic boundary conditions to avoid unphysical artefacts. For instance, if the boundary box is too small, the head of a macromolecule, such as the hemagglutinin, may interact with its own tail, which is extremely unrealistic. Also, if Columbic interactions are involved, the system must be electrostatically neutral; otherwise, the total charge becomes infinite due to the endless replication of the system associated with the PBC. Ions could be added to the solvation box to neutralize the system. Even if the latter is neutral, ions, such as sodium and chloride, may be added to reproduce the ionic strength of the solvent in which the macromolecule evolves. Due to the periodic nature of the boundary conditions, duplicate interactions may appear. It is customary to apply the minimum image convention in which such duplicate interactions are not allowed. If the structure is too large, the solvation may be limited to a specific region of interest: for example, a binding site or a channel while implicit solvation may be utilized for the remaining part of the macromolecule. There are various solvation models [49–52]. Generally, their parameters are adjusted to reproduce the enthalpy of vaporization and density of water.

The solvation box increases the complexity of the simulation. Indeed, most of the computational effort is directed toward simulating the solvent. Nevertheless, dielectric screening, electrostatic effects, and free energy, among others, may only be simulated through explicit solvation, which, consequently, is amply justified [53] though one should notice that explicit solvation does not allow for the simulation of the solvent viscosity.

If the macromolecule is solvated in an ionic solution, the electrostatic potential ϕ may be obtained with a greater accuracy by solving the Poisson-Boltzmann equation:

$$\frac{\partial}{\partial \mathbf{r}} \cdot \left(\varepsilon(\mathbf{r}) \frac{\partial \phi(\mathbf{r})}{\partial \mathbf{r}} \right) = -\rho(\mathbf{r}) - \sum_{k=1}^{K} q_i c_i^{\infty} \lambda(\mathbf{r}) \exp(-\beta q_i \phi(\mathbf{r})), \tag{44}$$

where $\rho(\mathrm{r})$ represents the charge density of the solute (the macromolecule), qi is the ionic charge, c^{∞}_{i} is the ionic concentration far from the solute, $\varepsilon(\mathrm{r})$ is the dielectric permittivity, and $\lambda(\mathrm{r})$ is an accessibility factor. The Poisson-Boltzmann equation is often used in modelling implicit solvation. It may be solved efficiently with finite element methods (FEM) [54].

In the next section, we show how to solve Newton's equation to obtain the trajectory of a macromolecule.

Integration of Newton's Equations: When Technical Details Matter

To obtain the trajectory associated with a given macromolecule, one should solve the corresponding Newton's equations. This is appropriate, since the trajectories are assumed to follow the laws of classical mechanics [55]. In this section we present a completely general approach from which most finite difference algorithms may be derived.

Newton's equation and their generalization, Hamilton's equations, present two important characteristic: they are time-reversible and any infinitesimal volume in phase space (space of all coordinates and momenta) is conserved with time. The latter property is known as the Liouville theorem [56]:

$$d^{3N}p(0)\, d^{3N}q(0) = d^{3N}p(t)\, d^{3N}q(t). \tag{45}$$

This is another formulation of the conservation of the total energy of the system. Any numerical algorithm must enforce these two properties at all times to be physically realistic and consequently relevant. When aiming to derive a finite different algorithm that complies with these requirements, we start with the Liouville operator defined as

$$iL \equiv \sum_{\alpha=1}^{3N} \left[\frac{\partial \mathcal{H}}{\partial p^{\alpha}} \frac{\partial}{\partial q_{\alpha}} - \frac{\partial \mathcal{H}}{\partial q_{\alpha}} \frac{\partial}{\partial p^{\alpha}} \right], \tag{46}$$

where $i = \sqrt{-1}$. This operator allows recasting Hamilton's equations in the form:

$$q^{3N}(t) = \exp[iLt]\, q^{3N}(0), \tag{47}$$

which is known as the Liouville equation. In Cartesian coordinates the Liouville operator becomes

$$iL = iL_1 + iL_2,$$
$$iL_1 \equiv \sum_{i=1}^{N} \frac{\mathbf{p}_i}{m_i} \cdot \frac{\partial}{\partial \mathbf{r}_i}; \qquad iL_2 \equiv \sum_{i=1}^{N} \left(-\frac{\partial \mathcal{U}(\mathbf{r}^N)}{\partial \mathbf{r}_i} \right) \cdot \frac{\partial}{\partial \mathbf{p}_i}. \tag{48}$$

Because the two parts of the Liouville operator do not commute $L_1 L_2 \neq L_2 L_1$, one must rely on the Trotter theorem [5] to develop the argument of the exponential:

$$\exp[iL\Delta t] = \exp[i(L_1 + L_2)\Delta t] \approx \exp\left[iL_2 \frac{\Delta t}{2}\right] \exp[iL_1 \Delta t] \exp\left[iL_2 \frac{\Delta t}{2}\right] + \mathcal{O}(\Delta t^3). \tag{49}$$

Then, if each exponential is approximated in terms of a truncated Taylor expansion, the Liouville equation becomes

$$\mathbf{r}_i(t + \Delta t) = 2\mathbf{r}_i(t) - \mathbf{r}_i(t - \Delta t) - \frac{1}{m_i} \frac{\partial \mathcal{U}(\mathbf{r}^N(t))}{\partial \mathbf{r}_i} \Delta t^2 + \mathcal{O}(\Delta t^4) \Bigg|_{i=1,\dots,N}, \tag{50}$$

which is the well-known Verlet algorithm [1]. Note that other common algorithms, such as the Leap Frog algorithm and the reference system propagator algorithm (RESPA), may be derived following a similar approach [6, 7]. The Liouville operator may be partitioned in more than one way. For instance, the potential may be divided into short- and long-range potentials (e.g., Coulombs and van der Waal). Since long-range potentials tend to change more slowly than short-range potentials, a larger time increment may be used for the former.

If this procedure is repeated, a multiple time-step algorithm may be designed. Once the trajectory has been calculated, the functional important motions may be separated from the random thermal motion by performing a principal component analysis (PCA), which expresses the dominant modes as linear combinations of the underlying motion [57].

The algorithms described in this section are only valid when the total energy is conserved, that is, in the microcanonical ensemble. In next section we show how to extend this approach to the experimentally more realistic canonical and isobaric-isothermal ensembles.

NON-HAMILTONIAN MOLECULAR DYNAMICS OR HOW TO REPRODUCE REALISTIC EXPERIMENTAL CONDITIONS DURING MOLECULAR DYNAMICS SIMULATIONS

To simulate realistic experimental conditions, the MD simulations must be performed either at constant volume and temperature or at constant pressure and temperature. Unfortunately, the canonical and the isobaric-isothermal ensembles do not conserve the total energy. The conservation of a quantity, such as the volume or the pressure, requires a constant exchange of energy in between the macromolecular system and the surrounding heat bath. Such a process, in which the Liouville theorem is not valid, may be described in terms of non-Hamiltonian molecular dynamics [58], as will be detailed here. Firstly, we describe the general approach, which is subsequently applied to the canonical ensemble.

General Approach

To apply the abovementioned general method, we must complement the N Cartesian positions and momenta associated with the constituent atoms with M additional generalized coordinates and momenta. The set of all coordinates is collectively denoted by

$$\boldsymbol{\eta} = \left[\mathbf{r}^N \mid \mathbf{p}^N \mid q^M \mid p_q^M\right]^T . \tag{51}$$

It follows that the original Hamiltonian must be modified to include the generalized coordinates as well as an adjustable parameter ℓ:

$$\mathscr{H}\left(\mathbf{r}^N,\mathbf{p}^N\right) \longrightarrow \mathscr{H}'\left(\mathbf{r}^N,\mathbf{p}^N,q^M,p_q^M;\ell\right). \tag{52}$$

Unfortunately, the theory does not specify a particular form for the modified Hamiltonian. Because of the non-Hamiltonian nature of the dynamics, the Liouville theorem does not apply, which means that elementary volume elements are not conserved in phase space. The reason for this is that the modified dynamics have altered the geometry of the phase space from an Euclidean (flat) geometry to a Riemannian (curved) geometry. In Riemannian geometry, the Liouville theorem becomes [59]

$$\sqrt{\left|\det \mathbf{G}\left(\boldsymbol{\eta}(t)\right)\right|}\,d\boldsymbol{\eta}(t) = \sqrt{\left|\det \mathbf{G}\left(\boldsymbol{\eta}(0)\right)\right|}\,d\boldsymbol{\eta}(0), \tag{53}$$

where $G(\eta)$ is the metric associated with the Riemannian space, and det is the determinant of the matrix. We are only interested in the determinant of this matrix since only the latter is involved in the microcanonical partition function associated with the non-Hamiltonian system. The latter may be obtained directly from

$$\sqrt{\left|\det \mathbf{G}\left(\boldsymbol{\eta}(t)\right)\right|} = \exp\left[-\int_0^t \frac{\partial}{\partial \boldsymbol{\eta}(\tau)}\cdot \dot{\boldsymbol{\eta}}(\tau)\,d\tau\right] \tag{54}$$

The number of microstates associated with the non-Hamiltonian system may be written as

$$\Omega\left(\mathbf{r}^N,\mathbf{p}^N,q^M,p_q^M;\ell\right) = \Omega_0 \int d\boldsymbol{\eta}\sqrt{\left|\det \mathbf{G}(\boldsymbol{\eta})\right|}\prod_{k=1}^{M}\delta\left(\Lambda_k(\boldsymbol{\eta}) - C_k\right), \tag{55}$$

where $d\boldsymbol{\eta}\sqrt{|\det \mathbf{G}(\boldsymbol{\eta})|}$ is the invariant volume element in extended phase space and $\Lambda_k(\boldsymbol{\eta}) - C_k$ is a function associated with a conservation law (for instance, the total energy, or the momentum associated with the barycentre of the system). If we integrate or marginalize the additional coordinates and momenta, we obtain a function that depends solely on the coordinates and momenta of the constituent atoms

$$\Omega\left(\mathbf{r}^N,\mathbf{p}^N,q^M,p_q^M;\ell\right) \xrightarrow[\int dq^M dp_q^M]{} \Omega'\left(\mathbf{r}^N,\mathbf{p}^N;\ell\right). \tag{56}$$

The adjustable parameter is chosen in such a way that the marginalized microcanonical partition function is equal to the partition function of interest (for instance, the canonical (constant volume) partition function)

$$\Omega\left(\mathbf{r}^N,\mathbf{p}^N,q^M,p_q^M;\ell\right) \xrightarrow[\int dq^M dp_q^M]{} \Omega'\left(\mathbf{r}^N,\mathbf{p}^N;\ell\right). \tag{57}$$

In the next section, we further clarify these notions by applying the general approach to the canonical ensemble.

Molecular Dynamics at NVT

In this section we outline the method to obtain the Hamiltonian that describes the canonical ensemble. This Hamiltonian may be substituted in the Liouville operator to obtain a finite difference equation. The modified Hamiltonian, called the Nosé-Hoover chain Hamiltonian [29, 60–62], is defined as

$$\mathscr{H}_{\mathrm{NHC}} = \mathscr{H}\left(\mathbf{r}^N,\mathbf{p}^N\right) + \sum_{i=1}^{M}\frac{p_{q_i}^2}{2\mu_{q_i}} + \ell\frac{q_1}{\beta} + \sum_{i=2}^{M}\frac{q_i}{\beta}, \tag{58}$$

where

$$\mathscr{H}\left(\mathbf{r}^N,\mathbf{p}^N\right) = \sum_{i=1}^{N-1}\frac{p_i'^2}{2m_i'} + \frac{p_\odot^2}{2m_\odot} + \mathscr{U}\left(\mathbf{r}'^N\right) \tag{59}$$

is the Hamiltonian of the macromolecule in barycentric coordinates, the primed variables are the barycentric coordinates (relative to the centre of mass), and $\odot$ refers to the barycentre. If we assume that the energy of the modified Hamiltonian and the barycentre momentum are conserved (isolated system), we have, for the microcanonical partition function,

$$\Omega_{\mathrm{NHC}} = \Omega_0 \int dp_q^M dq^M d\mathbf{p}'^{N-1} dp_{\odot} d\mathbf{r}'^{N-1} \times \exp\left[(3N-2)\, q_1 + \sum_{i=2}^{M} q_i\right] \delta\left(\mathcal{H}_{\mathrm{NHC}} - C_1\right) \times \delta\left(e^{q_1} p_\Omega - C_2\right). \quad (60)$$

If we marginalise the additional coordinate and momenta and choose $\ell = 3N$, the marginalised microcanonical partition function becomes equal to the canonical partition function, which means that the modified Hamiltonian describes the dynamics of the canonical ensemble

$$\Omega'\left(\mathbf{r}^N, \mathbf{p}^N\right)_{\mathrm{NHC}}\Big|_{\ell=3N} \equiv Q(N, V, T). \quad (61)$$

In the next section we focus on the constrained molecular dynamics.

Constrained Molecular Dynamics or How to Reduce the Computational Complexity

Macromolecules, such as the influenza hemagglutinin, possess many degrees of freedom. A typical influenza hemagglutinin is constituted of approximately 23.000 atoms, which means the macromolecules have typically 138.000 degrees of freedom in phase space (atomic positions and momenta). Unsurprisingly, the number of degrees of freedom increases dramatically if the structure is solvated. To reduce the computational complexity, it may be advantageous to impose constraints on certain degrees of freedom. Since the hydrogen atoms are light, they tend to follow the motion of heavier atoms quasi-instantaneously. Consequently, it is customary to fix their bonding length to reduce the computational burden. Let us assume that we have K holonomic (that depends only on time and coordinates) constraints:

$$\sigma_k(t) = \left\|\mathbf{r}_i(t) - \mathbf{r}_j(t)\right\|^2 - d_k^2 \equiv 0. \quad (62)$$

The constraints are enforced in the equations of motion with the method of Lagrange multipliers. Each constraint becomes a potential to which a variable is attached called a Lagrange multiplier λ_k:

$$m_i \ddot{\mathbf{r}}_i(t) = -\frac{\partial \mathcal{U}\left(\mathbf{r}^N(t)\right)}{\partial \mathbf{r}_i} + \sum_{k=1}^{K} \lambda_k \frac{\partial \sigma_k}{\partial \mathbf{r}_i}\Bigg|_{i=1,\ldots,N}. \quad (63)$$

Since the constraints are holonomic, they must be enforced at all times:

$$\sigma_k(t + \Delta t) \equiv 0. \quad (64)$$

If we perform a truncated Taylor development of the previous equation, we may demonstrate that such a condition remains valid if a correction is applied to the Lagrange multiplier at each time step:

$$\tilde{\lambda}_k(t + \Delta t) = \tilde{\lambda}_k(t) + \delta\tilde{\lambda}_k, \quad (65)$$

where $\tilde{\lambda}_k(t) = (\Delta t^2/2)\lambda_k$ is the value of the Lagrange multiplier at time t, $\delta\tilde{\lambda}_k$, is the correction, $\tilde{\lambda}_k(t+\Delta t)$ is the value of the Lagrange multiplier time $t + \Delta t$, and

$$\mathbf{A}\delta\tilde{\boldsymbol{\lambda}} \approx -\boldsymbol{\sigma}\left(\mathbf{r}^N(t + \Delta t)\right), \quad (66)$$

which are defined as

$$\mathbf{A} = \left[\mathbf{A}_{lk}\right] = \left[\sum_{i=1}^{N} \frac{1}{m_i} \frac{\partial \sigma_l\left(\mathbf{r}^N(t + \Delta t)\right)}{\partial \mathbf{r}_i} \cdot \frac{\partial \sigma_k\left(\mathbf{r}^N(t)\right)}{\partial \mathbf{r}_i}\right], \quad \delta\tilde{\boldsymbol{\lambda}} = \left[\delta\tilde{\lambda}_k\right]; \quad \boldsymbol{\sigma} = \left[\sigma_k\right]. \quad (67)$$

In this the linear equation governing the corrections to the Lagrange multipliers in which r(t) are the constrained atomic coordinates at time t. Consequently, to obtain the correction, one must solve the linear equation associated with the correction.

Algorithms distinguish themselves by the approach they use to solve this equation. For the RATTLE algorithm, this equation is solved analytically [63]. For the SHAKE algorithm [64], only the diagonal elements of the matrix are considered to reduce the complexity of the calculation, while for the MSHAKE algorithm [65] the whole matrix is solved with the LU decomposition. If the constraints are periodical, they must be expressed in terms of quaternions in which case the algorithm becomes QSHAKE [66].

In the next section, we consider a hybrid approach based on both molecular dynamics and Monte Carlo simulations.

Hybrid Monte Carlo Dynamics or How to Wisely Accelerate the Dynamics

In order to obtain realistic molecular dynamics simulations, the time increment must be kept small. Usually, the value of the time increment is chosen around 1 femtosecond (10^{-15} seconds), which is typically one to two orders of magnitude smaller than the time scale associated with the fastest molecular event, the bound-length vibration [48]. If one desires to explore slow molecular events, the time required to run the simulation may become rapidly prohibitive. For instance, the rotation of buried side chains takes about 10^{-4} to 1 second to complete, while an helix-coil transition may take as long as 10^4 seconds. The situation might be even worse with macromolecular docking. Naturally, the time increment could be increased but the corresponding trajectory becomes rapidly unphysical.

This problem may be partially solved with an approach Eric Paquet and Herna L. Viktor, "Molecular Dynamics, Monte Carlo Simulations, and Langevin Dynamics: A Computational Review," BioMed Research International, vol. 2015, Article ID 183918, 18 pages, 2015. doi:10.1155/2015/183918.

indicated by its name, HMC is a combination of MD and MC. The time evolution of the macromolecule is calculated with standard MD but with larger time increments. The outcome of a time iteration is accepted only if it is physically feasible. The feasibility of the outcome is assessed with a standard MC acceptance probability rule:

$$A\left(\mathbf{r}^N(t), \mathbf{p}^N(t) \longrightarrow \mathbf{r}^N(t + \Delta t), \mathbf{p}^N(t + \Delta t)\right) = \min\left\{1, \exp\left[-\beta\left(\mathcal{H}\left(\mathbf{r}^N(t + \Delta t), \mathbf{p}^N(t + \Delta t)\right) - \mathcal{H}\left(\mathbf{r}^N(t), \mathbf{p}^N(t)\right)\right)\right]\right\}. \quad (68)$$

Note that there is an importance subtlety associated with HMC. Because Newton's equations are time reversible (any trajectory may be reversed by inverting the direction of the motion), the transition probability must also be so [68]. That implies that the transition probability must be chosen so that

$$T\left(\mathbf{r}^N(t),\mathbf{p}^N(t) \longrightarrow \mathbf{r}^N(t+\Delta t),\mathbf{p}^N(t+\Delta t)\right) = T\left(\mathbf{r}^N(t+\Delta t),-\mathbf{p}^N(t+\Delta t) \longrightarrow \mathbf{r}^N(t),-\mathbf{p}^N(t)\right). \quad (69)$$

Otherwise, the HMC may generate unphysical results.

In the next section, we explore other ways that may be used to assess slow processes.

Computational Alchemy or How to Virtually Explore Experimentally Hidden Processes

Molecular dynamics is not limited to real physical potentials and equilibrium processes. Introducing an external, nonphysical, time-dependent, and possibly position-dependent potential may accelerate the dynamics of a slow process. This is called steered molecular dynamics or computational alchemy [69]. For instance, a force may be applied to a hemagglutinin to study potential conformational changes or to a drug (ligand) to analyze the docking process in between a drug and a neuraminidase. An example of how this could be achieved is by applying a random force (both in terms of amplitude and direction) to the barycentre of the ligand [70]. If the displacement of the ligand is above a certain threshold, the force is reapplied without modification. Otherwise, a new random force is generated. The whole process is repeated until the desired outcome is obtained. Such an approach is often combined with an immersive virtual environment to better visualize the outcome of the simulation [69]. In the next section, we explore an alternative to molecular dynamics and Monte Carlo simulations, namely, stochastic dynamics.

Molecular Dynamics and Influenza

There have been numerous applications of molecular dynamics in studies related to the influenza virus. For instance, molecular dynamics has been utilized for rational drug design for the influenza neuraminidase [7]. It has been instrumental to illustrate that the electrostatic funnel directs binding of Tamiflu to the influenza N1 neuraminidase [8]. Using molecular dynamics, it has been revealed that HR1039 is a potent inhibitor of the 2009 A (H1N1) influenza neuraminidase [9]. In addition, molecular dynamics has been valuable to show the role played by the BM2 channel in proton conductance and drug resistance for the influenza virus B [10]. Using molecular dynamics has made possible the design of inhibitor targeting drug-resistant mutants of the influenza A virus M2. Finally, conformational analysis of peptides and lipids of the influenza hemagglutinin fusion peptide in micelles and bilayer would not have been possible without the aid of molecular dynamics [14].

LANGEVIN DYNAMICS, SELF-GUIDE LANGEVIN DYNAMICS, AND SELF-GUIDED MOLECULAR DYNAMICS: TOWARD A BETTER SAMPLING OF THE CONFORMATIONAL SPACE

The dynamics of a macromolecular system is entirely determined by the potential $\mathcal{U}(\mathbf{r}^N)$ associated with the process. For computational and practical reasons, this potential is virtually always an approximation of the real physical potential. Stochastic (random) dynamics attempts to bridge the gap in between the real and the approximate potential. Stochastic dynamics does not attempt to define the real potential but a general correction, which is independent on the particular details of the real potential. In other words, stochastic dynamics attempt to take into account the neglected degrees of freedom to obtain more realistic simulations, especially in the case of conformational sampling [71].

In this framework, it is assumed that each particle is under the influence of the potential $\mathcal{U}(\mathbf{r}^N)$ and of a heat bath formed by the remaining $N-1$ particles. Under general conditions using the Mori-Zwanzig theory [72] (one may obtain the same result by assuming that each particle is minimally coupled to a harmonic heat bath formed from the other particles), it is possible to demonstrate that the particles obey the generalized Langevin equation (GLE) [73]:

$$m_i\ddot{\mathbf{r}}_i(t) = -\frac{\partial \mathcal{U}\left(\mathbf{r}^N(t)\right)}{\partial \mathbf{r}_i} - \int_0^t d\tau \dot{\mathbf{r}}_i(\tau)\gamma_i(t-\tau) + \boldsymbol{\xi}_i(t)\Bigg|_{i=1,\ldots,N}, \quad (70)$$

where $\gamma_i(t-\tau)$ is the memory kernel or dynamic friction and $\xi_i(t)$ is a stochastic term with mean:

$$\langle \boldsymbol{\xi}_i(t)\rangle = 0 \quad (71)$$

and covariance:

$$\left\langle \boldsymbol{\xi}_i(0)\boldsymbol{\xi}_j(t)\right\rangle = 2m_i k_B \gamma_i T\delta(t)\delta_{ij}\mathbf{I}. \quad (72)$$

Consequently, the correction is formed of two terms: a friction term, which introduces an artificial viscosity, and a stochastic term, which takes into account the unknown nature of the correction. One should notice that the frictional term depends on the previous history of the trajectory (Markovian process). The friction term is important in obtaining realistic simulations, as it takes into account the viscosity of the solvent (a feature, which is absent from both MD and MC). If one assumes that the frictional term is constant (no history), one obtains the celebrated Langevin equation (LE):

$$\dot{\mathbf{p}}_i(t) = -\frac{\partial \mathcal{U}\left(\mathbf{r}^N(t)\right)}{\partial \mathbf{r}_i} - \gamma_i\dot{\mathbf{p}}_i(t) + \boldsymbol{\xi}_i(t)\Bigg|_{i=1,\ldots,N}. \quad (73)$$

In that particular case, γ_i is simply called the frictional or damping term. The Langevin equation improves conformational sampling over standard molecular dynamics.

Conformational sampling may still be further improved if the trajectory history is reintroduced into the model [71].

To take history into account, a history-dependent guiding term $\mathbf{\Gamma}_i(t)$ is added to the Langevin equation:

$$\dot{\mathbf{p}}_i(t) = -\frac{\partial \mathcal{U}\left(\mathbf{r}^N(t)\right)}{\partial \mathbf{r}_i} - \gamma_i \dot{\mathbf{p}}_i(t) + \boldsymbol{\xi}_i(t) + \lambda_i \mathbf{\Gamma}_i(t)\Bigg|_{i=1,\ldots,N}. \tag{74}$$

This term could be defined in various manners. In self-guided Langevin dynamics (SGLD), the history-dependent guiding term is defined as the time average of the momentum over the last ℓ iterations:

$$\begin{aligned} {}^{\mathrm{SGLD}}\mathbf{\Gamma}_i(t) &= \gamma_i \overline{\mathbf{p}_i(t)}\Big|_{t\in[t-\ell\Delta t,t]} \\ &= \left(1-\frac{\Delta t}{\ell\Delta t}\right)\mathbf{\Gamma}_i(t-\Delta t) + \frac{\Delta t}{\ell\Delta t}\gamma_i m_i \dot{\mathbf{r}}_i\left(t-\frac{\Delta t}{2}\right). \end{aligned} \tag{75}$$

When, for self-guided molecular dynamics (SGMD), the history-dependent guiding term is defined as the time average of the potential plus its self-time average,

$$\begin{aligned} {}^{\mathrm{SGMD}}\mathbf{\Gamma}_i(t) &= \overline{-\frac{\partial \mathcal{U}\left(\mathbf{r}^N(t)\right)}{\partial \mathbf{r}_i} + \lambda_i \mathbf{\Gamma}_i(t)}\Bigg|_{t\in[t-\ell\Delta t,t]} \\ &= \left(1-\frac{\Delta t}{\ell\Delta t}\right)\mathbf{\Gamma}_i(t-\Delta t) \\ &\quad + \frac{\Delta t}{\ell\Delta t}\left(-\frac{\partial \mathcal{U}\left(\mathbf{r}^N(t)\right)}{\partial \mathbf{r}_i} + \lambda_i \mathbf{\Gamma}_i(t)\right). \end{aligned} \tag{76}$$

The guiding term is unphysical (as opposed to the memory kernel) and does not conserve energy. To reestablish energy conservation, an additional term $\varsigma(t)\mathbf{p}_i(t)$, proportional to the momentum, must be added to the SGLD equation:

$$\begin{aligned} \dot{\mathbf{p}}_i(t) &= -\frac{\partial \mathcal{U}\left(\mathbf{r}^N(t)\right)}{\partial \mathbf{r}_i} - \gamma_i \dot{\mathbf{p}}_i(t) \\ &\quad + \boldsymbol{\xi}_i(t) + \lambda_i \mathbf{\Gamma}_i(t) - \varsigma(t)\mathbf{p}_i(t)\Big|_{i=1,\ldots,N}. \end{aligned} \tag{77}$$

This term ensures the conservation of energy, providing that the running constant (t) obeys

$$\varsigma(t) = \frac{\sum_{i=1}^N \lambda_i \mathbf{\Gamma}_i(t)\cdot\dot{\mathbf{r}}_i(t)}{\sum_{i=1}^N \mathbf{p}_i(t)\cdot\dot{\mathbf{r}}_i(t)}. \tag{78}$$

These equations may be easily integrated if they are recast under the form of Wiener processes [74]. The integration requires two independent Gaussian random variables.

Langevin dynamics has been used in many applications. For instance it has been instrumental to unravel the bilayer conformation of the fusion peptide of the influenza virus hemagglutinin [75]. Further, the use of Langevin dynamics aided researchers to characterize the loop dynamics and ligand recognition in human- and avian-type influenza neuraminidase [76].

In the next section, we address complexes stability such as an interaction between a drug and neuraminidase or that between monoclonal antibodies and hemagglutinin.

FREE ENERGY, BINDING, AND COMPLEXES OR A MEASURE OF STABILITY

The interaction in between a small molecule (drug) or an antibody with an antigen (hemagglutinin or neuraminidase) may result in a complex. It follows that, often, the stability of such a complex must be assessed. Indeed, if the complex is stable, the small molecule or the antibody may efficiently neutralize the corresponding antigen. Free energy [53] is a measure of such stability. Consequently, the calculation of free energy is addressed in the next section.

Thermodynamic Integration

The thermodynamics integration method [77] requires two states to determine the free energy: an unbound state $\mathcal{I}$ and a bound state $\mathcal{J}$. From them, a parametric potential is defined as a linear superposition of the two potentials:

$$\mathcal{U}\left(\mathbf{r}^N;\lambda\right) = (1-\lambda)\,\mathcal{U}_{\mathcal{I}}\left(\mathbf{r}^N\right) + \lambda\,\mathcal{U}_{\mathcal{J}}\left(\mathbf{r}^N\right). \tag{79}$$

A corresponding parametric partition function is also defined as follows:

$$Z(N,V,T;\lambda) = \int d\mathbf{r}^N \exp\left[-\beta\,\mathcal{U}\left(\mathbf{r}^N;\lambda\right)\right]. \tag{80}$$

Such a procedure is called adiabatic switching. The parametric derivative of the free energy may be obtained from the definition of the free energy:

$$\frac{\partial F(N,V,T;\lambda)}{\partial\lambda} = -\frac{1}{\beta}\frac{\partial}{\partial\lambda}\ln Z(N,V,T;\lambda). \tag{81}$$

It follows that the difference of free energy in between the bound and the unbound state may be evaluated from

$$F_{\mathcal{J}} - F_{\mathcal{I}} = \int_0^1 d\lambda \left\langle \frac{\partial\,\mathcal{U}\left(\mathbf{r}^N;\lambda\right)}{\partial\lambda}\right\rangle_{\mathcal{U}(\mathbf{r}^N;\lambda)}. \tag{82}$$

In the next section, we introduce a method based on irreversible work.

Nonequilibrium Method

In this section we describe an approach inspired from steered molecular dynamics (Section 4.5). As opposed to thermodynamic integration, the nonequilibrium method does not require the bound state. This state is reached as a result of the irreversible work effectuated by the synthetic time-dependent Hamiltonian. The nonequilibrium method is based on Jarzinski's equality and irreversible work [78]. Jarzinski's equality may be derived from a time-dependent Hamiltonian ensemble average and the Liouville theorem. From this equality, the difference in free energy in between a bound and an unbound state may be calculated from the ensemble average of the irreversible work applied to the system to bring the unbound state into a bound state:

$$F_{\mathcal{J}} - F_{\mathcal{I}} = -kT\ln\left\langle \exp\left[-\beta\,\mathcal{W}_{\mathcal{I}\mathcal{J}}\left(\mathbf{r}^N,\mathbf{p}^N\right)\right]\right\rangle_{\mathcal{I}}, \tag{83}$$

where $\mathcal{W}_{\mathcal{I}\mathcal{J}}(\mathbf{r}^N,\mathbf{p}^N)$, the irreversible work, is given by

$$\mathcal{W}_{\mathcal{I}\mathcal{J}}\left(\mathbf{r}^N,\mathbf{p}^N\right) = \int_{t_{\mathcal{I}}}^{t_{\mathcal{J}}} d\tau\,\frac{\partial\,\mathcal{H}_{\mathcal{I}\mathcal{J}}\left(\mathbf{r}^N,\mathbf{p}^N,t\right)}{\partial\tau}, \tag{84}$$

and where the time-dependent Hamiltonian $\mathcal{H}_{\mathcal{I}\mathcal{J}}(\mathbf{r}^N, \mathbf{p}^N, t)$ is equal to the sum of the unbound macromolecular Hamiltonian $\mathcal{H}_{\mathcal{I}}(\mathbf{r}^N, \mathbf{p}^N)$ and of an external arbitrary time-dependent potential $\mathcal{U}_{\mathcal{I}\mathcal{J}}(\mathbf{r}^N; t)$, which brings the macromolecular system from an unbound state $\mathcal{I}$ to a bound state $\mathcal{J}$:

$$\mathcal{H}_{\mathcal{I}\mathcal{J}}\left(\mathbf{r}^N, \mathbf{p}^N, t\right) = \mathcal{H}_{\mathcal{I}}\left(\mathbf{r}^N, \mathbf{p}^N\right) + \mathcal{U}_{\mathcal{I}\mathcal{J}}\left(\mathbf{r}^N; t\right). \tag{85}$$

The ensemble average of the irreversible work is performed with the unbound Hamiltonian:

$$\begin{aligned}\left\langle \exp\left[-\beta \mathcal{W}_{\mathcal{I}\mathcal{J}}\left(\mathbf{r}^N, \mathbf{p}^N\right)\right]\right\rangle_{\mathcal{I}} &= \left(\int d\mathbf{p}^N d\mathbf{r}^N \exp\left[-\beta \mathcal{W}_{\mathcal{I}\mathcal{J}}\left(\mathbf{r}^N, \mathbf{p}^N\right)\right] \right.\\ &\quad \left. \times \exp\left[-\beta \mathcal{H}_{\mathcal{I}}\left(\mathbf{r}^N, \mathbf{p}^N\right)\right]\right)\\ &\quad \times \left(Z_{\mathcal{I}}(N, V, T)\right)^{-1}.\end{aligned} \tag{86}$$

The next two sections present methods, which evaluate the free energy based on a subset of their generalized coordinates.

Blue Moon Ensemble Approach

Thermodynamic integration and the nonequilibrium method both exploit all the coordinates and momenta associated with the macromolecular system to evaluate the free energy. Correspondingly, one may evaluate the free energy from a subset called the reaction coordinates (e.g., bound length and bound angle) of the generalized coordinates such as the generalized coordinates directly associated with the binding process. This considerably reduces the complexity of the calculation but introduces a certain level of approximation, as most coordinates are neglected. In order not to clutter the notation, we will use only one reaction coordinate while keeping in mind that the generalization of more than one coordinates is immediate.

The probability that a reaction coordinate q_α has a value s is

$$\Pr(s) = \frac{\int d\mathbf{p}^N d\mathbf{r}^N \exp\left[-\beta \mathcal{H}\right] \delta\left(q_\alpha\left(\mathbf{r}^N\right) - s\right)}{Z(N, V, T)}, \tag{87}$$

where

$$q_\alpha = q_\alpha\left(\mathbf{r}^N\right) \tag{88}$$

is the representation of the reaction coordinates in terms of the Cartesian coordinates. The free energy associated with this coordinate is by definition

$$F(s) = -k_B T \ln \Pr(s). \tag{89}$$

If we marginalize (integrate) the momenta and if we express the potential in terms of the generalized coordinates, we obtain for the free energy [79]

$$F(q) = F(s_i) + \int_{s_i}^{q} \left\langle \frac{\partial \mathcal{U}\left(\mathbf{r}^N\left(q^{3N}\right)\right)}{\partial q_\alpha} - k_B T \frac{\partial \ln \det \mathbf{J}}{\partial q_\alpha} \right\rangle_q ds, \tag{90}$$

where

$$\begin{aligned}\langle O \rangle_q &= \frac{1}{Z(N, V, T)\left\langle \delta\left(q_\alpha - s\right)\right\rangle}\\ &\quad \times \int dq^{3N} O\\ &\quad \times \exp\left[-\beta\left(\mathcal{U}\left(\mathbf{r}^N\left(q^{3N}\right)\right)\right.\right.\\ &\quad \left.\left. - k_B T \ln \det \mathbf{J}\right)\right] \delta\left(q_\alpha - s\right),\end{aligned} \tag{91}$$

and where the Jacobian matrix of the transformation is

$$\mathbf{J} = \frac{\partial \mathbf{r}^N}{\partial q^{3N}}. \tag{92}$$

The main disadvantage of this approach is that all Cartesian coordinates must be converted into generalized coordinates, although only a subset of them (the reactions coordinates) are effectively exploited in the calculation of the free energy. This inefficiency is addressed in the next section.

Umbrella Sampling and Weighted Histogram Methods

When using umbrella sampling [80, 81] and the weighted histogram method [82] for the blue moon ensemble approach, the free energy is estimated from the reaction coordinates. The range (codomain and image) associated with the reaction coordinates is divided into a set of windows or intervals. A reaction coordinate distribution is associated with each window. The unbiased estimator for the probability of occurrence of a reaction coordinate within a given window is assumed to be the product of a biased Gaussian probability distribution and a biased correction based on an umbrella potential:

$$\Pr(q) \approx \frac{1}{\sqrt{2\pi\sigma_k^2}} \exp\left[-\frac{\left(q - \bar{q}_k\right)^2}{2\sigma_k^2}\right] \exp\left[-\beta W\left(q, s^{(k)}\right)\right], \tag{93}$$

where

$$W\left(q, s^{(k)}\right) = \frac{1}{2}\kappa\left(q - s^{(k)}\right)^2 \tag{94}$$

is the harmonic umbrella potential and $s^{(k)}$ is the center of a given window, while $\bar{q}_k$ and σ_k^2 are, respectively, the average and the variance of all the realizations of the reaction coordinates within window k. By definition, the derivative of the free energy associated with a particular window is equal to

$$\frac{dF_k}{dq} = -\frac{k_B T}{\Pr(q)} \frac{d\Pr(q)}{dq} = \frac{k_B T}{\sigma_k^2}\left(q - \bar{q}_k\right) - \kappa\left(q - s^{(k)}\right). \tag{95}$$

It is assumed that the derivative of the total free energy is a combination of the derivatives of the windows' free energy:

$$\frac{dF}{dq} = \sum_{k=1}^{K} C_k(q) \frac{dF_k}{dq}, \tag{96}$$

and that the weighting coefficients are normalized

$$\sum_{k=1}^{K} C_k(q) = 1. \tag{97}$$

The full free energy profile may be extracted with numerical integration. In the next section, we describe another approach for the calculation of the free energy, which introduces a bias potential to better sample the reaction coordinates.

Metadynamics

As opposed to the previous methods, which were based on a single molecular dynamics simulation, the metadynamics approach involved two molecular dynamics simulations: the usual one in Cartesian coordinates under the influence of the real physical potential and a second one, the meta-simulation, which is performed under the influence of a biased potential. The bias potential is defined as

$$\mathcal{U}_G\left(\mathbf{r}^N\right) = W \sum_{t=\tau_G,2\tau_G,\dots} \exp\left[-\sum_{\alpha=1}^{M} \frac{\left(q_\alpha\left(\mathbf{r}^N(t)\right) - q_\alpha\left(\mathbf{r}_G^N(t)\right)\right)^2}{2\Delta q^2}\right], \tag{98}$$

where W is a constant, τ_G is a time interval; $r^N(t)$ is the time evolution (trajectory) of the complete set of Cartesian coordinates up to time t under the action of the real physical potential $U(r^N)$, and $r^N_G(t)$ is the trajectory of the system under the influence of both the physical potential and the biased potential $U(r^N) + U_G(r^N)$. From an analysis based on the Langevin equation, it is possible to demonstrate that the free energy may be obtained from

$$F\left(q^M\right) \approx W \sum_{t=\tau_G,2\tau_G,\dots} \exp\left[-\sum_{\alpha=1}^{M} \frac{\left(q_\alpha - q_\alpha\left(\mathbf{r}_G^N(t)\right)\right)^2}{2\Delta q^2}\right], \tag{99}$$

in which the reaction coordinates are expressed in a function of the metacoordinates. The proof of this equation is beyond the scope of this review and may be found in [83]. As for umbrella sampling, only the reaction coordinates need to be expressed in terms of the Cartesian coordinates.

The use of free energy calculations has provided insights into the susceptibility of antiviral drugs against the E119G mutant of the 2009 influenza A (H1N1) neuraminidase [26]. It has also been utilized to show the impact of calcium on the N1 influenza neuraminidase [25].

In the next section, we explain how molecular dynamics may be combined with macromolecular docking to assert the dynamics of a macromolecular complex, such as a viral hemagglutinin, when interacting with a broadly neutralizing monoclonal antibody.

BINDING AND DOCKING OR COMPUTATIONAL PHARMACOLOGY AND VACCINOLOGY

From the outset, one may assume that the relative pose of the constituent of a complex may be obtained through molecular dynamics without resorting to macromolecular docking. Although this is, in principle, feasible, it is usually not realistic in practice. Indeed, the time scale associated with macromolecular docking is usually out of reach for most molecular dynamics simulations [48]. Larger time steps may be taken if a hybrid Monte Carlo approach is followed (Section 4.4), but most of the time, the increase is insufficient. For this reason macromolecular docking remains the favorite method to determine the relative pose in between a receptor and a ligand. (Note that macromolecular docking is beyond the scope of this paper, but a comprehensive algorithmic review may be found in [84]).

The abovementioned observation does not imply that MD is irrelevant for macromolecular docking. On the contrary, once the pose in between the receptor and the ligand has been determined through macromolecular docking, the dynamics of the resulting complex may be explored with MD simulations. Various dynamical aspects of the complex may be studied. These include the study of the conformational space associated with the receptor [85], binding path analysis [56], examining the structural changes associated with induced fit, studying the flexibility of the docking region, analyzing transient binding, and aiding in structural refinement of the complex [86] as well as the design of drugs [87] and antibodies (see Figure 2) with optimal kinetic properties. MD may also be employed prior to macromolecular docking to generate conformations, which become the initial point for static docking. This is particularly beneficial if the ligand binds to conformations that rarely occur (relax complex method). Nevertheless, standard MD should not be utilized to explore low frequency motions associated with large conformational transformations. This is due to its limited ability to explore conformational change. Langevin dynamics may be more suited in that particular case. Also, from a computational point of view, multiple copies of the ligand may simultaneously interact with the receptor to search, in parallel, for the best docking region. There should be no interaction in between the copies. Molecular dynamics may also be utilized to evaluate the structural accessibility of a highly conserved epitope [4].

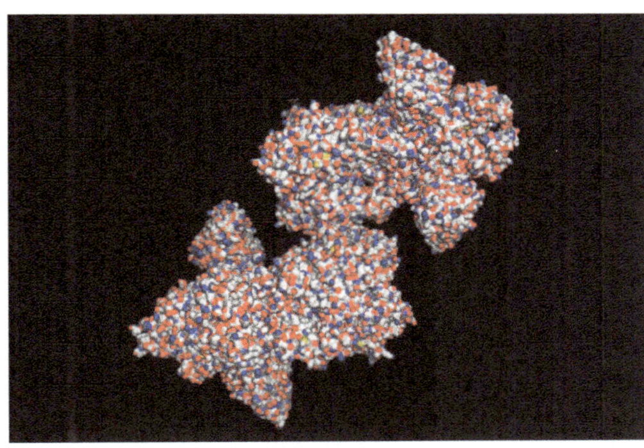

Figure 2: Molecular surface of influenza hemagglutinin (A/Viet Nam/1203/2004 (H5N1)) in complex with a broadly neutralizing antibody F10.

In the next section, we briefly explore the relationship in between quantum mechanics and molecular dynamics.

PATH INTEGRAL OR WHEN QUANTUM MECHANICS MEETS MOLECULAR DYNAMICS

The calculations that we have performed so far are based on classical and semiclassical approximations of quantum mechanics. Though being valid, these approximations are not as accurate as their quantum-mechanical counterpart

[88]. However, quantum-mechanical (QM) methods are often computationally prohibitive. For that reason, quantum-mechanical methods are restricted to a small number of coordinates of interest (reaction coordinates) and the residual coordinates are analyzed with molecular dynamics (the so-called hybrid quantum-mechanical molecular dynamics). Quantum mechanical methods are outside the scope of this review. More details may be found in [88]. Here, we limit ourselves to the connection in between quantum-mechanical methods and molecular dynamics. Quantum-mechanical methods are also based on the partition function formalism, which in the quantum case takes the form:

$$Q(N,V,T) = \operatorname{tr}\exp\left[-\beta\widehat{\mathscr{H}}\right], \tag{100}$$

where tr is the trace operator and $\widehat{\mathscr{H}}$ is the quantum Hamiltonian operator, which may be obtained from the classical Hamiltonian as follows:

$$\mathscr{H}\left(\mathbf{r}_i, \frac{\hbar}{i}\frac{\partial}{\partial\mathbf{r}_i}\right) \longrightarrow \widehat{\mathscr{H}}. \tag{101}$$

By extensively using the following identities:

$$\int dx\,|x\rangle\langle x| = \hat{I}; \qquad \int dp\,|p\rangle\langle p| = \hat{I};$$
$$\langle x \mid p\rangle = \exp\left[\frac{\hbar}{i}px\right], \tag{102}$$

and by making a certain number of approximations (Born-Oppenheimer approximation, Boltzmann statistics, and adiabatic approximation [83], etc.), it may be demonstrated that the partition function

$$Q(N,V,T) = \oint \mathscr{D}\mathbf{r}_1(\tau)\cdots\mathscr{D}\mathbf{r}_N(\tau) \times \exp\left[-\frac{1}{\hbar}\int_0^{\beta\hbar} d\tau \frac{1}{2}\sum_{i=1}^{N} m_i\dot{\mathbf{r}}_i^2(\tau) + \mathscr{U}\left(\mathbf{r}^N(\tau)\right)\right] \tag{103}$$

is equal to the path integral over all possible closed paths of the exponential of the classical Hamiltonian. The notation $\mathrm{Dr}(\tau)$ means that we must integrate $r_i(\tau)$ over all possible paths or trajectories. The resulting approach is known as path integral molecular dynamics [89]. Equation (103) involves only the atomics coordinates, but it is possible to introduce the conjugate momenta without altering the partition function by adding a harmonic potential to the Hamiltonian:

$$Q(N,V,T) = \oint \mathscr{D}\mathbf{r}_1(\tau)\cdots\mathscr{D}\mathbf{r}_N(\tau)\mathscr{D}\mathbf{p}_1(\tau)\mathscr{D}\mathbf{p}_N(\tau) \times \exp\left[-\frac{1}{\hbar}\int_0^{\beta\hbar} d\tau \sum_{i=1}^{N}\frac{\mathbf{p}_i^2}{2m_i'}(\tau)\right] \times \exp\left[-\frac{1}{\hbar}\int_0^{\beta\hbar}\sum_{i=1}^{N}\frac{1}{2}m_i\omega_K^2\mathbf{r}_i^2(\tau) + \mathscr{U}\left(\mathbf{r}^N(\tau)\right)\right], \tag{104}$$

where

$$m_i' = \frac{m_iK}{(2\pi\hbar)^2}; \quad \omega_K = \frac{\sqrt{K}}{\beta\hbar}; \quad K \longrightarrow \infty. \tag{105}$$

The two equations are entirely equivalent. If we develop the notation, we obtain

$$Q(N,V,T) \approx \int \prod_{i=1}^{N} d\mathbf{r}_i^{(1)}\cdots d\mathbf{r}_i^{(K)} d\mathbf{p}_i^{(1)}\cdots d\mathbf{p}_i^{(K)} \times \exp\left[-\beta\sum_{k=1}^{K}\sum_{i=1}^{N}\frac{\left(\mathbf{p}_i^{(k)}\right)^2}{2m_i'} + \frac{1}{2}m_i\omega_K^2\left(\mathbf{r}_i^{(k+1)} - \mathbf{r}_i^{(k)}\right)^2 + \frac{1}{K}\mathscr{U}\left(\mathbf{r}_1^{(k)}\cdots\mathbf{r}_N^{(k)}\right)\right]\Bigg|_{\mathbf{r}_i^{(K+1)}=\mathbf{r}_i^{(1)}\forall i}. \tag{106}$$

This path integral may be solved with the following infinite set of molecular dynamics equations:

$$\dot{\mathbf{r}}_i^{(k)} = \left.\frac{\mathbf{p}_i^{(k)}}{m_i'}\right|_{k=1,\ldots,\infty},$$
$$\dot{\mathbf{p}}_i^{(k)} = -m_i\omega_K^2\left(2\mathbf{r}_i^{(k)} - \mathbf{r}_i^{(k+1)} - \mathbf{r}_i^{(k-1)}\right) - \frac{1}{K}\left.\frac{\partial\mathscr{U}\left(\mathbf{r}_1^{(k)}\cdots\mathbf{r}_N^{(k)}\right)}{\partial\mathbf{r}_i^{(k)}}\right|_{k=1,\ldots,\infty}. \tag{107}$$

Consequently, QM (at the approximation level indicated earlier) may be formulated in terms of an infinite (in practice, large) number of MD equations.

These equations may also be solved with an MC approach in which the position is sampled from a Gaussian distribution (the harmonic distribution in the path integral), and the new position is accepted or rejected according to a Metropolis rule based on the potential $\mathscr{U}$. The sampling is made difficult by the fact that the positions are not statistically independent. To make them independent, the positions are expended in terms of their Fourier transform or their staging variables. When the staging variables are used, the partition function becomes

$$Q(N,V,T) \approx \int \prod_{i=1}^{N} d\mathbf{u}_i^{(1)}\cdots d\mathbf{u}_i^{(K)} d\mathbf{p}_i^{(1)}\cdots d\mathbf{p}_i^{(K)} \times \exp\left[-\beta\sum_{k=1}^{K}\sum_{i=1}^{N}\frac{\left(\mathbf{p}_i^{(k)}\right)^2}{2m_i'^{(k)}} + \frac{1}{2}m_i^{(k)}\omega_K^2\left(\mathbf{u}_i^{(k)}\right)^2 + \frac{1}{K}\mathscr{U}\left(\mathbf{r}_1^{(k)}(\mathbf{u}_1),\ldots,\mathbf{r}_N^{(k)}(\mathbf{u}_N)\right)\right]\Bigg|_{\mathbf{u}_i^{(K+1)}=\mathbf{u}_i^{(1)}\forall i}, \tag{108}$$

where

$$\mathbf{r}_i^{(1)} = \mathbf{u}_i^{(1)},$$

$$\mathbf{r}_i^{(k)} = \mathbf{u}_i^{(1)} + \sum_{l=k}^{K} \frac{k-1}{l-1} \mathbf{u}_i^{(k)}, \quad k = 2, \ldots, K,$$

$$m_i^{(1)} = 0; \quad m_i^{(k)} = \frac{k}{k-1} m_i, \quad k = 2, \ldots, K,$$

$$m_i'^{(1)} = m_i; \qquad m_i'^{(k)} = m_i^{(k)}. \tag{109}$$

As a result, each normal mode $\mathbf{u}_i^{(k)}$ may be sampled from an independent Gaussian distribution and the positions may be reconstructed from their corresponding staging transformations. For instance, path integrals have been instrumental to underline the importance of quantum effects in the influenza neuraminidase [90].

CONCLUSIONS

The conformational space and the dynamics associated with macromolecules and macromolecular complexes, as well as their structural stability, may be explored and asserted through molecular dynamics and Monte Carlo simulations. Although full knowledge of the algorithms involved is not absolutely required, to obtain meaningful simulations, it is impressible to understand the conditions, approximations, and hypotheses on which they are based, as well as their limitations and potentialities.

Although most simulations are currently limited to individual structures or small organisms and complexes (a few millions atoms), it is conceivable that in the near future large viruses, such as the influenza, may be entirely simulated in silico over long periods of time. That would open the door to an essentially in silico design and lead to a better assessment of toxicity, specificity, and efficiency. Such a phenomenon has occurred before in aerodynamics in which computational fluid dynamics has essentially replaced wind tunnel experiments in the design of large aeroplane fuselages and wings.

REFERENCES

1. G. Morra and G. Colombo, "Relationship between energy distribution and fold stability: insights from molecular dynamics simulations of native and mutant proteins," Proteins: Structure, Function and Genetics, vol. 72, no. 2, pp. 660–672, 2008. · ·
2. R. Alhadeff, D. Assa, P. Astrahan, M. Krugliak, and I. T. Arkin, "Computational and experimental analysis of drug binding to the Influenza M2 channel," Biochimica et Biophysica Acta—Biomembranes, vol. 1838, no. 4, pp. 1068–1073, 2014. ·
3. H. Dong, G. Fiorin, W. F. Degrado, and M. L. Klein, "Exploring histidine conformations in the M2 channel lumen of the influenza a virus at neutral pH via molecular simulations," Journal of Physical Chemistry Letters, vol. 4, no. 18, pp. 3067–3071, 2013. · ·
4. C. F. Reboul, G. R. Meyer, B. T. Porebski, N. A. Borg, and A. M. Buckle, "Epitope flexibility and dynamic footprint revealed by molecular dynamics of a pMHC-TCR complex," PLoS Computational Biology, vol. 8, no. 3, Article ID e1002404, 2012. · ·
5. B. Tarus, C. Chevalier, C. A. Richard, B. Delmas, C. Primo, and A. Slama-Schwok, "Molecular dynamics studies of the nucleoprotein of influenza a virus: role of the protein flexibility in RNA binding," PLoS ONE, vol. 7, no. 1, Article ID e30038, 2012. · ·
6. R. E. Amaro, R. V. Swift, L. Votapka, W. W. Li, R. C. Walker, and R. M. Bush, "Mechanism of 150-cavity formation in influenza neuraminidase," Nature Communications, vol. 2, no. 1, article 388, 2011. · ·
7. L. Le, "Incorporating molecular dynamics simulations into rational drug design: a case study on influenza a neuraminidases," in Bioinformatics, H. Pérez-Sánchez, Ed., chapter 7, pp. 159–184, InTech, 2012. ·
8. L. Le, E. H. Lee, D. J. Hardy, T. N. Truong, and K. Schulten, "Molecular dynamics simulations suggest that electrostatic funnel directs binding of Tamiflu to influenza N1 neuraminidases," PLoS Computational Biology, vol. 6, no. 9, Article ID e1000939, 2010. · ·
9. Y.-T. Wang, C.-H. Chan, Z.-Y. Su, and C.-L. Chen, "Homology modeling, docking, and molecular dynamics reveal HR1039 as a potent inhibitor of 2009 A(H1N1) influenza neuraminidase," Biophysical Chemistry, vol. 147, no. 1-2, pp. 74–80, 2010. · ·
10. Y. Zhang, H. Shen, M. Zhang, and G. Li, "Exploring the proton conductance and drug resistance of BM2 channel through molecular dynamics simulations and free energy calculations at different pH conditions," Journal of Physical Chemistry B, vol. 117, no. 4, pp. 982–988, 2013. · ·
11. N.-J. Deng, W. Zheng, E. Gallicchio, and R. M. Levy, "Insights into the dynamics of HIV-1 protease: a kinetic network model constructed from atomistic simulations," Journal of the American Chemical Society, vol. 133, no. 24, pp. 9387–9394, 2011. · ·
12. J. Wang, C. Ma, G. Fiorin et al., "Molecular dynamics simulation directed rational design of inhibitors targeting drug-resistant mutants of influenza A virus M2," Journal of the American Chemical Society, vol. 133, no. 32, pp. 12834–12841, 2011. · ·
13. S. S. Chang, H. J. Huang, and C. Y. C. Chen, "High performance screening, structural and molecular dynamics analysis to identify H1 inhibitors from TCM Database@Taiwan," Molecular BioSystems, vol. 7, no. 12, pp. 3366–3374, 2011. · ·
14. P. Lagüe, B. Roux, and R. W. Pastor, "Molecular dynamics simulations of the influenza hemagglutinin fusion peptide in micelles and bilayers: conformational analysis of peptide and lipids," Journal of Molecular Biology, vol. 354, no. 5, pp. 1129–1141, 2005. · ·
15. C. Siebena, R. Kappelb Zhuc, A. Wozniakd, et al., "Influenza virus binds its host cell using multiple dynamic interactions," Proceedings of the National Academy of Sciences, vol. 109, no. 34, pp. 13626–13631, 2012. ·

16. Z. Yang, Y. Nie, G. Yang, Y. Zu, Y. Fu, and L. Zhou, "Synergistic effects in the designs of neuraminidase ligands: analysis from docking and molecular dynamics studies," Journal of Theoretical Biology, vol. 267, no. 3, pp. 363–374, 2010. · ·

17. T. Rungrotmongkol, M. Malaisree, N. Nunthaboot, P. Sompornpisut, and S. Hannongbua, "Molecular prediction of oseltamivir efficiency against probable influenza a (H1N1-2009) mutants: molecular modeling approach," Amino Acids, vol. 39, no. 2, pp. 393–398, 2010. · ·

18. D. Pan, H. Sun, C. Bai et al., "Prediction of zanamivir efficiency over the possible 2009 Influenza A (H1N1) mutants by multiple molecular dynamics simulations and free energy calculations," Journal of Molecular Modeling, vol. 17, no. 10, pp. 2465–2473, 2011. · ·

19. G. G. Dodson, D. P. Lane, and C. S. Verma, "Molecular simulations of protein dynamics: new windows on mechanisms in biology," EMBO Reports, vol. 9, no. 2, pp. 144–150, 2008. · ·

20. C. Sieben, C. Kappel, R. Zhu et al., "Influenza virus binds its host cell using multiple dynamic interactions," Proceedings of the National Academy of Sciences of the United States of America, vol. 109, no. 34, pp. 13626–13631, 2012.

21. D. Vlachakis, A. Karozou, and S. Kossida, "3D molecular modelling study of the H7N9 RNA-dependent RNA polymerase as an emerging pharmacological target," Influenza Research and Treatment, vol. 2013, Article ID 645348, 9 pages, 2013. ·

22. C. Peri, P. Gagni, F. Combi, et al., "Rational epitope design for protein targeting," ACS Chemical Biology, vol. 8, no. 2, pp. 397–404, 2013. · ·

23. D. R. Flower, K. Phadwal, I. K. MacDonald, P. V. Coveney, M. N. Davies, and S. Wan, "T-cell epitope prediction and immune complex simulation using molecular dynamics: state of the art and persisting challenges," Immunome Research, vol. 6, no. 2, article S4, 2010. · ·

24. D. Xu, E. I. Newhouse, R. E. Amaro et al., "Distinct glycan topology for avian and human sialopentasaccharide receptor analogues upon binding different hemagglutinins: a molecular dynamics perspective," Journal of Molecular Biology, vol. 387, no. 2, pp. 465–491, 2009. · ·

25. M. Lawrenz, J. Wereszczynski, R. Amaro, R. Walker, A. Roitberg, and J. A. McCammon, "Impact of calcium on N1 influenza neuraminidase dynamics and binding free energy," Proteins, vol. 78, no. 11, pp. 2523–2532, 2010. · ·

26. P. Pan, L. Li, Y. Li, D. Li, and T. Hou, "Insights into susceptibility of antiviral drugs against the E119G mutant of 2009 influenza A (H1N1) neuraminidase by molecular dynamics simulations and free energy calculations," Antiviral Research, vol. 100, no. 2, pp. 356–364, 2013. · ·

27. K. A. Fichthorn and W. H. Weinberg, "Theoretical foundations of dynamical Monte Carlo simulations,"The Journal of Chemical Physics, vol. 95, no. 2, pp. 1090–1096, 1991. · ·

28. S. Duane, A. D. Kennedy, B. J. Pendleton, and D. Roweth, "Hybrid Monte Carlo," Physics Letters B, vol. 195, no. 2, pp. 216–222, 1987. · ·

29. G. J. Martyna, D. J. Tobias, and M. L. Klein, "Constant pressure molecular dynamics algorithms," The Journal of Chemical Physics, vol. 101, no. 5, pp. 4177–4189, 1994. · ·

30. M. Shu, Z. Lin, Y. Zhang, Y. Wu, H. Mei, and Y. Jiang, "Molecular dynamics simulation of oseltamivir resistance in neuraminidase of avian influenza H5N1 virus," Journal of Molecular Modeling, vol. 17, no. 3, pp. 587–592, 2011. · ·

31. N. Metropolis, A. W. Rosenbluth, M. N. Rosenbluth, A. H. Teller, and E. Teller, "Equation of state calculations by fast computing machines," The Journal of Chemical Physics, vol. 21, no. 6, pp. 1087–1092, 1953. · ·

32. D. Frenkel, G. C. A. M. Mooij, and B. Smit, "Novel scheme to study structural and thermal properties of continuously deformable molecules," Journal of Physics: Condensed Matter, vol. 4, no. 12, pp. 3053–3076, 1992. · ·

33. J. I. Siepmann and D. Frenkel, "Configurational-bias Monte Carlo: a new sampling scheme for flexible chains," Molecular Physics, vol. 75, pp. 59–70, 1992.

34. P. Minary, M. E. Tuckerman, and G. J. Martyna, "Dynamical spatial warping: a novel method for the conformational sampling of biophysical structure," SIAM Journal on Scientific Computing, vol. 30, no. 4, pp. 2055–2083, 2007. · ·

35. T. Huber and W. F. van Gunsteren, "SWARM-MD: searching conformational space by cooperative molecular dynamics," Journal of Physical Chemistry A, vol. 102, no. 29, pp. 5937–5943, 1998. · ·

36. J. Zhang, "An effective simulated annealing refined replica exchange Markov chain Monte Carlo algorithm for the infectious disease model of H1N1 influenza pandemic," World Journal of Modelling and Simulation, vol. 7, no. 1, pp. 29–39, 2011. ·

37. J. I. Siepmann and D. Frenkel, "Configurational-bias Monte Carlo: a new sampling scheme for flexible chains," Molecular Physics, vol. 75, pp. 59–70, 1992. ·

38. J. P. Ulmschneider and W. L. Jorgensen, "Monte Carlo backbone sampling for polypeptides with variable bond angles and dihedral angles using concerted rotations and a Gaussian bias," The Journal of Chemical Physics, vol. 118, no. 9, pp. 4261–4271, 2003. · ·

39. K. Lindorff-Larsen, P. Maragakis, S. Piana, M. P. Eastwood, R. O. Dror, and D. E. Shaw, "Systematic validation of protein force fields against experimental data," PLoS ONE, vol. 7, no. 2, Article ID e32131, 2012. · ·

40. A. K. Harrisa, J. R. Meyersona, Y. Matsuokac et al., "Structure and accessibility of HA trimers on intact 2009 H1N1 pandemic influenza virus to stemregion-specific neutralizing antibodies," Proceedings of the National Academy of Sciences, vol. 110, pp. 4592–4597, 2009.

41. S. Toxvaerd, "Hamiltonians for discrete dynamics," Physical Review E, vol. 50, no. 3, pp. 2271–2274, 1994. · ·

42. K. Vanommeslaeghe, E. Hatcher, C. Acharya, et al., "CHARMM general force field: a force field for drug-like molecules compatible with the CHARMM all-atom additive biological force fields," Journal of Computational Chemistry, vol. 31, no. 4, pp. 671–690, 2010. · ·

43. B. R. Brooks, C. L. Brooks III, A. D. Mackerell Jr. et al., "CHARMM: the biomolecular simulation program," Journal of Computational Chemistry, vol. 30, no. 10, pp. 1545–1614, 2009. · ·

44. D. Baker and A. Sali, "Protein structure prediction and structural genomics," Science, vol. 294, no. 5540, pp. 93–96, 2001. · ·

45. A. Özen, T. Haliloğlu, and C. A. Schiffer, "Dynamics of preferential substrate recognition in HIV-1 protease: redefining the substrate envelope," Journal of Molecular Biology, vol. 410, no. 4, pp. 726–744, 2011. · ·

46. J. Nocedal and S. J. Wright, Numerical Optimisation, Springer, New York, NY, USA, 2006.

47. S. Kirkpatrick, C. D. Gelatt, and M. P. Vecchi, "Optimization by simulated annealing," Science, vol. 220, pp. 671–680, 1983.

48. R. O. Dror, M. O. Jensen, D. W. Borhani, and D. E. Shaw, "Exploring atomic resolution physiology on a femtosecond to millisecond timescale using molecular dynamics simulations," Journal of General Physiology, vol. 135, no. 6, pp. 555–562, 2010. · ·

49. M. Feig and C. L. Brooks III, "Recent advances in the development and application of implicit solvent models in biomolecule simulations," Current Opinion in Structural Biology, vol. 14, no. 2, pp. 217–224, 2004. · ·

50. W. Im, S. Bernèche, and B. Roux, "Generalized solvent boundary potential for computer simulations,"The Journal of Chemical Physics, vol. 114, no. 7, pp. 2924–2937, 2001. · ·

51. E. Gallicchio and R. M. Levy, "GBNP, an analytic implicit solvent model suitable for molecular dynamics simulations and high-resolution modeling," Journal of Computational Chemistry, vol. 25, no. 4, pp. 479–499, 2004. · ·

52. M. W. Mahoney and W. L. Jorgensen, "A five-site model for liquid water and the reproduction of the density anomaly by rigid, nonpolarizable potential functions," The Journal of Chemical Physics, vol. 112, no. 20, pp. 8910–8922, 2000. · ·

53. C. Chipot and A. Pohorille, Eds., Free Energy Calculations, Springer, Berlin, Germany, 2007.

54. T. J. Chung, Computational Fluid Dynamics, Cambridge University Press, New York, NY, USA, 2010.

55. M. Tuckerman, B. J. Berne, and G. J. Martyna, "Reversible multiple time scale molecular dynamics," The Journal of Chemical Physics B, vol. 380, pp. 665–677, 1992.

56. C. Dellago, P. G. Bolhuis, and P. L. Geissler, "Transition path sampling," in Advances in Chemical Physics, vol. 132, chapter 1, 2002. ·

57. A. E. Garcia, "Large-amplitude non-linear motions in proteins," Physical Review Letters, vol. 68, Article ID 2696, 1992.

58. M. E. Tuckerman, Y. Liu, G. Ciccotti, and G. J. Martyna, "Non-Hamiltonian molecular dynamics: Generalizing Hamiltonian phase space principles to non-Hamiltonian systems," The Journal of Chemical Physics, vol. 115, no. 4, pp. 1678–1702, 2001. · ·

59. G. S. Ezra, "On the statistical mechanics of non-Hamiltonian systems: the generalized Liouville equation, entropy, and time-dependent metrics," Journal of Mathematical Chemistry, vol. 35, no. 1, pp. 29–53, 2004. · · View at Zentralblatt MATH · ·

60. S. Nosé, "A unified formulation of the constant temperature molecular dynamics methods," The Journal of Chemical Physics, vol. 81, no. 1, pp. 511–519, 1984. · ·

61. Y. Liu and M. E. Tuckerman, "Generalized Gaussian moment thermostatting: a new continuous dynamical approach to the canonical ensemble," The Journal of Chemical Physics, vol. 112, no. 4, pp. 1685–1700, 2000. · ·

62. G. J. Martyna, M. L. Klein, and M. Tuckerman, "Nosé-Hoover chains: the canonical ensemble via continuous dynamics," The Journal of Chemical Physics, vol. 97, no. 4, pp. 2635–2643, 1992. · ·

63. H. C. Andersen, "Rattle: a "velocity" version of the shake algorithm for molecular dynamics calculations," Journal of Computational Physics, vol. 52, no. 1, pp. 24–34, 1983. · ·

64. J.-P. Ryckaert, G. Ciccotti, and H. J. C. Berendsen, "Numerical integration of the cartesian equations of motion of a system with constraints: molecular dynamics of n-alkanes," Journal of Computational Physics, vol. 23, no. 3, pp. 327–341, 1977. · ·

65. V. Kraeutler, W. F. van Gunsteren, and P. H. Huenenberger, "A fast SHAKE algorithm to solve distance constraint equtions for small molecules in molecular dynamics simulations," Journal of Computational Chemistry, vol. 22, pp. 501–508, 2001. ·

66. T. R. Forester and W. Smith, "SHAKE, rattle, and roll: efficient constraint algorithms for linked rigid bodies," Journal of Computational Chemistry, vol. 19, no. 1, pp. 102–111, 1998. · ·

67. J. A. Izaguirre and S. S. Hampton, "Shadow hybrid Monte Carlo: an efficient propagator in phase space of macromolecules," Journal of Computational Physics, vol. 200, no. 2, pp. 581–604, 2004. · ·

68. G. J. Martyna, M. E. Tuckerman, D. J. Tobias, and M. L. Klein, "Explicit reversible integrators for extended systems dynamics," Molecular Physics, vol. 87, no. 5, pp. 1117–1157, 1996. · ·

69. B. K. Mai, M. H. Viet, and M. S. Li, "Top leads for swine influenza A/H1N1 virus revealed by steered molecular

dynamics approach," Journal of Chemical Information and Modeling, vol. 50, no. 12, pp. 2236–2247, 2010. · ·

70. M. Klvana, M. Pavlova, T. Koudelakova, et al., "Pathways and mechanisms for product release in the engineered haloalkane dehalogenases explored using classical and random acceleration molecular dynamics simulations," Journal of Molecular Biology, vol. 392, no. 5, pp. 1339–1356, 2009. · ·

71. X. Wu and B. R. Brooks, "Self-guided Langevin dynamics simulation method," Chemical Physics Letters, vol. 381, no. 3-4, pp. 512–518, 2003. · ·

72. R. W. Zwanzig, "Nonlinear generalized Langevin equations," Journal of Statistical Physics, vol. 9, no. 3, pp. 215–220, 1973. · ·

73. S. Izvekov and G. A. Voth, "Modeling real dynamics in the coarse-grained representation of condensed phase systems," Journal of Chemical Physics, vol. 125, no. 15, Article ID 151101, 2006. · ·

74. E. Vanden-Eijnden and G. Ciccotti, "Second-order integrators for Langevin equations with holonomic constraints," Chemical Physics Letters, vol. 429, no. 1–3, pp. 310–316, 2006. · ·

75. Q. Huang, C.-L. Chen, and A. Herrmann, "Bilayer conformation of fusion peptide of influenza virus hemagglutinin: a molecular dynamics simulation study," Biophysical Journal, vol. 87, no. 1, pp. 14–22, 2004. · ·

76. R. E. Amaro, X. Cheng, I. Ivanov, D. Xu, and J. A. McCammon, "Characterizing loop dynamics and ligand recognition in human- and avian-type influenza neuraminidases via generalized born molecular dynamics and end-point free energy calculations," Journal of the American Chemical Society, vol. 131, no. 13, pp. 4702–4709, 2009. · ·

77. J. G. Kirkwood, "Statistical mechanics of fluid mixtures," The Journal of Chemical Physics, vol. 3, no. 5, pp. 300–313, 1935. · · View at Zentralblatt MATH ·

78. S. Park, F. Khalili-Araghi, E. Tajkhorshid, and K. Schulten, "Free energy calculation from steered molecular dynamics simulations using Jarzynski›s equality," The Journal of Chemical Physics, vol. 119, no. 6, pp. 3559–3566, 2003. · ·

79. M. Sprik and G. Ciccotti, "Free energy from constrained molecular dynamics," The Journal of Chemical Physics, vol. 109, no. 18, pp. 7737–7744, 1998. · ·

80. J. Kästner and W. Thiel, "Bridging the gap between thermodynamic integration and umbrella sampling provides a novel analysis method: umbrella integration," Journal of Chemical Physics, vol. 123, no. 14, Article ID 144104, 2005. · ·

81. G. M. Torrie and J. P. Valleau, "Nonphysical sampling distributions in Monte Carlo free-energy estimation: umbrella sampling," Journal of Computational Physics, vol. 23, no. 2, pp. 187–199, 1977. · ·

82. S. Kumar, D. Bouzida, R. H. Svendsen, P. A. Kollman, and J. M. Rosenber, "The weighted histogram analysis method for free-energy calculations on biomolecules. 1. The method," Journal of Computational Chemistry, vol. 13, p. 1011, 1992. ·

83. A. Laio, A. Rodriguez-Fortea, F. L. Gervasio, M. Ceccarelli, and M. Parrinello, "Assessing the accuracy of metadynamics," Journal of Physical Chemistry B, vol. 109, no. 14, pp. 6714–6721, 2005. · ·

84. E. Paquet and H. L. Viktor, "Macromolecular structure comparison and docking: an algorithmic Review,"Current Pharmaceutical Design, vol. 19, no. 12, pp. 2183–2193, 2013. · ·

85. M. Raab and I. Tvaroška, "The binding properties of the H5N1 influenza virus neuraminidase as inferred from molecular modeling," Journal of Molecular Modeling, vol. 17, no. 6, pp. 1445–1456, 2011. · ·

86. C. T. T. Su, C. Schönbach, and C.-K. Kwoh, "Molecular docking analysis of 2009-H1N1 and 2004-H5N1 influenza virus HLA-B*4405-restricted HA epitope candidates: implications for TCR cross-recognition and vaccine development," BMC Bioinformatics, vol. 14, supplement 2, article S21, 2013. · ·

87. H. Liu and X. Yao, "Molecular basis of the interaction for an essential subunit PA-PB1 in influenza virus RNA polymerase: Insights from molecular dynamics simulation and free energy calculation," Molecular Pharmaceutics, vol. 7, no. 1, pp. 75–85, 2010. · ·

88. M. Dominik and J. Hutter, Ab Initio Molecular Dynamics: Basic Theory and Advanced Methods, Cambridge University Press, Cambridge, UK, 2009.

89. M. Ceriotti, M. Parrinello, T. E. Markland, and D. E. Manolopoulos, "Efficient stochastic thermostatting of path integral molecular dynamics," Journal of Chemical Physics, vol. 133, no. 12, Article ID 124104, 2010. · ·

90. F. Perrucio, L. Ridder, and A. J. Mulholland, "Quantum-mechanicl/molecular-mechanical methods in medicinal chemistry," in Quamtum Medicinal Chemistry, P. Carloni and F. Alber, Eds., Wilely, Weinheim, Germany, 2003.

CHAPTER 11

MOLECULAR DYNAMICS SIMULATION OF PROTEINS: A BRIEF OVERVIEW

Sachin Patodia[1], Ashima Bagaria[1] and Deepak Chopra[2]

[1]Centre for Converging Technologies, University of Rajasthan, Jaipur 302004, Rajasthan, India

[2]Department of Chemistry, Indian Institute of Science Education and Research, Bhopal 462066, India

ABSTRACT

MD simulation has become an essential tool for understanding the physical basis of the structure of proteins and their biological functions. During the current decade we witnessed significant progress in MD simulation of proteins with advancement in atomistic simulation algorithms, force fields, computational methods and facilities, comprehensive analysis and experimental validation, as well as integration in wide area bioinformatics and structural/systems biology frameworks. In this review, we present the methodology on protein simulations and recent advancements in the field. MD simulation provides a platform to study protein–protein, protein–ligand and protein–nucleic acid interactions. MD simulation is also done with NMR relaxation timescale in order to get residual dipolar coupling and order parameter of protein molecules.

INTRODUCTION

MD methods were originally devised within the theoretical physics community during the 1950s and in 1957 when Alder and Wainwright [1] performed the first MD simulation using a hard-sphere model. The first macromolecular simulation, of BPTI, was done in 1975 with a crude molecular mechanics potential, and lasted for only 9.2 ps [2]. MD simulations mimic the physical motions of atoms in the protein molecule present in the actual environment. The atoms are allowed to interact for a certain period of time, which will help to compute their trajectory in and around the protein molecule. Simulation provides great detail of information about individual motion of atoms as a function of time. The potentials used in simulation procedures are approximate and under the control of the user by removing and altering specific contribution terms, their role in determining a given property can be examined.

Molecular Mechanics force fields are the cornerstone of biomolecular simulations, being used to compute the potential energy of a system of particles. The role of solvent is very important in simulation to determine the internal motion [3] of proteins at different temperatures, particularly below the glass transition [4] temperature, since experimentally it may be sometimes difficult to capture the dynamics associated with the internal motion of proteins. Molecular dynamics simulations provide connection between structure and dynamics by enabling the study of the conformational energy landscape accessible to protein molecules [5]. In recent years, some widely used

Citation: Patodia S, Bagaria A, Chopra D (2014) Molecular Dynamics Simulation of Proteins: A Brief Overview. J Phys Chem Biophys 4:166. doi:10.4172/2161-0398.1000166.

MD simulation packages such as NAMD [6], GROMACS [7], and AMBER [8], have all substantially improved their algorithmic sophistication and parallel performance, being able to deliver up to ~10-100 ns/day/workstation/cluster [9]. MD simulation provides an alternate approach to the study of protein dynamics at NMR relaxation time scales. MD simulation studies are done at NMR timescales to calculate order parameter [10] and residual dipolar coupling [11] of proteins. Residual dipolar coupling provides information about the relative orientation of the parts of the protein that are present far apart in the structure. NMR spectroscopy enables the measurement of order parameters that gives an atomistic description of fluctuations in protein structure over pico- and nanoseconds [12]. Contrast between NMR spectroscopy and MD simulations can be used to interpret experimental results [13] and to improve the quality of simulation related force fields and integration methods [14]. Over time a large number of ingenious alternative approaches to classical MD simulation have been developed such as Monte-Carlo sampling of conformational space, simulated annealing, steered MD, hybrid Quantum Mechanics/ Molecular Mechanics (QM/MM), coarse-grained dynamics, Brownian dynamics, normal vibration modes analysis, molecular docking simulations and other non-dynamic methods, all leading to spectacular applications and developments in biomolecular simulation.

FORCE FIELD IN PROTEIN SIMULATION

Force fields provide information about the potential energy of a system of particles. Force field parameters are obtained from experimental and quantum mechanical studies of small molecules, and it is postulated that such parameters may be transferred to desired larger molecules. Force field function comprises bonded and non-bonded interaction terms. Bonded interactions include harmonic oscillator energy of bond lengths, bond angles, and sometimes improper dihedrals (hard terms) and torsional dihedral angles (soft terms, sometimes including improper dihedrals), while Van der Waals interactions and electrostatic interactions contribute to non-bonded interactions. It is worth mentioning that Van der Waals interaction terms, described by a Lennard-Jones (6-12) potential function, include only dispersion or London interactions between transient dipoles, while Keesom interactions (between permanent dipoles) and Debye (induced dipole) interactions are included in the electrostatic (Coulomb) term, without explicit development into charge distribution momenta. Following is the most widely used energy function [15]:

$$V = \underbrace{\underbrace{\sum_{bonds} \frac{1}{2} k_b (r - r_0)^2 + \sum_{angles} \frac{1}{2} k_\theta (\theta - \theta_0)^2 + \sum_{\substack{improper \\ dihedrals}} \frac{1}{2} k_\xi (\xi - \xi_0)^2}_{hard\ terms} + \underbrace{\sum_{dihedrals} K_\phi [1 + \cos(n\phi - \delta)]}_{soft\ terms}}_{bonded\ terms}$$

$$+ \underbrace{\sum_{\substack{atom \\ pairs\ i,j}} \left(\frac{1}{4\pi\varepsilon_0} q_i q_j r_{ij}^{-1} + A_{ij} r_{ij}^{-12} - B_{ij} r_{ij}^{-6} \right)}_{non-bonded\ terms} \qquad (1)$$

In equation 1, potential energy V is a function of the Cartesian coordinate set R, specifying the positions of all atoms, from which the internal coordinates for bond lengths (r), bond angles (θ), dihedral angles (φ) and interparticle distances (r_{ij}) are calculated. There are multiple force fields available, and selection of the most adequate force field for a specific application is essential in simulation studies. For biomolecular simulations widely used molecular force fields include AMBER03 [16], AMBER94 [17], AMBER96 [18], CHARMM27 [19], OPLS-AA [20], GROMOS87 [21], GROMOS96 [22].

SIMULATION METHODOLOGY

There are multiple steps involved in simulation and Figure 1 shows some of the most important ones. MD simulation starts with the knowledge of the potential energy of the system with respect to its position coordinates. The first derivative of the potential function to the position coordinates helps to compute the force acting on individual atoms of the system. The essential steps involved in the MD simulations of proteins are as follows.

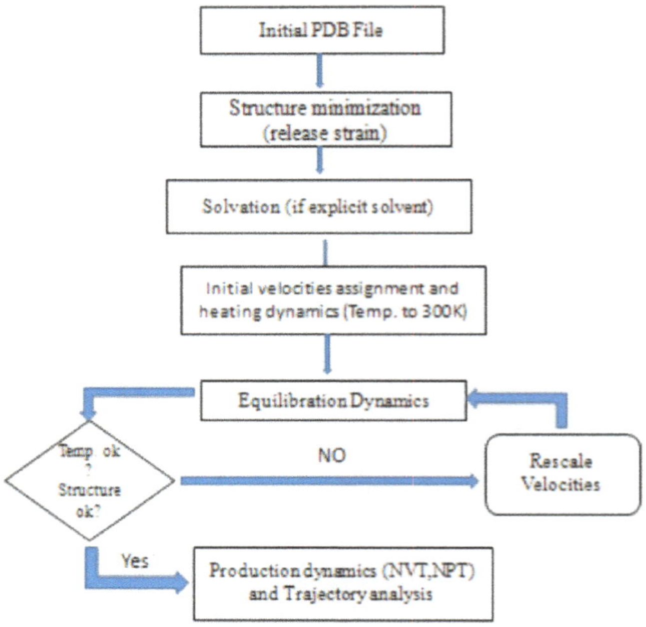

Figure 1: Flow chart depicting steps involved in MD simulation of a protein.

Simulation Environment

Protein simulation is done to replicate the experimental conditions, so several parameters for the different physical conditions are considered (such as pressure, temperature). In general the protein simulation is done in canonical ensemble (NVT), particularly the initial equilibration steps, or isothermal-isobaric (NPT) ensemble. For simulation, the protein molecules should be kept in the unit cell and solvated with explicit solvent. There is a broad range of explicit water models available and most popular of these models include TIP3P, TIP4P [23], TIP5P [24], SPC, and SPC/E [25]. The aforementioned water models are determined by Quantum mechanics-driven Molecular mechanics and validated by experimental methods. Water models are used to imitate the specific nature and complexity of molecule hydration, including orientation of solvent dipoles and effective electrostatic shielding, subtle hydrogen bond network rearrangements, clathration of hydrophobic surfaces, and accompanying changes in entropy. Unfortunately, due to limited time resolution of MD simulations and the intricate quantum nature of hydrogen bonds, most simulation environments do not treat them explicitly, including instead an average energy contribution to the non-bonded terms and using shake algorithms for solvent hydrogens repositioning. On the other hand, the use of implicit solvent models seeks to approximate the solute potential of the mean force, which determines the statistical weight of solute conformations, and is obtained by averaging over the solvent degrees of freedom [26]. To avoid polarization of the simulation ensemble, the total charge of the system should be kept zero by replacing certain solvent molecules with ions. To avoid interaction problems at system boundary, constrained spherical boundary models for solute and solvent can be applied, with specific corrections for density and polarization effects or the highly popular approach of cubic or rectangular periodic boundary conditions (PBC) consisting of repetitions of the system in the 26 adjacent unit cells. When a molecule leaves the system on one side, its equivalent image enters from the adjacent system on the opposite side, leading to conservation of mass and number of particles. Molecules within image systems are used for computation of long-range non-bonded interactions, and the Ewald summation has been directly applied in standard solvated periodic boundary simulations of biomolecular systems to compute the electrostatic interaction in the system [27].

Energy Minimization

This step involves finding the global minimum energy with respect to the position of side chains atoms that represents the geometry of the particular arrangements of atoms in which the net attractive force on each atom reaches a maximum. There are various methods to compute the minimum energy but most widely used methods are steepest descent and conjugate gradient. Steepest descent method is one of several first-order iterative descent methods and utilizes the gradient of the potential energy surface. It directly relates to the forces in the Molecular mechanical description of molecular systems, to guide a search path toward the nearest energy minimum [28]. An essential issue is correcting the protonation state of titratable residues. This task can be performed using either free energy of perturbation (FEP) MD simulations or with continuum electrostatics models such as finite differences Poisson-Boltzmann (FDPB) or protein dipole-Langevin dipole (PDLD).

Heating the System and Equilibration

In heating phase, initial velocities (at 0 K) are assigned to each atom of the system during energy minimization and Newton's equations of motion that represent the time evolution of system are numerically integrated. At short predefined intervals, new velocities are assigned corresponding to a slightly higher temperature and the simulation is allowed to continue until desired temperature is achieved. Force constrains on different subdomains of the simulation system are gradually removed as structural tensions dissipate by heating. Thermalization is usually performed at constant volume with Langevin dynamics.

Equilibration stage is used to equilibrate kinetic and potential energies means distribute the kinetic energy "pumped" into the system during heating among all degrees of freedom. In explicit solvent simulation, protein positions are fixed and waters move accordingly. Once the solvent is equilibrated, the constraints on the protein can be removed and the whole system (protein + solvent) can evolve in time.

Production Phase

Production phase is the last step of the simulation methodology to remove constraints on protein. It is carried out for desired time scale to generate trajectory of protein molecule in compliance with particular equilibrium conditions such as NVT, NPT and NVE. The timescale can be varied from several hundred picoseconds to microseconds or more. To avoid large trajectory artifacts during long runs, a 2D grid correction map of backbone φ and ψ parameters obtained from alanine, proline and glycine dipeptide surfaces (CMAP correction) has been included in more recent versions of CHARMM protein parameter files.

Analysis

In this step, stored coordinates and velocities of the system are used for further analysis. MD simulations can help to visualize and understand conformational changes at an atomic level when combined with visualization software which can display the structural parameters of interest in a time dependent way. MD simulation routinely calculates the following quantities 1) Time average structure 2) Radius of Gyration 3) Total energy of system 4) RMSD difference between two structures 5) interface related terms like order parameter, density of groups, electrostatic potential [29] . As an example we present a 3-nanosecond MD simulation of Protein-tyrosine phosphatase 1B (PTP1B) and inhibitor SNA complex [30]. Figure 2A shows that the total energy of PTP1B complexed with inhibitor decreases more sharply, and Figure 2B shows that the energy of uncomplexed PTP1B does not change too much. Root mean square deviation (RMSD) values of PTP1B (uncomplexed) and PTP1B-inhibitor (complexed) become stabilized after 1700 ps and 500 ps equilibrations, as shown in Figure 3. There is larger fluctuation in Radius of gyration in uncomplexed PTP1B relative to PTP1B complexed with inhibitor, as shown in Figure 4.

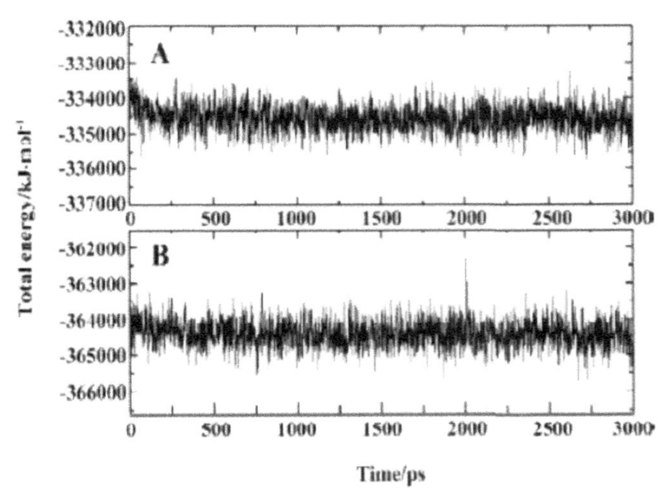

Figure 2: Total energy of (A) the uncomplexed PTP1B and (B) PTP1B-SNA complex as a function of simulation time for 3 –ns MD simulation.

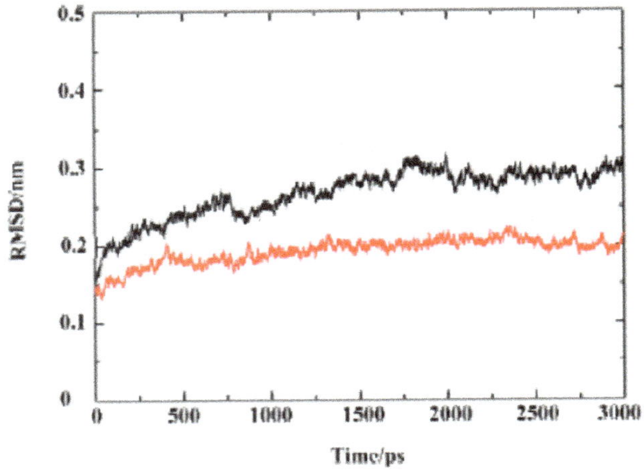

Figure 3: Time dependence of the RMS deviation from the crystal structure of the uncomplexed PTP1B (black) and the PTP1B-SNA complex (red) for all atoms over the 3-ns MD simulation.

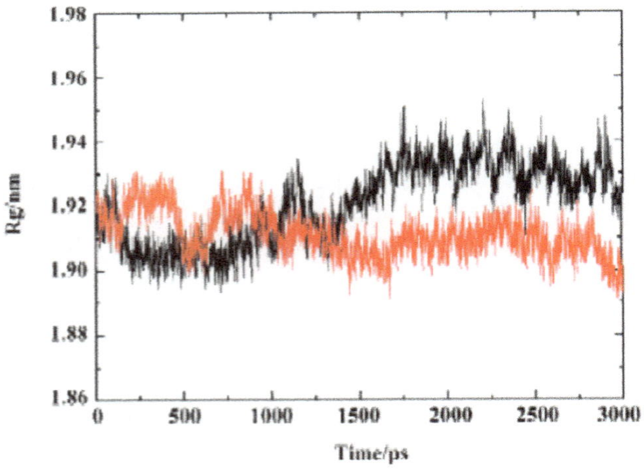

Figure 4: Radius of gyration as a function of time of uncomplexed PTP1B (black) and the PTP1B-SNA complex (red) during individual 3-ns MD simulation.

CONCLUSIONS

In the current scenario, advancement in algorithms of MD simulation and computational facilities helps us to explore bigger biological systems for a longer timescale. A typical simulation of system size about 10^5-10^6 atoms for several nanoseconds simulation will probably require 10^6-10^7 time steps and is expected to take a couple of days on workstation/cluster. During the period of simulation run it produces gigabytes of data for further analysis and visualization. Simulation analysis with visualization software enables investigation of motions and conformational changes that have functional inference and yields information that is not available through any other means. MD simulation results can be validated by protein dynamics by NMR and use to improve parameterization of the force fields. MD simulations yield great level of information but the challenge is to establish its validation. Molecular Dynamics simulation is more effective when it is coupled with experimental results on protein function at routinely basis, which play an essential role in validating and improving the simulations [31].

ACKNOWLEDGEMENTS

SP and DC gratefully acknowledge the IISER-BHOPAL team for research facilities and infrastructure.

REFERENCES

1. Alder BJ, Wainwright TE (1957) Phase Transition for a Hard Sphere System. J Chem Phys 27: 1208.
2. Artymiuk PJ, Blake CC, Grace DE, Oatley SJ, Phillips DC, et al. (1979) Crystallographic studies of the dynamic properties of lysozyme. Nature 280: 563-568.
3. Vitkup D, Ringe D, Petsko GA, Karplus M (2000) Solvent mobility and the protein 'glass' transition. Nat Struct Biol 7: 34-38.
4. Hagen SJ, Hofrichter J, Eaton WA (1995) Protein reaction kinetics in a room-temperature glass. Science 269: 959-962.
5. Karplus M, Mc Cammon JA (2002) Molecular dynamics simulations of biomolecules. Nat Struct Biol 9: 646-652.
6. Phillips JC, Braun R, Wang W, Gumbart J, Tajkhorshid E, et al. (2005) Scalable molecular dynamics with NAMD. J Comput Chem 26: 1781-1802.
7. Pronk S, Pall S, Schulz R, Larsson P, Bjelkmar P, et al. (2013) GROMACS 4.5: a high-throughput and highly parallel open source molecular simulation toolkit. Bioinformatics 29: 845-854.
8. Case DA, Cheatham TE 3rd, Darden T, Gohlke H, Luo R, et al. (2005) The Amber biomolecular simulation programs. J Comput Chem 26: 1668-1688.
9. Dror RO, Dirks RM, Grossman JP, Xu H, Shaw DE (2012) Biomolecular simulation: a computational microscope for molecular biology. Annu Rev Biophys 41: 429-452.
10. Tolman JR, Al-Hashimi HM, Kay LE, Prestegard JH (2001) Structural and dynamic analysis of residual dipolar coupling data for proteins. J Am Chem Soc 123: 1416-1424.
11. Shinji Sunada, Nobuhiro Go (1995) Calculation of nuclear magnetic resonance order parameter in proteins by normal mode of analysis. J Chem Phys 104: 4768-4774.
12. Maragakis P, Lindorff-Larsen K, Eastwood MP, Dror RO, Klepeis JL, et al. (2008) Microsecond molecular dynamics simulation shows effect of slow loop dynamics on backbone amide order parameters of proteins. J Phys Chem B 112: 6155-6158.
13. Best RB, Vendruscolo M (2004) Determination of protein structures consistent with NMR order parameters. J Am Chem Soc 126: 8090-8091.
14. Giovanni Lipari, Attila Szabo, Ronald M Levy (1982) Protein dynamics and NMR relaxation: comparison of simulations with experiment. Nature 300: 197–198.
15. Brooks BR, Bruccoleri RE, Olafson BD, States DJ, Swaminathan S, et al. (1983) CHARMM: A Program for Macromolecular Energy, Minimization, and Dynamics Calculations. J Comput Chem 4: 187–217.

16. Duan Y, Wu C, Chowdhury S, Lee MC, Xiong G, et al. (2003) A point-charge force field for molecular mechanics simulations of proteins based on condensed-phase quantum mechanical calculations. J Comput Chem 24: 1999-2012.
17. Cornell WD, Cieplak P, Bayly CI, Gould IR, Merz KM, et al.,(1995) A Second Generation Force Field for the Simulation of Proteins, Nucleic Acids, and Organic Molecules. J Am Chem Soc 117: 5179-5197.
18. Kollman PA (1996) Advances and Continuing Challenges in Achieving Realistic and Predictive Simulations of the Properties of Organic and Biological Molecules. Acc Chem Res 29: 461-469.
19. Foloppe N, MacKerell AD (2000) All-atom empirical force field for nucleic acids: I. Parameter optimization based on small molecule and condensed phase macromolecular target data. J Comput Chem 21: 86–104.
20. Kaminski GA, Friesner RA (2001) Evaluation and Reparametrization of the OPLS-AA Force Field for Proteins via Comparison with Accurate Quantum Chemical Calculations on Peptides. J Phys Chem B 105: 6474–6487.
21. Oostenbrink C, Villa A, Mark AE, van Gunsteren WF (2004) A biomolecular force field based on the free enthalpy of hydration and solvation: the GROMOS force-field parameter sets 53A5 and 53A6. J Comput Chem 25: 1656-1676.
22. Schuler LD, Daura X, van Gunsteren WF (2001) An improved GROMOS96 force field for aliphatic hydrocarbons in the condensed phase. J Comput Chem 22: 1205–1218.
23. Jorgensen WL, Chandrasekhar J, Madura JD, Impey RW, Klein ML (1983) Comparison of simple potential functions for simulating liquid water. J Chem Phys 79: 926-935.
24. Mahoney MW, Jorgensen WL (2001) A five-site model for liquid water and the reproduction of the density anomaly by rigid, nonpolarizable potential functions. J Chem Phys 112: 8910-8922.
25. Berendsen HJC, Grigera JR, Straatsma TP (1987) The missing term in effective pair potentials. J Phys Chem 24: 6269-6271.
26. Roux B, Simonson T (1999) Implicit solvent models. Biophys Chem 78: 1-20.
27. Cheatham TE, Miller JH, Fox T, Darden PA, Kollman PA (1995) Molecular Dynamics Simulations on Solvated Biomolecular Systems: The Particle Mesh Ewald Method Leads to Stable Trajectories of DNA, RNA, and Proteins. J Am Chem Soc 117: 4193-4194.
28. Daggett V (2006) Protein folding-simulation. Chem Rev 106: 1898-1916.
29. Hess B, van Der Spoel D, Lindahl E (2010) Gromacs user manual version 4.5. 4. University of Groningen, Netherland.
30. Liu GX, Tan JZ, Niu CY, Shen JH, Luo XM, et al. (2006) Molecular dynamics simulations of interaction between protein-tyrosine phosphatase 1B and a bidentate inhibitor. Acta Pharmacol Sin 27: 100-110.
31. Case DA (2002) Molecular dynamics and NMR spin relaxation in proteins. Acc Chem Res 35: 325-331.

CHAPTER 12 MOLECULAR DYNAMICS SIMULATION TO INVESTIGATE THE INTERACTION OF ASPHALTENE AND OXIDE IN AGGREGATE

Rui Li[1], Hui Du[2], Zepeng Fan[1] and Jianzhong Pei[1]

[1]School of Highway, Chang'an University, Xi'an 710064, China

[2]School of Transportation Engineering, Southeast University, Nanjing 210096, China

ABSTRACT

The asphalt-aggregate interface interaction (AAI) plays a significant role in the overall performances of asphalt mixture, which is caused due to the complicated physicochemical processes and is influenced by various factors, including the acid-base property of aggregates. In order to analyze the effects of the chemical constitution of aggregate on the AAI, the average structure $C_{65}H_{74}N_2S_2$ is selected to represent the asphaltene in asphalt and magnesium oxide (MgO), calcium oxide (CaO), aluminium sesquioxide (Al_2O_3), and silicon dioxide (SiO_2) are selected to represent the major oxides in aggregate. The molecular models are established for asphaltene and the four oxides, respectively, and the molecular dynamics (MD) simulation was conducted for the four kinds of asphaltene-oxide system at different temperatures. The interfacial energy in MD simulation is calculated to evaluate the AAI, and higher value means better interaction. The results show that interfacial energy between asphaltene and oxide reaches the maximum value at 25°C and 80°C and the minimum value at 40°C. In addition, the interfacial energy between asphaltene and MgO was found to be the greatest, followed by CaO, Al_2O_3, and SiO_2, which demonstrates that the AAI between asphalt and alkaline aggregates is better than acidic aggregates.

INTRODUCTION

Studies have shown that asphalt pavement damage phenomenon has a strong dependency upon asphalt-aggregate interfacial nature. If the bond between asphalt and aggregate is weak, then the interface between asphalt and aggregate is prone to failure, which results in the early damage of pavement. Therefore, many methods were used to study the interface between asphalt and aggregate in different perspectives. Elphingstone [1] proposed that the surface free energy is an effective approach to predict the fatigue and moisture properties of mixture, and adhesion process of aggregate plays an important role. Tan et al. [2] tested the phase angle and complex shear modulus of matrix asphalt and asphalt mastic by dynamic shear rheometer, and the asphalt-aggregate interaction capabilities were evaluated by K.Ziegel-B parameter. Cheng et al. [3] measured the surface free energy of different asphalt and aggregates and calculated the adhesion work of asphalt and reactive aggregates with and without water. Bhasin et al. [4] proposed quantifying the free

Citation: Rui Li, Hui Du, Zepeng Fan, and Jianzhong Pei, "Molecular Dynamics Simulation to Investigate the Interaction of Asphaltene and Oxide in Aggregate," Advances in Materials Science and Engineering, vol. 2016, Article ID 3817123, 10 pages, 2016. doi:10.1155/2016/3817123.

energy of the surface of asphalt-aggregate adhesion energy, and the results have shown that the same type of aggregate and bitumen adhesion between the energies has significant difference. To sum up, on the asphalt-aggregate interface studies have been largely analyzed from the experimental point of view. However, the experimental results were affected by many factors, in the dynamic shear rheological experiments; powder-binder ratio had greater impact on the test results, and how to determine the best powder-binder ratio is a problem. While aggregate composition is complex, powder, and mixed into the asphalt, mastic asphalt does not characterize asphalt-aggregate interface behavior. In the surface energy calculation, measurement results on the final great influence, different methods may result in different surface energy, resulting in a larger difference in the results.

In recent years, with the improvement of computer hardware and the perfection of related theory, molecular dynamics (MD) have become the effective means to study a complex system from the molecular level [5], the effective methods for the deep understanding of asphaltene adsorption characteristics, and the interaction mechanism of the interface from the microscopic view. At present, the domestic and foreign scholars mostly from the angle of test studied asphalt with aggregate interface area properties, and the computer simulation research is less. As a result, the MD method asphalt-aggregate interface behavior of asphalt mixture's research has the vital significance.

Based on the molecular dynamics calculation, we can simulate the asphalt-aggregate interfacial behavior. In four components of asphalt, asphaltene molecular mass is bigger and has stronger polarity and the asphaltenes content has a great impact on the properties of asphalt [6]. Asphaltene is fused aromatic ring system as the core, with a number of different alkyl chains linked structure, with N, S, and O heteroatoms, not soluble in nonpolar molecules of the corresponding normal alkanes soluble in benzene and toluene matter [7]. The composition, structure, and morphology of the asphaltene are very complex. In the study, the average molecular model is often used to replace the chemical structure [8]. Rogel [9] proposed a representative average asphaltene molecular structure model for the study of this paper. According to the composition of aggregate analysis, aggregates content of more oxides was analyzed, the remaining content of less neglected ones. The asphalt-oxide interface model based on dynamics simulation of the interfacial energy and adsorption configuration, which can quantitatively point out the ability of the main oxides in asphalt and aggregate interface interaction, for further understanding the mechanism of adhesion between asphalt and aggregate, is of great significance.

COMPUTER SIMULATION THEORY

Basic Theory

Calculations were conducted with Accelrys Materials Studio software package in this study and this method relies mainly on Newtonian mechanics to simulate the movement of molecular systems, the samples are taken in different states of the molecular system, and the configuration of the system is integrated. The thermodynamic quantities and other macroscopic properties are further calculated based on the results of the integration configuration.

COMPASS is selected for force field simulation, and the potential energy function form is shown as follows:

$$E_{\text{total}} = \sum_{\text{bond}} E_b(b) + \sum_{\text{angle}} E_\theta(\theta) + \sum_{\text{dihedral}} E_\phi(\phi) + \sum_{\text{out-of-plane}} E_\chi(\chi) + \sum_{\text{cross}} E(b,\theta,\phi) + E_{\text{coulomb}} + E_{\text{vdw}}. \tag{1}$$

In (1), the preceding five letters are bonding interaction of which the potential energy function contains bond lengths, bond angles, dihedral angles, plane, and cross-term; b, θ, ϕ, and χ are expressed as bond lengths, bond angles, dihedral angles, and off-plane vibration; the next two letters are nonbonding interactions which are electrostatic E_{coulomb} and van der Waals interaction E_{vdw}, respectively. E_{coulomb} and E_{vdw} are calculated by the following:

$$E_{\text{coulomb}} = \sum_{i>j} \frac{q_i q_j}{r_{ij}}, \tag{2}$$

$$E_{\text{vdw}} = \sum \varepsilon_{ij} \left[2\left(\frac{r_{ij}^0}{r_{ij}}\right)^9 - 3\left(\frac{r_{ij}^0}{r_{ij}}\right)^6 \right]. \tag{3}$$

In (2) and (3), i and j represent different atoms; q is atomic charge; r is the distance of different atomic; ε is potential well depth.

The interface model is composed of three layers. The first layer is oxide, the second layer is asphaltene, and the third layer is vacuum, eliminating the impact of the cyclical interaction between the particles resulting from double counting.

The molecular dynamics simulation of the canonical ensemble (NVT) is completed by Discover module. According to the set temperature, the temperature is controlled by Andersen thermostat, the initial rate of each molecule is randomly generated by the Maxwell-Boltzmann distribution, Newton's equation of motion is calculated by Velocity Verlet algorithm, and the van der Waals interaction and Coulomb interaction are calculated by Charge Group method.

Interfacial Energy Calculation

Interfacial energy is also known as the interaction energy, the calculation method is similar to the surface energy, and the difference is that the interfacial energy reflects the bonding strength between the two materials. The interface may be between the same materials or different materials; the interfacial energy of the same material can be theoretically equal to the surface energy. The interfacial energy can be calculated as shown in Figure 1.

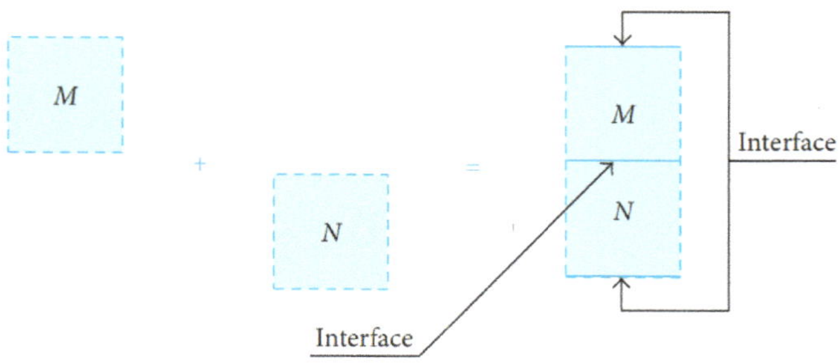

Figure 1: Schematic diagram of the interfacial energy.

In Figure 1, to calculate the interfacial energy between materials M and N, the interface model is constructed as shown in the right of Figure 1.The model is calculated in temperature T and system NVT by molecular dynamics until the system reaches a steady state; the energy of the model is E_{total}. The interface models M and N are constructed as shown in the left of Figure 1 and calculated by molecular dynamics; the energy of the two models is E_M and E_N, respectively. If the area of the combination surface is A, the interfacial energy is calculated as follows:

$$\gamma_{int} = \frac{(E_M + E_N - E_{total})}{2A}. \tag{4}$$

RESULTS AND DISCUSSION

Asphaltene Molecular Structure Model

Since the composition and structure of the asphaltene are very complex, Rogel et al. built average molecular structure, which was standardized samples studied in the Strategic Highway Research Program (SHRP). The average structure is shown in Figure 2.

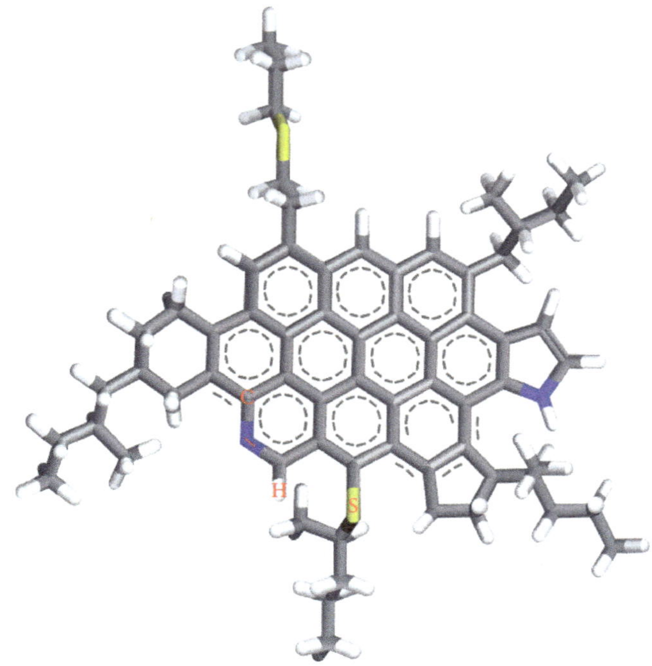

Figure 2: Average molecular structure of asphaltene.

Because asphaltenes are highly polydisperse molecules, the construction is highly complex. So in the research, we used one of the average molecules. For the simulation, we chose the COMPASS force field. For the asphaltene, the target density was set to 1 g/cm^3, constructing the configuration of asphaltene system. In order to analyze the temperature dependence of the asphaltene system, six different temperatures were chosen: 10°C (283.15 K), 25°C (298.15 K), 40°C (313.15 K), 60°C (333.15 K), 80°C (353.15 K), and 100°C (373.15 K), and the results are shown in Figure 3.

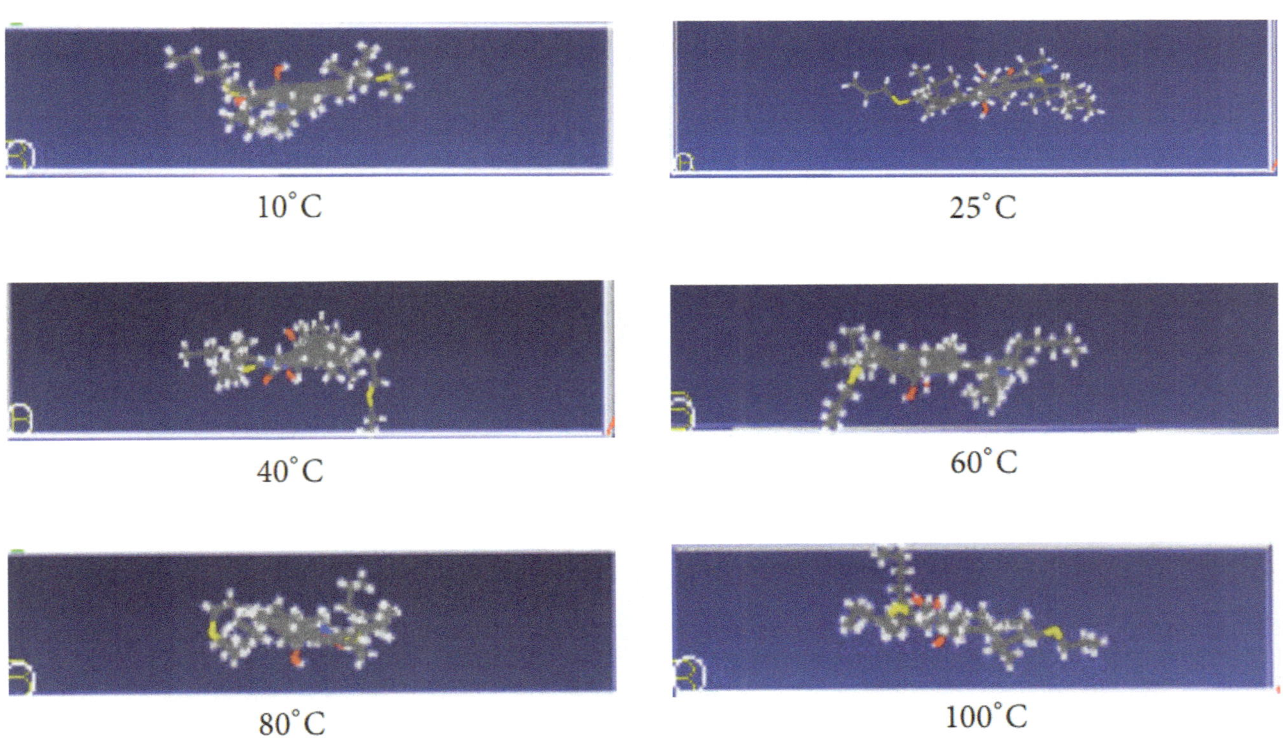

Figure 3: The configuration of asphaltene molecules in different temperature.

Oxide Molecular Structure Model

Selection of Representative Oxides

To simulate aggregate, we used 10 chemical compositions of aggregate (*M*1, *M*2, . . . , *M*10) suggested by Ozkahraman and Is¸ik [10] (as seen in Figure 4). The chemical composition of aggregate is $SiO_2 > CaO > Al_2O_3 > MgO > Fe_2O_3 > NaO > K_2O$.

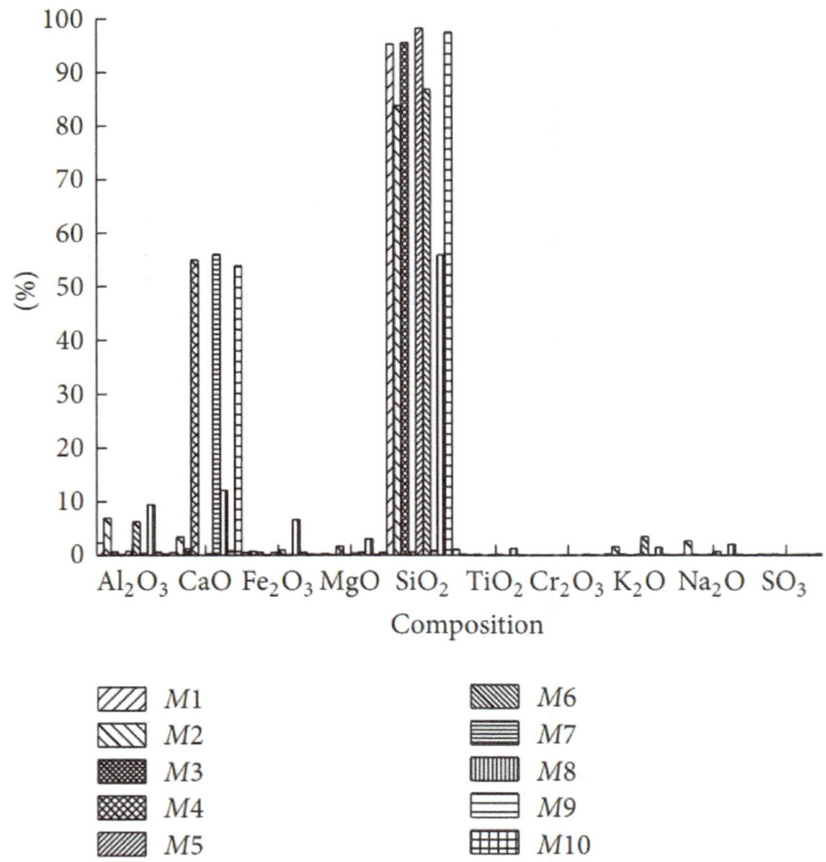

Figure 4: The chemical composition of aggregates.

Therefore, the simulation is mainly focused on the four oxides which are SiO_2, CaO, Al_2O_3, and MgO.

Building Oxide Molecular Cell

By the molecular dynamics software that comes with the unit cell model of oxide molecules, molecular cell structure of oxide is formed after importing the unit cell model, as shown in Figure 5.

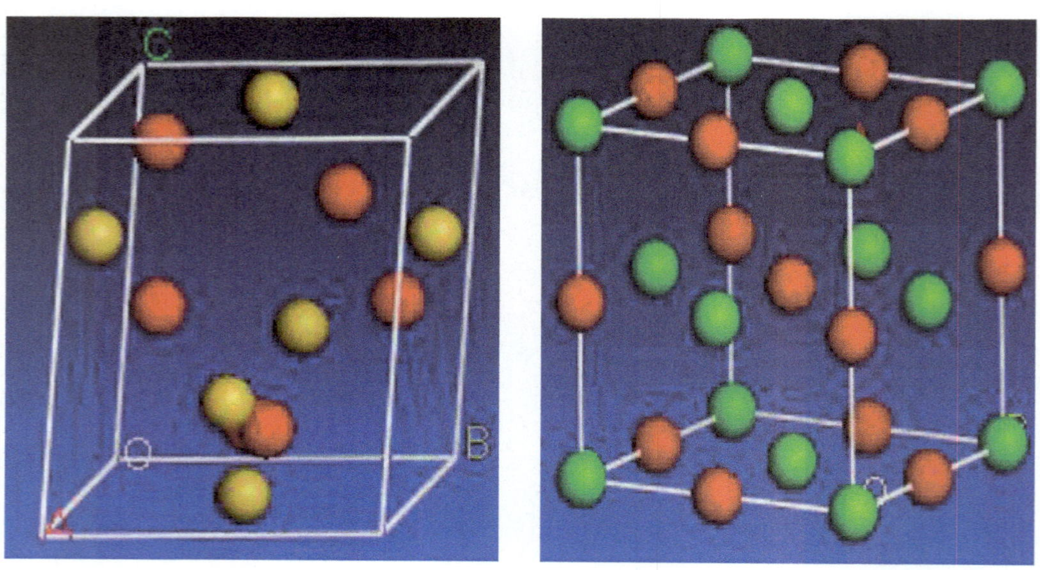

SiO_2 CaO

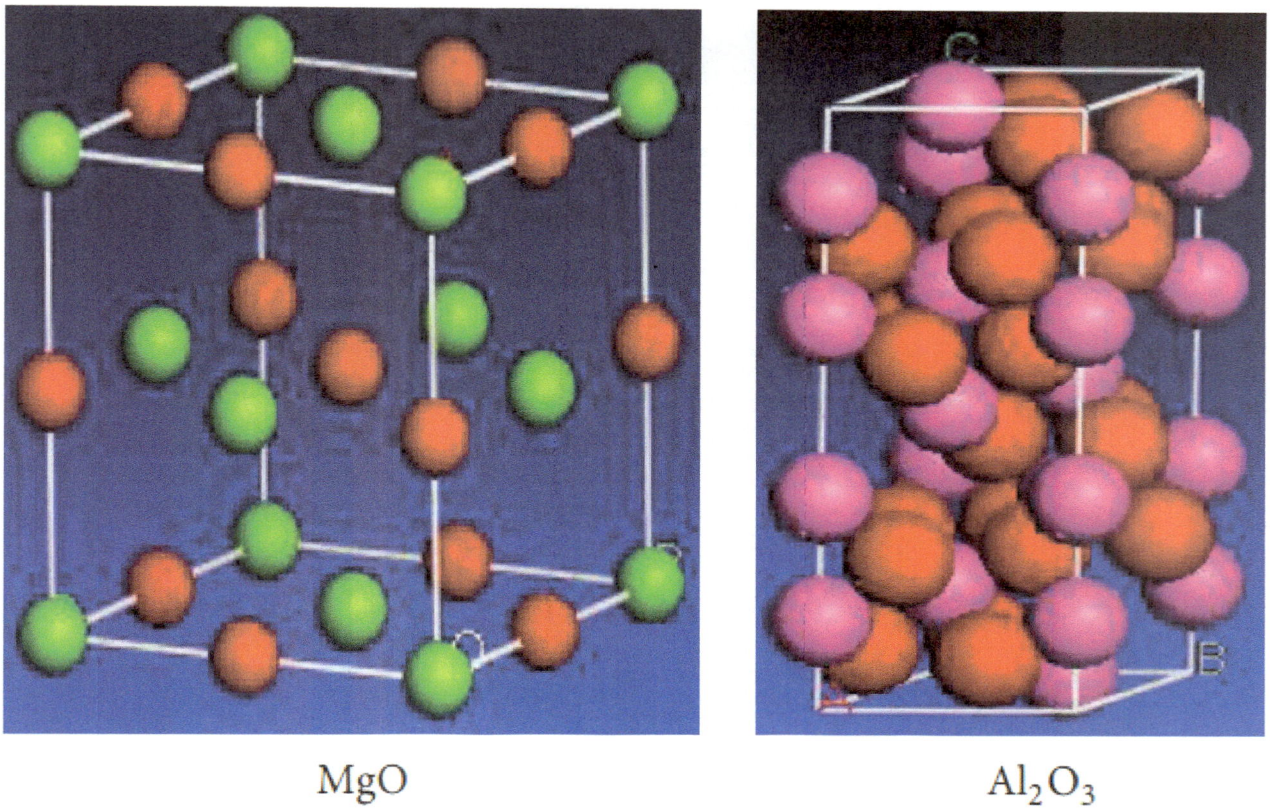

Figure 5: Four kinds of oxide molecular cell structure.

Building Oxide Supercell

Based on the molecular cell above, the cell is intercepted by section 0 0 1. Secondly, the force field is assigned to each atom of cell. In the paper, the force field is COMPASS. The energy of cell is minimized by the Minimize module of Discover. Finally, the oxide supercell structure is built by the supercell module of Build which is composed of 5 × 5 matrix. For example, the process of building supercell structure of SiO_2 is shown in Figure 6.

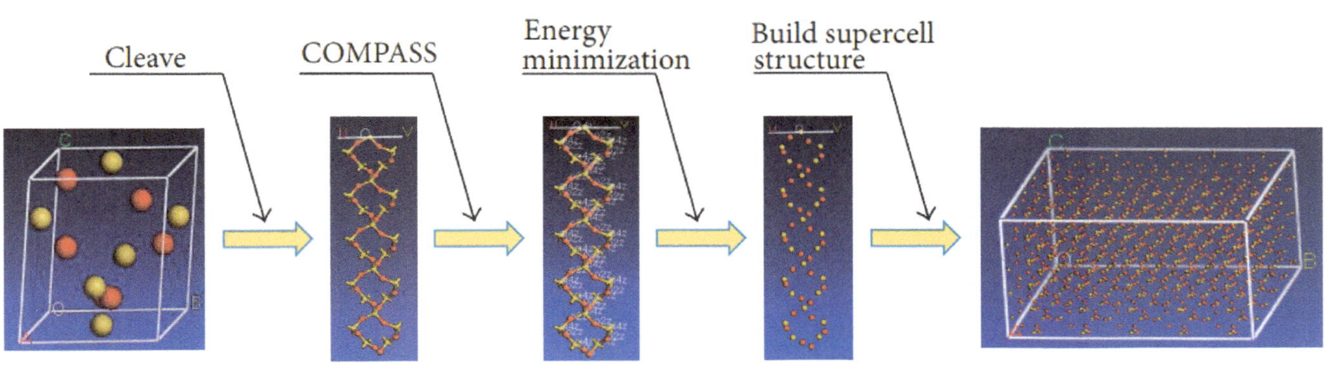

Figure 6: The process of building supercell structure.

The Interaction of Asphaltene and Oxide

Creating of Asphaltene/Oxide Interface

Interface model is composed of three layers. Layer 1 is oxide, layer 2 is asphaltene, and the last layer is 30 Å thick vacuum layer. According the depiction earlier, the three-layer interface structure was formed. For example, the asphaltene/SiO_2 interface model of 25°C is shown in Figure 7.

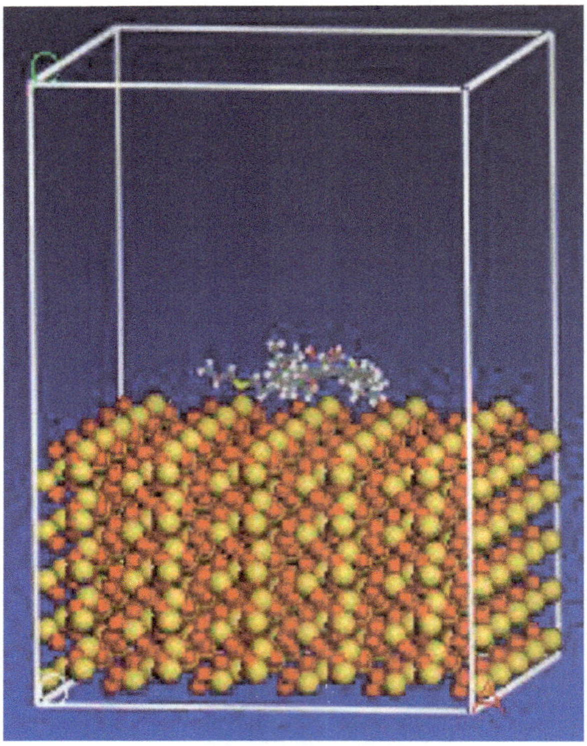

Figure 7: Asphaltene/SiO_2 interface model.

Dynamic Simulation Interface of Asphaltene/Oxide

As shown in the interface model in Figure 7, molecular dynamic simulation was then conducted in order to equilibrate systems further. The temperature is set and maintained constant by Andersen thermostat; the initial rate of molecular cell structure is randomly generated by the Maxwell-Boltzmann distribution. The Velocity Verlet algorithm is used to solve Newton's equation of motion. The van der Waals interaction and the Coulomb interaction are calculated by Charge Group method. The intermolecular interaction is corrected by an average density approximation. The system was simulated under NVT conditions with velocity rescaling, which was used for 2000 ps with a time step of 1.0 fs to dissipate high energy interactions between molecules.

MD simulation of four-group interface model was performed for temperatures of 10°C (283.15 K), 25°C (298.15 K), 40°C (313.15 K), 60°C (333.15 K), 80°C (353.15 K), and 100°C (373.15 K). The results are shown in Figures 8–15.

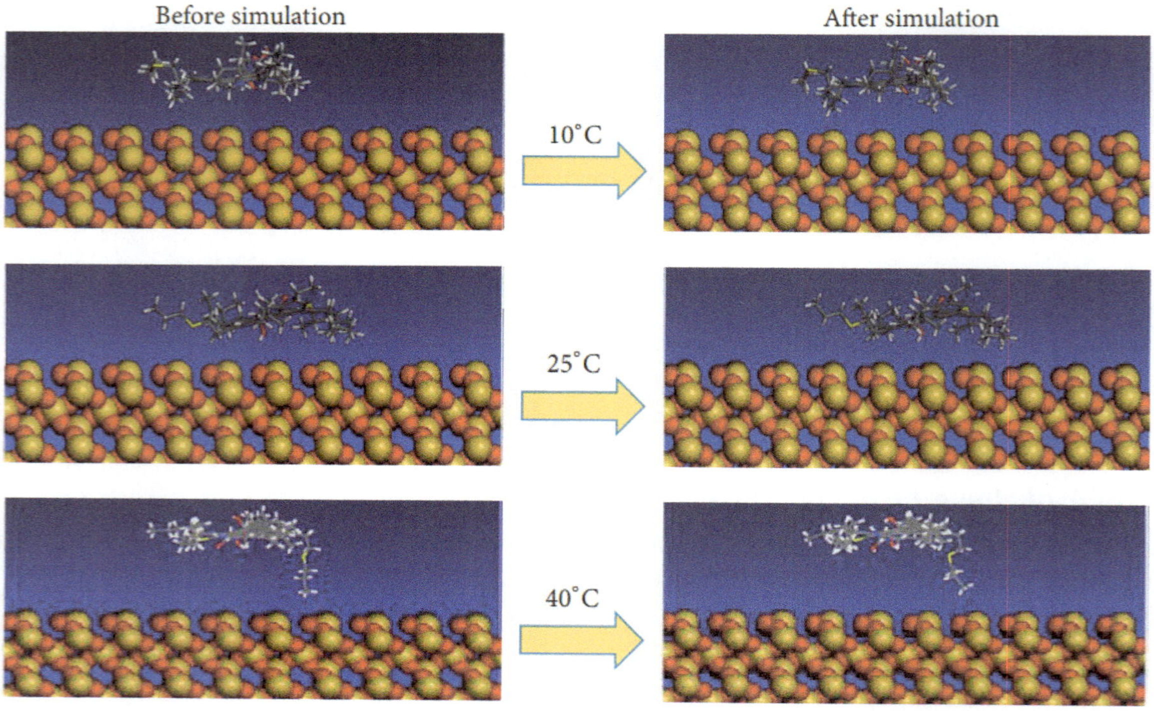

Figure 8: The interface configuration of asphaltene/SiO_2 (10–40°C).

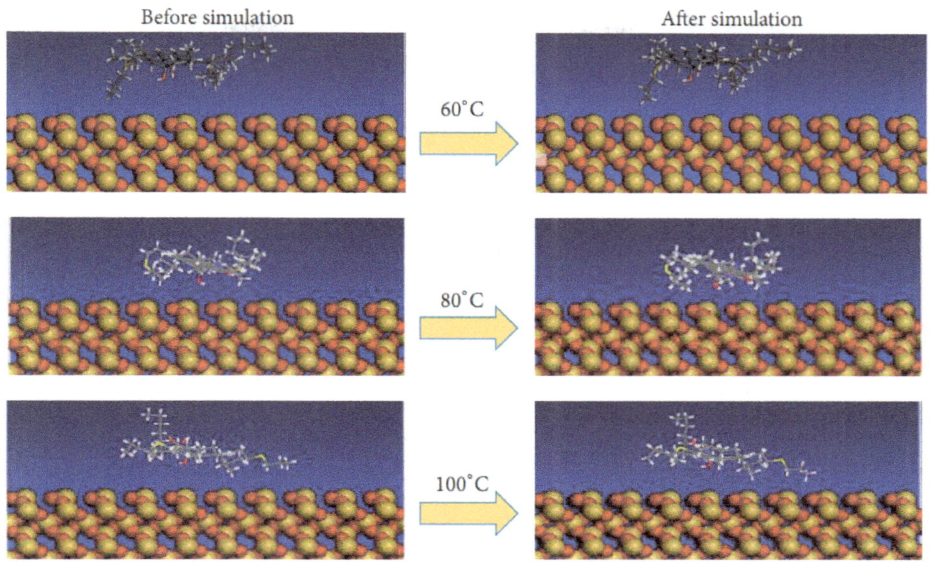

Figure 9: The interface configuration of asphaltene/SiO_2 (60–100°C).

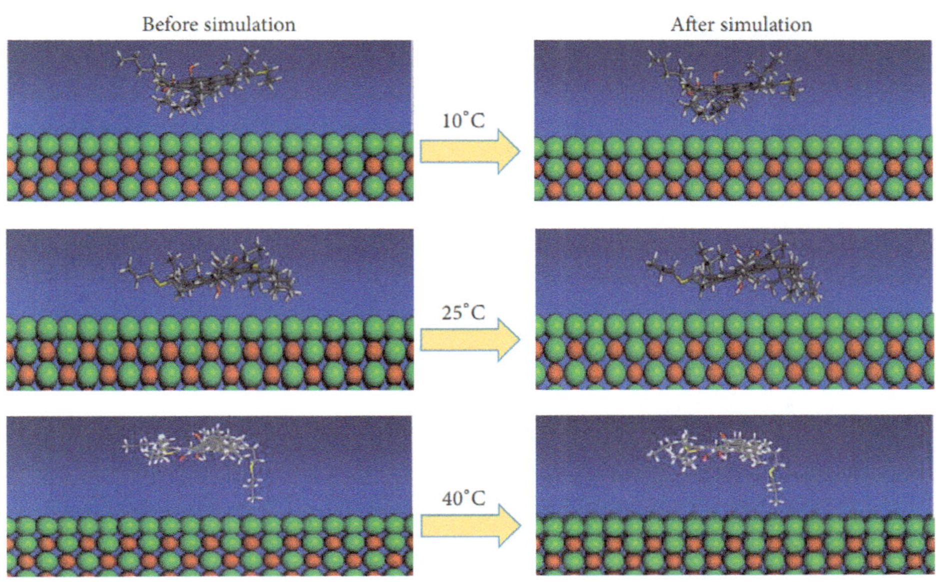

Figure 10: The interface configuration of asphaltene/CaO (10–40°C).

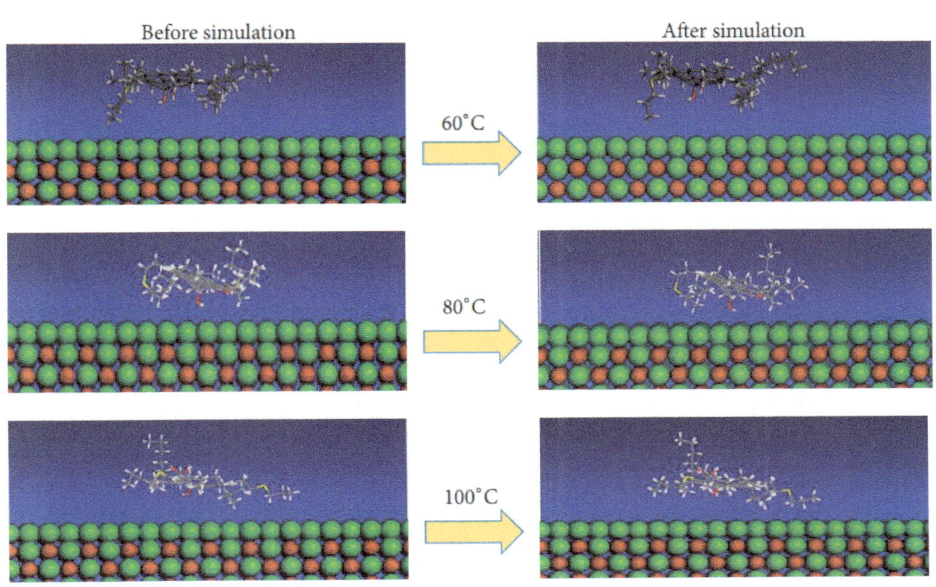

Figure 11: The interface configuration of asphaltene/CaO (60–100°C).

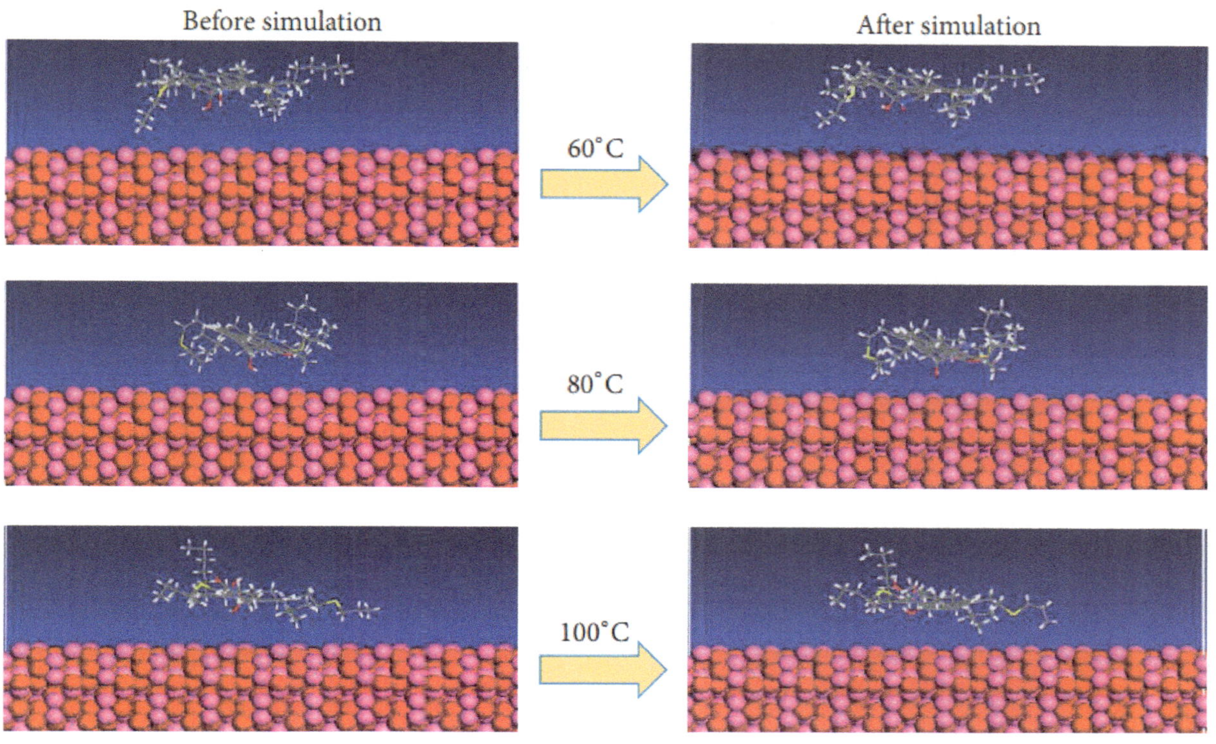

Figure 15: The interface configuration of asphaltene/Al_2O_3 (60–100°C).

As shown in the four interface configurations, the asphaltene molecular structures change in varying degree after the simulation. The asphaltene molecular structure moves closer to the oxide surface. Comparing the four kinds of interface configuration, it is difficult to evaluate the interaction between asphaltene and oxides accurately. For the fixed interface configuration, it is almost impossible to distinguish whether the interaction changes with temperature. Therefore, the interfacial energy is introduced to evaluate the interaction more precisely.

The Calculation of Interfacial Energy

The interfacial energy can be calculated according to the following:

$$\gamma_{int} = \frac{\left(E_{polymer} + E_{surface} - E_{total}\right)}{2A}. \qquad (5)$$

The four-interface model size is 49.000 × 49.000 × 26.38067 Å. So the interfacial energy equation can be simplified as follows:

$$\gamma'_{int} = \left(E_{polymer} + E_{surface} - E_{total}\right). \qquad (6)$$

After the molecular dynamics simulation, the four kinds of interfacial energy are shown in Tables 1 and 2.

Table 1: Asphaltene/SiO_2 and asphaltene/CaO

	Temp. (°C)	E_{total} (kJ/mol)	$E_{polymer}$ (kJ/mol)	$E_{surface}$ (kJ/mol)	$E_{Interface}$ (kJ/mol)
Asphaltene/SiO_2	10	$9.03E+06$	$3.63E+02$	$9.03E+06$	$2.22E+01$
	25	$9.03E+06$	$3.83E+02$	$9.03E+06$	$2.09E+02$
	40	$9.03E+06$	$3.73E+02$	$9.03E+06$	$3.25E+01$
	60	$9.03E+06$	$3.72E+02$	$9.03E+06$	$3.88E+01$
	80	$9.03E+06$	$3.83E+02$	$9.03E+06$	$1.36E+02$
	100	$9.03E+06$	$3.85E+02$	$9.03E+06$	$1.31E+02$
Asphaltene/CaO	10	$-1.57E+06$	$3.71E+03$	$-1.57E+06$	$8.99E+02$
	25	$-1.57E+06$	$3.69E+03$	$-1.57E+06$	$9.63E+02$
	40	$-1.57E+06$	$3.68E+03$	$-1.57E+06$	$7.44E+02$
	60	$-1.57E+06$	$3.87E+03$	$-1.57E+06$	$8.51E+02$
	80	$-1.57E+06$	$3.83E+03$	$-1.57E+06$	$1.00E+03$
	100	$-1.57E+06$	$3.81E+03$	$-1.57E+06$	$8.13E+02$

Table 2: Asphaltene/MgO and asphaltene/Al_2O_3

	Temp. (°C)	E_{total} (kJ/mol)	$E_{polymer}$ (kJ/mol)	$E_{surface}$ (kJ/mol)	$E_{Interface}$ (kJ/mol)
Asphaltene/MgO	10	−2.61*E* + 06	3.71*E* + 03	−2.61*E* + 06	1.67*E* + 03
	25	−2.61*E* + 06	3.79*E* + 03	−2.61*E* + 06	1.79*E* + 03
	40	−2.61*E* + 06	3.73*E* + 03	−2.61*E* + 06	1.61*E* + 03
	60	−2.61*E* + 06	3.76*E* + 03	−2.61*E* + 06	1.61*E* + 03
	80	−2.61*E* + 06	3.76*E* + 03	−2.61*E* + 06	1.68*E* + 03
	100	−2.61*E* + 06	3.81*E* + 03	−2.61*E* + 06	1.63*E* + 03
Asphaltene/Al_2O_3	10	−2.01*E* + 07	3.75*E* + 03	−2.01*E* + 07	4.61*E* + 02
	25	−2.01*E* + 07	3.72*E* + 03	−2.01*E* + 07	4.32*E* + 02
	40	−2.01*E* + 07	3.92*E* + 03	−2.01*E* + 07	1.19*E* + 02
	60	−2.01*E* + 07	3.98*E* + 03	−2.01*E* + 07	3.10*E* + 02
	80	−2.01*E* + 07	3.97*E* + 03	−2.01*E* + 07	6.43*E* + 02
	100	−2.01*E* + 07	3.87*E* + 03	−2.01*E* + 07	2.71*E* + 02

The interfacial energy of asphaltene/SiO2 varies from 2.22*E* + 01 to 2.09*E* + 02 kJ/mol within the temperature range of 10∘ C to 100∘ C; the calculated variation ranges of asphaltene/CaO, asphaltene/MgO, and asphaltene/Al_2O_3 within the same temperature range are 7.44*E* + 02 to 1.00*E* + 03 kJ/mol, 1.61*E* + 03 to 1.79*E* + 03 kJ/mol, and 1.19*E* + 02 to 6.43*E* + 02 kJ/mol, respectively. Further, the interfacial energy was plotted in Figure 16.

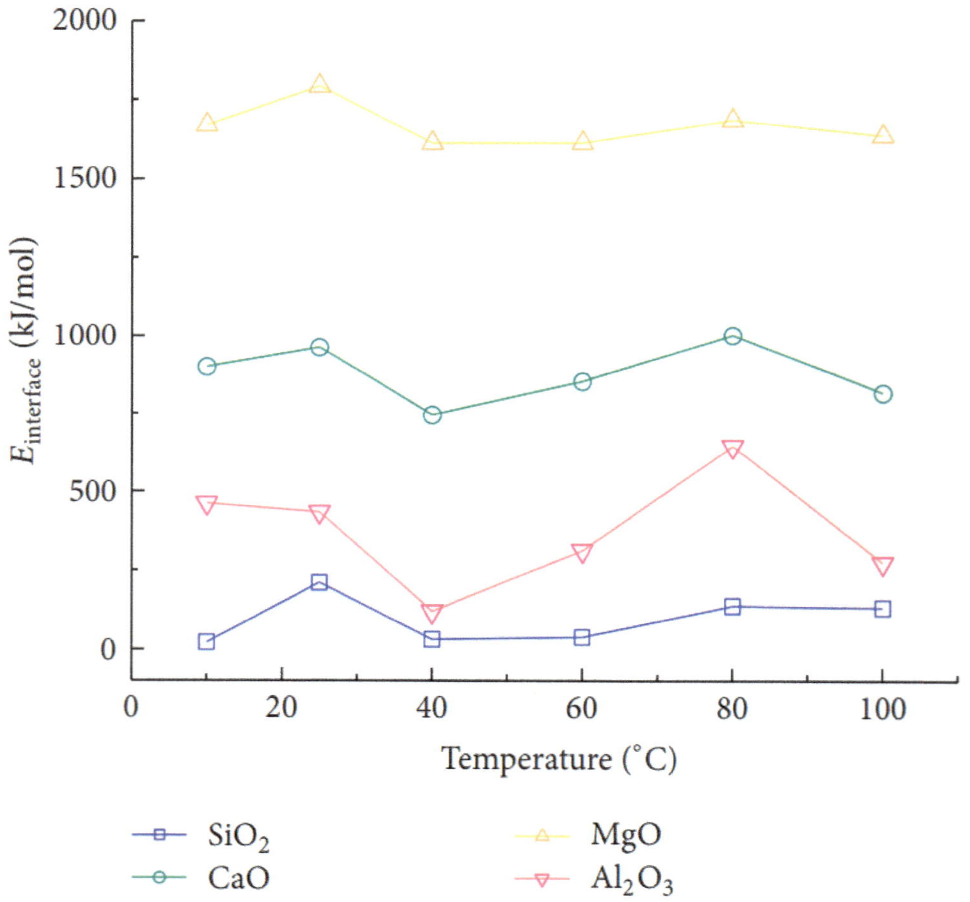

Figure 16: The interfacial energy with different temperature.

As shown in Figure 16, the four kinds of interfacial energy reach the extreme twice at 25°C and 80°C which indicates stronger interaction between asphaltene and oxides. Within the temperature range of 10°C to 100°C, the interaction capability ranking of the four oxides to asphaltene is as follows: MgO > CaO > Al_2O_3 > SiO_2.

CONCLUSION

The interaction and the interfacial energy between asphaltene and four kinds of oxides (MgO, CaO, Al_2O_3, and SiO_2) under temperature range from 10°C to 100°C were analyzed by virtue of molecular dynamic simulations and the following conclusions can be made: the interaction between asphaltene and oxides reaches strongest level at the temperature of 25°C and 80°C and reaches weakest level at the temperature of 40°C. It could be concluded according to the calculation results of the interfacial energy that the interfacial energy y between asphaltene and four kinds of oxides ranges from $2.22E+01$ kJ/mol to $1.79E+03$ kJ/mol within the temperature range of 10∘ C to 100∘ C. In this temperature range, the interaction capability ranking of the four oxides to asphaltene is as follows: MgO > CaO > Al_2O_3 > SiO_2.

ACKNOWLEDGMENTS

This work is supported by the National Natural Science Foundation of China (Grant nos. 51408048 and 51378037), the Key Program for International Science and Technology Cooperation Projects of Shaanxi province (Grant nos. 2012KW-09 and 2014KW10-03), the China Postdoctoral Science Foundation (Grant nos. 2015M570805 and 2015M582592), the Fundamental Research Funds for the Central Universities of Chang'an University (Grant nos. 310821153502 and 310821152005), and the Research Fund for the Doctoral Program of Higher Education of China (Grant no. 20120205120010).

REFERENCES

1. G. M. Elphingstone, Adhesion and cohesion in asphalt-aggregate systems [Ph.D. thesis], Texas A&M University, College Station, Tex, USA, 1997.
2. Y.-Q. Tan, J.-T. Wu, X.-M. Li, and L. Ji, "Evaluation indexes for interaction capability of asphalt and aggregate," Journal of Harbin Institute of Technology, vol. 41, no. 7, pp. 81–84, 2009. ·
3. D. Cheng, D. N. Little, R. L. Lytton, and J. C. Holste, "Use of surface free energy properties of the asphalt-aggregate system to predict moisture damage potential," Journal of the Association of Asphalt Paving Technologists, vol. 71, pp. 59–88, 2002.
4. A. Bhasin, E. Masad, D. Little, and R. Lytton, "Limits on adhesive bond energy for improved resistance of hot-mix asphalt to moisture damage," Transportation Research Record, vol. 1970, pp. 3–13, 2006. ·
5. R. Notman and T. R. Walsh, "Molecular dynamics studies of the interactions of water and amino acid analogues with quartz surfaces," Langmuir, vol. 25, no. 3, pp. 1638–1644, 2009. · ·
6. K. J. Leontaritis, J. O. Amaefule, and R. E. Charles, "Systematic approach for the prevention and treatment of formation damage caused by asphaltene deposition," in Proceedings of the International Symposium on Formation Damage Control, pp. 383–395, Lafayette, La, USA, February 1992.
7. Y.-F. Li, G.-W. Lu, W. Sun, et al., "Study on the molecular dynamics of petroleum-derived asphaltene aggregate," Acta Petrolei Sinica (Petroleum Processing Section), vol. 23, no. 4, pp. 25–31, 2007.
8. P. Srivastava, W. G. Chapman, and P. E. Laibinis, "Molecular dynamics simulation of oxygen transport through n-alkanethiolate self-assembled monolayers on gold and copper," The Journal of Physical Chemistry B, vol. 113, no. 2, pp. 456–464, 2009. · ·
9. E. Rogel, "Studies on asphaltene aggregation via computational chemistry," Colloids and Surfaces A: Physicochemical and Engineering Aspects, vol. 104, no. 1, pp. 85–93, 1995. · ·
10. H. T. Ozkahraman and E. C. Işik, "The effect of chemical and mineralogical composition of aggregates on tensile adhesion strength of tiles," Construction and Building Materials, vol. 19, no. 4, pp. 251–255, 2005. · ·

CHAPTER 13

MOLECULAR DYNAMICS SIMULATIONS OF THE BACTERIAL URAA H+-URACIL SYMPORTER IN LIPID BILAYERS REVEAL A CLOSED STATE AND A SELECTIVE INTERACTION WITH CARDIOLIPIN

Antreas C. Kalli[1], Mark S. P. Sansom[1], Reinhart A. F. Reithmeier[2]

[1] Department of Biochemistry, University of Oxford, Oxford, United Kingdom

[2] Department of Biochemistry, University of Toronto, Toronto, Canada

ABSTRACT

The *Escherichia coli* UraA H^+-uracil symporter is a member of the nucleobase/ascorbate transporter (NAT) family of proteins, and is responsible for the proton-driven uptake of uracil. Multiscale molecular dynamics simulations of the UraA symporter in phospholipid bilayers consisting of: 1) 1-palmitoyl 2-oleoyl-phosphatidylcholine (POPC); 2) 1-palmitoyl 2-oleoyl-phosphatidylethanolamine (POPE); and 3) a mixture of 75% POPE, 20% 1-palmitoyl 2-oleoyl-phosphatidylglycerol (POPG); and 5% 1-palmitoyl 2-oleoyl-diphosphatidylglycerol/cardiolipin (CL) to mimic the lipid composition of the bacterial inner membrane, were performed using the MARTINI coarse-grained force field to self-assemble lipids around the crystal structure of this membrane transport protein, followed by atomistic simulations. The overall fold of the protein in lipid bilayers remained similar to the crystal structure in detergent on the timescale of our simulations. Simulations were performed in the absence of uracil, and resulted in a closed state of the transporter, due to relative movement of the gate and core domains. Anionic lipids, including POPG and especially CL, were found to associate with UraA, involving interactions between specific basic residues in loop regions and phosphate oxygens of the CL head group. In particular, three CL binding sites were identified on UraA: two in the inner leaflet and a single site in the outer leaflet. Mutation of basic residues in the binding sites resulted in the loss of CL binding in the simulations. CL may play a role as a "proton trap" that channels protons to and from this transporter within CL-enriched areas of the inner bacterial membrane.

INTRODUCTION

Crystal structures of membrane proteins, particularly from bacterial sources, are being determined at an increasing rate (http://blanco.biomol.uci.edu/mpstruc/). The membrane protein is typically isolated and crystallized in the presence of detergents and indeed, non-native detergent molecules are often found tightly associated with the membrane protein. It is therefore important to relate the crystal structure back to the state of the protein in a native lipid bilayer [1,2].

Molecular dynamics (MD) simulations of membrane proteins in increasingly complex lipid bilayers [3] can now be applied with confidence to build dynamic models of membrane proteins in a native milieu [4,5]. A key question to be explored is to what extent the crystal structure determined in the presence

Citation: Kalli AC, Sansom MSP, Reithmeier RAF (2015) Molecular Dynamics Simulations of the Bacterial UraA H+-Uracil Symporter in Lipid Bilayers Reveal a Closed State and a Selective Interaction with Cardiolipin. PLoS Comput Biol 11(3): e1004123. doi:10.1371/journal.pcbi.1004123.

of detergents is the same as the structure of the membrane protein in a lipid bilayer. For example, the influenza M2 protein transmembrane domain shows significant structural differences between detergent-containing crystals and the structure in lipid bilayers determined by solid state NMR [1]. Furthermore, there are a number of examples of membrane protein structures that contain tightly associated lipids bound to specific sites on the protein [6,7]. MD simulations allow us to explore molecular motions and the dynamic interactions between lipids and proteins, providing a complementary approach to the temporal and spatially averaged static structures determined by X-ray crystallography. Recent studies using MD simulations have identified cardiolipin (CL) binding sites in the cytochrome bc_1transporter [8] and in the cytochrome c oxidase [9], and PIP_2 binding sites in Kir channels [10,11]. MD simulations were also used to identify interactions of lipids with the aquaporin family [12,13] and other proteins [14,15]. Recently MD simulations were also used to study the conformation of proton-driven transporters LacY [16] and UpaA [17]. So, how close is the structure of a membrane protein in a lipid bilayer determined by MD simulations to the original structure of the protein determined by X-ray crystallography in the presence of detergents? Can we use MD simulations to detect specific lipid-protein interactions not often found in crystal structures and study their dynamics?

We explored these questions using the *E. coli* UraA H^+-uracil symporter, the crystal structure of which was determined with bound uracil in the detergent n-nonyl-β-D-glucopyranoside (NG;Fig. 1). UraA is a member of the nucleobase/ascorbate transporter NAT family of proteins (also known as nucleobase-cation symporter-2 [NCS2]) which can be found in all species [18]. The majority of the NAT proteins are responsible for the uptake of xanthine, uric acid or uracil with H^+ in the case of most bacteria and Na^+ in the case of the mammalian transporters [17,18]. In mammals, NAT proteins are also responsible for the transport of vitamin C [18]. Including UraA, there are ten members of the NCS2 family of transporters in *E. coli*. Extensive mutagenesis and *in silico* studies identified a conserved 11-residue NAT motif in the NAT/NCS2 family ([Q/E/P-N-x-G-x-x-x-x-[RKG]) that is crucial for the function of its members. Mutation of the conserved residues of the motif in the *Aspergillus nidulans* uric acid/xanthine permease, UapA and the *E. coli* xanthine permease XanQ (YgfO) have shown that residues within the NAT motif contribute to the substrate specificity [19]. Extensive cysteine-scanning mutagenesis studies on XanQ have identified key functional residues, including Gln324 and Asn325 in the NAT motif within TM10 [19].

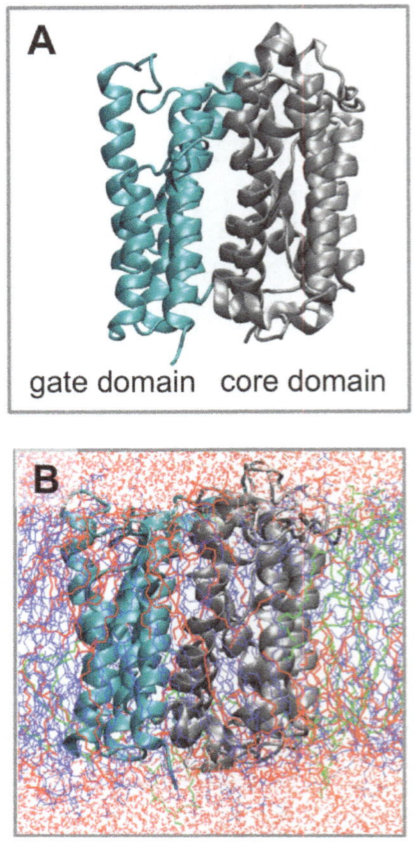

Figure 1. Structure of UraA. A. The core and the gate domains as suggested by *Lu et al.* [20] are shown in grey and cyan respectively. B. Snapshot from the end of one of the UraA-AT simulations. The same colors for UraA as in A are used. The POPE, POPG and CL are shown in blue, red and green respectively. The water solvent is shown in red. C. Sequence of UraA. The secondary structure is shown below the sequence. The same colors for the core and the gate domains as in A are used.

The 2.8 Å crystal structure of the UraA uracil transporter (Fig. 1), the first member of the NCS2 family to be solved, revealed a monomer protein with 14 transmembrane segments with TM segments 1–7 and TMs 8–14 having an inverted topology [20]. Uracil was located in a central cavity open via a water-filled passage to the cytosolic side of the membrane, bound between a pair of short anti-parallel β-strands in TM3 and 10 via a predominant interaction with an invariant Glu241 [20]. The crystal structure was obtained in the detergent n-nonyl-β-d-glucopyranoside with one inverted detergent molecule associated with the protein with the glucose head group of the detergent donating a hydrogen bond to the carbonyl in uracil, thereby stabilizing its binding.

In this study a serial multiscale MD simulation approach [21] was used to study the dynamics of the UraA (3QE7), simulated without bound uracil or detergent in a native-like lipid bilayer, enabling us to explore its specific interactions with the lipids. To this end, a lipid bilayer containing POPC lipids was self-assembled around UraA and then simulations in pure POPC, pure POPE and a POPE (75%), POPG (20%) and CL (5%) mixture to mimic the native bacteria membrane in the coarse-grained and atomistic resolutions were performed. Simulations in the absence of substrate (and bound detergent) resulted in a closed state of the protein. The various lipid bilayers that we have used allowed us to study the differences in the conformational states of UraA in different lipid environments. We discovered a preferential interaction of anionic lipids PG and CL with UraA. In particular, two CL sites were identified on UraA on the inner leaflet and a single CL site in the outer leaflet. The inner association involved a preferential interaction of the phosphate oxygen of the CL head group with R299 and with R4 near the cytosolic exit site of the transporter at one site, and with R265 and K109 at the second site. CL also showed a preferential interaction with UraA on the outer leaflet involving K321 and W212 in the third site. *In silico* mutation of these basic residues led to loss of CL binding highlighting the importance of an electrostatic interaction in the association of this anionic lipid with UraA.

Table 1. Summary of the principal simulations

Simulation	Description	Bilayer Concentration	Duration
UraA-CG	UraA (PDB: 3QE7)	POPE(75%)/POPG(20%)/CL (5%)	7 x 1 μs, 3 x 10 μs
UraAmut-1-CG	UraA [mutations: R4A, R299A, K321A]	POPE(75%)/POPG(20%)/CL (5%)	10 x 1 μs
UraAmut-2-CG	UraA [mutations: K157A, R222A, K321A, K381A, K390A]	POPE(75%)/POPG(20%)/CL (5%)	10 x 1 μs
UraAmut-3-CG	UraA [mutations: R4A, R11A, K62A, K109A, K110A, R177A, R265A, R299A, R351A, K357A, K362A, K407A]	POPE(75%)/POPG(20%)/CL (5%)	10 x 1 μs
UraA-AT	UraA	POPE(75%)/POPG(20%)/CL (5%)	2 x 0.1 μs 2 x 0.5 μs

METHODS

Coarse-Grained Molecular Dynamics Simulations

Coarse-grained molecular dynamics (CG-MD) simulations were performed using the MARTINI force field [22]. In the MARTINI CG model used [22,23] four heavy atoms (i.e. non-hydrogen) are represented as one CG particle. Four CG particle types exist: charged (Q), non-polar (N), apolar (I) and polar (P). Additional particle subtypes exist for assigning H-bond capabilities. Each amino acid is represented by the aforementioned particles representing the alpha carbon and the various portions of the side chain [22,23]. An elastic network was applied to all backbone particles using a cut-off distance of 7 Å to model secondary and tertiary structure. The use of an elastic network model (ENM) to maintain protein tertiary structure during the CG-MD simulations restricts any major conformational changes of the protein. However, during the subsequent atomistic simulations the ENM is removed. In the CG-MD simulations the original MARTINI water model [22,23] was used, without addition of antifreeze particles to any of the systems.

The starting structure of the UraA transporter, without the bound uracil or single detergent molecule, was taken from the protein coordinates of 2.8 Å crystal structure deposited in the PDB (3QE7). The CG protein was centered in a 14 x 14 x 12 nm³ simulation box of sufficient size to accommodate the 429-residue protein monomer (45 kDa) and about 400 phospholipid molecules. UraA was oriented to align the transport path of the protein with the z axis of the simulation box with the bottom of the structure facing the cytosol and the top the periplamic space. UraA contains 14 transmembrane segments and both the N- and C-termini face the cytosol.

Insertion of UraA in a Lipid Bilayer

POPC lipids were placed randomly in the simulation box and the system was solvated with CG water molecules and NaCl ions to ~150 mM were added to neutralize the system. In particular 296 Na^+ and 176 Cl^- ions were added to neutralize the -84*e* charge of the POPG lipids, the -42*e* charge of the CL and the +6*e* net charge of the protein. These ions were used to neutralize the systems, and no specific interactions between the ions and the lipids or the protein were observed. The system was energy-minimized using the steepest descent algorithm embedded in GROMACS. Subsequently, a CG-MD simulation for 100 ns was run which allowed the lipids to self-assemble forming a bilayer around the membrane protein. Note that during the self-assembly simulation the protein Cα atoms were restrained so the protein backbone could not move, while the side chains were mobile. At the end of the 100 ns self-assembly CG-MD simulation with UraA now in a bilayer, the protein restraints were removed and further 1 μs of production simulation with the protein free to diffuse in the bilayer was run. This allowed the protein to optimize both its position in the lipid bilayer (e.g. optimize the protein displacement

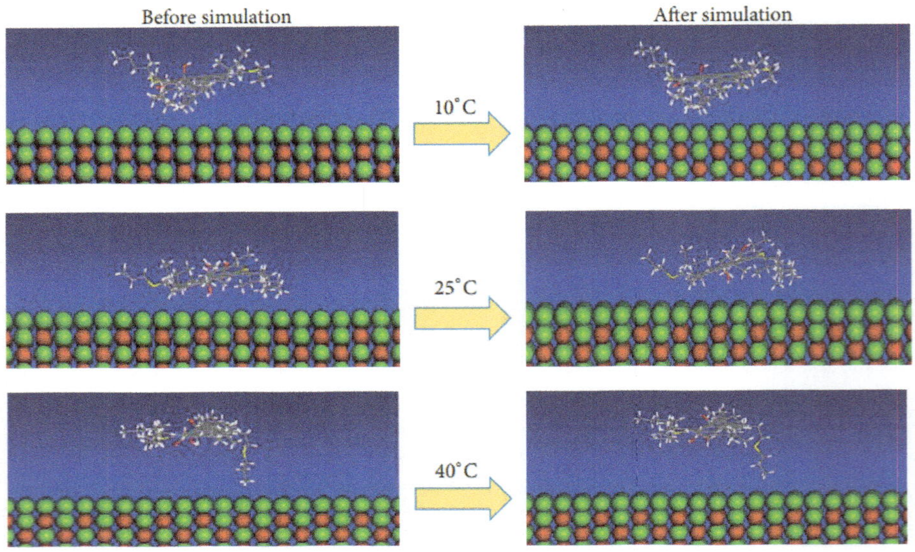

Figure 12: The interface configuration of asphaltene/MgO (10–40°C).

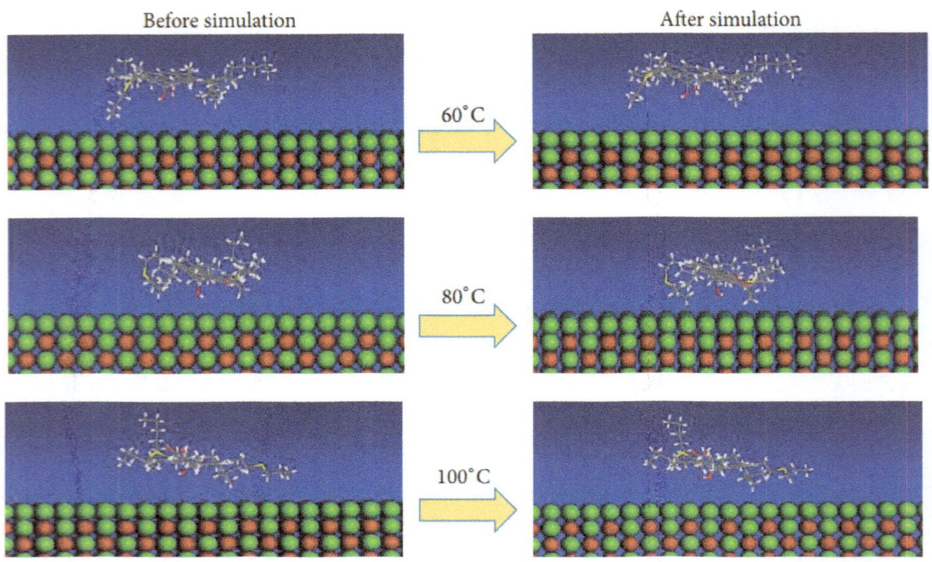

Figure 13: The interface configuration of asphaltene/MgO (60–100°C).

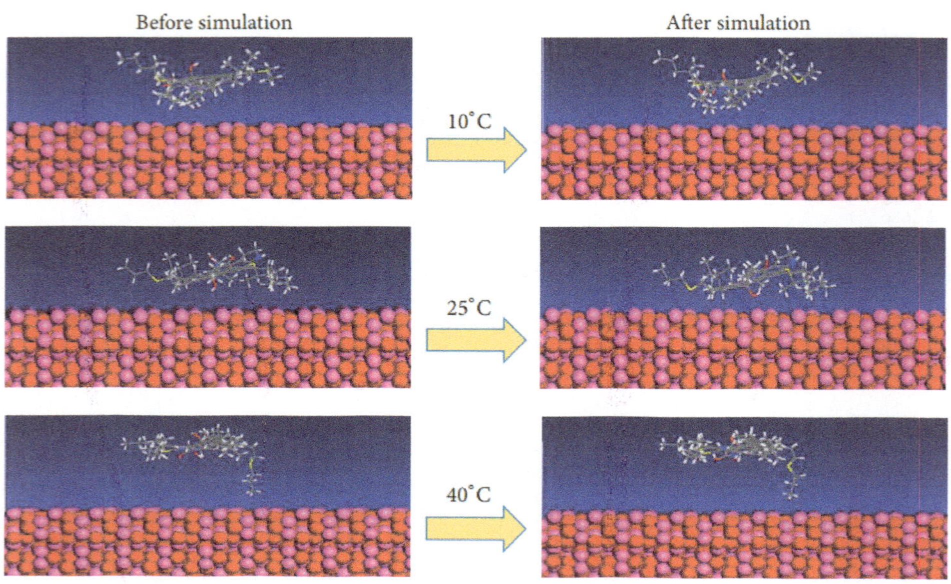

Figure 14: The interface configuration of asphaltene/Al_2O_3 (10–40°C).

and tilt angle relative to the bilayer) and its interaction with surrounding lipids.

Lipid Exchange

At the end of the unrestrained 1 μs simulation in the POPC bilayer the protein was re-centered in the simulation box and the POPC lipids were exchanged for 100% POPE, or 75% POPE, 20% POPG, 5% CL [24] using a locally written script (see Table 1). The script randomly replaces one POPC molecule (in the case of POPE or POPG) or two POPC molecules (in the case of CL, which consists of two diacylphosphatidic acid moieties connected by a central glycerol) with the desired number of POPE and/or CL molecules in each leaflet thus creating a new bilayer where the lipids are randomly distributed in the two leaflets in the desired concentrations. After lipid exchange the systems were energy minimized followed by 5 ns of equilibration at 323K with the protein Cα atoms restrained. It should be emphasized that in the latter bilayer the lipids were randomly distributed between the two leaflets. Seven independent repeat production simulations for 1 μs each and three independent repeat production simulations for 10 μs each were performed starting with different initial configurations (i.e. with the same initial position but with different initial velocities) to study the dynamic interactions of UraA with native lipids. For the pure POPE bilayer only 1 simulation was performed (for 1 μs).

Mutant Forms of UraA

Simulations using the aforementioned system with UraA in the 75% POPE, 20% POPG, 5% CL bilayer were also performed using mutated forms of the protein to study the role of specific residues in binding lipids (see Table 1 for more information). Additionally, removing basic residues from the CL binding sites of UraA facing the outer bilayer leaflet (UraAmut-2-CG), facing the inner bilayer leaflet (UraAmut-3-CG) or a combination of the two (UraAmut-1-CG) allowed us to probe if CL molecules bound cooperatively to the three CL binding sites. All mutations were performed using Modeller [25,26]. In particular, in UraAmut-1-CG residues R4 and R299 in site 1 and K321 in site 3 were mutated to alanines. In UraAmut-2-CG residues K157, R222, K321, K381, and K390 in or close to site 1 were mutated to alanines. These mutations removed the CL binding site in the outer leaflet of the bilayer. In UraAmut-3-CG residues R4, R11, R177, R299, R351, R357 and R362 in or close to site 1 and K62, K109, K110, R265 and K407 in or close to site 2 were mutated to alanines. These mutations removed the CL binding sites in the region of UraA facing the inner leaflet of the bilayer. Finally, simulations with mutations restricted to each of the three CL-binding sites: i) K321A, ii) K109A/R265A and iii) R4A/R299A were also performed. Note that in many cases more than one basic residue in the CL binding site was mutated to alanine in order to change the net positive charge of the CL binding site to neutral. This allowed us to prevent the CL molecules from interacting with other positive residues in the CL binding site, thus compensating for the loss of the main interacting residues. In these simulations, after the exchange of the lipids, the wild-type form of UraA was replaced by the mutated form and 10 individual repeat production simulations of 1 μs each were performed for each mutant.

CG-MD Simulations

All CG-MD simulations were performed using GROMACS 4.5 (www.gromacs.org). A Berendsen thermostat (coupling constant of 1.0 ps; reference temperature 310 K for POPC and POPE lipid bilayer and 323 K for the more physiological lipid mixture) and barostat (coupling constant of 1.0 ps; reference pressure 1 bar) were used. The integration time step was 20 fs. Lennard–Jones and Coulombic interactions were shifted to zero between 9 and 12 Å, and 0 and 12 Å, respectively, following the original MARTINI parameterization.

Atomistic Molecular Dynamics Simulations

Coarse-grained (CG) to atomistic (AT) conversion was performed using the CG2AT protocol as described previously by Stansfeld *et al.* [27]. In this approach the lipids are reconstructed by aligning AT lipid fragments with the corresponding CG fragments from an energy-minimized library of atomistic lipid conformations. The protein backbone is reconstructed by the CG backbone trace using PULCHRA [28] and the side chains using Modeller [25,26]. Atomistic molecular dynamics (AT-MD) simulations were performed using the GROMOS96 53a6 force field that has been widely used in simulation studies of membrane proteins. The Parrinello–Rahman barostat [29] and the V-rescale thermostat [30] were used for pressure and temperature coupling, respectively. The LINCS algorithm [31] was used to constrain bond lengths, and the Particle Mesh Ewald (PME) algorithm [32] was used to model long-range electrostatic interactions. Lipid parameters for POPE and CL were from [33] and for POPG were from [34]. All atomistic systems were equilibrated for 2.5 ns with the protein backbone Cα atoms restrained, followed by unrestrained atomistic MD simulations from 100 to 500 ns with a time step of 2 fs. Note that the removal of uracil is necessitated by the lack of accurate simulation parameters for this ligand.

Analyses

All the analysis was performed using GROMACS [35,36], VMD [37] and locally-written scripts. Structures were visualized using VMD [37] and PyMol (http://www.pymol.org). Analysis of the lifetime of interactions between the different lipid types and the 3 CL binding sites was performed by calculating distances (d_{ij}) between each binding site (i) and the head group of each of the 3 different lipid types (j). To define a contact a cut-off distance (δ) of 0.7 nm for POPE and POPG was used. For the CL lipids instead of using the standard 0.8 nm cutoff distance we have evaluated the optimal cutoff distance (δ) over a range of distances from 0.8 to 1.6 nm. This was done to reflect the larger headgroup of the CL molecules and additionally to take into consideration "rattling in a cage" motion of CL molecules. This motion of CL molecules has been reported in other CG-MD studies of these lipids [8]. Whilst the calculation of the lifetime of interactions is sensitive to the cutoff distance used, the trend that CL molecules form more prolonged interactions with CL binding site 1 is retained for all cutoff distances used. In particular, the longest lifetime of interactions of the CL molecules binding to CL site 1 for simulation 3 varies from ~200 ns to ~800 ns when varying δ. Similarly, for CL binding site 2 it varies from ~200 ns to ~450 ns, and for CL binding site 3 varies from ~100 ns to ~300 ns. A comparable sensitivity of the residence time has been observed to other studies which used CG-MD simulations to probe the CL interactions with membrane proteins [8]. Note that in all Figures/Tables we show the results obtained using a cutoff distance of 1.1 nm to define a contact for CL lipids. An occupancy variable $\zeta_{ij}(t)$ was determined, defined as:

$$\zeta_{ij}(t) = 1 \text{ if } d_{ij}(t) \leq \delta \text{ or}$$

$$\zeta_{ij}(t) = 0 \text{ if } d_{ij}(t) > \delta$$

Thus if the distance between the lipid and the CL binding site was lower that the cut-off distance the occupancy was 1 otherwise the occupancy was 0.

Using this measure of occupancy the stretches of time for which a site was continuously occupied by a specific lipid type were determined. For this analysis the trajectory was sampled every 0.3 ns, and we did not perform any smoothing of our raw data to remove any fast binding/unbinding events. This was because we concerned that smoothing of our trajectories would prevent us from capturing the faster binding/unbinding events of POPE and POPG lipids compared to CL, and additionally it has been suggested that application of different smoothing times of a trajectory may yield significantly different results [8]. For all lipid types we have observed some fast binding/unbinding events, but for our analysis we have focused on the most extended continuous periods of interactions for each lipid. Finally, note that we present all of our results without any normalization in terms of the number of lipids in the bilayer. To define the CL binding sites we have used two residues in each site for which our contact analysis had revealed that were predominantly involved in the interactions with the lipids (site 1: R4 and R299; site 2: K109 and R265; site 3: K321 and W212). Note that using a two-residue representation of the CL binding site often allows more than one lipid to occupy the CL binding site. After we have extracted the periods of time for which a binding site was continuously occupied by a specific lipid type, we have used these data to calculate how many changes of single lipids of the same lipid type occurred in the CL binding site during the time of which the site was continuously occupied by a specific lipid type. It should be noted, however, that the nature of the CG-MD simulations is such that calculated lifetimes of interactions should be viewed as indicative rather than absolute predictions.

In order to quantify the enrichment of anionic lipids (i.e. POPG and CL) around the protein we have merged the trajectories from the three extended CG-MD simulations, and have then calculated the mean density of each of the lipid types on the bilayer xy plane. The density was calculated for each leaflet separately. A region with a radius of 3 nm from the protein center of mass (representing the region adjacent to the protein) and a region with a radius of 3.5 to 6.0 nm from the protein center of mass (representing the bulk region of the bilayer) were selected. The change in the density of each lipid type between the region adjacent to the protein and the bulk region of the bilayer was calculated by dividing the mean density of the lipids in the annulus close to the protein over the mean density of the lipids in the annulus representing the bulk region:

$$S = \frac{mean\ density(0-3\ nm)}{mean\ density(3.5-6\ nm)}$$

Therefore if the value of S is above 1 then it means that there is enrichment of the specific lipid type around the protein whereas if S is < 1 it suggests that the density of the lipids in the bulk region is higher compared to the density of the lipids close to the protein. Note that for the CL molecules, because of their preference for the CL binding sites, we have also calculated S using only the mean density of CL molecules in the CL binding sites.

RESULTS

Cardiolipin (CL) Binding Sites on the UraA Transporter

To study the preferential interaction of UraA with the anionic lipids in a membrane, the crystal structure of UraA (3QE7) was assembled in a lipid bilayer that resembles its native environment in the bacterial inner membrane. In particular, UraA in a lipid bilayer containing POPE (75%), POPG (20%) and CL (5%) was subjected to MD simulations at coarse-grained and atomistic resolutions (see Methods and Table 1 for more information). Note that in all simulations the uracil molecule was absent, as was the detergent molecule present in the crystal structure, producing an empty carrier. A key goal of this study was to explore the conformation of the empty carrier, which is an intermediate in the transport mechanism.

Analysis of the lipid densities around the protein during the CG-MD simulations revealed a high density of acidic lipids in the vicinity of the protein (Fig. 2). In particular, we have identified three main preferential sites in UraA that interact with the CL. In all CL binding sites lysine or arginine residues interact with the phosphate groups of the lipid head groups. These sites are shown in Fig. 2 and Fig. 3 and were reproducible in the repeat CG-MD simulations performed with the wild-type UraA. One of these sites (site 3) is located in the extracellular side of the protein at residue K321, which forms the main interactions with the phosphate group of CLs (formed mainly by TM helix 11, 12 and the unstructured region between helices 7 and 8). Other residues in this CL binding site that form significant interactions with the CL headgroup are V202 to T204 and I208 to W212 (Fig. 2 and Fig. 3). The remaining two CL binding sites (sites 1 and 2) are located in the cytosolic side of the protein, which unlike the UraA extracellular part, has many lysine and arginine residues in the lipid-exposed area of the protein creating a net positive surface. CL binding site 1 is located around residues R4, A5, I6, P13, F176, R177, G178 and A296 to V300 (TM helices 1, 7 and the loop region between helices 10 and 11 that contains R299). CL binding site 2 is located around residues C61, K62, I107 to A111 and R265 (formed by TM helices 2, 4 and 9). Other dynamic interactions of CLs with other parts of the protein were also observed. Calculation of the average number of CLs in contact with the protein during the simulations suggested that 2–3 CLs were in contact most of the time with the protein. Visualization of the final snapshots from all the individual repeat CG-MD simulations (Fig. 3) showed that in the vast majority of the simulations one CL lipid was associated with each CL binding site. To test if the interactions between the CL molecules and UraA are retained on longer timescales, three simulations (to 10 μs) have also been performed (see Table 1). Analysis of the interactions showed that the CL binding sites were retained over this time period.

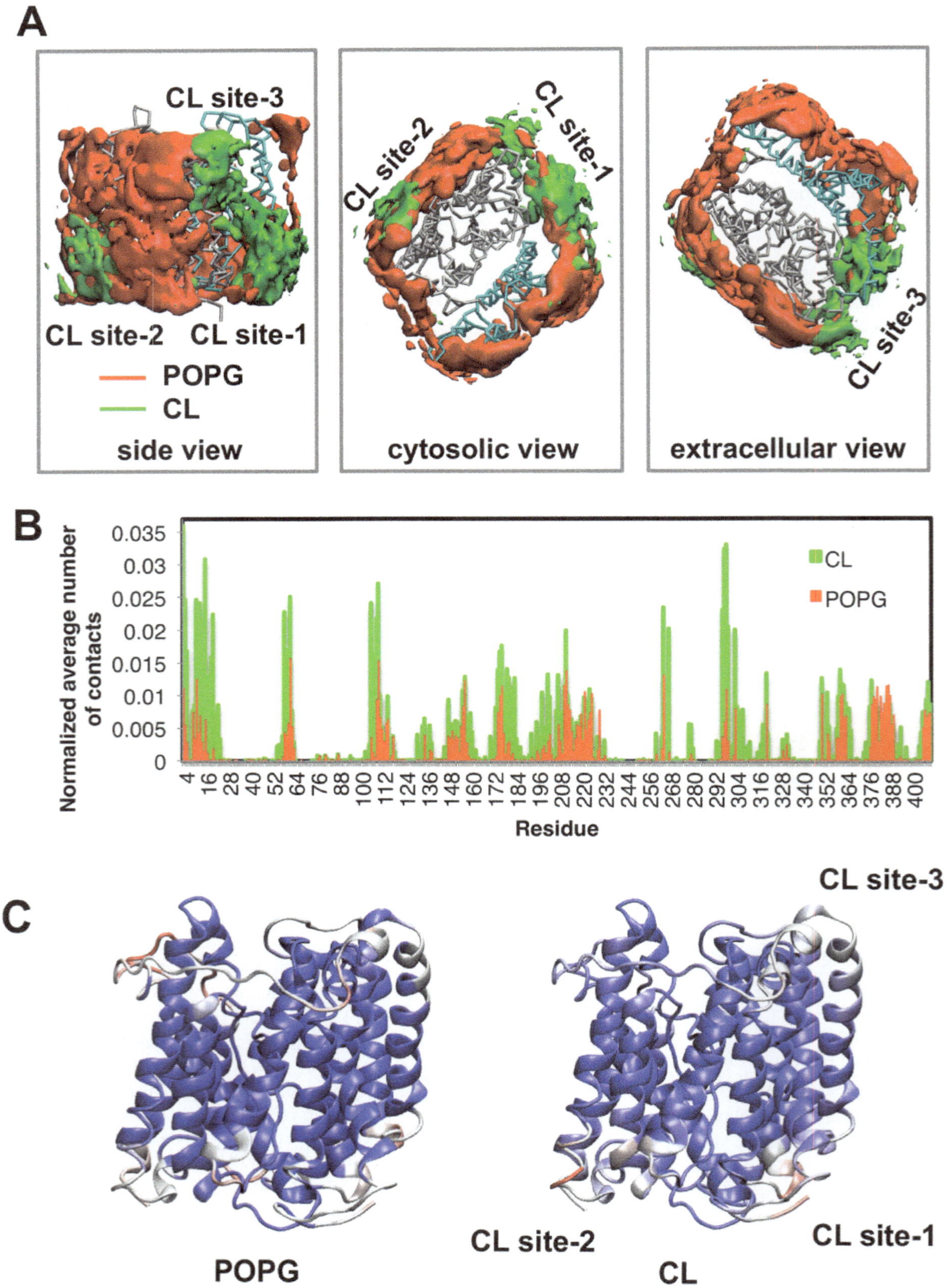

Figure 2. Interactions of UraA with lipids. A. Occupancy plots showing the probability of occurrence of cardiolipin (CL; green) and of the POPG (red) lipids around UraA. The occupancy was calculated as the average over all repeat coarse-grained simulations of the wild type UraA (UraA-CG in Table 1). Three different views are shown: a side view (left) showing the three CL binding sites, a cytosolic view facing the inner leaflet (middle) showing CL sites 1 and 2, and an extracellular view facing the outer leaflet of the bilayer (right) showing CL site 3. Normalized average number of contacts (using a cut-off distance of 7 Å for POPE and POPG and 8 Å for CL molecules) between the UraA and the head groups of POPG and CL lipids in the bilayer (across all repeats of the extended UraA-CG simulations; see Table 1). For the normalization, the number of contacts of a residue with a lipid type was divided by the number of lipids, the number of frames and the ratio of cutoff volumes. C. The number of contacts from one of the CG-MD simulations was mapped on the UraA crystal structure. Blue indicates a low number, white indicates a medium number and red a large number of contacts.

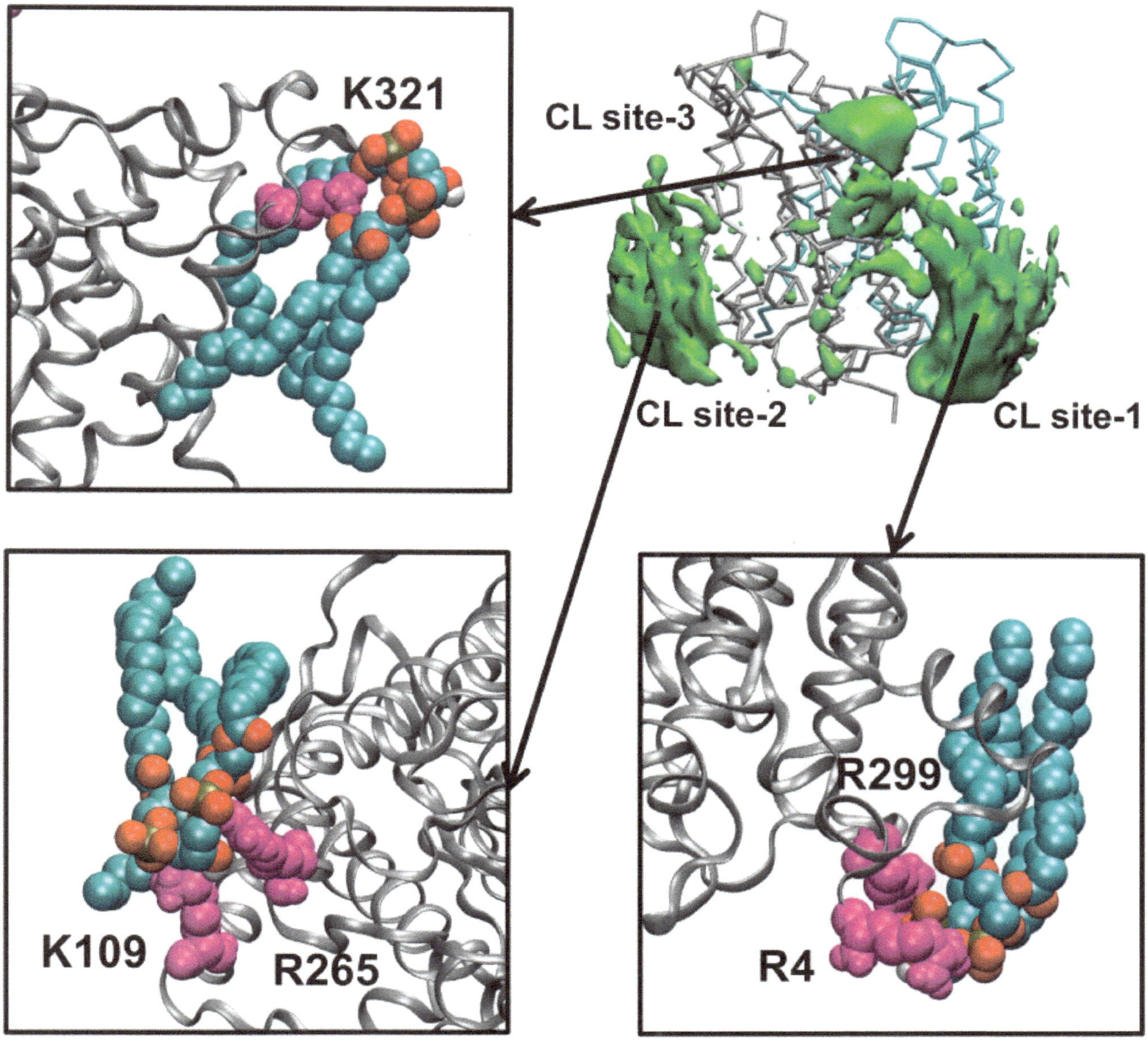

Figure 3. CL binding sites. Atomistic simulations of UraA in a 75% POPE/ 20% POPG/ 5% CL bilayer. Snapshots from the specific interactions between the CL and UraA are shown for the three CL binding sites (shown in green). The protein is represented as grey ribbons. The positively charged residues that form the main interactions with the CL in each CL binding site are shown in magenta. CL is shown in VDW format. Note that for this Figure we show snapshots with only CL lipids bound in the CL binding sites.

Fig. 3 shows the CL interactions with arginine and/or lysine residues in the CL binding sites identified above. In the CG-MD simulations the interactions of CL molecules with UraA are mediated by interactions of the negatively charged phosphate atom of the CL headgroup with polar part of the protein and hydrophobic interactions of the acyl chains with the non-polar belt of the membrane protein. The most common residues involved in the interaction with the phosphates are arginine and lysine as has been found in other simulations [12]. In the CL binding site 1, CL interacts with R299 located in a short loop connecting TM helix 10 and 11 and/or R4 in the mobile N-terminus of the protein near the exit site of the transporter. The CL interaction site in the outer leaflet involves K321 located at the N-terminus of the short helix connecting TM helices 11 and 12 and W212 at the N-terminus of TM helix 7. Hydrophobic interactions also play a role in creating lipid binding grooves in UraA. CL tails interact with the hydrophobic part of the protein located below the arginine and lysine residues identified above. The surface of UraA contains a number of hydrophobic grooves that accommodate the lipid acyl chains. The surface created by TM helices 2, 4, 9 and 11 is quite smooth. There are hydrophobic grooves between TM helices 1 and 6, TM helices 5 and 6, and between TM helices 13 and 14 and the rest of the protein. The simulations indicate that while the head groups of the lipids are immobilized by interaction with basic residues in the protein, the acyl chains are quite dynamic.

Interestingly, in all individual CG-MD simulations a high density of POPG molecules was also observed around the protein. The POPG lipids interact with the majority of the protein lipid exposed surface however basic residues form the highest number of interactions with POPG (Fig. 2.). This creates an acidic annulus around the protein, which may be essential in UraA function by hoarding protons required for transport. Similar annuli of anionic lipids have also been observed around TM helices [38] and around the TM domains of receptors e.g. EGFR [39]. Simulations studies have also shown that the mobility of lipids in an annulus up to ~3 nm from the membrane surface is smaller compared to the bulk region of the bilayer [40,41]. POPE can bind indiscriminately to also form a complete annulus of lipids around the protein. All the interacting residues with the CL, POPG and the POPE lipids are shown in Fig. 2. Calculation of the mean number of lipids in contact with the protein reveals that 2 - 3 CLs and ~ 8 POPG lipids are in contact with the protein.

To quantify the enrichment of anionic lipids around the protein we have calculated the ratio of the mean density of each lipid type in each leaflet in an annulus close to the protein and in the bulk region of the bilayer (S; see Methods for details of this analysis). If $S > 1$ then there is enrichment of the specific lipid type around the protein whereas if $S < 1$ it indicates that the mean density of the lipids in the bulk region is higher compared to the density of the lipids close to the protein. For POPG $S = 1.16$ for the outer leaflets and 1.33 for the inner leaflet demonstrating an enrichment of POPG around the protein. For CL, $S = 1.14$ for the inner leaflet, and 0.59 for the outer leaflet. However if we calculate S using only the density of the CL lipids in the CL binding sites (divided by the mean density in the bulk region) then the local value is $S = 5$ for the inner leaflet and $S = 2.3$ for the outer leaflet. This confirms the raised density of CL molecules in the discrete CL binding sites rather than the overall annulus, and thus demonstrates the preferential interaction of CL molecules with the three CL binding sites. For reference, for the zwitterionic lipid POPE $S = 0.63$ for the outer leaflet and 0.71 for the inner leaflet.

To examine in more detail the association of the different lipid types with the aforementioned CL binding sites we have approximated the lifetime of interactions between the different lipids and the 3 CL binding sites (see Methods, Fig. 4.). This analysis suggests that whilst all 3 lipid types are able to bind to the CL binding sites, the CL molecules form prolonged interactions. Therefore, when a CL binding site is not occupied by CL it is likely to be occupied mainly by POPG lipids or POPE lipids to a lesser extent. Additionally, this analysis suggests that all lipid types form more stable interactions with the CL binding site 1, followed by CL binding sites 2 and 3. In particular, in the CL binding site 1 CL molecules form the longest interactions, followed by POPG and POPE. In CL binding sites 2 and 3 the same pattern of interactions is observed however the time for which a lipid type continuously occupies a CL binding site is lower. Interestingly, when we calculated how many changes of single lipids of the same lipid type occurred in the CL binding site when the site is continuously occupied by a lipid type, we found that in most cases only one CL molecule interacts with the CL binding site whereas in the case of POPG and POPE lipids 3 to 7 changes of single lipids occurred in the CL binding site. This analysis suggests a faster exchange of the POPG and POPE lipids between the unbound and unbound state despite the fact that for this analysis we have used only those periods when a single lipid from the same lipid type occupies the CL binding site. Note that the stretches of times we have obtained in our study for CL binding sites 2 and 3 are short compared to those in a recent study by Arnarez *et al.* on the interactions of CL molecules with cytochrome bc_1 [8]. The lifetime of interactions for CL binding site 1 are comparable with the residence times at the lower occupancy CL sites of cytochrome bc_1, but not with the times for those sites with prolonged CL interactions with the cytochrome protein. Besides the fact that we have used a different protein, the lower residence times we have observed may be due to the fact that in our study we have POPE, POPG and CL whereas in the previous study only CL and POPC lipids were present, which may induce competition between the lipids especially between the anionic CL and POPG lipids. However, the CL binding sites identified in the cytochrome bc_1study contain multiple lysine and arginine residues which is similar to the sites we have identified on UraA. Poyry *et al.* also studied the cytochrome bc_1/CL interactions, using atomistic simulations, and also demonstrated that the interactions between CL lipids and cytochrome $bc_1$are mainly mediated by the interactions between the lipids and positively charged residues [42]. Other studies have shown that other anionic lipids interact with specific sites on proteins e.g. PIP_2 lipids to ion channels [10].

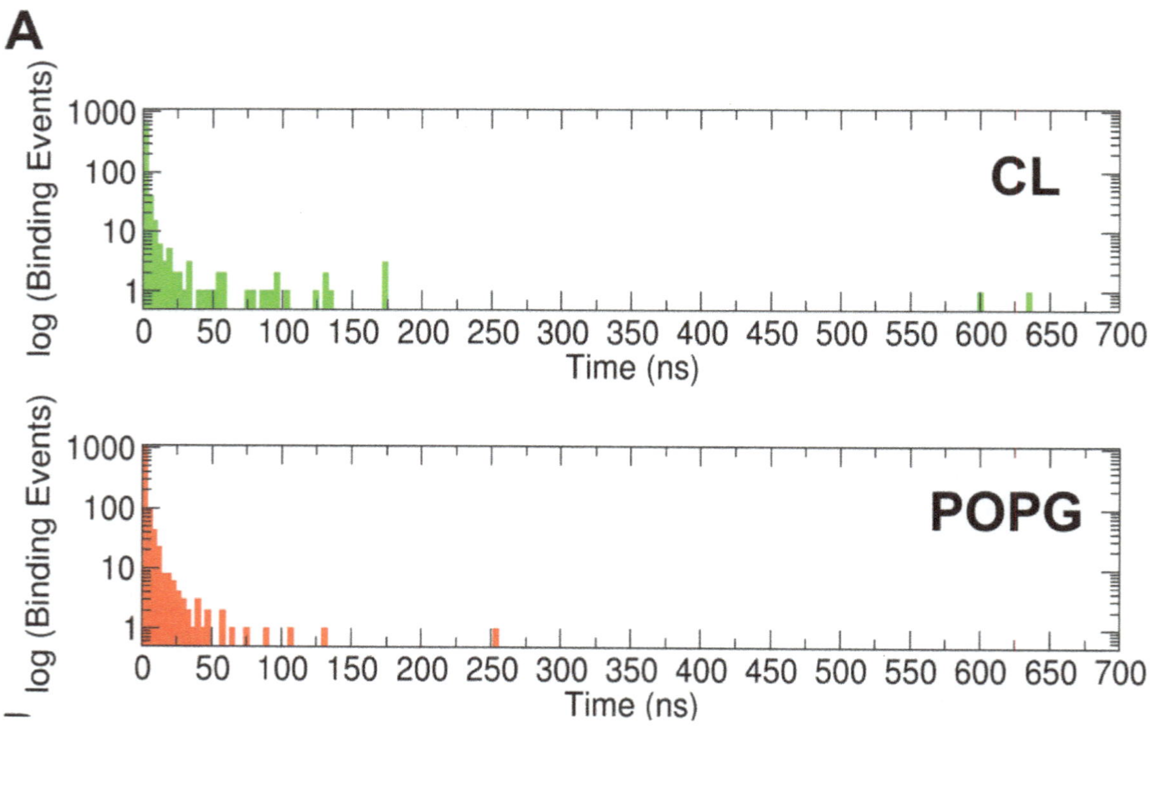

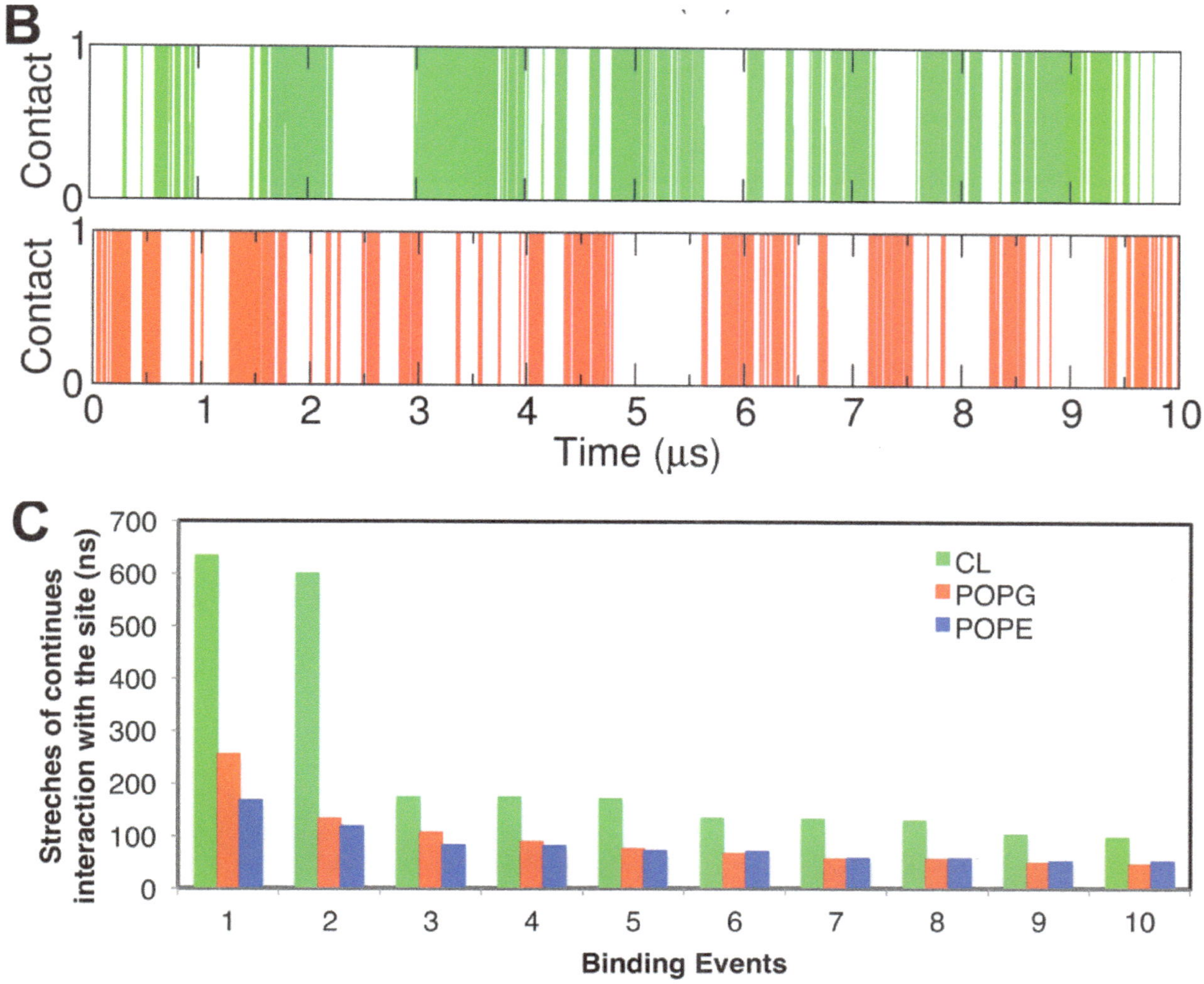

Figure 4. Interactions between the lipids and CL binding site 1. A. Histograms of binding event durations (defined as the time for which CL binding site 1 was continuously occupied by a lipid; see Methods for details). B. Interactions of UraA R299 residue with the POPG and CL molecules as a function of the simulation time for one of the CG simulations with the wild type UraA (10 μs simulation; UraA-CG in Table 1). An occupancy index of 1 is used when the lipids are in contact with the protein and 0 when the lipids are not in contact. The 10 longest stretches of continuous interactions between the different lipid types and CL binding sites 1.

We have also used the extended 10 μs simulations to examine the effect of the protein on the lateral diffusion of the POPG lipids and the CLs in the membrane. It is possible that the strong interactions between UraA and the aforementioned acidic lipids would influence the lipid diffusion rates. Indeed, the diffusion of the CL lipids in a bilayer with UraA decreases compared to a bilayer with the same lipid concentration but with no protein component. The diffusion of the POPG molecules is also somewhat slower in the system where the protein is present compared to a bilayer with the same lipid concentration but with no protein component, possibly due to their interactions with the protein. Interestingly, calculation of the mean diffusion coefficient for the lipids in the system without any protein component (pcpgcl-CG) reveals diffusion coefficients of $4.5 \pm 0.2 \times 10^{-7}$ cm^2/s for CL, $5.9 \pm 0.3 \times 10^{-7}$ cm^2/s for POPG and $6.2 \pm 0.2 \times 10^{-7}$ cm^2/s for POPE. This suggests that the four acyl chains of the CL and its larger cross-sectional diameter slow its diffusion about 1.5-fold relative to POPE or POPG. Comparison of the diffusion coefficients of POPE and POPG with those in a CG-MD study with a similar mixture of lipids (but not with the inclusion of CL lipids and with a higher number of proteins in the bilayer) suggests a somewhat lower diffusion coefficient for the POPG and POPE lipids in our system ($5.9 \pm 0.3 \times 10^{-7}$ cm^2/s for POPG in our system compared to $8.5 \pm 0.3 \times 10^{-7}$ cm^2/s for POPG in Goose *et al.* [41]). Short atomistic simulations by Rog *et al.* [43] and by Murzyn *et al.* [44], again with a POPE/POPG mixture, have suggested short timescale diffusion coefficients for POPG of ~3.5×10^{-7} cm^2/s. Recent microsecond scale simulations of a POPE/POPG mixture showed that the lipid diffusion coefficient is dependent on the NaCl concentration present, with a lipid lateral diffusion coefficient of 3.1×10^{-8} cm^2/s in the presence of 0.15 M NaCl [45].

To test the essential role of the basic residues in UraA in CL binding and to examine whether CL molecules bind cooperatively to the three CL binding sites, three different sets of *in silico*mutations (mutating basic residues in the CL binding sites of UraA facing the outer bilayer leaflet (UraAmut-2-CG), facing the inner bilayer leaflet (UraAmut-3-CG) or a combination of the two (UraAmut-1-CG)) were performed (Table 1) and the CG-MD simulations were rerun with the mutated UraA in the lipid mixture. All the simulations with the mutated forms of UraA were initiated by replacing the wild-type protein with the mutated form of the protein after the exchange of lipids step as described in the Methods section. Fig. 5 shows the distribution of CLs in the simulations with the mutated form of the protein. In the UraAmut-1-CG system, mutation of basic residues in CL site-1 and CL site-3 (R4A and

R299A in site 1 and K321A in site 3) resulted in the loss of these two CL binding sites. In this case only the non-mutated CL side-2 was observed. Similarly in the UraAmut-2-CG (K157A, R222A, K321A, K381A, K390A in site 1) and UraAmut-3-CG (R4A, R11A, R177A, R299A, R351A, R357A and R362A in site 1 and K62A, K109A, K110A, R265A and K407A in site 2) systems mutation in the CL binding site 3 and CL binding sites 1 and 2 respectively resulted in the loss of the mutated CL binding sides. To test perhaps more "biological" mutations we have also performed additional sets of simulations with the following *in silico* changes: (i) K321A, (ii) K109A/R265A, and (iii) R4A/R299A. In all cases the simulations resulted in the loss of binding of CL at the mutated site. In systems (i) and (ii) in addition to the strong interactions of the CL lipids with the non-mutated CL binding sites, interactions of CL molecules with R222 and K381 in (i) and R11, K62, K64 and R351 in (ii) were also observed. This augments our previous observation that basic residues in the three CL binding sides regulate the CL/UraA interactions and that CL bind to the 3 CL binding sites independently. Note that when both CL binding sides in the cytosolic part of the protein were mutated the association of the POPG lipids with UraA in the inner leaflet was also reduced. Convergence analysis of the simulations showed that all simulations were converged after 4 or 5 repeat simulations.

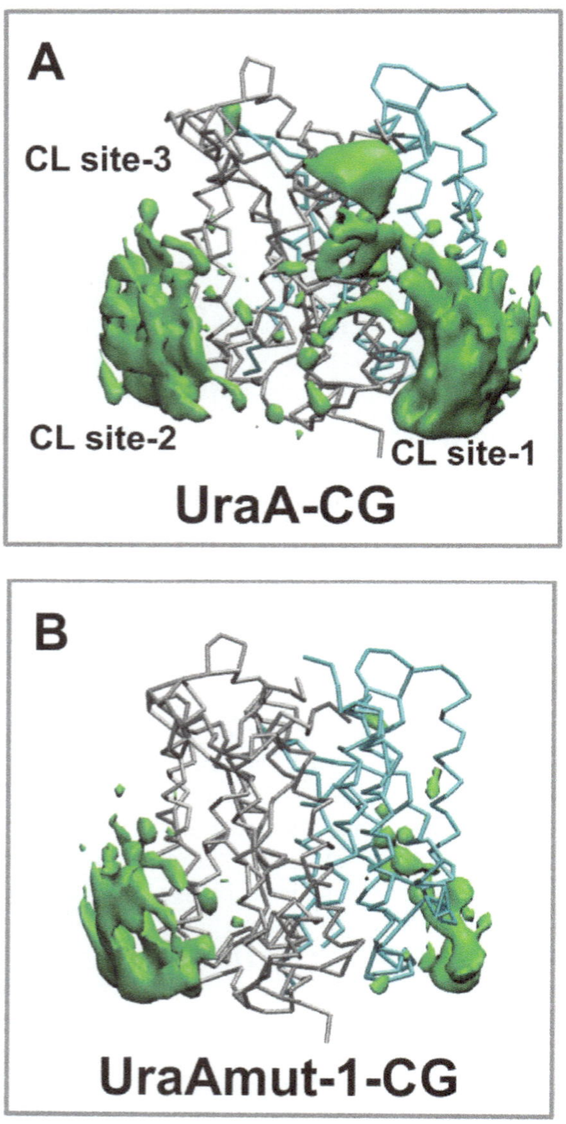

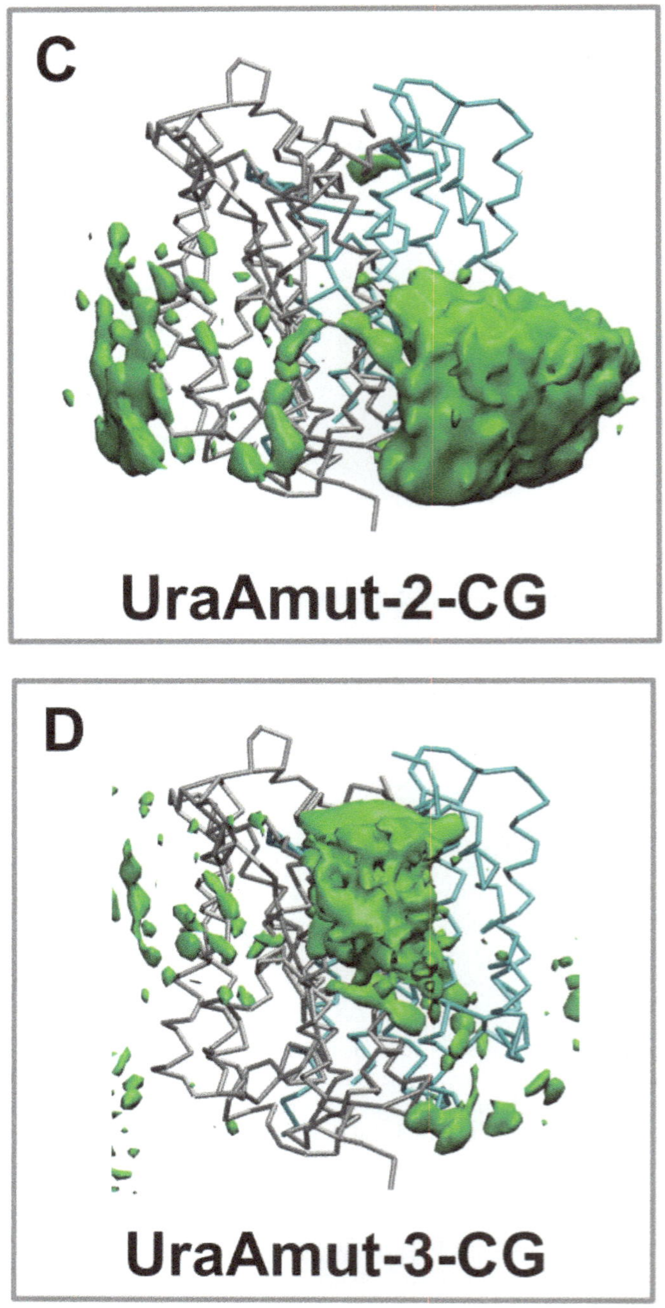

Figure 5. Mutations on the CL binding sites. Occupancy plots for the location of cardiolipins (CL; green) around UraA. The occupancy was calculated as the average over the multiple coarse-grained simulations of the wild type and the mutated forms of UraA (Mut 1: UraAmut-1-CG, Mut 2: UraAmut-2-CG and Mut 3: UraAmut-3-CG; see Table 1).

Atomistic Simulations of UraA

To study the possible conformational changes within the protein and the preferential interaction of UraA with the different lipids in the bilayer the final snapshots from four of the ten repeat CG-MD simulations with the WT UraA in the POPE/POPG/CL lipid mixture were converted to an atomistic representation (see Methods for more details). This is a standard protocol used in previous studies where interactions between a protein and lipid molecules were studied. In all the atomistic snapshots CLs were bound in the aforementioned CL binding sites. Atomistic simulations (2 x 100 ns and 2 x 500 ns) were then run. Calculation of the root mean square displacement

(RMSD) of the protein during the simulation revealed a sharp increase of the protein RMSD at the beginning of the simulations to ~0.35 nm. The RMSD for the longer AT-MD simulations subsequently increased to final values of ~0.5 nm. This increase is due to large fluctuations of the unstructured regions of UraA. Indeed, calculation of the route mean square fluctuation (RMSF) showed that core secondary structure elements have RMSFs between ~0.1 and ~0.2 nm whereas unstructured regions have RMSFs between ~0.4 to ~0.6 nm. Alignment of the crystal structure with the final snapshot of the extended simulations demonstrates that the integrity of all 14 transmembrane helices was maintained during the simulation, however the positions and angles of the helices that form the gate domain changed from the crystal structure (Fig. 6). Calculation of the transport path of the protein at the end of all atomistic simulations, using the program HOLE, revealed that the protein moved from an inward facing conformation that initially bound uracil (and a single detergent) to a closed conformation collapsing the water-filled passage to the cytosol (Fig. 6). This is presumably due to the fact that the simulations were performed in the absence of uracil. Alignment of the gate and the core domains from the end of the four atomistic simulations with the crystal structure reveals no significant conformational changes within the domains themselves. This suggests that in the absence of the uracil molecule a rigid body movement of the cytosolic part of the gate domain toward the cytosolic part of the core domain shifts the protein conformation to a close state. Simulations of the UraA protein in a simpler POPC or POPE lipid bilayer resulted in the same shift of the protein from the inward facing to a closed conformation, showing that the conformation change was not dependent upon a specific lipid composition. During the simulation no ion flux is observed. However occasionally we observe a small flux of water molecules.

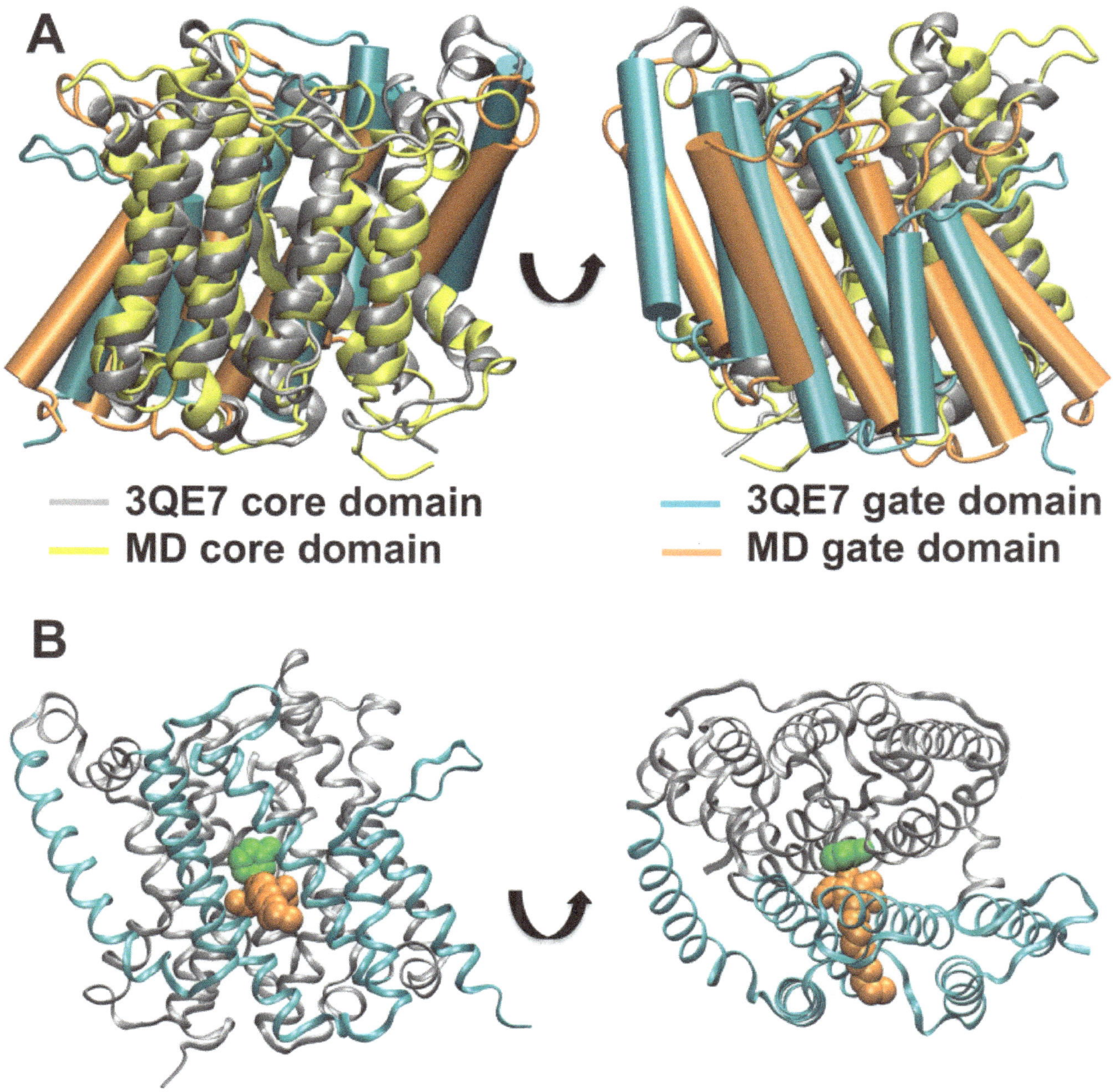

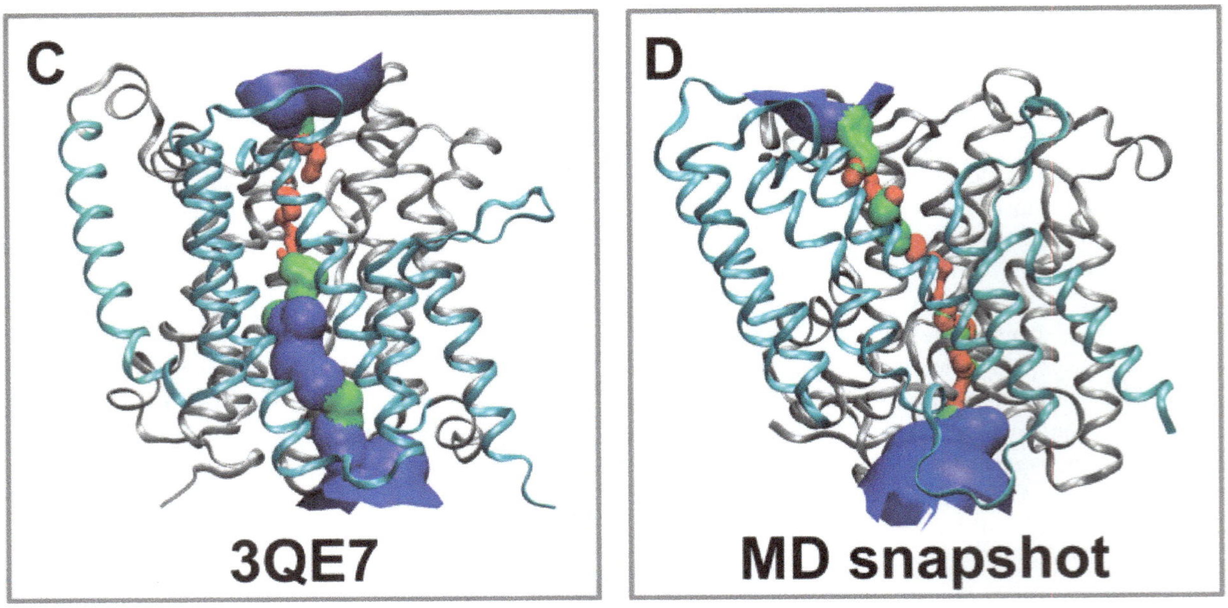

Figure 6. Closed state of UraA. A. Alignment of the core domain of the final snapshot of one of the atomistic simulations (UraA-AT) with the core domain of the crystal structure. B. Ribbon structure of UraA showing the location of the bound uracil (green) and NG detergent (orange). C,D. Pore lining surfaces of the proteins aligned in A. Red is used when the pore radius is lower than 1.15 Å, green when the pore radius is between 1.15 and 2.3 Å and blue when the pore radius is larger than 2.3 Å. Note the loss of water cavity (blue) from the inward-facing passage during the simulation to produce a closed conformational state of UraA.

Closer examination of the interactions between CL and the protein at the atomistic level suggests stable interactions between the CL oxygen atoms and the protein for the complete duration of the simulations (Fig. 3). The contacts made between UraA and CL in the corresponding CG-MD simulations were maintained during the atomistic simulations. In particular CL head groups can be seen to form significant interactions with the R4 and K299 residues in CL binding site 1, K109 and R265 residues in the CL binding site 2 and K321 residues in CL binding site 3 (Fig. 3). The CL tails form dynamic interactions with the hydrophobic part of the CL binding sites. Additionally, the annulus of the CL and POPG lipids around the proteins is retained on the timescale of our AT-MD simulations.

DISCUSSION

Cardiolipin Binding Sites in UraA

The major finding of our study is the identification of three CL binding sites in UraA. That these were not seen in the crystal structure is perhaps not surprising given that the purification/crystallization procedures extracts the majority of any bound lipids and that CL is a relatively minor (5%) component of the *E. coli* inner membrane. Indeed, no phospholipids could be detected in purified samples (~1 mg protein/ml) of His-tagged UraA prepared using the detergent dodecylmaltoside and Ni-affinity chromatography. CL has some unique physiochemical properties [46]. It consist of two phosphatidic acid linked together by a central glycerol. The free hydroxyl on the glycerol can form a hydrogen bond with the neighboring phosphate oxygens. This results in the formation of an acid-anion and a CL molecule with a single net negative charge with one phosphate having a typical low pKa (~2–3) and the second phosphate having a much higher pKa in the physiological pH range [47]. CL can thus act as a ready proton acceptor and donor. CL also has a very small head-group compared to the size of its hydrophobic region with four acyl chains. This allows CL molecules to pack closely together to form clusters (e.g. CL "lipid rafts") in mitochondrial and bacterial membranes. The cross-sectional area of CL is decreased upon binding of counter-ions like sodium due to reduction in electrostatic repulsion. The CL binding site identified in the extracellular site of UraA is located close to the transport path implying the CL may be important for providing the proton required for transport, as suggested for other proton transporters [8,48]. The inner CL binding sites may also have active role in the proton transport, by buffering protons as they exit the protein, however further investigation is required to identify the exact role of these CL binding sites in transport. Calculation of the binding energy of CL in cytochrome c oxidase by Marrink *et al.* using extended CG-MD simulations (100 μs) revealed a free energy of association of ~-30 kJ/mol for the main CL binding sites [9]. A quantitative measure of the association of lipids with specific membrane proteins may be given by EPR and fluorescence quenching methods [49]. Note that no associated lipids were reported in the crystal structure of UraA [20]. Other techniques such as crosslinking often require the use of chemically-modified lipids that have different properties from native lipids used in this study. The residues which are involved in the UraA/CL interactions are expected to be exposed to the lipid environment when the protein is in the outward open state and thus it is possible that CLs also associate with UraA when the protein is in the outward open conformation. However structural data for UraA in the outward open state are needed to be able to study this computationally.

UraA is a member of the nucelobase/ascorbate transporter (NAT) family, of which there are 10 family members in *E. coli*. We have shown that certain arginine and lysine residues in UraA are required for CL binding. How conserved are these residues in the NAT family in *E. coli*? Alignment of all 10 members of the NAT family in *E. coli* suggests that an arginine residue at the UraA R4 position is conserved in 4 out 10 members of the family. However with the exception of YbbY all other members of the family have a conserved basic residue in the vicinity of R4 (6 residues away). Inspection of UraA suggests that this conserved positive residue is also located in the CL binding site 1 described in this study. Additionally, all the members of the NAT family found in *E. coli* have a basic residue in the same position as R299 or in the same loop 1–3 residues distally. Interestingly, residue R299 is found at the end of the NAT-signature motif. This is a motif which is conserved in the NAT/NC2 family of proteins ([Q/E/P-N-x-G-x-x-x-x-[RKG]) and it was shown to be important in the function of UapA [50]. In particular R299 corresponds to R417 on UapA, the fungal uric acid/xanthine permease from *Aspergillus*, and mutation of this residue affect the kinetics and specificity of UapA for uric acid [50]. Additionally, cysteine scanning mutagenesis of the xanthine permease XanQ (YgfO) residue G333, which again corresponds to the UraA R299 residue, suggests that it plays a crucial role to the specificity of XanQ for different substrates [51]. Furthermore, replacement of certain basic residues located within TM segments in XanQ (K164, K249, R341, R385) with cysteine resulted in functional transporters but with significant decreases in xanthine uptake [52]. Mutation of basic residues (R285, R286, R296) in the cytosolic loop connected TM 8 and 9 showed modest decreases in transport [53]. The K298C mutant showed ~50% diminished initial xanthine uptake. K298 is located at an equivalent position one residue proximal to R265 in UraA at the beginning of TM9, while R285/286 in XanQ are lysine residues in UraA.

In CL binding site 2 a basic residue in the same position as R265 is present with the exception of UapA, XanQ, XanP and PurP. Note, however, that XanQ, XanP and PurP proteins have an adjacent positive residue to the R265 position whereas UapA has a basic amino acid 3 residues distal from R265. K109 is poorly conserved in the *E. coli* NAT/NCS2 family, however basic residues are located in the same loop 3–5 residues distally in all family members. Finally, K321 is conserved only in some family members (RutG, UacT) but not others. CL is not found in eukaryotic plasma membranes and thus key basic residues responsible for CL interaction in bacteria are not expected to be conserved in these NAT family members.

CL plays an essential role in the assembly and functioning of mitochondrial and bacterial membranes [54]. Yeast mutants deficient in CL biosynthesis have impaired mitochondrial function [55]. The ADP/ATP translocator is associated with CL clusters in the inner mitochondrial membrane along with respiratory sub-complexes [56] and this transporter is dependent upon CL for optimal activity [57]. The mitochondrial enzyme creatine kinase induces clustering of CL [58] and in turn, CL promotes the association of ATP synthase into ribbon-like structures in the mitochondrial membrane [59].

CL exists in clusters in bacterial membranes in regions of high curvature such as in "mini-cells" [60] and at the cell poles [61,62]. ProP, a H^+-proline symporter, is localized to CL clusters at the cell poles of *E. coli* [63]. Phospholipids interact with membrane proteins during their insertion into the membrane acting as chaperones to assist in their folding [64]. Indeed CL interacts tightly with the bacterial SecYEG translocon where it creates a high-affinity binding site for the SecA ATPase and localizes the complex to the cell poles [65]. This suggests that CL may be involved in membrane protein folding and insertion into the bilayer during biosynthesis. CL may also act as a "proton trap" through its acid-anion structure and is commonly associated with proteins involved in oxidative phosphorylation, associating with sub-complexes and providing a source of protons for the ATP synthase [66]. CL is found associated with sub-complexes of the respiratory chain such as in the bacterial photoreaction center [67,68] and the yeast cytochrome bc1 complex [6,69]. The association of CL with proton-driven symporters like UraA may point to a similar role by providing a source of buffered protons in the vicinity of these transporters. As in the case of ProP, CL may also provide a mechanism to cluster and localize proton-driven transporters to the cell poles in bacteria. Proton may be channeled directly between CL and the proton-binding site in the center of the transport protein via a proton highway composed in part of acidic residues. For example, CL is located close to the exit site for protons in cytochrome oxidase [66]. A similar role of CL has been also identified recently for the cytochrome c oxidases [9].

A Closed State for UraA

The second major finding of this study is the observation of a closed state of UraA, resulting from removal of the substrate uracil and a stabilizing non-native detergent molecule. Membrane proteins in general can tolerate changes in membrane lipid composition, which occur commonly in micro-organism as they adapt to changes in their environment [2]. Any changes in conformation that occurred during the simulation may be due to lipid interactions or loss of the substrate, or a combination of both factors. We found a similar change in the UraA structure in three different lipid environments (POPC, POPE and the lipid mixture), suggesting that lipids did not greatly influence the protein motions in moving from an empty to a closed state conformation. This suggests that the CL binding is not required for the transition from the inward facing to the closed state.

There is a water-filled passage leading from the bound uracil to the cytosol, partly occluded by the bound detergent. The protein can respond in two ways: 1, by filling in the space with water or 2, by changing conformation to close the space. The final atomistic structure of UraA in a bilayer is in a closed state conformation. This change occurred regardless of the lipid composition and thus is due to performing the simulations with the empty carrier. Thus, the MD simulation produced a closed state of the carrier, i.e. a state that may not be readily accessible by crystallography or other structural methods and that is an intermediate in the transport mechanism.

Membrane Proteins in Lipid Bilayers

Molecular dynamics simulations provide a powerful tool to analyze the structure and dynamics of membrane proteins in lipid bilayers of defined composition [4]. A molecular dynamics simulation of LacY, in various lipids identified specific interactions between the lipid head-groups and sites on the protein [70]. Recent studies also identified cardiolipin (CL) binding sites in the cytochrome bc_1 transporter [8] and in the cytochrome c oxidase [9] using CG-MD simulations. Similar studies identified specific interactions between cholesterol [15] or PIP lipids [10,11] with integral membrane proteins. The inner membrane of gram-negative bacteria, like*E. coli*, is composed of approximately 75% PE, 20% PG and 5% CL under normal laboratory growth conditions at 37°C [71]. In this simulation we found an annulus of anionic lipids around the UraA protein, followed by concentric rings spaced ~5 Å apart defined by the cross-sectional diameter of the lipid. Rings are discernable out to 20 Å from the protein surface. This suggests that the protein restricts the mobility of the annular lipids, a property of membrane proteins noted for some time [72,73]. The flexible acyl chains can fill in gaps in the protein's hydrophobic surface and the head groups can mediate more specific interactions with the polar parts of the protein [74,75]. Molecular simulations indicate that the association of annular lipids, with aquaporin for example can be quite dynamic [12]. Tightly-bound lipids have been identified in a number of X-ray crystal structures of membrane proteins showing that there are specific binding sites for lipids on the surface of some membrane proteins, which may assist in their folding or functioning [7]. These interactions involve tight binding of the acyl chains to the hydrophobic surface of the protein as well as polar/ionic interactions involving the head groups. The different simulations run for UraA indicate that CL is preferentially associated with the protein, although this lipid was only present at a 5% content. Indeed, positively-charged residues on hydrophilic loops of the protein can engage anionic lipids in a dynamic fashion, the lipids moving on and off the protein over the overall ~40 μs time course of the simulation. PE although in excess does not interact in a stable manner with UraA, likely due to its zwitterionic nature. The dynamic nature of the lipid-protein interactions observed with UraA may allow the protein to adapt to different lipid environments that the organisms like *E. coli* might encounter [2].

LIMITATIONS

It is important to consider possible limitations of simulations discussed in this study. The use of a CG-MD based approach implies some approximations in the protein and in the lipids. It has been shown, however, that CG-MD simulations correctly predict interactions of CL [8] and of other lipids (e.g. PIP_2) with integral membrane proteins [10–15]. Conversion of the final snapshot of four individual CG-MD simulation to a fully atomistic representation and extended AT-MD simulation runs confirm the stability of the identified CL binding sites and gives more detail information about the positioning of the CL lipid tails. The use of an elastic network model within the CG-MD simulations restricts any possible conformational changes of the protein. The use of a serial multiscale approach, with subsequent atomistic simulations, in part addresses this limitation allowing us to observe a closed state of the protein. Despite seen some conformational changes within the protein more extended simulations and/or exchange sampling techniques (e.g., dynamic importance sampling (DIMS) [76] or self-guided Langevin dynamics [77] are likely to be needed to more fully address possible conformational changes (e.g. outward facing conformation). Another limitation which arises due to the intrinsic approximations of the CG representation is that quantities such as the durations of protein/lipid interactions or diffusion coefficients of lipids are likely to be influenced by the coarse-graining of the protein and to be sensitive to the exact protocol used. Such analyses, however, provide a semi-quantitative measure of the stability of protein-lipid interactions observed in the CG-MD simulations. The removal of the uracil from the binding site produced a novel closed state of UraA, which may form part of the transport cycle. In future studies it will be important both to include the uracil molecule and possibly extend the time frames to capture the binding and movement of substrate across the membrane and the various conformational state of the transport protein in real time. It will also of course be important to test e.g. the predictions concerning CL binding sites and key residues via experimental mutagenesis studies.

AUTHOR CONTRIBUTIONS

Conceived and designed the experiments: ACK MSPS RAFR. Performed the experiments: ACK. Analyzed the data: ACK. Contributed reagents/materials/analysis tools: ACK MSPS RAFR. Wrote the paper: ACK MSPS RAFR.

REFERENCES

1. Cross TA, Sharma M, Yi M, Zhou H-X (2011) Influence of solubilizing environments on membrane protein structures. Trends Biochem Sci 36: 117–125. doi: 10.1016/j.tibs.2010.07.005. pmid:20724162
2. Sanders CR, Mittendorf KF (2011) Tolerance to changes in membrane lipid composition as a selected trait of membrane proteins. Biochemistry 50: 7858–7867. doi: 10.1021/bi2011527. pmid:21848311
3. Niemelä PS, Hyvönen MT, Vattulainen I (2009) Atom-scale molecular interactions in lipid raft mixtures. Biochim Biophys Acta Biomembranes 1788: 122–135. doi: 10.1016/j.bbamem.2008.08.018. pmid:18817748
4. Lindahl E, Sansom MSP (2008) Membrane proteins: molecular dynamics simulations. Curr Opin Struct Biol 18: 425–431. doi: 10.1016/j.sbi.2008.02.003. pmid:18406600
5. Stansfeld PJ, Sansom MSP (2011) Molecular simulation approaches to membrane proteins. Structure 19: 1562–1572. doi: 10.1016/j.str.2011.10.002. pmid:22078556
6. Palsdottir H, Hunte C (2004) Lipids in membrane protein structures. Biophys Acta Biomembranes 1666: 2–18. pmid:15519305 doi: 10.1016/j.bbamem.2004.06.012
7. Hunte C (2005) Specific protein-lipid interactions in membrane proteins. Biochem Soc Trans 33: 938–942. pmid:16246015 doi: 10.1042/bst20050938
8. Arnarez C, Mazat J-P, Elezgaray J, Marrink S-J, Periole X (2013) Evidence for cardiolipin binding sites on the membrane-exposed surface of the Cytochrome bc1. J Am Chem Soc 135: 3112–3120. doi: 10.1021/ja310577u. pmid:23363024
9. Arnarez C, Marrink SJ, Periole X (2013) Identification of cardiolipin binding sites on cytochrome c oxidase at the entrance of proton channels. Sci Rep 3. doi: 10.1038/srep03544. pmid:24384914
10. Schmidt MR, Stansfeld PJ, Tucker SJ, Sansom MSP (2012) Simulation-based prediction of phosphatidylinositol 4,5-bisphosphate binding to an ion channel. Biochemistry 52: 279–281. doi: 10.1021/bi301350s. pmid:23270460
11. Stansfeld PJ, Hopkinson R, Ashcroft FM, Sansom MSP (2009) PIP2-binding site in Kir channels: definition by multiscale biomolecular simulations. Biochemistry 48: 10926–10933. doi: 10.1021/bi9013193. pmid:19839652

12. Stansfeld Phillip J, Jefferys Elizabeth E, Sansom Mark SP (2013) Multiscale simulations reveal conserved patterns of lipid interactions with aquaporins. Structure 21: 810–819. doi: 10.1016/j.str.2013.03.005. pmid:23602661

13. Aponte-Santamaría C, Briones R, Schenk AD, Walz T, de Groot BL (2012) Molecular driving forces defining lipid positions around aquaporin-0. Proc Natl Acad Sci USA 109: 9887–9892. doi: 10.1073/pnas.1121054109. pmid:22679286

14. Koivuniemi A, Vuorela T, Kovanen PT, Vattulainen I, Hyvönen MT (2012) Lipid exchange mechanism of the cholesteryl ester transfer protein clarified by atomistic and coarse-grained simulations. PLoS Comput Biol 8: e1002299. doi: 10.1371/journal.pcbi.1002299. pmid:22253581

15. Sengupta D, Chattopadhyay A (2012) Identification of cholesterol binding sites in the Serotonin1A Receptor. J Phys Chem B 116: 12991–12996. doi: 10.1021/jp309888u. pmid:23067252

16. Stelzl LS, Fowler PW, Sansom MSP, Beckstein O (2014) Flexible gates generate occluded intermediates in the transport cycle of LacY. J Mol Biol 426: 735–751. pmid:24513108 doi: 10.1016/j.jmb.2013.10.024

17. Kosti V, Lambrinidis G, Myrianthopoulos V, Diallinas G, Mikros E (2012) Identification of the substrate recognition and transport pathway in a Eeukaryotic member of the Nucleobase-Ascorbate Transporter (NAT) family. PLoS ONE 7: e41939. doi: 10.1371/journal.pone.0041939. pmid:22848666

18. Gournas C, Papageorgiou I, Diallinas G (2008) The nucleobase-ascorbate transporter (NAT) family: genomics, evolution, structure-function relationships and physiological role. Mol BioSys 4: 404–416. doi: 10.1039/b719777b. pmid:18414738

19. Frillingos S (2012) Insights to the evolution of Nucleobase-Ascorbate Transporters (NAT/NCS2 family) from the Cys-scanning analysis of xanthine permease XanQ. Int J Biochem Mol Biol 3: 250–272. pmid:23097742

20. Lu F, Li S, Jiang Y, Jiang J, Fan H, et al. (2011) Structure and mechanism of the uracil transporter UraA. Nature 472: 243–246. doi: 10.1038/nature09885. pmid:21423164

21. Ayton GS, Noid WG, Voth GA (2007) Multiscale modeling of biomolecular systems: in serial and in parallel. Curr Opin Cell Biol 17: 192–198. doi: 10.1016/j.sbi.2007.03.004

22. Monticelli L, Kandasamy SK, Periole X, Larson RG, Tieleman DP, et al. (2008) The MARTINI coarse-grained force field: extension to proteins. J Chem Theory Comput 4: 819–834. doi: 10.1021/ct700324x

23. Marrink SJ, Risselada HJ, Yefimov S, Tieleman DP, de Vries AH (2007) The MARTINI force field: coarse grained model for biomolecular simulations. J Phys Chem B 111: 7812–7824. pmid:17569554 doi: 10.1021/jp071097f

24. Dahlberg M, Maliniak A (2010) Mechanical Properties of Coarse-Grained Bilayers Formed by Cardiolipin and Zwitterionic Lipids. J Chem Theory Comput 6: 1638–1649. doi: 10.1021/ct900654e

25. Sali A, Blundell TL (1993) Comparative protein modelling by satisfaction of spatial restraints. J Mol Biol 234: 779–815. pmid:8254673 doi: 10.1006/jmbi.1993.1626

26. Fiser A, Šali A (2003) Modeller: Generation and refinement of homology-based protein structure models. In: Carter Charles W. Jr, Robert MS, editors. Meths Enzymol: Academic Press. pp. 461–491.

27. Stansfeld PJ, Sansom MSP (2011) From coarse grained to atomistic: a serial multiscale approach to membrane protein simulations. J Chem Theory Comput 7: 1157–1166. doi: 10.1021/ct100569y

28. Rotkiewicz P, Skolnick J (2008) Fast procedure for reconstruction of full-atom protein models from reduced representations. J Comput Chem 29: 1460–1465. doi: 10.1002/jcc.20906. pmid:18196502

29. Parrinello M, Rahman A (1981) Polymorphic transitions in single crystals: a new molecular dynamics method. J Appl Phys 52: 7182–7190. doi: 10.1063/1.328693

30. Bussi G, Donadio D, Parrinello M (2007) Canonical sampling through velocity rescaling. J Chem Phys 126: -. pmid:17614595 doi: 10.1063/1.2408420

31. Hess B, Bekker H, Berendsen H, Fraaije J (1997) LINCS: a linear constraint solver for molecular simulations. J Comput Chem 18: 1463–1472. doi: 10.1002/(sici)1096-987x(199709)18:12<1463::aid-jcc4>3.3.co;2-l

32. Darden T, York D, Pedersen L (1993) Particle mesh Ewald: an Nlog(N) method for Ewald sums in large systems. J Chem Phys 98: 10089–10092. doi: 10.1063/1.464397

33. Piggot TJ, Holdbrook DA, Khalid S (2011) Electroporation of the E. coli and S. Aureus membranes: Molecular dynamics simulations of complex bacterial membranes. J Phys Chem B 115: 13381–13388. doi: 10.1021/jp207013v. pmid:21970408

34. Kukol A (2009) Lipid models for united-atom molecular dynamics simulations of proteins. J Chem Theory Comput 5: 615–626. doi: 10.1021/ct8003468

35. Hess B, Kutzner C, van der Spoel D, Lindahl E (2008) GROMACS 4: algorithms for highly efficient, load-balanced, and scalable molecular simulation. J Chem Theory Comput 4: 435–447. doi: 10.1021/ct700301q

36. David Van Der S, Erik L, Berk H, Gerrit G, Alan EM, et al. (2005) GROMACS: fast, flexible, and free. J Comput Chem 26: 1701–1718. pmid:16211538 doi: 10.1002/jcc.20291

37. Humphrey W, Dalke A, Schulten K (1996) VMD: visual molecular dynamics. J Mol Graph 14: 33–38. pmid:8744570 doi: 10.1016/0263-7855(96)00018-5

38. Koldsø H, Sansom MSP (2012) Local lipid reorganization by a transmembrane protein domain. J Phys Chem Lett 3: 3498–3502. doi: 10.1021/jz301570w

39. Abd Halim KB, Koldsø H, Sansom MSP (2014) Interactions of the EGFR juxtamembrane domain with PIP2-containing lipid bilayers: Insights from multiscale

molecular dynamics simulations. Biochim Biophys Acta—General Subjects.

40. Niemela PS, Miettinen MS, Monticelli L, Hammaren H, Bjelkmar P, et al. (2010) Membrane proteins diffuse as dynamic complexes with lipids. J Am Chem Soc 132: 7574–7575. doi: 10.1021/ja101481b. pmid:20469857

41. Goose JE, Sansom MSP (2013) Reduced lateral mobility of lipids and proteins in crowded membranes. PLoS Comput Biol 9: e1003033. doi: 10.1371/journal.pcbi.1003033. pmid:23592975

42. Pöyry S, Cramariuc O, Postila PA, Kaszuba K, Sarewicz M, et al. (2013) Atomistic simulations indicate cardiolipin to have an integral role in the structure of the cytochrome bc1 complex. Biochim Biophys Acta—Bioenergetics 1827: 769–778. doi: 10.1016/j.bbabio.2015.01.009. pmid:25626174

43. Róg T, Murzyn K, Pasenkiewicz-Gierula M (2003) Molecular dynamics simulations of charged and neutral lipid bilayers: treatment of electrostatic interactions. Acta Biochimica Polonica 50: 789–798. pmid:14515159

44. Murzyn K, Róg T, Pasenkiewicz-Gierula M (2005) Phosphatidylethanolamine-Phosphatidylglycerol Bilayer as a Model of the Inner Bacterial Membrane. Biophys J 88: 1091–1103. pmid:15556990 doi: 10.1529/biophysj.104.048835

45. Hong C, Tieleman DP, Wang Y (2014) Microsecond molecular dynamics simulations of lipid mixing. Langmuir 30: 11993–12001. doi: 10.1021/la502363b. pmid:25237736

46. Lewis RNAH, McElhaney RN (2009) The physicochemical properties of cardiolipin bilayers and cardiolipin-containing lipid membranes. Biochim Biophys Acta Biomembranes 1788: 2069–2079. doi: 10.1016/j.bbamem.2009.03.014. pmid:19328771

47. Kates M, Syz J-Y, Gosser D, Haines T (1993) pH-dissociation characteristics of cardiolipin and its 2′-deoxy analogue. Lipids 28: 877–882. pmid:8246687 doi: 10.1007/bf02537494

48. Lange C, Nett JH, Trumpower BL, Hunte C (2001) Specific roles of protein–phospholipid interactions in the yeast cytochrome bc1 complex structure. EMBO J 20: 6591–6600. pmid:11726495 doi: 10.1093/emboj/20.23.6591

49. Marsh D (2008) Electron spin resonance in membrane research: protein–lipid interactions. Methods 46: 83–96. doi: 10.1016/j.ymeth.2008.07.001. pmid:18672066

50. Koukaki M, Vlanti A, Goudela S, Pantazopoulou A, Gioule H, et al. (2005) The Nucleobase-ascorbate Transporter (NAT) signature motif in UapA defines the function of the purine translocation pathway. J Mol Biol 350: 499–513. pmid:15953615 doi: 10.1016/j.jmb.2005.04.076

51. Karatza P, Panos P, Georgopoulou E, Frillingos S (2006) Cysteine-scanning analysis of the nucleobase-ascorbate transporter signature motif in YgfO permease of escherichia coli: Gln-324 and Asn-325 are essential, and Ile-329–Val-339 from an α-helix. J Biol Chem 281: 39881–39890. pmid:17077086 doi: 10.1074/jbc.m605748200

52. Karena E, Frillingos S (2009) Role of intramembrane polar residues in the YgfO xanthine permease: HIS-31 and ASN-93 are crucial for affinity and specificty, and ASP-304 and GLU-272 are irreplaceable. J Biol Chem 284: 24257–24268. doi: 10.1074/jbc.M109.030734. pmid:19581302

53. Mermelekas G, Georgopoulou E, Kallis A, Botou M, Vlantos V, et al. (2010) Cysteine-scanning analysis of helices TM8, TM9a, and TM9b and intervening loops in the YgfO xanthine permease: a carboxyl group is essential at ASP-276. J Biol Chem 285: 35011–35020. doi: 10.1074/jbc.M110.170415. pmid:20802252

54. Schlame M (2008) Thematic review series: Glycerolipids. Cardiolipin synthesis for the assembly of bacterial and mitochondrial membranes. J Lipid Res 49: 1607–1620. pmid:18077827 doi: 10.1194/jlr.r700018-jlr200

55. Jiang F, Ryan MT, Schlame M, Zhao M, Gu Z, et al. (2000) Absence of cardiolipin in the crd1 null mutant results in decreased mitochondrial membrane potential and reduced mitochondrial function. J Biol Chem 275: 22387–22394. pmid:10777514 doi: 10.1074/jbc.m909868199

56. Claypool SM, Oktay Y, Boontheung P, Loo JA, Koehler CM (2008) Cardiolipin defines the interactome of the major ADP/ATP carrier protein of the mitochondrial inner membrane. J Cell Biol 182: 937–950. doi: 10.1083/jcb.200801152. pmid:18779372

57. Hoffmann B, Stöckl A, Schlame M, Beyer K, Klingenberg M (1994) The reconstituted ADP/ATP carrier activity has an absolute requirement for cardiolipin as shown in cysteine mutants. J Biol Chem 269: 1940–1944. pmid:8294444

58. Epand RF, Tokarska-Schlattner M, Schlattner U, Wallimann T, Epand RM (2007) Cardiolipin clusters and membrane domain formation induced by mitochondrial proteins. J Mol Biol 365: 968–980. pmid:17097675 doi: 10.1016/j.jmb.2006.10.028

59. Acehan D, Malhotra A, Xu Y, Ren M, Stokes David L, et al. (2011) Cardiolipin affects the supramolecular organization of ATP synthase in mitochondria. Biophys J 100: 2184–2192. doi: 10.1016/j.bpj.2011.03.031. pmid:21539786

60. Koppelman C-M, Den Blaauwen T, Duursma MC, Heeren RMA, Nanninga N (2001) Escherichia coli minicell membranes are enriched in cardiolipin. J Bacteriol 183: 6144–6147. pmid:11567016 doi: 10.1128/jb.183.20.6144-6147.2001

61. Renner LD, Weibel DB (2011) Cardiolipin microdomains localize to negatively curved regions of Escherichia coli membranes. Proc Natl Acad Sci USA 108: 6264–6269. doi: 10.1073/pnas.1015757108. pmid:21444798

62. Huang KC, Mukhopadhyay R, Wingreen NS (2006) A Curvature-mediated mechanism for localization of lipids to bacterial poles. PLoS Comput Biol 2: e151. pmid:17096591 doi: 10.1371/journal.pcbi.0020151

63. Romantsov T, Guan Z, Wood JM (2009) Cardiolipin and the osmotic stress responses of bacteria. Biochim Biophys Acta Biomembranes 1788: 2092–2100. doi: 10.1016/j.bbamem.2009.06.010. pmid:19539601

64. Dowhan W, Bogdanov M (2012) Molecular genetic and biochemical approaches for defining lipid-dependent membrane protein folding. Biochim Biophys Acta Biomembranes 1818: 1097–1107. doi: 10.1016/j.bbamem.2011.09.013. pmid:21945882

65. Gold VAM, Robson A, Bao H, Romantsov T, Duong F, et al. (2010) The action of cardiolipin on the bacterial translocon. Proc Natl Acad Sci USA 107: 10044–10049. doi: 10.1073/pnas.0914680107. pmid:20479269

66. Haines TH, Dencher NA (2002) Cardiolipin: a proton trap for oxidative phosphorylation. FEBS Letters 528: 35–39. pmid:12297275 doi: 10.1016/s0014-5793(02)03292-1

67. McAuley KE, Fyfe PK, Ridge JP, Isaacs NW, Cogdell RJ, et al. (1999) Structural details of an interaction between cardiolipin and an integral membrane protein. Proc Natl Acad Sci USA 96: 14706–14711. pmid:10611277 doi: 10.1073/pnas.96.26.14706

68. Wakeham MC, Sessions RB, Jones MR, Fyfe PK (2001) Is there a conserved interaction between cardiolipin and the type II bacterial reaction center? Biophys J 80: 1395–1405. pmid:11222300 doi: 10.1016/s0006-3495(01)76112-7

69. Lange C, Hunte C (2002) Crystal structure of the yeast cytochrome bc1 complex with its bound substrate cytochrome c. Proc Natl Acad Sci USA 99: 2800–2805. pmid:11880631 doi: 10.1073/pnas.052704699

70. Lensink MF, Govaerts C, Ruysschaert J-M (2010) Identification of specific lipid-binding sites in integral membrane proteins. J Biol Chem 285: 10519–10526. doi: 10.1074/jbc.M109.068890. pmid:20139086

71. Ballesta JPG, Schaechter M (1971) Effect of shift-down and growth inhibition on phospholipid metabolism of escherichia coli. J Bacteriol 107: 251–258. pmid:4327511

72. Lee AG (2011) Biological membranes: the importance of molecular detail. Trends Biochem Sci 36: 493–500. doi: 10.1016/j.tibs.2011.06.007. pmid:21855348

73. Contreras F-X, Ernst AM, Wieland F, Brugger B (2011) Specificity of intramembrane protein-lipid interactions. Cold Spring Harbor Perspect Biol 3: a004705. doi: 10.1101/cshperspect.a004705

74. Lee AG (2005) How lipids and proteins interact in a membrane: a molecular approach. Mol Biosyst 1: 203–212. pmid:16880984 doi: 10.1039/b504527d

75. Hite RK, Li Z, Walz T (2010) Principles of membrane protein interactions with annular lipids deduced from aquaporin-0 2D crystals. EMBO J 29: 1652–1658. doi: 10.1038/emboj.2010.68. pmid:20389283

76. Perilla JR, Beckstein O, Denning EJ, Woolf TB (2011) Computing ensembles of transitions from stable states: Dynamic importance sampling. J Comput Chem 32: 196–209. doi: 10.1002/jcc.21564. pmid:21132840

77. Wu X, Brooks BR (2003) Self-guided Langevin dynamics simulation method. Chem Phys Lett 381: 512–518. doi: 10.1016/j.cplett.2003.10.013

78. Baker NA, Sept D, Joseph S, Holst MJ, McCammon JA (2001) Electrostatics of nanosystems: Application to microtubules and the ribosome. Proc Natl Acad Sci USA 98: 10037–10041. pmid:11517324 doi: 10.1073/pnas.181342398

79. DeLano WL (2002) The PyMOL molecular graphics system. http://www.pymol.org.

80. von Heijne G (1992) Membrane protein structure prediction: Hydrophobicity analysis and the positive-inside rule. J Mol Biol 225: 487–494. pmid:1593632

CHAPTER 14

ATOMISTIC MONTE CARLO SIMULATION OF LIPID MEMBRANES

Daniel Wüstner[1] and Heinz Sklenar[2]

[1]Department of Biochemistry and Molecular Biology, University of Southern Denmark, Odense M DK-5230, Denmark

[2]Theoretical Biophysics Group, Max Delbrück Center for Molecular Medicine, Robert-Rössle-Str. 10, Berlin D-13125, Germany

ABSTRACT

Biological membranes are complex assemblies of many different molecules of which analysis demands a variety of experimental and computational approaches. In this article, we explain challenges and advantages of atomistic Monte Carlo (MC) simulation of lipid membranes. We provide an introduction into the various move sets that are implemented in current MC methods for efficient conformational sampling of lipids and other molecules. In the second part, we demonstrate for a concrete example, how an atomistic local-move set can be implemented for MC simulations of phospholipid monomers and bilayer patches. We use our recently devised chain breakage/closure (CBC) local move set in the bond-/torsion angle space with the constant-bond-length approximation (CBLA) for the phospholipid dipalmitoylphosphatidylcholine (DPPC). We demonstrate rapid conformational equilibration for a single DPPC molecule, as assessed by calculation of molecular energies and entropies. We also show transition from a crystalline-like to a fluid DPPC bilayer by the CBC local-move MC method, as indicated by the electron density profile, head group orientation, area per lipid, and whole-lipid displacements. We discuss the potential of local-move MC methods in combination with molecular dynamics simulations, for example, for studying multi-component lipid membranes containing cholesterol.

INTRODUCTION

Membrane Simulations and the Time-Scaling Problem

Analysis of the molecular mechanisms underlying membrane lipid organization is of crucial importance for understanding the function of membranes in living cells. Recent interest has focused on the biophysical characterization of model membranes to explore the physical principles governing the behavior of biological membranes. For example, lipid phase behavior and separation was investigated in model membranes by fluorescence microscopy, neutron scattering and NMR, while cholesterol dependent lipid order in endoplasmic reticulum membranes was determined by EPR spectroscopy [1–5]. Despite a large increase in computational power, atomistic membrane simulations lack behind this experimental progress. This is largely due to the large time scale at which biologically relevant processes in membranes occur. While characteristic times for stretching of a C–C bond in a fatty acid tail are in the range of a few picoseconds, change of the lateral position of two lipids requires several nanoseconds, and transverse phospholipid migration in pure lipid bilayers takes place on a time scale of hours [6,7]. Similarly, important

Citation: Wüstner, D.; Sklenar, H. Atomistic Monte Carlo Simulation of Lipid Membranes. Int. J. Mol. Sci. 2014, 15, 1767-1803. · DOI: 10.3390/ijms15021767

biological processes like vesicle formation during intracellular transport require the collective dynamics of hundreds of membrane components for several seconds. Molecular detail, however, cannot be ignored when developing models of these processes, since small structural changes in the involved lipids can promote or antagonize membrane budding and fusion [8,9]. In addition, biological membranes are of extreme complexity by consisting of several hundred different lipid species, transmembrane proteins, like G-protein coupled receptors and peripheral proteins like prenylated ras or the glycosylphosphatidylinositol- (GPI-) anchored folate receptor [10,11]. There is also the subcortical actin playing an important role in membrane organization and dynamics [12–17], the extracellular oriented glycocalyx [18] and the well-established transbilayer phospholipid asymmetry [19]. This compositional complexity together with the dynamically varying solute and ion composition on both sides of the plasma membrane naturally call for many different modeling approaches being suitable for each level of description [20–23].

Molecular Dynamics Simulation of Lipid Membranes

Atomistic molecular dynamics (MD) simulation, which solves explicitly Newton's equations of motion, is now a very advanced and established technique being able to simulate the behavior of phospholipid membranes up to several hundred nanoseconds [7,24]. Important collective phenomena, such as thickness fluctuations and bilayer undulations, correlative lipid motion during diffusion, heat capacity changes due to increasing salt concentration, or cholesterol's impact on the lateral pressure profile in lipid membranes have been studied by this approach [24–27]. Recently, cholesterol translocation has been studied by atomistic MD simulation using the replica exchange approach [28]. In this method, multiple replicas of the membrane are simulated in parallel at various temperatures to sample rare events, such as cholesterol flip-flop [28,29]. By this approach, the overall simulation time, including the advancement of the simulation clock for each detected rare event, could be extended tremendously to 15 µs altogether [28]. Local interaction of membrane proteins and lipids as well as the impact of fluorescent probes on bilayer properties has also been investigated by atomistic MD simulations [30–34]. MD simulations were combined with ab-initio calculation and spectroscopy experiments to determine dynamics of fluorescent dyes in ground and excited states [35]. Finally, using a specially designed supercomputer [36], classical atomistic MD simulations of protein-lipid systems were extended into the µsec-range and used to study voltage gating of a membrane embedded ion channel, to decipher conformational changes of the epidermal growth factor receptor on ligand binding and to characterize specific lipid interactions of receptor subunits [37,38].

Together, classical atomistic MD simulations provide information about many relevant membrane processes in the time window of several fs to <~0.5 µs for a few hundred lipid molecules. Using improved sampling schemes, such as replica exchange MD or supercomputer architectures optimized for MD simulations, the simulation time can be extended by a factor of ten. However, even with further improvement of computer instrumentation, it can be expected that other highly relevant but slower phenomena like lipid domain formation or phospholipid flip-flop in membranes taking place in the range of milliseconds to seconds and/or involving larger membrane areas cannot be studied solely by the MD technique. This is a disadvantage of the inherent time dependence of the method, which requires time propagation steps of typically 1–2 fs to follow the fastest processes, such as intramolecular vibrations. As a consequence, membrane phenomena occurring on a large time scale in nature will require lengths of MD simulation runs not accessible by today's conventional computational power. Supercomputers, such as "Anton", developed by David E. Shaw's group, might provide a way to extend MD further towards the relevant mesoscopic scheme [36]. However, it is of large interest to develop alternative computational approaches reaching thermodynamic equilibrium of slow membrane processes within reasonable calculation time on conventional computer systems. One way is the coarse graining of the molecular lipid structure and force field, thereby allowing for efficient simulation of slow collective membrane phenomena, such as membrane domain formation, vesicle fusion, membrane crowding, or flip-flop of phospholipids and transbilayer coupling [39–41]. These advantages come to the price of neglecting molecular detail, such that fine differences in lipid structure, for example, between cholesterol and its various biosynthetic precursors [42], cannot be studied adequately by coarse-grained (CG) simulation. This is a rapidly growing field, and a detailed discussion of these developments is beyond the scope of this article. We will, though, briefly touch on this topic when we discuss the potential of local move Monte Carlo (MC) simulations for multiscale and CG modeling of lipid membranes in Section 5 of this article.

Monte Carlo Methods for Atomistic Simulation of Lipids and Other Biomolecules

As alternative to MD simulations of atomistic or CG molecular models, one can circumvent the explicit time dependence of the investigated processes and sample molecular conformations along a Markov chain in the configuration space based on their probability in a Boltzmann distribution [43]. This approach, which is at the core of the Metropolis MC method, samples different configurations from an ensemble of structures and should—According to the Ergodic theorem—Provide the same result as MD sampling of the time-evolution of a single structure [44]. Let X^N be the configuration vector comprising positional and rotational coordinates of an N-molecule system, i.e., $X^N = \{X_1, X_2, \ldots X_N\}$, average thermodynamic properties, such as the internal energy, U, of the system are expressed in a statistical sense as average according to:

$$\mathbf{U} = \int \ldots \int \mathbf{E}(X^N) \cdot \mathbf{P}(X^N) \mathbf{d}X^N = \left\langle \mathbf{E}(X^N) \right\rangle \tag{1}$$

Here, $E(X^N)$ is the configuration energy, while the probability of finding the system in state X^N is given by

$$\mathbf{P}(X^N) = \frac{\exp(-\boldsymbol{\beta} \cdot \mathbf{E}(X^N))}{\int \ldots \int \exp(-\boldsymbol{\beta} \cdot \mathbf{E}(X^N))\, \mathbf{d}X^N} \tag{2}$$

This high-dimensional configuration integral and thereby the thermodynamic average properties, such as the internal energy, cannot be evaluated by numerical integration. However, $P(X^N)$ can be evaluated by generating an irreducible Markov chain, of which the limiting distribution is the Boltzmann distribution:

$$\mathbf{P}(X^{N}) = \left\{ \mathbf{P}(X_1^{N}), \mathbf{P}(X_2^{N}), \ldots \right\} \quad (3)$$

Here, P(XNi) is the Boltzmann probability of the ith configuration of the system; for a detailed discussion of the underlying statistical theory, see ref. [43–45]. The standard Metropolis MC method has been first applied to simulation of water and aqueous solutions of ions [46,47]. It was limited by slow convergence, i.e., reaching the thermodynamic equilibrium of the system given by the limiting distribution of the Markov chain (see above) in reasonable computational time. This problem hampered the application of the method to larger molecular systems.

Early attempts to improve sampling efficiency modified the underlying transition matrix, thereby biasing the moves in a certain direction. For example, smart MC developed by Ross et al. included forces into the MC sampling procedure, thereby increasing the likelihood of moves in the direction of forces acting on the simple molecule models [48]. A similar modification of the Markov chain has been introduced by Pangali et al. in their force-biased MC algorithm [49]. Computationally faster equilibration of solvated rigid molecules has been also achieved in such early studies using a preferred sampling method, where solvent molecules close to the solute surface have been moved with higher frequency [50]. Not only intermolecular forces but also the virial in the NPT-ensemble can be included in the acceptance criterion to increase the sampling efficiency [51]. In addition, the step size of the attempted moves can be adjusted according to the total interaction energy of the particle [52]. Another way of accelerating equilibration of the system is to shift to the grand canonical ensemble (GCE) and to insert/delete molecules as part of the sampling procedure. Mezei and co-workers developed cavity-biased GCE simulations in which MC moves are only attempted into intermolecular cavities of suitable size [53]. In addition, other methods have been developed, and this short overview is not comprehensive, but just serves the aim of making the reader familiar with the major challenges in applying MC methods to molecular simulations. For a more general introduction into various MC methods and enhanced sampling techniques, the interested reader might consult the text book by Allen and Tildesley [44] or two recent reviews [54,55]. Similarly, the fruitful application of the MC method to two-dimensional lattice and off-lattice simulations of simple membrane models are covered elsewhere [22].

While being a problem in simulating particles in dense fluids, the low acceptance rate of standard Markov chain MC methods has been even a larger hurdle in MC simulations of non-rigid molecules and macromolecular assemblies, such as proteins or lipid bilayers. The reason for this is that intramolecular moves often result in steric clashes associated with high energy and thereby rejection of the attempted move. For example, even a modest change of a torsion angle in the middle of a molecular chain can lead to large movement of the atoms at the end of the chain causing, for example, clashes between neighboring fatty acyl chains of phospholipids in the membrane [56–58]. Several strategies have been developed to overcome this problem and to improve sampling efficiency. Reptation moves were proposed and applied, among others, to long chain n-alkanes and polymethylenes [59,60]. Here, an atom is removed at the end of one chain, while another atom is simultaneously added at the end of an adjacent chain. End rotation is the simple change of the last torsion angle in a chain altering the position of the last atom, only [61,62]. Another straightforward local MC move set is the flip move, in which an inner (united) atom of the chain is picked at random and rotated around the adjacent bond by an angle sampled from a uniform distribution in the range of 20°–30° [63]. This changes a total of four torsion and two bond angles. These three sampling methods replace only one atom per MC move, which is in contrast to MC moves, involving the geometric reconstruction of a group of atoms. One approach of the latter category is the so-called configurational biased MC (CBMC). In this method, an end segment of a chain molecule is removed and the missing part is rebuilt sequentially with a bias towards avoiding overlaps with the unmoved neighbor sites. Thus, CBMC introduce biases in the sampling procedure which sample certain torsion moves preferentially based on additional information about the studied molecular system [58,64]. CBMC has been combined with intervening MD simulations to enhance equilibration of a bilayer of 100 molecules of dipalmitoylphosphatidylcholine (DPPC) [57,65]. The cavity-biased GCE method has been also employed in conjunction with CBMC for simulation of a membrane patch of dimyristoylphosphatidylcholine (DMPC) resulting in improved sampling compared to MD runs of comparable length [66]. Permeation of small molecules across DMPC bilayers, as well as the effect of cholesterol on membrane permeability, has been studied by atomistic MC simulation in the torsion angle space with fixed bond length and bond angles [67–69]. Intramolecular rotations as well as whole lipid rotations were performed in an extension-biased scheme [67].

However, these approaches suffer from rather slow convergence, as small changes in torsion angles can be associated with large changes in configurational energy due to van der Waals contacts of distant atoms and steric clashes [70]. Much effort was, therefore, devoted to the development of local move sets with constant bond lengths and bond angle constraints for more efficient equilibrium ensemble simulations. The theoretical background of such work is provided by the classical paper of Go and Scheraga (1970) [71], where a general formalism for ring and chain closure in the space of dihedral angles (the "torsion angle space") was developed. The basic idea of this type of local or "window" moves is to change first one torsion angle (the driver torsion) followed by adjustment of the next six dihedral angles in the chain in a way that keeps the remaining chain (i.e., from torsion angle □ = 7 on) unchanged. This closure method is inspired by the inverse kinematic problem in robotics, where a robotic arm joint by several links must be placed in such a manner that a desired position of an end-effector is provided [72]. To ensure chain connectivity, the closure problem in torsion angle space is solved by sequential application of rotation matrices under the constraint of recreating the correct distance between the last atom of the moved fragment and the second fixed fragment of the chain [71,73–75]. The approach introduced by Go and Scheraga (1970) has been adapted to the use in MC simulations of chain molecules by the concerted rotation (CR) method, which ensured detailed balance by taking the different volume elements of the coordinate systems into account [76]. A number of special CR variants were developed and successfully tested in MC simulations of polypeptide structures [70,74,75,77,78] and lipid bilayers [79], and the CR algorithm was implemented as an option of the MC module for the CHARMM program [80]. Another solution to the local MC move problem was presented by Pant and Theodorou (1995) with their "end bridging" (EB) and "internal re-bridging" (IRB) moves [81]. As in the original CR method [76], the geometric problem of chain closure in torsion angle space was formulated as determining the coordinates of a moved trimer relative to two specially confined dimers [81]. In the EB method, the fixed dimers belong to two different chains, while in the IRB method the dimers belong to the same molecular chain. Thereby, the IRB method is equivalent to the CR move set, just with two driver torsions, one for each of the fixed segments. In both bridging moves, a set of nine equations is numerically solved, which links the Cartesian coordinates of the atoms of the bridging trimer to the constrained bond lengths. Here, bond lengths are defined between the connected adjacent atoms and as the distance between the second neighbors, essentially fixing the bond angles in the molecule [81]. The EB method alters the chain length distribution and is therefore suitable to simulate polydisperse polymer melts [81,82]. In these systems, proper weighting and simulation in the semigrand ensemble using EB, reptation, and end-rotation moves guarantee the correct polydispersity, i.e., chain length distribution for the polymer model [81]. The EB method becomes more accurate for long chains, but cannot be used for lipid membranes, which is for the following reasons. First, EB moves alter the chain lengths, which is inappropriate for phospholipid bilayers, and, second, the performance of the EB moving scheme drops dramatically if the chains are ordered, as it happens to be in lipid membranes [62]. The bridging scheme of the original study [81] has been later extended for monodisperse systems using the double bridging moves [62]. It was also extended to internal segments of arbitrary lengths and combined with CBMC moves to systematically re-growth the internally moved segment [83,84].

Importantly, the torsion angle changes during a CR move are correlated such that the volume element in the Boltzmann factor $\exp(-E/kT)d\square_1 \ldots d\square_6$ is not volume preserving. This necessitates inclusion of a Jacobian, J, in the Metropolis acceptance criterion as first shown by Dodd et al. [76]. The Jacobian accounts for the different volume elements in the Cartesian space versus the space of torsion angles and constraint variables. More generally, using any definition of generalized coordinates for describing the configuration space of a molecular system requires a transformation of the volume element in the configuration integral of the partition function. A subset of the new coordinates is usually considered as constraints (like bond lengths and angles in the CR procedure). Thus, the transformation of the volume element from the canonical Cartesian space into a new set of transformed coordinates reads:

$$d\Gamma = \prod_{i=1,N} dx_i dy_i dz_i = J \prod_{i=1,M} dq_i \prod_{j=1,3N-M} dc_j \quad (4)$$

where $\{q_i\}$ is the subset of independent generalized coordinates, and $\{c_i\}$ are the constrained variables. The Jacobian J is the functional determinant of the transformation and is typically calculated as the inverse:

$$J^{-1} = \left| \frac{\partial(q_1,\cdots,q_M,c_1,\cdots,c_{3N-M})}{\partial(\mathbf{r}_1,\cdots,\mathbf{r}_N)} \right| \quad (5)$$

Importance sampling, according to the Metropolis method, basically evaluates the integral over configurations of a subset of the configurational space in Cartesian coordinates. Therefore, the Jacobian has to be included in the Metropolis acceptance criterion to account for the coordinate transformation from cartesian to internal coordinates in the closure procedure. The Jacobian enters the acceptance criterion for the trial move m → n according to:

$$p^{acc}_{m\to n} = \min\left\{1, \frac{p^{sel}_{n\to m}\cdot e^{-E/kT}\cdot J(n)}{p^{sel}_{m\to n}\cdot e^{-E/kT}\cdot J(m)}\right\} \quad (6)$$

Here, pseln→m and paccm→n are the probabilities for selecting and accepting the move, respectively [56]. Omitting the Jacobians from the acceptance criterion results in wrong torsion angle distributions since it violates the criterion of microscopic reversibility of the moves. The latter is required to ensure detailed balance and thereby to sample the Boltzmann distribution correctly. All CR variants in the torsion angle space have in common that complex non-linear chain closure equations have to be solved, with at most 16 solutions in the general case. Hence, computationally demanding numerical procedures are involved that do not readily provide the full set of solutions. To overcome this problem, Mezei (2003) devised a method to choose that solution of the closure problem, which is closest to the original conformation before attempting the move. Making this choice eliminated all moves which are unlikely to proceed [79]. To maintain microscopic reversibility, it was tested whether the reverse rotation applied leads again to the original conformation, which was called the "reverse proximity criterion" [79].

Like the other CR methods, this approach is limited to moves in the torsion angle space, since bond lengths and angles are kept constant in the simulation. Fixing bond angles, however, is not in accordance with MD simulations and results in a rough, artificial energy landscape. For example fixing bond angles increases the energy barrier for a cis–trans conversion of alkanes significantly making this constraint less suitable for membrane simulations [78]. The problems encountered by using fixed bond angles in protein systems were first addressed by Bruccoleri and Karplus (1985) who modified the original CR method by allowing for limited bond angle variations [85]. Definition of a modified (φ,ψ)-torsion potential to compensate for otherwise increased torsional barriers also allowed for partially removing the problems caused by fixed bond angles [70]. These problems were finally solved by extending the CR approach to include flexible bond angles in the closure algorithm [78]. In the latter method called concerted rotations with angles (CRA), the mathematical formalism for solving the kinematic problem in torsion angle space was modified by replacing three torsions around dihedral angles with changing bond angles. The overall derivation of the closure equations is very similar to the classical CR method including the associated Jacobians. The pre-rotation, however, was additionally restricted by a Gaussian bias to ensure small structural changes likely being accepted during each move [78]. The algebraic expressions become simpler in the CRA method with only two branches in the polynomial equation for solving the closure problem compared to four branches in the original CR method. It was shown that the CRA local moving scheme results in faster equilibration and more soft energy landscape [78,86,87]. In addition to sequential rotations and rebridging, the chain closure problem has been formulated as a fixed-end move for polypeptide chains, in which a crankshaft rotation is performed between distant alpha carbons [88]. This local move type does also alter the bond angles adjacent to the respective alpha carbons and was found to be more efficient than the original CR method. Related crankshaft-type local move sets, in concert with extensive use of side chain structure libraries, have been repeatedly applied to conformational sampling of flexible loops in proteins, but to review this rapidly growing field in detail is beyond the scope of this article [89–91]. Very recently, the geometric problem of CR has been rephrased in such a way that an exact analytical solution for the chain closure could be derived instead of the tedious numerical procedure [92]. Having this analytical solution, the authors could express the coupling between pre- and postrotation as linear transformation allowing for construction of a probability distribution controlling all necessary degrees of freedom (DOF). More precisely, the interdependencies between the various DOF during the local move could be expressed as correlations in a multivariate Gaussian distribution. This information could be used during the move to avoid imbalances in angular variations of pre- and postrotations, a problem from which even the CRA method suffered [78,92]. The so-called "concerted rotations involving self-consistent proposals" (CRISP) move was combined with an implicit solvent description and was found to sample the conformational ensemble of ubiquitin and other proteins more efficiently than the CRA move set [92]. The CRISP move along with CRA, crankshaft, pivot, and other MC move sets has been implemented in a program suite called "Phaistos" for rapid conformational sampling of proteins in implicit solvents [93].

We have recently presented a very different local-move set MC method, the chain breakage/closure (CBC) algorithm, which solves closure equations in the bond/torsion angle space using the constant bond length approximation (CBLA) [94]. The CBC equations provide two simple analytical solutions to the closure problem. Jacobi factors were analytically derived for chain molecules as well as for the ribose ring of DNA molecules and included in the Metropolis acceptance criterion. The method was found to be computationally very efficient and highly suitable to simulate sequence-specific DNA structure and drug docking to DNA [95,96]. It set the stage for further important developments and application in the field of modeling nucleic acids [97,98]. To take advantage of this approach in MC simulations of phospholipid membranes, local move sets within the CBC framework were adapted and defined for describing structural variations of DPPC molecules. In the following, we describe how this can be used to simulate single DPPC molecules and DPPC membranes in an implicit solvent description. We give a detailed explanation of implementation of local MC moves within the CBC framework into lipid and membrane simulations. This will make the reader familiar with the technical details and the challenges of this type of molecular simulations. In particular, we describe the sequential moving of individual units in each DPPC molecule, the set up and details of the bilayer simulation and the analysis of the data. We demonstrate that local moves in the bond/torsion angle space can provide fast structural equilibration of DPPC molecules in an implicit solvent, and we describe several measures for assessing this equilibration. We also show the rapid, in terms of computation time, transition of a crystalline-like DPPC bilayer into the fluid state, and discuss the limitations of the used solvent description. We do not claim, however, that the presented data for the membrane simulation resemble comprehensive sampling of the conformational space of all the lipids in the bilayer, as this would require much more extensive simulations and a more elaborated implicit solvent model, both, clearly beyond the scope of this review article.

SIMULATION DETAILS AND ANALYSIS OF DPPC STRUCTURE AS MONOMER AND IN THE BILAYER

Molecular Model and Assignment of Moved and Dependent Variables

For some aspects of the lipid molecule, we used a united atom representation of DPPC recently developed for the AMBER 4.1. Force field [99]. This DPPC parametrization combines methyl groups of the fatty acyl chains and methylen groups at the nitrogen and in the fatty acid tails, respectively, into one united atom, thereby reducing the total number of atoms from 120 to 50 per molecule (Figure 1). For the fatty acyl chains, the Ryckaert-Bellemann potential was applied as described [100,101]. A branched main chain of atoms in the DPPC molecule was defined leaving the united CH_3 atoms (atoms # 44–46) on the head group nitrogen (atom # 1), the oxygen atoms (atoms # 47,48) on the headgroup phosphate (atom # 5), and the carbonyl atoms (atoms # 49,50) as external atoms (light grey in Figure 1, see preceding paragraph). From the 3N DOF, where N = 50 (number of (united) atoms) in the molecule, we have, due to CBLA and introduction of additional constraints, a reduction from 150 to 88 DOF. Figure 1 shows the local move sets defined for the variation of DPPC structures. The chain breaking fragments of the CBC algorithm are reduced to single atoms (red circles). After Cartesian moves of such atoms, the chemically bound neighbor atoms are repositioned by using the closure equations of CBC in the frame of CBLA. Crankshaft rotations of these atoms define additional DOF. Chemical moieties at chain ends or attached to the chain are moved by using classical internal coordinates (bond and torsion angles). All moves, including volume moves, were performed in a sequential manner including sequential updating of the energy. This has been shown to fulfil the balance condition and thereby to sample correctly from the Boltzmann distribution [102].

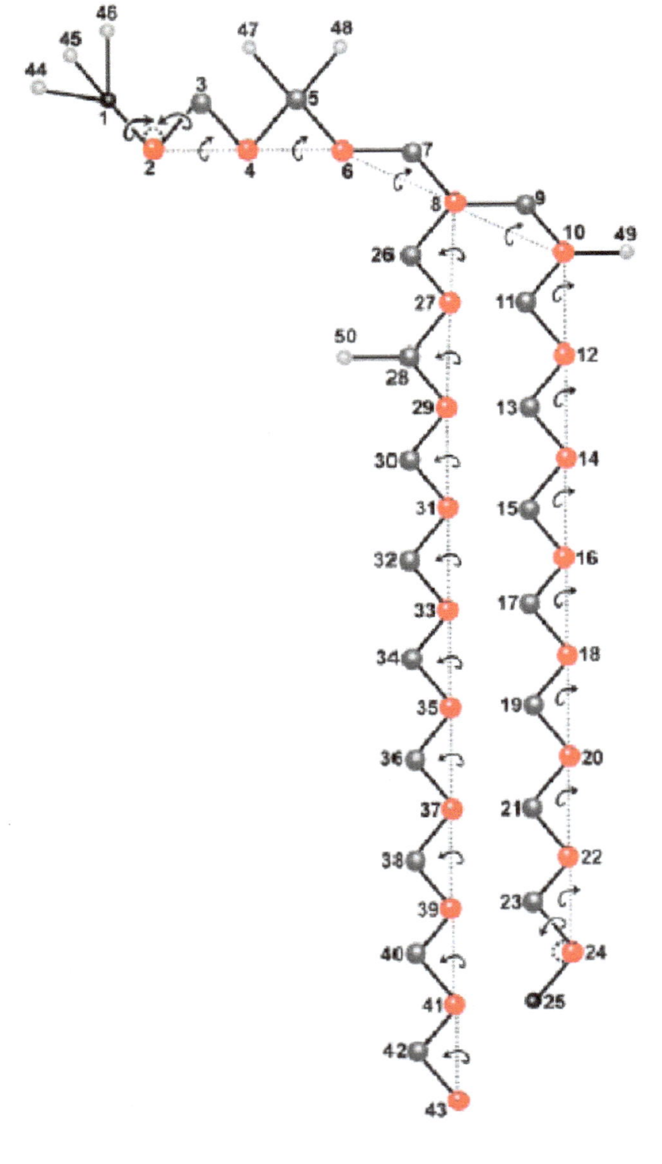

Figure 1. Molecular model of DPPC for atomistic Monte Carlo simulation. A united atom representation of dipalmitoylphosphatidylcholine (DPPC), which neglects hydrogen atoms, is employed reducing the number of atoms per DPPC molecule to 50 [81].

Start Configuration, Boundary Conditions, and Solvent Treatment

A DPPC molecule was built using in-house modeling software. Values for bond angles, torsion angles, and bond lengths were taken from their equilibrium values defined in [99]. These values yielded straight fatty acyl chains of DPPC and a molecular conformation close to but not identical to the crystal structure previously defined for the closely related DMPC (PDB ID XP4; Figure 1) [103]. For one-lipid simulations, one MC cycle comprised changing all 88 DOF. Efficiency of energy calculations was guaranteed by calculating energy differences for each moved fragment with respect to all other atoms in the same lipid prior and after the local MC move. For the bilayer, the energy calculation included additionally the non-bonded interactions to other lipids (see below). The DPPC bilayer is approximately described by a two-dimensional periodic model with a cubic elementary cell, and the simulations were done under the condition of constant pressure of 1 atmosphere. This means that, in addition to structural moves, random variations of the volume have to be included, and the energetic contribution of the volume change adds to the structural intra- and intermolecular energy. The resulting enthalpy decides on the acceptance of the move by the Metropolis algorithm. We started with the crystalline-like conformation for 64 DPPC molecules (compare Figure 2A, "start" and Figure 5A). Cut-offs used in calculation of pair-wise (atom-atom) energies were defined by the minimum image convention [44]. For 6–12-Lennard Jones and electrostatic interactions a cut-off of 10 Å and a shell with a non-bonded list of 10 Å was generated. This list was updated every 10 cycles. That is, interaction of groups at a distance between 0 and 10 Å were calculated every cycle, while those of groups at a distance between 10 and 20 Å were stored in the pair list, similar as described previously [44,101]. Volume moves were attempted in every MC cycle for the bilayer by isotropic adjustment of the box length. For that purpose, the center-of-mass distance between all DPPC molecules was adjusted in each MC cycle. To account for effects of the solvent, an implicit description of dielectric properties of the water phase close to the bilayer/solvent interface was chosen. Thus, electrostatics was treated by employing a sigmoidal damping function for the dielectric constant □(r) in the Coulomb term of the AMBER 4.1. force field function [104]. This is based on a function originally proposed by Hingerty et al. [105], and refitted by Lavery et al. [106], for large distances from the interface:

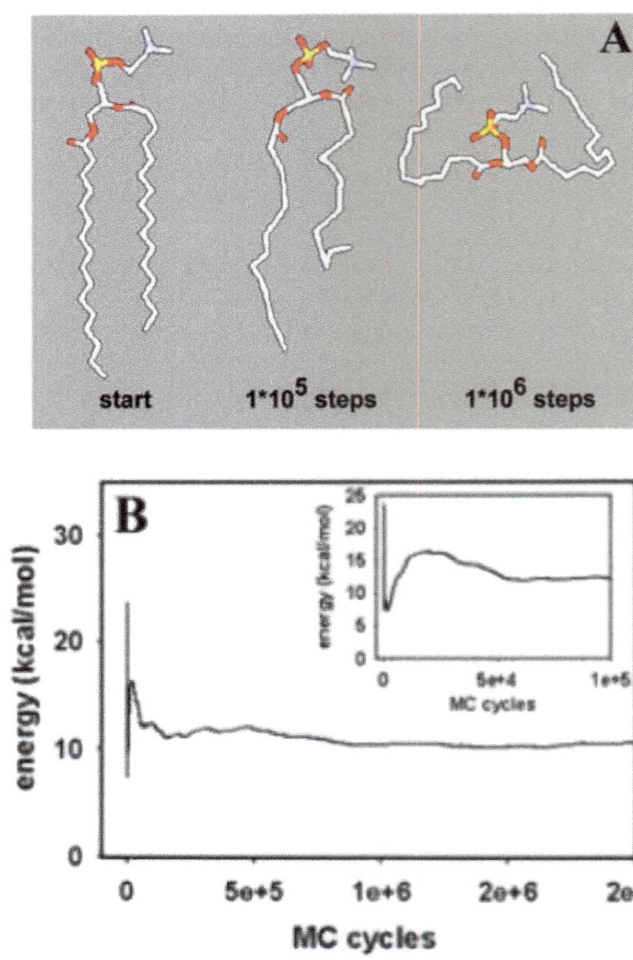

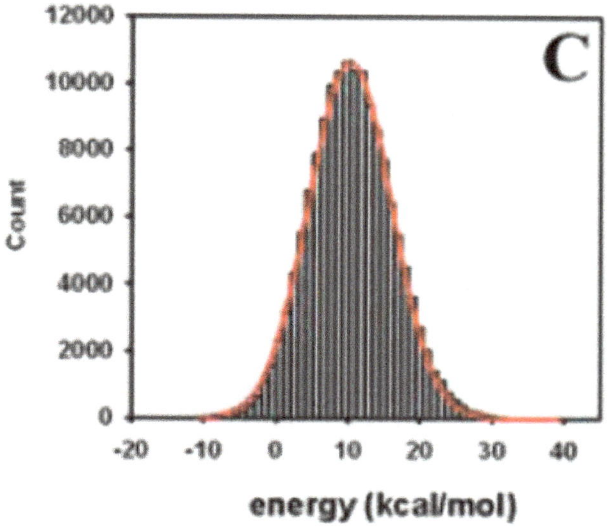

Figure 2. Snapshot and energy of a single-molecule Monte Carlo simulation. (**A**) starting conformation (start) and representative snapshots of a single molecule MC simulation are shown after $1\cdot\times 10^5$ and $1\cdot\times 10^6$ MC steps. Large conformational moves of the fatty acyl chains and the lipid head group can be observed; (**B**) mean system energy equivalent to the conformational energy calculated from the AMBER force field during the simulation [85]. The conformational energy becomes stable already after about 750,000 MC steps. The inset shows the initial phase of energy equilibration with a maximum after about 20,000 MC steps; (**C**) histogram of the conformational energy of the last one million MC steps (grey bars) overlaid with a fit to a Gaussian function of the form $f(E) = A\cdot\exp\left(\left(E_i - \langle E\rangle\right)/\sqrt{2}\cdot\sigma\right)^2\right)$ providing the mean energy at equilibrium, E, and the standard deviation, σ, as a measure of fluctuations around the mean value (red line). See text for further explanation.

$$\varepsilon(r) = d - \frac{d-1}{2}\cdot\left((rs)^2 + 2rs + 2\right)\cdot e^{-rs} \tag{7}$$

where d is the plateau value and s the slope for the dielectric permeability. This function models the change of dielectric permittivity due to solvation of the phospholipid molecules. It, thereby, accounts for the overall effect of solvent (water) reorientation close to the solute surface. Simulations were run on an AMD PC with an Athlon XP 2700+ processor having 2.2 GHz clock frequency. On this machine, 10^5 steps of MC simulation of a DPPC bilayer consisting of 64 molecules using the implicit solvent model required ten days of simulation time.

Evaluation of DPPC Conformations and Bilayer Structure

All trajectory analysis was performed using in-house developed Fortran77-based software. Systems energy and entropy were calculated from the trajectories of obtained conformations of the DPPC molecule. A histogram of molecule energies was generated for the one-lipid and bilayer simulation and was fitted to a simple Gaussian model using SigmaPlot 4.0 (SPSS Inc., Chicago, IL, USA). To calculate the entropy associated with the conformational ensemble of the single DPPC molecule, the following formula was used [107]:

$$S = k_b\cdot\ln\left(\mathbf{1} + \frac{k_b\cdot T\cdot e^2}{\hbar}\cdot \mathbf{M}\odot\mathbf{C}\right) \tag{8}$$

Here, $\mathbf{1}$ is the unity matrix, $\mathbf{M}$ the mass matrix containing atomic masses in the diagonal, k_b is the Boltzmann constant, e is Euler's number, and $\mathbf{C}$ is the symmetrical covariance matrix whose elements σ_{ij} are given by:

$$\sigma_{ij} = \left\langle (x_i - \langle x_i\rangle)\cdot(x_j - \langle x_j\rangle)\right\rangle \tag{9}$$

The respective mean value in angular brackets is the average of all positions of an atom in Cartesian space calculated for a given number of MC steps. For example, the covariance matrix after 20,000 MC steps provides the variances and covariances of 20,000 conformations around the respective mean value separately calculated for each x-, y-, z-coordinate of the atoms in the DPPC molecule. Non-vanishing off-diagonal elements (covariances) indicate correlations between structural fluctuations of adjacent atom coordinates. The atomic Cartesian coordinates are given by x_1, ... x_{3N}, and the matrix $\mathbf{C}$ has the dimension 150 × 150 for a single DPPC molecule. Values for the matrix $\mathbf{C}$ were calculated according to Equation (9), exported as text-file and plotted using the open-source image analysis software ImageJ (developed at the U.S. National Institutes of Health and available on the Internet at http://rsb.info.nih.gov/ij) in tagged image file (TIFF) format. The conformational entropy was calculated according to Equation (8), above, as function of MC steps (see Results section, Figure 3). The basic assumption here is that fluctuations are normal-distributed around the mean allowing for estimating covariances from a multidimensional Gaussian distribution [107–109]. The electron density profile was calculated by placing a Gaussian distribution of the electrons on each atomic center with the variance being equal to the van der Waals radius for each configuration and averaged over all configurations [110]. The P-N-vector was calculated as the angle between the P–N bond in the lipid head groups with the bilayer normal, and the distribution was calculated for all molecules in the bilayer and averaged over all configurations [99,111]. Head group torsions for the single lipid and the DPPC membrane were calculated using the same notation as described [99,103]. To assess the lateral diffusion of DPPC in the bilayer plane, the mean square displacement (MSD) was calculated according to:

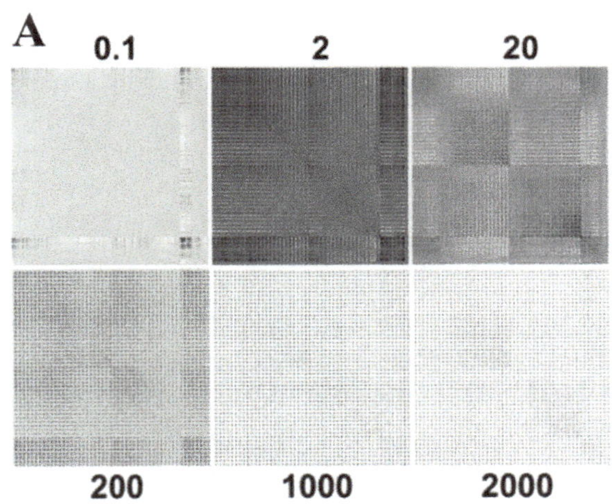

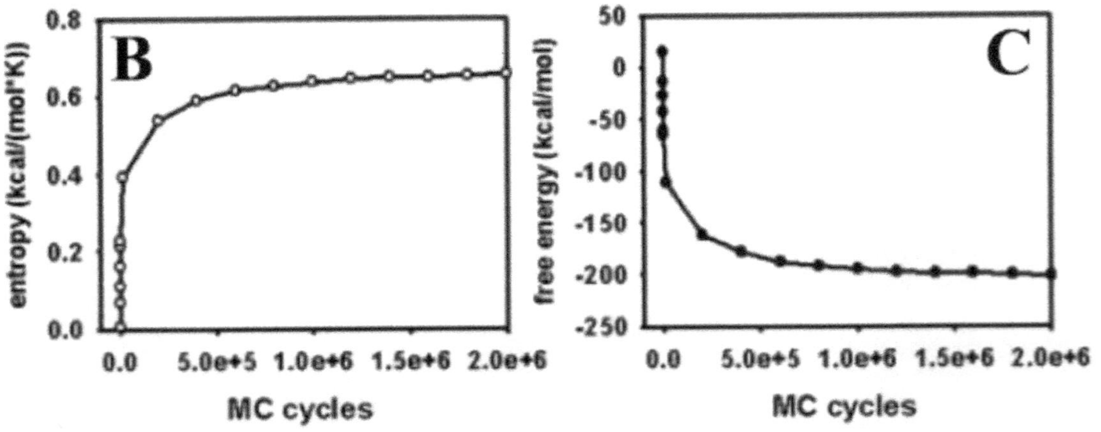

Figure 3. Covariance, conformational entropy and free energy of a single DPPC molecule. (**A**) from trajectories of single molecule simulations the covariance matrix of atomic positions was calculated and plotted as a 150 × 150 matrix for 100 (=0.1), 2,000 ("2"), 20,000 ("20"), 200,000 ("200"), one million ("1000"), and two million ("2000") MC steps, respectively. Dark and light spots indicate high and low values of (co-)variances, respectively. The matrix values do not change grossly after 1 million MC steps with low off-diagonal values (i.e., covariances). This indicates equilibration of the structural sampling and absence of significant correlations of fluctuations of adjacent atoms in the DPPC molecule; (**B**) the conformational entropy was calculated from the mass-weighted covariance matrix after a given number of MC steps, as described in the text; (**C**) the conformational free energy was calculated as F = E − T·S with E and S being the conformational (internal) energy and entropy, respectively. The temperature of the simulation was 323 K (i.e., 50 °C).

$$MSD = \frac{1}{N}\sum_{i=1}^{N}\sum_{s_0<S-s}\left|\vec{r}(s_0)-\vec{r}(s_0+s)\right|^2 \quad (10)$$

Here, N is the total number of lipids, $\vec{r}$ is the position vector of the center of mass of each lipid and s is the running index for the MC steps. An equivalent to a lateral diffusion coefficient, D, was estimated from the slope of the averaged MSD after an initial phase of box size adjustment in the constant NPT-ensemble simulation (see Results section) [27].

RESULTS AND DISCUSSION

Efficient Monte Carlo Sampling of Conformations of a Single Lipid Molecule

We started our simulations with a single DPPC molecule, as it allows us to assess structural equilibration of the basic unit of the membrane and to perform MC sampling of the whole conformational space. This is important, as thermodynamic properties calculated from bilayer simulations always average over individual lipids, which do not necessarily explore their configurational space individually. Thus, in order to assess equilibrium properties of the used molecular DPPC model, we performed first simulations of single DPPC molecules. This allows us to determine the evolution of structural correlations during MC sampling and thereby to assess the effect of the molecular constraints used in the description of DPPC (e.g., the CBLA and the move sets of external and closure atoms; see Figure 1 and Section 2.1., above). In addition, the properties of an equilibrated single DPPC structure can be directly compared with the same properties of DPPC in the bilayer assembly (see below and Figure 4 vs. Figure 6). Thereby, we can gather information about the degree of equilibration of the membrane during our simulation as well as about eventual structural confinement of DPPC in the bilayer. Both aspects are discussed below. Large structural moves in relatively short simulation times on a normal desktop computer were made possible by using the CBC moving scheme (Figure 2A). The conformational energy dropped, followed by a rise towards a local maximum before a stable value was reached after about 50,000 MC steps (Figure 2B). It can be concluded that the potential energy, E, for a single DPPC molecule is in the range of its equilibrium value after about 100,000 MC steps, while complete equilibration required about 750,000 steps. Fluctuations of E around the mean after obtaining a stable plateau value can be well approximated by a normal distribution (Figure 2C). This is a first indication that a thermodynamic equilibrium has been reached, since the probability density function (PDF) of the potential energy, $p_k(E)$, should be well approximated by a Gaussian function at equilibrium [112]. The PDF, $p_k(E)$, is the product of the density of states, n(E), a rapidly increasing function, and the Boltzmann weight factor (an exponentially decreasing function, as defined in Equations (2) and (3)) providing the bell-shaped energy distribution with maximum at the average conformational energy of the molecule at the given temperature [89,112]. From the non-linear regression one obtains an average conformational energy of 10.26 kcal/mol associated with a standard deviation as a measure for fluctuation of the conformational energy around the mean of SD = 5.71 kcal/mol. By squaring that value and multiplying with $\beta = 1/(k_b \cdot T)$ with T = 323 K, we can determine the molar heat capacity of a single DPPC molecule in the canonical ensemble to c_V = 9.43 cal/mol·K [112]. For the united atom representation with 150 DOF and assuming for the sake of simplicity that the heat capacity ratio, γ, can be calculated like for an ideal gas (i.e., γ = 1 + 2/DOF = 1.013~1), we would have approximately the same value for the heat capacity at constant pressure, c_p. As we use the CBLA and thereby ignore stretching vibrations of the bonds, these values are presumably smaller than those obtainable by quantum mechanical calculations. However, it serves as starting point for comparison with the bilayer simulation (see below).

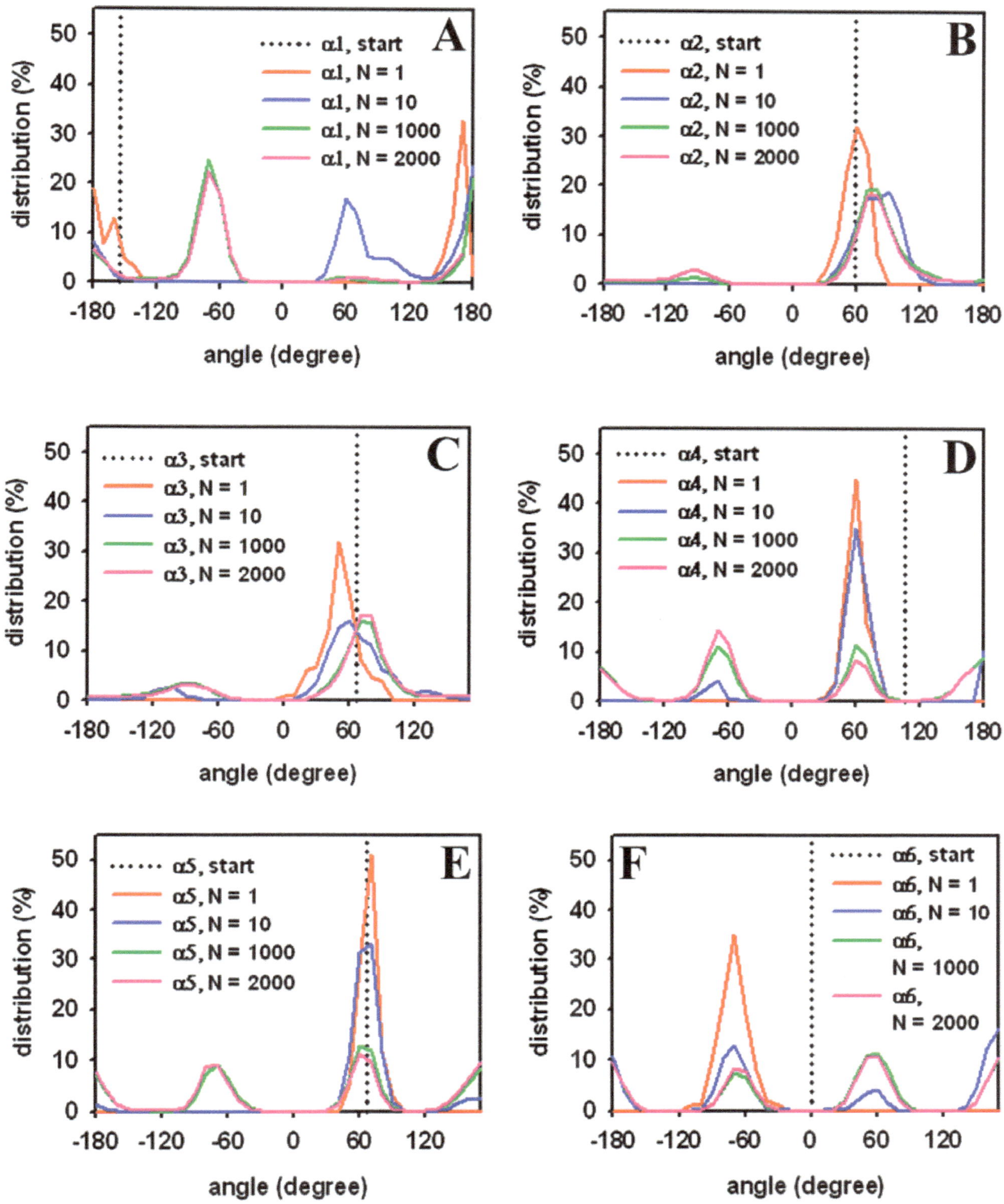

Figure 4. Distribution of DPPC head group dihedral angles from a single-molecule simulation. Head group torsions were defined as in Vanderkooi et al. [84] with the starting value defined by the initial conformation (dotted lines, "start"; compare Figure 2A). The percentage of occupation was calculated as function of torsion angle (in degree) after 1000 ("N = 1", red line), 10,000 ("N = 10", blue line), one million ("N = 1000", green line), and two million ("N = 2000", pink line) MC steps, respectively. The dihedral angles are defined as (**A**) torsions around atom 6 and 7 (α_1); (**B**) atom 5 and 6 (α_2); (**C**) atom 4 and 5 (α_3); (**D**) atom 3 and 4 (α_4); (**E**) atom 2 and 3 (α_5); and (**F**) atom 1 and 2 (α_6), respectively (compare Figure 2 for atom numbering in the DPPC molecule).

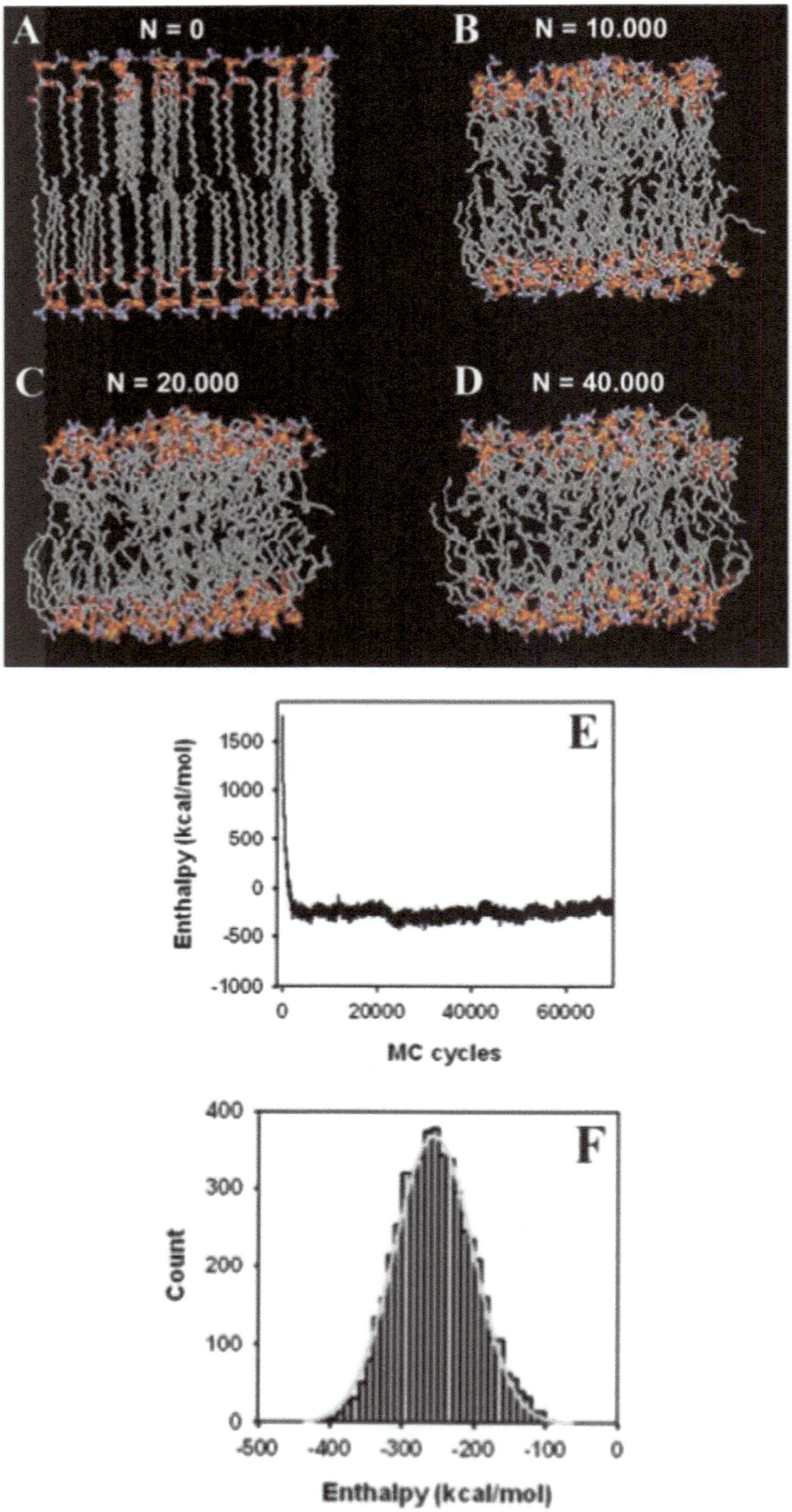

Figure 5. Simulation snapshot and system energy of a DPPC bilayer. The membrane simulation started from a crystalline bilayer consisting of 32 DPPC molecules with straight fatty acyl chains in each leaflet. Each molecule was rotated by a random rotation angle around the molecular long axis in the start configuration ((**A**), "N = 0"); (**B–D**) show snapshots after N = 10,000 (**B**); N =

20,000 (**C**); and N = 40,000 (**D**) MC steps, respectively. United atoms of methyl and methylen as well as carbon atoms are shown in grey, oxygen in red, nitrogen in blue, and phosphorus in yellow. Fatty acyl chains become increasingly disordered in course of the simulation; (**E**) the enthalpy of the bilayer in the implicit solvent was calculated after a given number of MC steps of the simulation performed in the constant NPT-ensemble. The temperature of the simulation was 323 K (i.e., 50 °C); (**F**) a histogram of the conformational energy of the last 40,000 MC steps (grey bars) overlaid with a fit to a Gaussian function of the form $f(E) = A \cdot \exp\left(\left(\left(E_i - \langle E \rangle\right)/\sqrt{2}\cdot\sigma\right)^2\right)$ provides the mean energy at equilibrium, E, and the standard deviation, σ, as a measure of fluctuations around the mean value (light grey line). See text for further explanation.

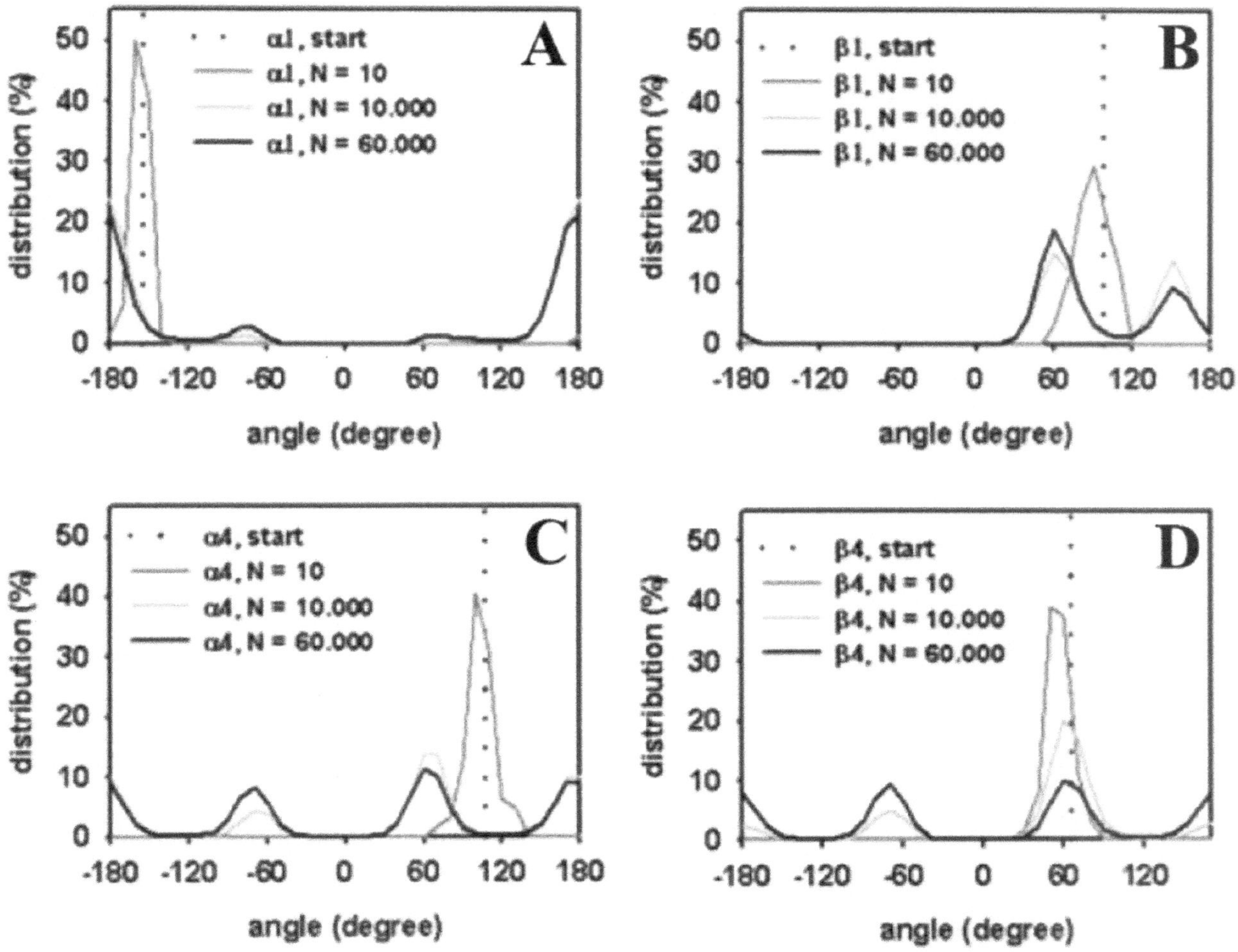

Figure 6. Head group torsions of DPPC in the membrane. Head group dihedral angles were defined as in Vanderkooi et al. [84], and as given in the legend to Figure 5. The starting value is defined by the initial conformation (dotted lines, "start"; compareFigure 2A). The percentage of occupation was calculated as function of torsion angle (in degree) after 10 ("N = 10", dark grey line), 10,000 ("N = 10,000", light grey line), 60,000 ("N = 60,000", black line) MC steps, respectively. (**A**) shows the torsion around the bond connecting atoms 6 and 7 (α_1); (**B**) around the bond connecting atoms 8 and 9 (β_1) (**C**) around the bond connecting atoms 3 and 4 (α_4) and (**D**) is the torsion angle distribution around the bond connecting atoms 11 and 12 (β_4).

It is possible that the energy landscape for the lipid molecule is quite flat, meaning that a large number of different conformations are associated with very similar total energy. To obtain an independent proof for successful sampling of a representative subset of the conformational ensemble by our MC method, the conformational entropy was calculated. According to the Boltzmann theorem, the entropy is a measure of the probability of the sampled conformational subensemble and should, therefore, tend towards a maximum after long simulations leading to a thermodynamic equilibrium [44]. The conformational entropy can be calculated from the covariances of atomic positions, as outlined in Section 2.3, and Equations (8) and (9), above [107,108]. A stable plateau of the entropy would be equivalent to a covariance matrix with more or less constant values. In Figure 3A, the covariance matrix is plotted for all 50 atoms of the single DPPC molecule after various MC steps. It is apparent from the figure that values of the matrix do not change after one million MC steps, indicating that atomic fluctuations are in equilibrium. Some correlations persist in the equilibrated structure, as indicated by non-vanishing covariances after one and two million MC steps. This is likely a consequence of two facts; first, we have performed the calculation with Cartesian coordinates, which will naturally result in correlations between the x-, y-,z-directions for a given atomic position [113], and second, in the chosen geometry of the DPPC molecule, some groups are always moved in concert. This applies to the groups being defined as external atoms, such as the methyl groups at the nitrogen in the phosphatidylcholine head group or the carbonyl groups together with the fatty acyl chains (light grey spheres in Figure 1) and to the closure atoms belonging to each atom moved directly in Cartesian space (dark grey and red spheres in Figure 1, respectively) In addition, the local move MC set defined by the CBC equations naturally introduces correlations in the moves, which is the reason for the local and efficient sampling but also for the necessity of including the Jacobian into the acceptance criterion (see Equations (4)–(6) in Section 1.3. and [91]). Stable values of the covariance matrix after prolonged simulation is, thus, an important condition for ensuring reliable sampling of many independent conformations by the CBC algorithm.

The Schlitter entropy calculated from the mass-weighted covariances reaches a plateau after about 1,000,000 MC steps; i.e., it needs longer simulations to equilibrate the conformational entropy than the energy, E (compare Figures 2B and 3B) [107]. From the entropy, S, and the conformational energy, E, of the molecule, the free energy, F, can be calculated according to $F = E - T \cdot S$, where T is the temperature at which the simulation was performed [44]. As shown in Figure 3C, the conformational free energy reaches a plateau value after about one million MC steps indicating that the system has reached a thermodynamic equilibrium. The simulation time required for stabilization of the conformational free energy of a DPPC molecule is largely determined by the term $T \cdot S$, i.e., by the time for equilibration of the molecular entropy. The simulation time for obtaining constant values of the conformational entropy could be best described ($\chi^2 = 0.9999$) by a tri-exponential function with fractional half-times of 385 MC steps for the first phase, 11.179 MC steps for the second phase, and 266.595 MC steps for the third phase, respectively (not shown). The first and third phase contributed, each, to about 1/3 of the decrease in entropy, while the second phase contributed half of the entropy drop. Thus, the conformational entropy of our DPPC model is equilibrated by the CBC move set to about 83% of the stable plateau value with a half-time of less than 12,000 MC steps. The rather slow equilibration of the remaining one third of the entropy (the third phase of the fit) is not found for the evolution of conformational energy of the molecule, which became stable already after about 750,000 MC steps (compare Figures 2B and 3B). This suggests that many different conformations of DPPC are associated with a similar energy. Please note, that the absolute value of the entropy calculated from the covariance matrix is just an estimate based on the assumption that the PDF of atomic positions is a multivariate Gaussian distribution. Thus, the reached plateau values of entropy and, consequently, free energy might be different in case that non-Gaussian PDF's underlie the fluctuation of atomic positions during the simulation. Together, calculation of conformational entropy and free energy, both reaching stable plateau values after about 1.5×10^6 MC steps, indicates that our atomistic MC algorithm efficiently samples an ensemble of lipid conformations, thereby, faithfully, representing the configurational integral [44,55].

Next, the distribution of head group torsions was analyzed from a trajectory of MC simulated DPPC. The start value is given as a dotted line similar to a delta function, and only the alpha-torsions following the definition of Vanderkooi are shown (Figure 4) [103]. Already, after 1000 MC steps, a broad distribution of torsions angles close to a local torsion energy minimum is found for the DPPC molecule (red lines). This indicates large structural moves in a small number of steps, reflecting the sampling efficiency of the algorithm. After 10,000 MC steps, the second minimum starts to be occupied (blue lines), while, after one million steps, the local minima are occupied to a similar extent (green lines). No further change in these values was observed after two million MC steps (violet lines), suggesting that, after one million MC steps, the sampled ensemble represents the equilibrium state for a single DPPC molecule. Analysis of head group torsions therefore supports that the one-molecule system rapidly reaches a thermodynamic equilibrium when sampled by our MC algorithm. Note that this simulation takes only about ten minutes on a Pentium PC.

Large Conformational Moves Mediate the Transition from a Crystalline to a Fluid Bilayer

Phospholipids are naturally associated in membrane assemblies, and it is therefore important to prove that the MC algorithm faithfully simulates lipid membranes as well. Having shown rapid equilibration of an isolated DPPC molecule does not prove that densely packed DPPC molecules in a bilayer arrangement can be efficiently simulated. Our starting structure is a crystalline-like DPPC bilayer with straight fatty acyl chains and identical conformations for the 64 lipid molecules that comprise our system (see Section 2.2 and Figure 5A). We wanted to determine whether our local-move MC technique is able to produce conformational moves large enough to create a structural transition from a crystalline to a fluid-like state in the bilayer. As shown in Figure 5B, already, after 10,000 MC steps, the high molecular order typical for the crystal-like structure is lost and acyl chains of individual PC molecules show large conformational differences. Individual molecules are tightly packed, which is due to the fact that the box size in the NPT-ensemble simulation is rapidly adjusted (see also the accompanying video sequence). After 20,000 MC cycles, fatty acyl chains became more disordered (Figure 5C), while after 40,000 MC steps, the acyl chains of the phospholipids show large structural variation compared to the starting configuration. These changes indicate that the bilayer system made a transition to a fluid lipid bilayer. The simulated membrane model system is driven in a state of slight undulations, as suggested by the wave-like appearance of the head group regions (Figure 5D). This mesoscopic organisation has previously been described for long MD simulations of fluid DPPC membranes [25]. However, due to the simple solvent representation used in our MC simulation, we cannot rule out that these phenomena are caused or at least influenced by the simple solvent description. Starting from the crystalline-like ordered structure shown in Figure 5A, we determined next whether our MC algorithm leads to equilibration of the DPPC bilayer in terms of system enthalpy. As shown in Figure 5E, the system enthalpy drops to the equilibrium value in less than 10,000 MC steps, which are simulated in about one day on a Pentium PC. Thus, no energy minimization of the bilayer is required as often performed prior to extensive MD simulations (see [114] as example). A plateau value around −250 kcal/mol is obtained, which pertains stable during the simulation run. The PDF of the system enthalpy, $p_k(H)$, is well approximated by a Gaussian function with mean −256.7 and SD = 54.1 kcal/mol (Figure 5F) [112]. Using again the relation between fluctuation in energy (or enthalpy, as the membrane simulation was done in the constant NPT-ensemble; see section 2.2, above) and the heat capacity, we can derive a value of $c_p = 821.63$ cal/mol·K~0.822 kcal/mol·K. This can be used for a consistency check of one-lipid and bilayer simulation, as follows. Given the proportionality between heat capacity and system size [112], a value of c_p = 0.6 kcal/mol·K (i.e., 64 times the value for the single lipid) would be expected. As the one-lipid simulation resembles individual DPPC molecules dissolved in an aqueous solution, one has to consider the additional heat capacity change due to transfer of lipid molecules from water to the hydrophobic membrane phase (i.e., due to the hydrophobic effect [115]). This has been estimated to give $\Delta c_p \sim -0.1$ kcal/mol·K per fatty acid chain of 16 carbon length [116]. Accordingly, based on this consideration and the one-lipid simulation, we would expect for the bilayer a value of c_p = 0.4 kcal/mol·K. Experimental values for the heat capacity of DPPC bilayers are in this range (e.g., ~0.15 kcal/mol measured by Heimburg and colleagues and ~0.3 kcal/mol as determined by Blume's group, both using various forms of calorimetry [117,118]). Given the small bilayer patch, the simple implicit solvent representation, some uncertainty in the force field parameters and the used constrained bond lengths in our simulation, the value found here for the heat capacity is in satisfactory agreement with the experiment.

Rapid Transition of PC Headgroup Conformations by Local Move Set MC Sampling

To quantify the extent of equilibration of DPPC headgroups during transition of the crystalline to the fluid bilayer state, we calculated the torsion angle distribution for selected head group torsion angles as performed for the single DPPC molecule (compare Figures 4 and 6). After only 10 steps, head group torsion angles occupy a relatively broad distribution around the starting value (dark grey line). After 10,000 MC steps, values for all torsion angles reveal a local energy minimum that starts to be occupied, while after 60,000 MC steps, energy minima of some head group torsions were equally occupied (see Figure 6C,D). By comparing, for example, the torsion angle distribution for the α4 torsion angle of phospholipids in the bilayer (Figure 6C) with that of the single DPPC molecule (Figure 4D) it becomes obvious that a similar level of occupancy is reached. Note, that we averaged over the given number of MC steps for all 64 phospholipid molecules in case of the bilayer, providing better statistics for a shorter simulation time than in case of a single DPPC molecule. In other words, the simulation of a single molecule for one million MC steps gives a very similar distribution for the α4 torsion angle in a DPPC bilayer with 64 molecules simulated for ~60,000 MC steps.

In contrast, the distribution of the α1 torsion angle clearly differs for the single molecule (Figure 4A) and the bilayer (Figure 6A). This torsion around the bond between atoms O(6) and C(7) in the lipid head group (the numbers in brackets are according to Figure 1) is therefore likely hindered in the bilayer compared to the free movement in a single molecule. Conformational equilibration for the bilayer lipids was also found for the β-torsions in the DPPC head groups after 60,000 MC steps (Figure 6B,D). We cannot rule out, however, that significantly longer simulation times would be required to ensure equilibration of all torsions in the head group region, especially those with high energy barriers due to steric confinement at the interface.

Electron Density Profile, Area per Lipid and Lateral Displacement of DPPC Molecules

Next, we evaluated the electron density profile parallel to the z-axis in the laboratory frame. In Figure 7A, the electron density profile of the starting structure is compared to that of the structure after 60,000 MC steps. The peak-to-peak distance in the head group region, D_{HH}, a measure for bilayer thickness [6], is D_{HH} = 46 Å in the crystalline-like start structure and shows a characteristic double peak in the headgroup/glycerol region (grey line in Figure 7A). This double peak, as well as the value for D_{HH}, is in line with X-ray diffraction data and earlier MD simulation of DPPC gel phase membranes [119,120]. Slight inter-digitations of the terminal part of the fatty acyl chains resulted in a small increase in electron density in this region. After 60,000 MC steps, the typical symmetrical density profile of a fluid DPPC bilayer with lowest electron density in the bilayer center was found (black line in Figure 7A). This confirms the transition from the crystalline-like to the fluid bilayer state during the simulation. The equilibrated bilayer structure is thinner (D_{HH} = 36 Å) than what is found in experiments for fluid DPPC bilayers at 323 K, i.e., D_{HH} ~ 38 Å [6]. We assign this discrepancy to the simple solvent treatment in our simulation and to the united atom AMBER force field, which we used for DPPC. Interestingly, Smondyrev and Berkowitz, who developed this force field parameterization for DPPC [99], reported also a significantly lower peak-to-peak distance for DPPC in an MD simulation at 323 K with explicit water compared to the experimental value [99,121].

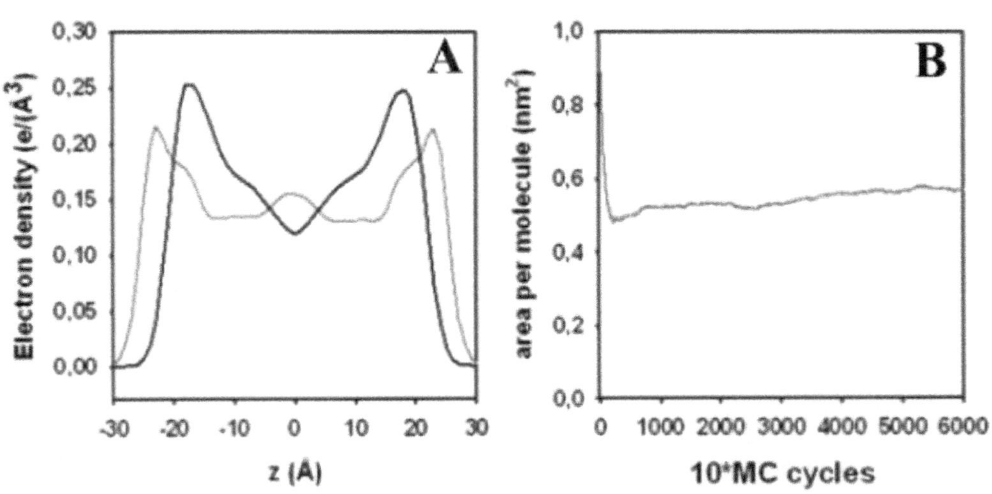

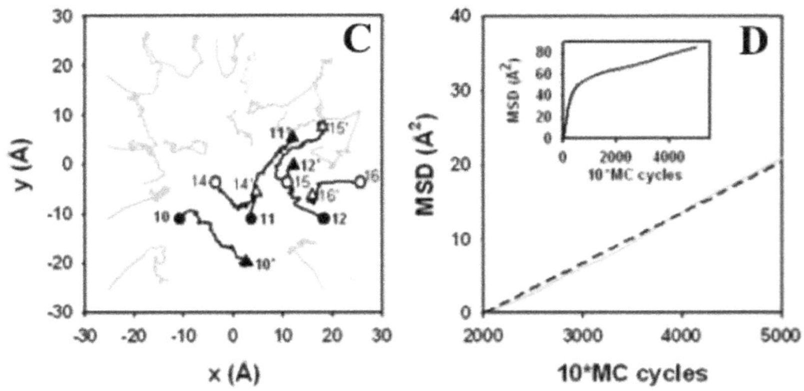

Figure 7. Electron density profile, area per lipid and lateral displacement in the bilayer. (**A**) the electron density profile was calculated for the starting configuration (grey line) and for the structure obtained after 60,000 MC steps (black line). One can clearly see that the membrane gets thinner with a large extent of fatty acyl chain disorder towards the bilayer center during the simulation; (**B**) the change of area per lipid from the crystalline start structure to the equilibrated value after 60,000 MC steps is shown; (**C**) a stroboscopic snapshot of selected lipid trajectories is shown for the upper leaflet during an MC simulation of 50,000 MC steps (lipid number X and X′ at the start, circles, and the end of the simulation, triangles). The position of the center of mass of the lipids is calculated. After adjustment of the box size causing initially large inward-directed "movement" of most lipids, the lipids show irregular "movement" as being characteristic for random walks (i.e., diffusion); (**D**) the mean square displacement (MSD) was calculated between 20,000 and 50,000 MC steps. The MSD is linear (light grey line) and can be well described by a linear fit (dashed black line), as characteristic for normal diffusion (see text for values). Inset shows the MSD as function of MC steps ("10*MC cycles") for the whole simulation.

Another criterion for the transition from the crystalline-like start structure of the bilayer to a fluid membrane is the change of the area per lipid during the MC simulation (Figure 7B). After an initial sharp drop from 1.02 nm^2 in the start structure to 0.53 nm^2 after 2200 MC steps, the bilayer expanded in the following 40,000 MC steps to reach a plateau value of 0.6 nm^2 area per lipid after 60,000 MC steps. Most of the initial drop is a consequence of the volume moves displacing whole lipids thereby resulting in tighter packing and a contraction of the simulation box. The subsequent increase in area per lipid is due to incremental mobility of the fatty acyl chains and their "melting" from the all-trans configuration [122,123]. In addition to intramolecular structural rearrangements during equilibration and conformational fluctuations at equilibrium, membrane phoshopholipids can exchange place by lateral diffusion in the bilayer plane. To determine whether our MC move scheme can cause significant lipid displacements in the membrane, the MSD of the center of mass of individual lipid molecules in the DPPC bilayer was calculated from the trajectory of an MC run and plotted as function of MC steps (Figure 7D). A stroboscopic snapshot of the center-of-mass position of tracked molecules in the upper membrane leaflet indicates that the DPPC molecules indeed perform large displacements during the simulation (Figure 7C). This confirms the fluid state of the DPPC bilayer reached after less than 30,000 MC steps. For example, lipid number 10 moves a distance of about 15 Å in the positive x-axis direction, while the lipid number 11 shows a similar displacement in y-axis direction. The MSD plot shows a steep increase for up to 10,000 MC steps, followed by a linear rise with much smaller slope after about 20,000 MC steps (Figure 7D, inset). While the initial rise again reflects primarily adjustment of the box size in our NPT-ensemble simulation as consequence of whole-lipid moves, the second linear phase indicates real Brownian-type diffusion. This part of the displacement curve can be fitted with a linear function (dashed line in Figure 7D). Following the Einstein relation for diffusion in a plane:

$$MSD(t)=4 \cdot D \cdot t \qquad (11)$$

an apparent diffusion coefficient of D = 0.68 Å²/MC step can be calculated. We note that sampling along a Markov chain in configurational space, especially in internal coordinates, does not provide real dynamic information about the investigated molecular system [44]. The MSD is defined as the second moment, i.e., the variance, of the PDF of the stochastic process underlying the observed time-dependent displacements. The PDF equals the step length distribution in tracking of randomly moving particles. Its time evolution is governed by the diffusion equation (i.e., it is the propagator of the studied diffusion process). The MSD calculated here as a function of MC steps cannot be used to extract physically meaningful self-diffusion constants of DPPC or to test various membrane diffusion models. It is just used here to assess the efficiency local move MC simulations in "lateral scrambling" of lipids as precondition for studying lipid mixtures by atomistic simulation.

Impact of the Implicit Solvent on Area per Lipid and Head Group Conformation

The advantage in sampling efficiency of atomistic Monte Carlo simulations of biomolecular structures can only become significant if an implicit solvent description is used. This is because the computational cost of calculating solute-solvent interactions for explicit solvent surmounts the gain in sampling efficiency by MC compared with other methods [41,65,87,124]. As electrostatic interactions play a fundamental role in stabilizing bilayer structure [125], we next assessed the impact of the solvent treatment in our simulation. To this end, we systematically varied the slope, s, of the distance-dependent dielectric constant (see Equation (7)) and determined the area per lipid and the angle between the P–N vector and the bilayer normal from the trajectories of separately run MC simulation, each having 60,000 MC cycles (Figure 8). Varying the slope, s, determined at which distance from the bilayer surface the dielectric constant approached the value found for water (i.e., ε = 78; Figure 8A,B). For example, for a slope, s = 0.65, the dielectric constant at a distance 10 Å from the membrane surface is ε = 74.69, which is close to the value in bulk water. In contrast, for a slope of s = 0.15, the dielectric constant at the same distance from the bilayer is only ε = 15.72, which resembles strong orientation of water molecules at the interface. There is a sigmoid dependence of εas function of the slope, s, for a given distance from the surface (Figure 8A). We found that the area per lipid depends on the slope of the distance-dependent dielectric constant also in a sigmoid manner, with values of ≥0.6 nm^2 for s > 0.5 (Figure 8C, mean ± SD). Thus, we chose a slope of s = 0.654 for all the simulations shown in Figures 5–7 to match experimental values as well as possible. The evolution of the area per lipid towards its equilibrium value was also found to depend on the slope value (not shown). The fluctuation around the mean value given by the respective SD of the area per lipid can be related to the bilayer compressibility modus K_A [123,126,127], for which we found in most simulations 260–430 mN/m. This is comparable to values found in equilibrated MD simulations [123,127].

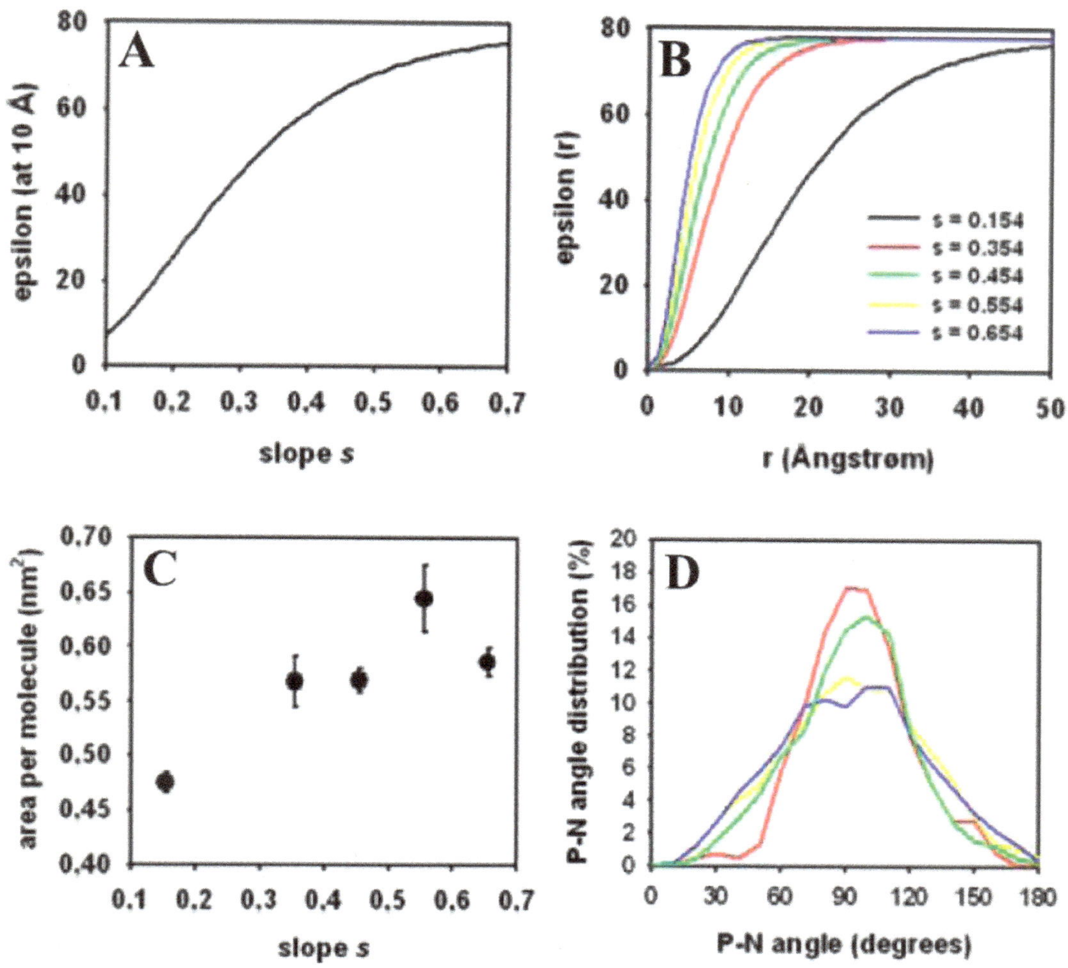

Figure 8. Impact of electrostatic interactions on membrane structure. (**A**) the distance-dependent dielectric constant is plotted as function of the slope, s, as defined inEquation (7) for a fixed distance from a charged group of 10 Å; (**B**) the distance-dependent dielectric constant is plotted as function of the distance from a charged group for various slope values of s = 0.154 (black line), s = 0.354 (red line), s = 0.454 (green line), s = 0.554 (yellow line), and s = 0.654 (blue line), respectively; (**C**) the area per lipid was calculated from the least 40,000 steps of separate MC runs performed with differing slope values of the dielectric constant. The mean ± SD is plotted as function of slope, s; (**D**) the head group orientation was inferred from the angle between the P–N vector and the bilayer normal and plotted for the separate MC runs. Increasing the slope, s, of the dielectric constant resulted in a broadening of the angle distribution (color coding is as for panel B). See text for more details.

The orientation of DPPC head groups was also affected by the parameter values chosen for the slope of the dielectric constant. For s = 0.154 and s = 0.354, the angle between P–N vector and bilayer normal showed a rather narrow peak at ~91° (Figure 8D, red and green curve, respectively). This constricted distribution is also found in simulations of DPPC bilayers in the gel phase [111], but also for certain lipid force field parameterizations [128]. It suggests that at low slope values of the dielectric constant the transition to the fluid bilayer state is incomplete in our simulations. For s = 0.554 and s = 0.654 we found a broad distribution of head group orientations with a maximum around 90°–100° (Figure 8D, yellow and blue curve, respectively). This has been similarly found in other MD simulations of DPPC membranes [99,128]. For s = 0.654 used in most of the simulations, there are two local maxima, one around 85° and one around 110°, while the overall distribution is rather broad (Figure 8D, blue curve). The maximum of the distribution suggests that DPPC head groups orient preferentially to cover the bilayer surface thereby efficiently shielding the hydrophobic acyl chains. In other words, this position of the P–N vector might have a function in covering the hydrophobic fatty acyl chains from contact to the aqueous environment [99,129]. Together, we found that the orientation of the P–N vector, similar as the bilayer thickness (see above), depend on the choice of the slope of the sigmoidal damping function in Equation (7) in Section 2.2. As this value is physically reasonable and produced the best agreement with experiments, we chose a value of s = 0.654 for this function in the simulation generating the results shown in Figures 5–7.

SUMMARY AND FUTURE IMPROVEMENTS OF THE PRESENTED LOCAL MC SIMULATION TECHNIQUE

In this study, we give a thorough introduction into atomistic MC simulations of lipid membranes. We describe the challenges and advantages of MC compared to other simulation methods and provide a concrete implementation of local move MC for efficient equilibration of lipid bilayer structures. While previous atomistic MC simulations of membranes use CBMC [65], cavity-biased moves [66], or local moves in the torsion angle space [67,79], our study shows the first implementation of an efficient move set in the torsion and bond angle space for lipid assemblies. As the CBC method allows large MC steps in the conformational space, it might overcome energetic barriers, which are not connected in a physical trajectory along phase space thereby enhancing the sampling efficiency. While this is true for other MC procedures as well, introducing flexible bond angles softens the energy landscape, which is particularly important for condensed systems of chain molecules, such as lipid bilayers. Some improvements of our method for future applications are as follows. First, we want to point out that the treatment of electrostatic interactions and the solvent description must be improved. Instead of the well-known double-cut offs and pair lists for calculating the long-ranging interactions, particle-mesh Ewald or reaction field methods might be used [125,127]. A better implicit solvent model could be implemented based on the Generalized Born model, the ABSINTH implicit solvent description or the field integrated electrostatic approach (FIESTA) [87,91,124,130,131]. Special care has to be taken in implementing such models to account for the multibody interactions, which can conflict with the necessary decomposition of the energy function into static and changing terms for MC simulations [91]. In any case, explicit water should be avoided as it will diminish any advantage due to rapid sampling by lengthy calculations of solute-solvent interactions. In fact, we have tried initial MC simulations with explicit water based on the SPC2 water model and found that the computation time per MC cycle was, depending on the exact moving scheme, between six- to twelve-fold that of simulation with the implicit solvent description used here (not shown). A suitable compromise could be to use a mixed explicit/implicit solvent description. A physiologically relevant salt, such as NaCl could be added as co-solute and modeled explicitly, while the continuous description of water is kept. This has been done previously in local move MC simulations of DNA using the CBC algorithm [95] and would be highly relevant for membrane simulations, as salts like NaCl have been shown to affect phospholipid dynamics in membranes, both in experiments and simulations [27].

Second, refined force field parameters could be used likely giving bilayer structures in better agreement with experiments [132–134], which might be even more important than a more elaborated treatment of long-range electrostatics [128]. In fact, this issue touches upon the similarity between the united atom description used here for part of the DPPC molecule and CG descriptions of lipid structures. The empirically defined interaction potentials in CG models of lipid structure often include the solvent in the interaction parameters thereby also allowing for an implicit solvent description (see Section 5.2 below). In some of these CG models, a cohesive force between the fatty acyl chains is introduced to ensure stability of the bilayer state [135]; see below. Together, this suggests that better force field parameters might be obtained for a united atom description of the DPPC lipid structure by taking the aim of an implicit solvent description into account during force field refinement.

Third, local move MC simulation of lipid membranes might be improvable by appropriate parallelization of the program code. Due to the sequential moving and updating scheme, MC simulations are traditionally very difficult to parallelize. The only routine, which in principle should be easy to parallelize is the energy calculation, for the whole system as well as for the energy difference prior to and after attempted local moves. One way of parallelizing the whole MC run is by implementing the replica-exchange technique, also called parallel tempering, into the MC algorithm. This method is useful for sampling rare events in MD and MC simulations (see an example for membranes in the Introduction, Section 1.2). As long as the energy distributions of the system simulated at various temperatures overlap, and the system size is not too large, replica-exchange MC simulation can be very efficient for overcoming low-energy barriers [89,136]. Translated into a MC simulation of membranes, the procedure could comprise parallel local move MC simulation of replicas of the bilayer system at various temperatures distributed over several processors. As the simulation proceeds, configurations are exchanged between the systems at random and become accepted by a Metropolis-like criterion, if the high-temperature simulation happens to have found one of the low-energy regions of conformational space [29,55,89,136]. As an example of this, a massively parallelized replica-exchange MC simulation of a CG model of a polymer has been performed on graphical processing units (GPUs) with an almost linear relation between number of replicas and acceleration of the computation time for the system [137]. Alternatively, the MC move set can be directly parallelized by introducing a suitable decomposition scheme in which moves are attempted in parallel for regions of the simulated system separated at a distance larger than the longest inter-pair interaction. This decomposition could be the checkerboard arrangement of interaction sites, which has been first used on spin systems [138] and very recently in many particle simulations of hard disks on GPUs [139]. For local move MC simulations of chain molecules, Uhlherr et al. (2002) used a stripe decomposition, in which randomly chosen stripes across a simulated polyethylene molecule were defined as active [61]. The molecular model was a united atom representation of polyethylene, in which only short range Lennard Jones interactions and harmonic bond stretching were considered. In the active regions, various local move MC steps, such as CR, end bridging, etc. were performed in parallel. Suitable neighbor list were defined to update interacting united atoms, while any displacement attempts out of the currently active stripes were rejected [61]. Despite this progress, we believe that local move MC simulations of dense systems with many chain-like molecules, such as lipid bilayers, will be very challenging to carry out on several processors in parallel. Atoms belonging to different molecules will hardly be found in one active stripe only, as fatty chains in the bilayer interior are intersecting and closely packed. Accordingly, many attempted moves out of a stripe will be rejected. In addition, the necessary incorporation of long-range electrostatic interactions makes the definition of stripes with intervening non-active zones difficult, especially when Ewald summation techniques are used. The same challenges of including all necessary interactions are also found in parallelized MD simulations of membranes in which system decomposition is used. These problems have been satisfactorily solved in, for example, the Gromacs program suite [131,140], and might be possible to overcome also for local move MC simulations. In addition, for increasing membrane size, another simpler parallelization scheme useful for one bilayer system might be worth to consider: the costly energy calculation could be spread over several GPUs, while the Metropolis routine (i.e., move, closure, and acceptance/rejection) could be run on the kernel-CPU. Indexing of interaction partners over an updatable pairlist would allow one to split the energy calculation efficiently over many processors/units.

RELATIONSHIP BETWEEN ATOMISTIC MC SIMULATIONS AND OTHER MEMBRANE MODELING TECHNIQUES

Combination of Atomistic MC and MD Simulations

The last two decades have witnessed an enormous progress of biomolecular simulations, in particular in the field of modeling lipid membranes. In this area, simulations can be very fruitful, since in most experimental techniques only bulk bilayer properties are assessed. In contrast, molecular simulations of membranes can provide structural detail of the underlying lipid and protein dynamics. The simulations classically performed by atomistic MD, however, are very time-consuming and cover only membrane patches of a size of a few nanometer. This naturally calls for several developments: first, hybrid simulations using MD and other simulation approaches. We suggest that the local move MC method presented here is specifically useful in combination with MD simulations of lipid membranes. MD simulations provide dynamic structural information being directly comparable to experimental data. Assessing membrane dynamics with confidence requires proper equilibration of the simulated molecular system with its inherently broad range of time scales. Metropolis MC simulations of macromolecular systems, in particular in case of local and connectivity altering move sets, lack temporal information, as sampling occurs along a phase space trajectory [44]. Only for MC simulations of particles with very small step sizes and for croft CG representations of peptide chains with extremely small changes of torsion angles as the sole implemnented MC moves, a correspondence between MC steps and physical time could be established [141,142]. This is because the Metropolis MC sampling is a stochastic process, governed by a Markovian master equation, which allows for comparison with Langevin dynamics in these special cases [141]. On the other hand, connectivity altering local-move set MC methods, such as the CBC, CR, or CRA algorithms allow for overcoming physically disconnected energy barriers in the conformational space, which can be very useful for speeding up the equilibration phase of MD simulations. In fact, in previous attempts of combined MC and MD simulation of membranes, MD runs were interrupted with CBMC sampling to significantly enhance equilibration of a DPPC bilayer [65] and to improve diffusive mixing in two-component PC bilayers [143]. A direct comparison of combined CBMC/MD versus pure MD simulation for several single-type PC membranes found the first method three times faster in terms of CPU time for equilibration of the bilayers [144]. Chiu and colleagues used that approach additionally for pre-equilibration of sphingomyelin membranes, starting from a DPPC structure [145] and in two-component bilayers made of DPPC and cholesterol [146], for the CBMC moves, explicit water was removed, and the dielectric constant was set to one, which is even simpler than our implicit solvent description [65]. In addition, in these and similar studies [147], CBMC moves are only applied to enhance sampling of the phospholipids but not of cholesterol. This is not suitable for determining the lateral and transverse membrane distribution of this important lipid component. As the CBC approach is very general, it will allow for implementing the method to steroid-like molecules. Thus, we believe that local move MC algorithms, as the one shown here, hold great promise for rapid pre-equilibration of multi-component membranes, especially in combination with classical or advanced MD techniques [28]. This is also indicated in a mixed MC/MD simulation of a polyalanine peptide in which MD runs, based on the velocity-Verlet algorithm, were alternated with CRA moves explained in Section 1.3, above [148]. The authors found significantly faster formation of secondary structures compared to MD simulations alone. Efficiency of MC moves in such studies might be further increased by using force-biased MC techniques being introduced in Section 1.3, above. These techniques consume additional computational efforts by requiring calculation of forces which enter the Metropolis acceptance criterion. This disadvantage is eventually more than compensated by the possibility to define a reasonable MC time step, which—upon appropriate weighting—can be compared to the time step in the MD part of the simulation [90]. Timoneva et al. (2010) showed, based on earlier work, that a modified form of force-biased MC called uniform-acceptance force-biased MC (UFMC) allows for physically realistic simulation of diffusion and phase transitions in various materials [149]. In UFMC all displacements are accepted, which is in contrast to the original force-biased MC method by Pangali and Rao (1979) [49], but the moved distance is determined by the forces acting on the particle. These forces, in turn, follow a Boltzmann distribution, as long as they are sufficiently small which is guaranteed by predefining a maximal displacement [90,149]. A modification of this approach called time-stamped UFMC has been presented by Mees et al. (2012), where the maximal displacement of particle i was weighted by the square root of the mass of the lightest particle divided by the mass of the particle i [150]. This allowed for defining an average time step per simulation being directly comparable to the time step in MD simulations. Local move MC for biomacromolecular systems, as defined in the CBC and other algorithms outlined above (Section 1.3), can in principal also be extended towards variants of force-biased MC like UFMC. This, however, might be complicated by the complexity of the move operation in internal coordinates, making the definition of a meaningful time step challenging. Even if the latter would turn out to be prohibitively difficult, force-biased MC variants of the presented connectivity altering MC move sets should results in accelerated sampling and ease combination with MD simulations, potentially compensating for the time-consuming force calculation.

Atomistic MC Simulations in Multi-Scale Modeling of Lipid Membranes

A second development in the field concerns multi-scale modeling of lipid membranes. Coarse-graining of the molecular structure of various lipid species has been used to simulate membrane phenomena beyond the reach of classical all-atom simulations including lipid assembly and vesicle formation [41]. A particular challenge in this area is the development and validation of the molecular model, which allows for simulating lipid structures under various conditions, for example temperature-induced phase transitions. In addition to appropriate grouping of atoms with similar physical properties into defined "CG beads" [135], CG simulations require adjustment of the inter-particle interaction potential. There is no way of grouping atoms and defining interaction potentials from first principles. Thus, the interaction potentials between CG beads are necessarily effective potentials, which need to be inferred by matching lipid partition coefficients between oil and water [39] or by inverse statistical mechanics using data from all-atom MD simulations of the same system. A procedure using the latter approach is the inverse Monte Carlo method developed by Lyubartsev and Laaksonen (1995) [151]. This method is based on a theorem by Henderson (1974), which states that the interaction potential between atoms or molecules can be inferred from the radial distribution function (RDF) of the system at thermodynamic equilibrium [152]. Inverse MC starts with a distance-dependent discretization of the Hamiltonian of the system, which allows for relating the Hamiltonian to the RDF and the pair interaction potential [151]. In practice, inverse MC determines effective pair potentials by repeatedly performing MC simulations of a predefined CG molecular system, each time with slightly varied short-range interactions. These simulations provide estimates of the RDF for each variation of the interaction potential. As initial guess for the effective pair potential, the potential of mean force might be used. The RDF of this CG molecular system calculated for each MC run is compared to the RDF calculated for the same system in all-atom simulations, the latter being considered as ground truth. By minimizing the distance between the MC-estimated RDF and the "true" RDF, non-linear optimization, such as a Newton-Raphson method, allows for determining effective pair potentials. The procedure stops when the RDF of the CG system resembles that of the all-atom simulated system to a satisfactorily degree. The MC simulations during the optimization procedure consider only simple Cartesian moves of the CG beads in the CG model. Once the CG lipid model is established, it is used for simulations of large membrane systems which are beyond the reach of classical all-atom MD simulations. Such CG simulations are typically performed by MD or MC simulations, the latter with Cartesian moves of the CG beads, i.e., the molecular sites [151,153]. We believe that local move MC simulations can be a very useful tool complementing the existing techniques for multiscale modeling of membranes in several ways. First, local move MC, for example in the CBC scheme developed here for the united atom representation of DPPC, can be applied to speed up the optimization procedure in inverse MC for obtaining the interaction potentials from the RDF (see above). Instead of simple Cartesian moves being prone for rejection due to steric clashes, the CBC moving scheme will allow for rapid exploration of the conformational space and consequently for reliable estimates of the RDF [151]. Second, local move MC might be applied directly to the simulations of the CG molecular model of a given phospholipid species. Indeed, in some implementations of CG membrane simulations, bond lengths are kept constant, while bond angle potentials between the CG beads are inferred from all-atom MD simulations [154]. Thus, connectivity altering local MC move sets, especially those which involve only three atoms per attempted move, as the CBC scheme discussed here, should be directly applicable to CG bilayer simulations and might result in faster equilibration compared to CG MD simulations.

ACKNOWLEDGMENTS

DW acknowledges funding from the Villum foundation and from a fellowship of the Max Delbrück Center for Molecular Medicine during the initial part of this work. DW and HS thank Remo Rohs (Molecular and Computational Biology Program, University of Southern California, Los Angeles, CA 90089, USA) for critical reading of the manuscript.

REFERENCES

1. Baumgart, T.; Hess, S.T.; Webb, W.W. Imaging coexisting fluid domains in biomembrane models coupling curvature and line tension. Nature 2003, 425, 821–824.
2. Korlach, J.; Schwille, P.; Webb, W.W.; Feigenson, G.W. Characterization of lipid bilayer phases by confocal microscopy and fluorescence correlation spectroscopy. Proc. Natl. Acad. Sci. USA 1999, 96, 8461–8466.
3. Gliss, C.; Randel, O.; Casalta, H.; Sackmann, E.; Zorn, R.; Bayerl, T. Anisotropic motion of cholesterol in oriented DPPC bilayers studied by quasielastic neutron scattering: The liquid-ordered phase. Biophys. J 1999, 77, 331–340.
4. Li, Y.; Ge, M.; Ciani, L.; Kuriakose, G.; Westover, E.J.; Dura, M.; Covey, D.F.; Freed, J.H.; Maxfield, F.R.; Lytton, J.; et al. Enrichment of endoplasmic reticulum with cholesterol inhibits sarcoplasmic-endoplasmic reticulum calcium ATPase-2b activity in parallel with increased order of membrane lipids: Implications for depletion of endoplasmic reticulum calcium stores and apoptosis in cholesterol-loaded macrophages. J. Biol. Chem 2004, 279, 37030–37039.
5. Oradd, G.; Westerman, P.W.; Lindblom, G. Lateral diffusion coefficients of separate lipid species in a ternary raft-forming bilayer: A Pfg-NMR multinuclear study. Biophys. J 2005, 89, 315–320.
6. Nagle, J.F.; Tristram-Nagle, S. Structure of lipid bilayers. Biochim. Biophys. Acta 2000, 1469, 159–195.
7. Tieleman, D.P.; Marrink, S.J.; Berendsen, H.J. A computer perspective of membranes: Molecular

dynamics studies of lipid bilayer systems. Biochim. Biophys. Acta 1997, 1331, 235–270.

8. De Figueiredo, P.; Doody, A.; Polizotto, R.S.; Drecktrah, D.; Wood, S.; Banta, M.; Strang, M.S.; Brown, W.J. Inhibition of transferrin recycling and endosome tubulation by phospholipase A2 antagonists. J. Biol. Chem 2001, 276, 47361–47370.
9. Sharma, D.K.; Brown, J.C.; Choudhury, A.; Peterson, T.E.; Holicky, E.; Marks, D.L.; Simari, R.; Parton, R.G.; Pagano, R.E. Selective stimulation of caveolar endocytosis by glycosphingolipids and cholesterol. Mol. Biol. Cell 2004, 15, 3114–3122.
10. Goodwin, J.S.; Drake, K.R.; Remmert, C.L.; Kenworthy, A.K. Ras diffusion is sensitive to plasma membrane viscosity.Biophys. J 2005, 89, 1398–1410.
11. Mayor, S.; Maxfield, F.R. Insolubility and redistribution of GPI-anchored proteins at the cell surface after detergent treatment. Mol. Biol. Cell 1995, 6, 929–944.
12. Fujimoto, L.M.; Roth, R.; Heuser, J.E.; Schmid, S.L. Actin assembly plays a variable, but not obligatory role in receptor-mediated endocytosis in mammalian cells. Traffic 2000, 1, 161–171.
13. Ganguly, S.; Pucadyil, T.J.; Chattopadhyay, A. Actin cytoskeleton-dependent dynamics of the human serotonin(1A) receptor correlates with receptor signaling. Biophys. J 2008, 95, 451–463.
14. Van Rheenen, J.; Jalink, K. Agonist-induced PIP(2) hydrolysis inhibits cortical actin dynamics: Regulation at a global but not at a micrometer scale. Mol. Biol. Cell 2002, 13, 3257–3267.
15. Kwik, J.; Boyle, S.; Fooksman, D.; Margolis, L.; Sheetz, M.P.; Edidin, M. Membrane cholesterol, lateral mobility, and the phosphatidylinositol 4,5-bisphosphate-dependent organization of cell actin. Proc. Natl. Acad. Sci. USA 2003, 100, 13964–13969.
16. Boucrot, E.; Saffarian, S.; Massol, R.; Kirchhausen, T.; Ehrlich, M. Role of lipids and actin in the formation of clathrin-coated pits. Exp. Cell Res 2006, 312, 4036–4048.
17. Mueller, V.; Ringemann, C.; Honigmann, A.; Schwarzmann, G.; Medda, R.; Leutenegger, M.; Polyakova, S.; Belov, V.N.; Hell, S.W.; Eggeling, C. STED nanoscopy reveals molecular details of cholesterol- and cytoskeleton-modulated lipid interactions in living cells. Biophys. J 2011, 101, 1651–1660.
18. Kultti, A.; Rilla, K.; Tiihonen, R.; Spicer, A.P.; Tammi, R.H.; Tammi, M. Hyaluronan synthesis induces microvillus-like cell surface protrusions. J. Biol. Chem 2006, 281, 15821–15828.
19. Pomorski, T.; Holthuis, J.C.; Herrmann, A.; van Meer, G. Tracking down lipid flippases and their biological functions.J. Cell Sci 2004, 117, 805–813.
20. Imparato, A.; Shillcock, J.C.; Lipowsky, R. Lateral and transverse diffusion in two-component bilayer membranes. Eur. Phys. J. E 2003, 11, 21–28.
21. Jülicher, F.; Lipowsky, R. Shape transformations of vesicles with intramembrane domains. Phys. Rev. E 1996, 53, 2670–2683.
22. Zuckermann, M.J.; Ipsen, J.H.; Miao, L.; Mouritsen, O.G.; Nielsen, M.; Polson, J.; Thewalt, J.; Vattulainen, I.; Zhu, H. Modeling lipid-sterol bilayers: Applications to structural evolution, lateral diffusion, and rafts. Methods Enzymol 2004,383, 198–229.
23. Ehrig, J.; Petrov, E.P.; Schwille, P. Near-critical fluctuations and cytoskeleton-assisted phase separation lead to subdiffusion in cell membranes. Biophys. J 2011, 100, 80–89.
24. Roark, M.; Feller, S.E. Molecular dynamics simulation study of correlated motions in phospholipid bilayer membranes. J. Phys. Chem. B 2009, 113, 13229–13234.
25. Lindahl, E.; Edholm, O. Mesoscopic undulations and thickness fluctuations in lipid bilayers from molecular dynamics simulations. Biophys. J 2000, 79, 426–433.
26. Falck, E.; Patra, M.; Karttunen, M.; Hyvönen, M.T.; Vattulainen, I. Impact of cholesterol on voids in phospholipid membranes. J. Chem. Phys 2004, 121, 12676–12689.
27. Böckmann, R.A.; Hac, A.; Heimburg, T.; Grubmüller, H. Effect of sodium chloride on a lipid bilayer. Biophys. J 2003,85, 1647–1655.
28. Choubey, A.; Kalia, R.K.; Malmstadt, N.; Nakano, A.; Vashishta, P. Cholesterol translocation in a phospholipid membrane. Biophys. J 2013, 104, 2429–2436.
29. Voter, A.F. Parallel replica method for dynamics of infrequent events. Phys. Rev. B 1998, 57, 13985–13988.
30. Hub, J.S.; Aponte-Santamaría, C.; Grubmüller, H.; de Groot, B.L. Voltage-regulated water flux through aquaporin channels in silico. Biophys. J 2010, 99, L97–L99.
31. Pöyry, S.; Cramariuc, O.; Postila, P.A.; Kaszuba, K.; Sarewicz, M.; Osyczka, A.; Vattulainen, I.; Róg, T. Atomistic simulations indicate cardiolipin to have an integral role in the structure of the cytochrome bc1 complex. Biochim. Biophys. Acta 2013, 1827, 769–778.
32. Hölttä-Vuori, M.; Uronen, R.L.; Repakova, J.; Salonen, E.; Vattulainen, I.; Panula, P.; Li, Z.; Bittman, R.; Ikonen, E. BODIPY-cholesterol: A new tool to visualize sterol trafficking in living cells and organisms. Traffic 2008, 9, 1839–1849.
33. Repáková, J.; Holopainen, J.M.; Karttunen, M.; Vattulainen, I. Influence of pyrene-labeling on fluid lipid membranes. J. Phys. Chem. B 2006, 110, 15403–15410.
34. Niemelä, P.S.; Miettinen, M.S.; Monticelli, L.; Hammaren, H.; Bjelkmar, P.; Murtola, T.; Lindahl, E.; Vattulainen, I. Membrane proteins diffuse as dynamic complexes with lipids. J. Am. Chem. Soc 2010, 132, 7574–7575.
35. Barucha-Kraszewska, J.; Kraszewski, S.; Jurkiewicz, P.; Ramseyer, C.; Hof, M. Numerical studies of the membrane fluorescent dyes dynamics in ground and excited states. Biochim. Biophys. Acta 2010, 1798, 1724–1734.
36. Shaw, D.E.; Maragakis, P.; Lindorff-Larsen, K.; Piana, S.; Dror, R.O.; Eastwood, M.P.; Bank, J.A.; Jumper, J.M.; Salmon, J.K.; Yibing, S.; et al. Atomic-

level characterization of the structural dynamics of proteins. Science 2010, 330, 341–346.

37. Jensen, M.Ø.; Jogini, V.; Borhani, D.W.; Leffler, A.E.; Dror, R.O.; Shaw, D.E. Mechanism of voltage gating in potassium channels. Science 2012, 336, 229–233.

38. Arkhipov, A.; Shan, Y.; Das, R.; Endres, N.F.; Eastwood, M.P.; Wemmer, D.E.; Kuriyan, J.; Shaw, D.E. Architecture and membrane interactions of the EGF receptor. Cell 2013, 152, 557–569.

39. Periole, X.; Marrink, S.J. The Martini coarse-grained force field. Methods Mol. Biol 2013, 924, 533–565.

40. Shillcock, J.C. Vesicles and vesicle fusion: Coarse-grained simulations. Methods Mol. Biol 2013, 924, 659–697.

41. Bennun, S.V.; Hoopes, M.I.; Xing, C.; Faller, R. Coarse-grained modeling of lipids. Chem. Phys. Lipids 2009, 159, 59–66.

42. Vainio, S.; Jansen, M.; Koivusalo, M.; Rog, T.; Karttunen, M.; Vattulainen, I.; Ikonen, E. Significance of sterol structural specificity. Desmosterol cannot replace cholesterol in lipid rafts. J. Biol. Chem 2006, 281, 348–355.

43. Metropolis, N.; Rosenbluth, A.W.; Rosenbluth, M.N.; Teller, E. Equation of state calculations by fast computing machines. J. Chem. Phys 1953, 21, 1087–1092.

44. Allen, M.P.; Tildesley, D.J. Computer Simulation of Liquids; Oxford University Press: Oxford, UK, 1987.

45. Frenkel, D. Introduction to Monte Carlo Methods. In Computational Soft Matter: From Synthetic Polymers to Proteins; Attig, N., Binder, K., Grubmüller, H., Kremer, K., Eds.; NIC Series; Lecture Notes: Jülich, Germany, 2004; Volume 23, pp. 29–60.

46. Mezei, M.; Swaminathan, S.; Beveridge, D.L. Convergence characteristics of Monte Carlo–Metropolis computer simulations on liquid water. J. Chem. Phys 1979, 71, 3366–3373.

47. Mezei, M.; Beveridge, D.L. Monte Carlo studies of the structure of dilute aqueous sclutions of Li^+, Na^+, K^+, F^-, and Cl^-. J. Chem. Phys 1981, 74, 6902–6910.

48. Rossky, P.J.; Doll, J.D.; Friedman, H.L. Brownian dynamics as smart Monte Carlo simulation. J. Chem. Phys 1978, 69, 4628–4633.

49. Pangali, C.; Rao, M.; Berne, B.J. On a novel Monte Carlo scheme for simulating water and aqueous solutions. Chem. Phys. Lett 1978, 55, 413–417.

50. Owicki, J.C.; Scheraga, H.A. Preferential sampling near solutes in Monte Carlo calculations on dilute solutions. Chem. Phys. Lett 1977, 47, 600–602.

51. Mezei, M. Virial-bias Monte Carlo methods. Efficient sampling in the (T,P,N) ensemble. Mol. Phys 1983, 48, 1075–1082.

52. Mezei, M.; Bencsath, K.A.; Goldman, S.; Singh, S. The detailed balance energy-scaled displacement Monte Carlo algorithm. Mol. Simul 1987, 1, 87–93.

53. Mezei, M. Grand-canonical ensemble Monte Carlo study of dense liquids: Lennard-Jones, soft spheres and water. Mol. Phys 1987, 61, 1207–1208.

54. Mitsutake, A.; Sugita, Y.; Okamoto, Y. Generalized-ensemble algorithms for molecular simulations of biopolymers.Biopolymers 2001, 60, 96–123.

55. Zuckermann, D.M. Equilibrium sampling in biomolecular simulations. Annu. Rev. Biophys 2011, 40, 41–62.

56. Mezei, M. On the Potential of Monte Carlo Methods for Simulating Macromolecular Assemblies. In Computational Methods for Macromolecules: Challenges and Applications, Third International Workshop for Methods for Macromolecular Modelling Conference Proceedings; Springer Verlag: Berlin/Heidelberg, Germany, 2002.

57. Scott, H.L. Modeling the lipid component of membranes. Curr. Opin. Struct. Biol 2002, 12, 495–502.

58. Siepmann, J.I.; Frenkel, D. Configurational bias Monte Carlo: A new sampling scheme for flexible chains. Mol. Phys1992, 75, 59–70.

59. Vacatello, M.; Avitabile, G.; Corradini, P.; Tuzi, A. A computer model of molecular arrangement in a n-paraffinic liquid. J. Chem. Phys 1980, 73, 548–552.

60. Boyd, R.H. An off-lattice constant-pressure simulation of liquid polymethylene. Macromolecules 1989, 22, 2477–2481.

61. Uhlherr, A.; Leak, S.J.; Adam, N.E.; Nyberg, P.E.; Doxastakis, M.; Mavrantzas, V.G.; Theodorou, D.N. Large scale atomistic polymer simulations using Monte Carlo methods for parallel vector processors. Comput. Phys. Commun 2002,144, 1–22.

62. Karayiannis, N.C.; Mavrantzas, V.G. Advanced Monte Carlo Methods for the Atomistic Simulation of Polymers with a Linear or a Non-Linear Molecular Architecture. In Multiscale Modelling of Polymer Properties; Laso, M., Perpéte, E.A., Eds.; Elsevier: Toronto, ON, Canada, 2006; pp. 31–67.

63. Mavrantzas, V.G.; Theodorou, D.N. Atomistic simulation of polymer elasticity: Calculation of the free energy of an oriented polymer melt. Macromolecules 1998, 31, 6310–6332.

64. Scott, H.L.; Jacobsson, E.; Subramanian, S. Simulations of lipid membranes with atomic resolution. Comput. Phys 1998,12, 328–334.

65. Chiu, S.W.; Clark, M.M.; Jakobsson, E.J.; Subramanian, S.; Scott, H.L. Application of combined Monte Carlo and Molecular Dynamics method to simulation of dipalmitoylphosphatidylcholine lipid bilayer. J. Comput. Chem 1999, 20, 1153–1164.

66. Jedlovszky, P.; Mezei, M. Grand canonical ensemble Monte Carlo simulation of a lipid bilayer using extension biased rotations. J. Chem. Phys 1999, 111, 10770–10773.

67. Jedlovszky, P.; Mezei, M. Orientational order of the water molecules across a fully hydrated DMPC bilayer: A Monte Carlo simulation study. J. Phys. Chem. B 2001, 105, 3614–3623.

68. Jedlovszky, P.; Mezei, M. Effect of cholesterol on the properties of phospholipid membranes. 2. Free energy profile of small molecules. J. Phys. Chem. B 2003, 107, 5322–5332.

69. Alinchenko, M.G.; Voloshin, V.P.; Medvedev, N.N.; Mezei, M.; Pártay, L.; Jedlovszky, P. Effect of cholesterol on the properties of phospholipid membranes. 4. Interatomic voids. J. Phys. Chem. B 2005, 109, 16490–16502.
70. Hoffmann, D.; Knapp, E.W. Polypeptide folding with off-lattice Monte Carlo dynamics: The method. Eur. Biophys. J1996, 24, 387–403.
71. Go, N.; Scheraga, H.A. Ring closure and local conformational deformations of chain molecules. Macromolecules 1970, 3, 178–187.
72. Manocha, D.; Canny, J.F. Efficient inverse kinematics for general 6R manipulators. IEEE Trans. Robot. Autom 1994, 10, 648–657.
73. Coutsias, E.A.; Seok, C.; Jacobson, M.P.; Dill, K.A. A kinematic view of loop closure. J. Comput. Chem 2004, 25, 510–528.
74. Wu, M.G.; Deem, M.W. Analytical rebridging Monte Carlo: Application to cis/trans isomerization in proline-containing. J. Chem. Phys 1999, 111, 6625–6632.
75. Dinner, A.R. Local deformations of polymers with nonplanar rigid main-chain internal coordinates. J. Comp. Chem2000, 21, 1132–1144.
76. Dodd, L.R.; Boone, T.D.; Theodorou, D.N. A concerted rotation algorithm for atomistic Monte Carlo simulation of polymer melts and glasses. Mol. Phys 1993, 78, 961–996.
77. Hoffmann, D.; Knapp, E.W. Folding pathways of a helix-turn-helix model protein. J. Phys. Chem. B 1997, 101, 6734–6740.
78. Ulmschneider, J.P.; Jorgensen, W.L. Monte Carlo backbone sampling for polypeptides with variable bond and dihedral angles using concerted rotations and a Gaussian bias. J. Chem. Phys 2003, 118, 4261–4271.
79. Mezei, M. Efficient Monte Carlo sampling for long molecular chains using local moves, tested on a solvated lipid bilayer. J. Chem. Phys 2003, 118, 3874–3879.
80. Hu, J.; Ma, A.; Dinner, A.R. Monte Carlo simulations of biomolecules: The MC module in CHARMM. J. Comput. Chem2006, 27, 203–216.
81. Pant, K.P.V.; Theodorou, D.N. Variable connectivity method for the atomistic Monte Carlo simulation of polydisperse poymer melts. Macromolecules 1995, 28, 7224–7234.
82. Mavrantzas, V.G.; Boone, T.D.; Zervopoulou, E.; Theodorou, D.N. End-bridging Monte Carlo: A fast algorithm for atomistic simulation of condensed phases og long polymer chains. Macromolecules 1999, 32, 5072–5096.
83. Uhlherr, A. Monte Carlo conformational sampling of the internal degrees of freedom of chain molecules. Macromolecules 2000, 33, 1351–1360.
84. Uhlherr, A.; Mavrantzas, V.G.; Doxastakis, M.; Theodorou, D.N. Directed bridging for fast atomistic Monte Carlo simulations of bulk polymers. Macromolecules 2001, 34, 8554–8568.
85. Bruccoleri, R.E.; Karplus, M. Chain closure with bond angle variations. Macromolecules 1985, 18, 2767–2773.
86. Ulmschneider, J.P.; Ulmschneider, M.B.; di Nola, A. Monte Carlo vs molecular dynamics for all-atom polypeptide folding simulations. J. Phys. Chem. B 2006, 110, 16733–16742.
87. Ulmschneider, J.P.; Jorgensen, W.L. Polypeptide folding using Monte Carlo sampling, concerted rotation, and continuum solvation. J. Am. Chem. Soc 2004, 126, 1849–1857.
88. Betancourt, M.R. Efficient Monte Carlo trial moves for polypeptide simulations. J. Chem. Phys 2005, 123, 174905:1–174905:7.
89. Okamoto, Y. Generalized-ensemble algorithms: Enhanced sampling techniques for Monte Carlo and molecular dynamics simulations. J. Mol. Graph. Model 2004, 22, 425–439.
90. Neyts, E.C.; Bogaerts, A. Combining molecular dynamics with Monte Carlo simulations: Implementations and applications. Theor. Chem. Acc 2013, 132, 1320.
91. Vitalis, A.; Pappu, R.V. Methods for Monte Carlo simulations of biomacromolecules. Annu. Rep. Comput. Chem 2009, 5, 49–76.
92. Bottaro, S.; Boomsma, W.; Johansson, K.E.; Andreetta, C.; Hamelryck, T.; Ferkinghoff-Borg, J. Subtle Monte Carlo updates in dense molecular systems. J. Chem. Theory Comput 2012, 8, 695–702.
93. Boomsma, W.; Frellsen, J.; Harder, T.; Bottaro, S.; Johansson, K.E.; Tian, P.; Stovgaard, K.; Andreetta, C.; Olsson, S.; Valentin, J.B.; et al. PHAISTOS: A framework for Markov Chain Monte Carlo simulation and inference of protein structure. J. Comput. Chem 2013, 34, 1697–1705.
94. Sklenar, H.; Wüstner, D.; Rohs, R. Using internal and collective variables in Monte Carlo simulations of nucleic acid structures: Chain breakage/closure algorithm and associated Jacobians. J. Comput. Chem 2006, 27, 309–315.
95. Rohs, R.; Bloch, I.; Sklenar, H.; Shakked, Z. Molecular flexibility in ab initio drug docking to DNA: Binding-site and binding-mode transitions in all-atom Monte Carlo simulations. Nucleic Acids Res 2005, 33, 7048–7057.
96. Rohs, R.; Sklenar, H.; Shakked, Z. Structural and energetic origins of sequence-specific DNA bending: Monte Carlo simulations of papillomavirus E2-DNA binding sites. Structure 2005, 13, 1499–1509.
97. Minary, P.; Levitt, M. Conformational optimization with natural degrees of freedom: A novel stochastic chain closure algorithm. J. Comput. Biol 2010, 17, 993–1010.
98. Sim, A.Y.L.; Minary, P.; Levitt, M. Modeling nucleic acids. Curr. Opin. Struct. Biol 2012, 22, 273–278.
99. Smondyrev, A.M.; Berkowitz, M.L. United atom force field for phospholipid membranes: Constant pressure molecular dynamics simulation of dipalmitoylphosphatidylcholine/water system. J. Comput. Chem 1999, 20, 531–545.
100. Ryckaert, J.-P.; Bellemans, A. Molecular dynamics of liquid n-butane near its boiling point. Chem. Phys. Lett 1975, 30, 123–125.

101. Berger, O.; Edholm, O.; Jähnig, F. Molecular dynamics simulations of a fluid bilayer of dipalmitoylphosphatidylcholine at full hydration, constant pressure, and constant temperature. Biophys. J 1997, 72, 2002–2013.
102. Manousiouthakis, V.I.; Deem, M.W. Strict detailed balance is unnecessary in Monte Carlo simulation. J. Chem. Phys1999, 110, 2753–2756.
103. Vanderkooi, G. Multibilayer structure of dimyristoylphosphatidylcholine dihydrate as determined by energy minimization. Biochemistry 1991, 30, 10760–10768.
104. Cornell, W.D.; Cieplak, P.; Bayly, C.I.; Gould, I.R.; Merz, K.M., Jr.; Ferguson, D.M.; Spellmeyer, D.C.; Fox, T.; Caldwell, J.W.; Kollman, P.A. A second generation force field for the simulation of proteins, nucleic acids, and organic molecules. J. Am. Chem. Soc 1995, 117, 5179–5197.
105. Hingerty, B.E.; Ritchie, R.H.; Ferrell, T.L.; Turner, J.E. Dielectric effects in biopolymers: The theory of ionic saturation revisited. Biopolymers 1985, 24, 427–439.
106. Lavery, R.; Sklenar, H.; Zakrzewska, K.; Pullman, B. The flexibility of the nucleic acids: (II). The calculation of internal energy and applications to mononucleotide repeat DNA. J. Biomol. Struct. Dyn 1986, 3, 989–1014.
107. Schlitter, J. Estimation of absolute and relative entropies of macromolecules using the covariance matrix. Chem. Phys. Lett 1993, 215, 617–621.
108. Andricioaei, I.; Karplus, M. On the calculation of entropy from covariance matrices of the atomic fluctuations. J. Chem. Phys 2001, 115, 6289–62812.
109. Altman, B.; Hughes, C.; Gerstein, M.B. Methods for displaying macromolecular structural uncertainty: Application to the globins. J. Mol. Graph 1995, 13, 142–152.
110. Tu, K.; Tobias, J.; Klein, M.L. Constant pressure and temperature molecular dynamics simulation of a fully hydrated liquid crystal phase dipalmitoylphosphatidylcholine bilayer. Biophys. J 1995, 69, 2558–2562.
111. Egberts, E.; Marrink, S.J.; Berendsen, H.J. Molecular dynamics simulation of a phospholipid membrane. Eur. Biophys. J1994, 22, 423–436.
112. Greiner, W.; Neise, L.; Stöcker, H. Thermodynamik und Statistische Mechanik; Verlag Harri Deutsch: Frankfurt, Germany, 1993; pp. 297–315.
113. Edholm, O.; Berendsen, H.J. Entropy estimation from simulation of non-diffusive systems. Mol. Phys 1984, 51, 1011–1028.
114. Moore,P.B.;Lopez,C.F.;Klein,M.L.Dynamicalproperties of a hydrated lipid bilayer from a multinanosecond molecular dynamics simulation. Biophys. J 2001, 81, 2484–2494.
115. Tanford, C. The Hydrophobic Effect. Formation of Micells and Biological Membranes; Wiley: New York, NY, USA, 1980.
116. Gill, S.J.; Wadsö, I. An equation of state describing hydrophobic interactions. Proc. Natl. Acad. Sci. USA 1976, 73, 2955–2958.
117. Grabitz, P.; Ivanova, V.P.; Heimburg, T. Relaxation kinetics of lipid membranes and its relation to the heat capacity.Biophys. J 2002, 82, 299–309.
118. Blume, A. Apparent molar heat capacities of phospholipids in aqueous dispersion. Effects of chain length and head group structure. Biochemistry 1983, 22, 5436–5442.
119. Wiener, M.C.; Suter, R.M.; Nagle, J.F. Structure of the fully hydrated gel phase of dipalmitoylphosphatidylcholine. Biophys. J 1989, 55, 315–325.
120. Tu, K.; Tobias, J.; Blasie, K.; Klein, M.L. Molecular dynamics investigation of the structure of a fully hydrated gel-phase dipalmitoylphosphatidylcholine bilayer. Biophys. J 1996, 70, 595–608.
121. Smondyrev, A.M.; Berkowitz, M.L. Molecular dynamics simulation of DPPC bilayer in DMSO. Biophys. J 1999, 76, 2472–2478.
122. Seelig, J. Deuterium magnetic resonance: Theory and application to lipid membranes. Q. Rev. Biophys 1977, 10, 353–418.
123. Lindahl, E.; Edholm, O. Spatial and energetic-entropic decomposition of surface tension in lipid bilayers from molecular dynamics simulations. J. Chem. Phys 2000, 113, 3882–3893.
124. Ulmschneider, J.P.; Ulmschneider, M.B.; di Nola, A. Monte Carlo folding of trans-membrane helical peptides in an implicit generalized born membrane. Proteins 2007, 69, 297–308.
125. Patra, M.; Karttunen, M.; Hyvönen, M.T.; Falck, E.; Lindqvist, P.; Vattulainen, I. Molecular dynamics simulations of lipid bilayers: Major artifacts due to truncating electrostatic interactions. Biophys. J 2003, 84, 3636–3645.
126. Feller, S.E.; Pastor, R.W. On simulating lipid bilayers with an applied surface tension: Periodic boundary conditions and undulations. Biophys. J 1996, 71, 1350–1355.
127. Anézo, C.; de Vries, A.H.; Höltje, H.D.; Tieleman, D.P.; Marrink, S.J. Methodological issues in lipid bilayer simulations. J. Phys. Chem. B 2003, 107, 9424–9433.
128. Poger, D.; Mark, A.E. Lipid bilayers: The effect of force field on ordering and dynamics. J. Chem. Theory Comput 2012,8, 4807–4817.
129. Smondyrev, A.M.; Berkowitz, M.L. Structure of dipalmitoylphosphatidylcholine/cholesterol bilayer at low and high cholesterol concentrations: Molecular dynamics simulation. Biophys. J 1999, 77, 2075–2089.
130. Sklenar, H.; Eisenhaber, F.; Poncin, M.; Lavery, R. Including solvent and counterion effects in the force field integrated electrostatic approach (FIESTA). In Theoretical Biochemistry and Molecular Biophysics; Beveridge, D.L., Lavery, R., Eds.; Adenine Press: Schenectady, NY, USA, 1990; pp. 317–335.
131. Larsson, P.; Lindahl, E. A high-performance parallel-generalized born implementation enabled by tabulated interaction rescaling. J. Comput. Chem 2010, 31, 2593–2600.

132. Chiu, S.W.; Pandit, S.A.; Scott, H.L.; Jakobsson, E. An improved united atom force field for simulation of mixed lipid bilayers. J. Phys. Chem. B 2009, 113, 2748–2763.

133. Piggot, T.J.; Pineiro, Á.; Khalid, S. Molecular dynamics simulations of phosphatidylcholine membranes: A comparative force field study. J. Chem.Theory Comput 2012, 8, 4593–4609.

134. Skjevik, Å.A.; Madej, B.D.; Walker, R.C.; Teigen, K. LIPID11: A modular framework for lipid simulations using amber.J. Phys. Chem. B 2012, 116, 11124–11136.

135. Wang, Z.J.; Deserno, M. A systematically coarse-grained solvent-free model for quantitative phospholipid bilayer simulations. J. Phys. Chem. B 2010, 114, 11207–11220.

136. Kofke, D.A. On the acceptance probability of replica-exchange Monte Carlo trials. J. Chem. Phys 2002, 117, 6911–6914.

137. Gross, J.; Janke, W.; Bachmann, M. Massively parallelized replica-exchange simulations of polymers on GPUs.Comput. Phys. Commun 2011, 182, 1638–1644.

138. Pawley, G.S.; Bowler, K.C.; Kenway, R.D.; Wallace, D.J. Concurrency and parallelism in MC and MD simulations in physics. Comput. Phys. Commun 1985, 37, 251–260.

139. Anderson, J.A.; Jankowski, E.; Grubb, T.L.; Engel, M.; Glotzer, S.C. Massively parallel Monte Carlo for many-particle simulations on GPUs. J. Comput. Phys 2013, 254, 27–38.

140. Pronk, S.; Páll, S.; Schulz, R.; Larsson, P.; Bjelkmar, P.; Apostolov, R.; Shirts, M.R.; Smith, J.C.; Kasson, P.M.; van der Spoel, D.; et al. GROMACS 4.5: A high-throughput and highly parallel open source molecular simulation toolkit.Bioinformatics 2013, 29, 845–854.

141. Kikuchi, K.; Yoshida, M.; Maekawa, T.; Watanabe, H. Metropolis Monte Carlo method as a numerical technique to solve the Fokker-Planck equation. Chem. Phys. Lett 1991, 185, 335–338.

142. Tiana, G.; Sutto, L.; Broglia, R.A. Use of the Metropolis algorithm to simulate the dynamics of protein chains. Phys. A2007, 380, 241–249.

143. Coppock, P.S.; Kindt, J.T. Atomistic simulations of mixed-lipid bilayers in gel and fluid phases. Langmuir 2009, 25, 352–359.

144. Chiu, S.W.; Jakobsson, E.; Subramaniam, S.; Scott, H.L. Combined Monte Carlo and molecular dynamics simulation of fully hydrated dioleyl and palmitoyl-oleyl phosphatidylcholine lipid bilayers. Biophys. J 1999, 77, 2462–2469.

145. Chiu, S.W.; Vasudevan, S.; Jakobsson, E.; Mashl, R.J.; Scott, H.L. Structure of sphingomyelin bilayers: A simulation study. Biophys. J 2003, 85, 3624–3635.

146. Chiu, S.W.; Jakobsson, E.; Mashl, R.J.; Scott, H.L. Cholesterol-induced modifications in lipid bilayers: A simulation study. Biophys. J 2002, 83, 1842–1853.

147. De Joannis, J.; Coppock, P.S.; Yin, F.; Mori, M.; Zamorano, A.; Kindt, J.T. Atomistic simulation of cholesterol effects on miscibility of saturated and unsaturated phospholipids: Implications for liquid-ordered/liquid-disordered phase coexistence. J. Am. Chem. Soc 2011, 133, 3625–3634.

148. Ribeiro, A.S.T.; de Alencastro, R.B. Mixed Monte Carlo/molecular dynamics simulations in explicit solvent. J. Comput. Chem 2012, 33, 901–905.

149. Timoneva, M.; Groenewegen, J.; Thijsse, B. Modeling diffusion and phase transitions by a uniform-acceptance force-biased Monte Carlo method. Phys. Rev. B 2010, 81, 144107:1–144107:13.

150. Mees, M.J.; Pourtois, G.; Neyts, E.C.; Thijsse, B.J.; Stesmans, A. Uniform-acceptance force-bias Monte Carlo method with time scale to study solid-state diffusion. Phys. Rev. B 2012, 85, 134301:1–134301:9.

151. Lyubartsev, A.P.; Laaksonen, A. Calculation of effective potentials from radial distribution fucntions: A reverse Monte Carlo approach. Phys. Rev. E 1995, 52, 3730–3737.

152. Henderson, R.L. A uniqueness theroem for fluid pair correlation functions. Phys. Lett 1974, 49A, 197–198.

153. Shelley, J.C.; Selley, M.Y.; Reeder, R.C.; Bandyopadhyay, S.; Moore, P.B.; Klein, M.L. Simulations of phospholipids using a coarse grain model. J. Phys. Chem. B 2001, 105, 9785–9792.

154. Sodt, A.J.; Head-Gordon, T. An implicit solvent coarse-grained lipid model with correct stress profile. In J. Chem. Phys; 2010; Volume 132, pp. 205103:1–205103:8.

CHAPTER 15

APPLICATION OF ENHANCED SAMPLING MONTE CARLO METHODS FOR HIGH-RESOLUTION PROTEIN-PROTEIN DOCKING IN ROSETTA

Zhe Zhang[1], Christina E. M. Schindler[1], Oliver F. Lange[2], Martin Zacharias[1]

[1] Physik-Department T38, Technische Universität München, James-Franck-Str. 1, 84748 Garching, Germany

[2] Biomolecular NMR and Munich Center for Integrated Protein Science, Department Chemie, Technische Universität München, Lichtenbergstr. 4, 85748 Garching, Germany

ABSTRACT

The high-resolution refinement of docked protein-protein complexes can provide valuable structural and mechanistic insight into protein complex formation complementing experiment. Monte Carlo (MC) based approaches are frequently applied to sample putative interaction geometries of proteins including also possible conformational changes of the binding partners. In order to explore efficiency improvements of the MC sampling, several enhanced sampling techniques, including temperature or Hamiltonian replica exchange and well-tempered ensemble approaches, have been combined with the MC method and were evaluated on 20 protein complexes using unbound partner structures. The well-tempered ensemble method combined with a 2-dimensional temperature and Hamiltonian replica exchange scheme (WTE-H-REMC) was identified as the most efficient search strategy. Comparison with prolonged MC searches indicates that the WTE-H-REMC approach requires approximately 5 times fewer MC steps to identify near native docking geometries compared to conventional MC searches.

INTRODUCTION

Protein-protein interactions are integral to many mechanisms of cellular activities, ranging from enzyme catalysis and inhibition to signal transduction and gene regulation. Atomic-level structural knowledge is essential to understand the function of protein-protein complexes in biological processes. However, experimental structure determination of protein-protein complexes is often difficult and for many interactions the corresponding complex structures are lacking [1,2]. Computational protein-protein docking methods can provide structural models of protein-protein interactions where experimental data is absent, of low-resolution or too sparse. Besides providing valuable structural biology information, high-resolution protein-protein docking can also help to explain binding affinities and specificities, the nature of the binding free energy funnel and effects of mutations. Furthermore, these techniques are essential for computational protein-protein interface design of the design of non-natural complexes [3–7]

Docking programs often employ a two-stage protocol [8,9]. First, the space of putative docking geometries is sampled broadly, keeping the partner structures rigid, which corresponds to six degrees of freedom. Second, structures are refined in one or multiple steps, typically employing partner structures at atomic resolution allowing for conformational changes of side chains and possibly also of the protein main chain. For the rigid-body stage, the application of grid-based

Citation: Zhang Z, Schindler CEM, Lange OF, Zacharias M (2015) Application of Enhanced Sampling Monte Carlo Methods for High-Resolution Protein-Protein Docking in Rosetta. PLoS ONE 10(6): e0125941. doi:10.1371/journal.pone.0125941.

fast Fourier transformation [10] or geometric hashing [11] can accelerate the search [8,9]. Alternatively, the search can also be performed efficiently at reduced resolution using a coarse-grained model of the protein structures [12,13]. Both, at the rigid-body search stage or during refinement, Monte Carlo (MC) methods can be very helpful [14–18].

For high-resolution refinement, most protocols require a starting configuration that is already fairly close to the native structure. The likely reason for this strong dependence on the starting structure is the energy landscape's ruggedness, which frustrates the sampling and renders the global energy minimum hard to reach [19]. Accordingly, the rationale behind typical two stage docking refinement protocols is that the initial docking predictions is likely to generate at least one structure close to the native. A justification of this assumption is given by the general hypothesis that the native conformation coincides with the global energy minimum at the bottom of a broad basin in a rugged energy landscape [20,21]. However, since initial docking stages often use a simplified energy function, the initial docking stage is routinely misled and does not actually produce sufficiently many near-native candidates for the subsequent refinement stage, especially when there is an alternative binding site with larger buried surface [14].

Focus of this work is to improve the sampling for high-resolution docking based on the MC approach. In the MC method, random translational and rotational moves or conformational changes of the partner structures are applied on the configuration in a step-wise manner using the Metropolis criterion for acceptance of a move. Advantages of the MC method are the generation of a physically relevant canonical ensemble of docking configurations, use of arbitrary energy functions that can contain discontinuities and for the possibility to include various levels of structural flexibility. However, an exhaustive high-resolution sampling of the conformational space with the MC method can be computationally demanding. In general, the docking success of MC docking is limited by the sampling of putative complex geometries and by the accuracy of the energy function used for scoring predicted complexes.

Parallel tempering or replica exchange techniques promise to overcome these challenges and have received wide-spread interest in recent years [22–25]. The general idea of parallel tempering is to simulate the system with M replicas at different temperatures and to frequently exchange configurations between neighboring replicas. The high temperature replicas sample broadly, whereas the low-temperature replicas allow precise exploration of deep energy minima. Due to the frequent exchanges between the (hot) broad sampling regime and the (cold) annealing regime, configurations are less likely to get trapped in local minima. A generalization of temperature replica exchange is to vary the Hamiltonian (H-REMC) among replicas [26], which allows, for instance, to blend between a smoothed van der Waals potential and a realistically hard formulation to allow overcoming of sampling barriers in molecular dynamics simulation [27,28]. For Rosetta, previous studies also showed that softening the Lennard-Jones repulsive term is beneficial and better suited for side-chain modeling and prediction [29,30]. Of course, it is possible to combine variation of temperature and Hamiltonian in multi-dimensional replica exchange approaches [31–35]. A bottleneck in using replica exchange is that to cover the same parameter range (temperature, or smoothness) the number of replicas required increases quickly with the number of degrees of freedom sampled. This is due to the fact, that to achieve efficient exchange between replicas, a substantial overlap between sampled energy levels of neighboring replicas is required [33,36].

Metadynamics is another popular enhanced sampling method, in which sampling is facilitated by a history-dependent biasing potential. It is constructed as the sum of Gaussian functions deposited along the trajectory in the collective variable space [37,38]. Choosing energy as collective variable gives rise to the so-called well-tempered ensemble (WTE) with much larger fluctuations in the sampled energies than the canonical ensemble [39]. This property of the WTE can be exploited to overcome the major bottleneck of temperature replica exchange discussed above. Since, the overlap of the energy distribution between neighboring replicas controls the exchange efficiency, using WTE drastically reduces the number of replicas required [39,40].

In the present study we have compared the efficiency of a standard MC protocol for high resolution protein-protein docking using RosettaDock and various extensions based on advanced sampling techniques. In particular, we tested four different protocols, standard Monte Carlo (MC), Temperature Replica Exchange Monte Carlo (REMC), well-tempered ensemble temperature Replica Exchange Monte Carlo (WTE-REMC), and well-tempered ensemble two dimensional Hamiltonian Replica Exchange Monte Carlo (WTE-H-REMC). The approaches were systematically evaluated on protein-protein complexes using unbound partner structures and starting in each case from the same starting placements. Overall best performance was achieved with the WTE-H-REMC method at the same computational cost compared to the alternative protocols.

METHODS

Energy scoring function and Starting Structure Generation

The standard all-atom energy function for RosettaDock as given by the weight-set docking [41] was used in all docking protocols. The docking energy function consists of van der Waals attractive and repulsive interactions, an implicit solvation term, hydrogen-bonding energies, a statistical residue-residue pairwise interaction term, a rotamer probability term and a pairwise electrostatic energy term [18,42]. For each target, the different docking simulation protocols were started from the same initial protein partner arrangements. The start geometries were based on unbound partner structures and one partner was initially separated relative to the position in the complex in a random direction by 15 Å and a random rotation by 60° relative to the bound geometry. Only geometries without steric overlap between partners were accepted. The ligand RMSD (L_rmsd, root mean square deviation of the mobile protein after best superposition of the receptor protein onto native complex structure) from the respective bound complex for all the targets was on average ~18 Å with slight variation depending on the shape and size of the binding partners (Fig 1). The initial placement corresponds to a scenario where the binding region is approximately known.

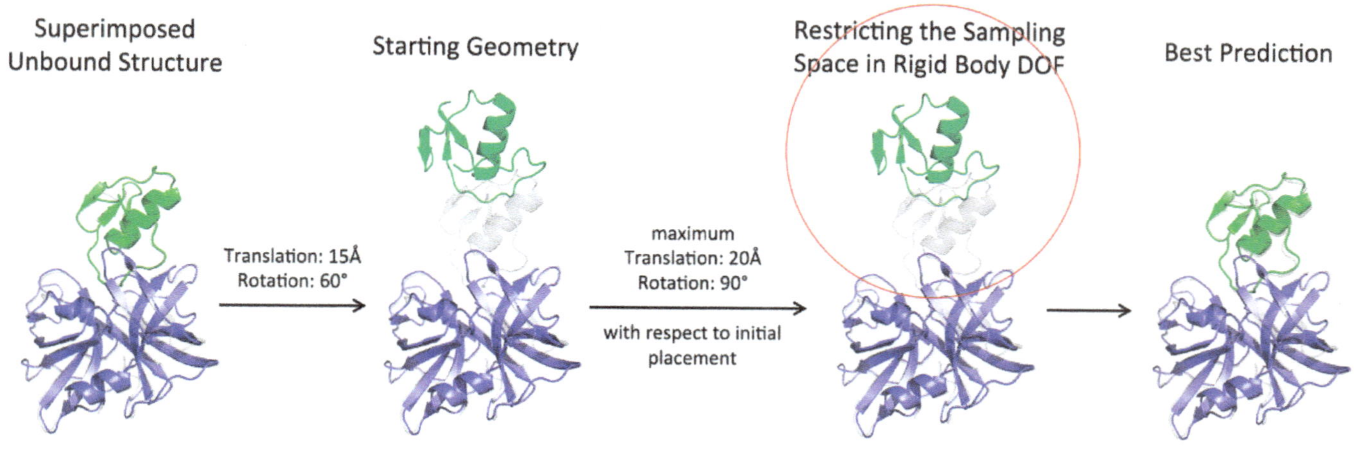

Figure 1. Docking refinement conditions. Each docking starting geometry was generated by an initial random translation of one unbound partner from the geometry in the complex by 15Å and random rotation of 60° (compare green displaced and grey cartoon representations). During the docking search translation and rotation of one partner with respect to the other was restricted relative to the starting geometry by 20Å and 90° (indicated by red circle), respectively.

Restricting the Sampling Space in Rigid Body Degrees Of Freedom

For rigid body moves, random translations drawn from Gaussian distribution are performed along all three axes, and the axis-angle notation is used to represent rotations [14]. In order to perform a local search in the vicinity of the starting geometry, the sampling space in the rigid body degrees of freedom was restricted with respect to the initial input conformation by a maximum translation of 20Å and maximum rotation of 90° (this exceeds the maximum displacement of the starting structure from the bound configuration, see previous paragraph and illustration in Fig 1). To avoid dissociation of the two binding partners, we also applied an encounter constraint, which acts on the distance between the center of mass of the two binding partners and only penalizes the sampled geometries if the two binding partners are too far apart [14].

General Settings

We have combined enhanced sampling techniques with Monte Carlo (MC) method to sample protein-protein docking with atomistic representation, and tested four protocols within Rosetta including standard MC, Temperature Replica Exchange Monte Carlo (REMC), well-tempered ensemble temperature Replica Exchange Monte Carlo (WTE-REMC), and well-tempered ensemble 2-dimensional Hamiltonian Replica Exchange Monte Carlo (WTE-H-REMC). In those docking approaches, rigid body displacements and side-chain optimization are accomplished by the rigid body mover UnbiasedRigidBodyPerturbNoCenterMover, and sidechain movers including JumpRotamerSidechainMover, PerturbRotamerSidechainMover and PerturbChiSidechainMover. Those movers are applied under the control of the Metropolis-Hastings framework. For each move, the MetropolisHastingsMover randomly applies one out of the four movers based on their sampling weights. Mover step size for UnbiasedRigidBodyPerturbNoCenterMover and PerturbChiSidechainMover are drawn from random Gaussian distributions. In the protocols with replica exchange, the magnitude for mover step size and sampling weight were modulated according to the replica level during initialization such that in the lower levels more frequent side-chain moves and fewer small rigid body moves were applied and in the higher levels less frequent side-chain moves and larger rigid-body moves. The magnitude of the step size and sampling weight were, however, kept fixed along the simulation in each replica. All the settings for the reference replica were made exactly the same as used in the standard MC protocol, and we denote these settings as reference settings. If not indicated otherwise, for all the protocols and on each target $2x10^6$ MC steps were employed. Snapshots are taken and stored every 1,000 steps. In the REMC protocols exchanges were attempted every 1,000 MC steps.

Monte Carlo and Asynchronous Parallel Tempering Protocol

For the standard MC docking protocol, 25 trajectories are run with temperature set to 0.15. At the end, about 25 x 2,000 sampled structures were collected for each target. The step size of translation and rotation for rigid-body moves are drawn from normal distributions with small mean value of 0.1Å and 1°. The sampling weights for UnbiasedRigidBodyPerturbNocenter, JumpRotamerSidechain, PerturbRotamerSidechain and PerturbChiSidechain are set to 0.5, 4, 6 and 10, respectively.

For the parallel tempering replica exchange (REMC) protocol, 13 temperature levels were drawn from geometric progressions ranging from 0.15 (reference) to 0.31. Two trajectories with the 13 replicas are run for each target. Exchanges were attempted between neighbor replicas every 1,000 steps. For all targets, good exchange rates (between 25% and 69% with median value 49%) are achieved and no further target dependent optimization was required. In replica exchange, it is common that the speed of each replica is not exactly the same. To avoid that the faster replica wait for the slower partner, we used an asynchronous exchange scheme. That is the faster replicas can perform more steps instead of waiting for its partner to reach the pre-defined exchange stride. Finally, the simulation will finish as soon as the slowest replica has reached the required step number.

Well-Tempered Parallel Tempering and Hamiltonian replica exchange protocol

We applied the well-tempered ensemble (WTE) technique with parallel tempering replica exchange Monte Carlo using a value of 5 for the tunable factor γ and reduced the temperature levels from 13 to 5 with the same range. The bin size for well-tempered ensemble technique to collect the history-dependent bias energy is set to two units of Rosetta docking energy. The resulting exchange rates are between 20% and 55% with median value 37%. For the WTE-H-REMC protocol we took advantage of the splitting of the van der Waals interactions into attractive and repulsive components in RosettaDock. It is represented with a modified Lennard-Jones 6–12 potential which includes a linear extrapolation in the repulsive part below the threshold of $0.6\sigma_{ij}$, where σ_{ij} is the sum of the atomic radii of atoms i and j. The atomic radii and energy well depth are taken from the CHARMM19 parameter set [13,18,43], and we denote this as "hard-rep". For the standard "soft-rep" in Rosetta, the atomic radii were either held fixed or scaled by a factor of 1.07 (typically for non-polar atoms) from the hard-rep radii, and the "switch point" for the linear extrapolation was selected empirically [29]. In the WTE-H-REMC protocol, we applied a 2-dimensional replica exchange, with the temperature as variable in the first dimension, and used the scaling factor for the soft Lennard-Jones repulsive term as the second dimension. The scaling factor allows linear interpolation of atomic radii and switch point between the hard-rep and soft-rep potentials (see above). In the Hamiltonian scaling dimension, we used five levels: hard_rep, soft 50%, soft 55%, soft 60% and soft 65%. In the temperature dimension five temperatures between 0.15 and 0.3 were used (in arbitrary units depending on the scaling of the Rosetta score), yielding a total of 25 replicas. Exchange between neighboring replicas is attempted every 1,000 steps along the two dimensions. Well-tempered ensemble technique was applied again to improve the exchange rate with tunable factor $\gamma = 5$. The bin size for well-tempered ensemble technique to collect the history-dependent bias energy is set to two units of the Rosetta docking energy. Note, that each replica accumulated separate history-dependent biasing potentials depending on the individual sampling history. Replica exchange rate ranged between 14% and 56% in the temperature dimension with median value 32%. In the scaling dimension, it increased along the shifting from soft to hard repulsive interaction, and ranged between 12% and 99% with median value 39%.

Implementation in Rosetta

Previously, we implemented replica exchange within the general Metropolis-Hastings framework of the Rosetta3 software package [14]. The replica exchange module is accessible through the RosettaScripts interface and can be combined with any conformational moves that are implemented as children of the ThermodynamicMover class. For rigid-body docking refinement, we have applied a rigid-body mover UnbiasedRigidBodyPerturbNoCenterMover, and sidechain movers including JumpRotamerSidechainMover, PerturbRotamerSidechainMover and PerturbChiSidechainMover. For detailed balance, UnbiasedRigidBodyPerturbNoCenterMover performs unbiased rotational and translational perturbations in the restricted space as described in the previous section, and sidechain movers provide the proposal density of the perturbation through implementation of the abstract virtual function compute_proposal_density() in the ThermodynamicMover interface. The acceptance of a move is decided by the Metropolis criterion [44].

Side-chain motion is applied on one randomly selected residue (among all the residues and all the residue types but Proline) each time. Continuous sampling of side chain chi angles are used instead of fixed rotamers in all the three sidechain movers. The angles are chosen according to the Dunbrack rotamer library 2010 probabilities [45]. PerturbChiSidechainMover does a perturbation on the side chain chi angles, either uniformly distributed or Gaussian distributed with a given magnitude. For JumpRotamerSidechainMover and PerturbRotamerSidechainMover, a rotamer is first selected randomly or selected such that it has the highest probability of proposing the old chi angles according to the Dunbrack rotamer library probabilities, respectively, then individual chi angles are chosen using Gaussian distributed random angles with the means and standard deviations from the Dunbrack rotamer library.

The well-tempered ensemble technique is implemented into the framework of MetropolisHastings as ThermodynamicObserver. It is applied with a certain time interval (here in the test stride = 1) and deposits the Gaussians to the bias energy with height of

$$W = \omega e^{-[V(s,t)/\Delta T]} \tau_G$$

where τ_G is the time interval or stride number, V(s, t) is the old bias energy in the energy bin where the current decoy's energy has dropped into, ω represents the initial bias deposition rate and $\Delta T = (1 - \gamma)T$, in which γ represents the tunable factor and T is the temperature in the simulation [37,39,40,46]. When well-tempered ensemble technique is applied, acceptance of a move or an exchange attempt is decided based on the total energy, which is the sum of the force field energy and the bias energy. For final analysis, only the force field scoring energy was used.

Construction of a Benchmark

The four protocols were first tested on 10 unbound targets (Table 1) from the benchmak4.0 set [47,48] with reasonable energy funnels using the RosettaDock scoring force field. This was checked by generating 1000 decoys with standard RosettaDock full protocol starting from the bound docking geometry (using unbound structures). These 10 targets belong to the group of "rigid body" docking cases with small changes of side chains associated with complex formation (according to the classification of the protein docking benchmark4.0 collection [47,48]). In addition, the standard MC protocol and the WTE_H_REMC protocols were also tested on another 10 unbound targets including one "difficult" (1JK9) and two "medium difficulty" (1MQ8 and 2CFH) targets (Table 1). The number of residues of the 20 targets range between 122 and 409.

Table 1. Test complex structures and partner structures

Complex	Cat.	Difficulty	Partner I	Nres 1	Partner II	Nres 2	RMSD (Å)	DASA (Å)
1EAW_A:B	E	rigid	1EAX_A	241	9PTI_	56	0.54	1866
1GCQ_B:C	O	rigid	1GRI_B	66	1GCP_B	56	0.92	1208
1KTZ_A:B	O	rigid	1TGK_	105	1M9Z_A	82	0.39	989
1PPE_E:I	E	rigid	1BTP_	223	1LU0_A	29	0.44	1688
1S1Q_A:B	O	rigid	2F0R_A	141	1YJ1_A	69	0.98	1288
2AYO_A:B	O	rigid	2AYN_A	337	2FCN_A	72	1.39	3027
2SNI_E:I	E	rigid	1UBN_A	274	2CI2_I	64	0.35	1628
3D5S_A:C	O	rigid	1C3D_A	294	2GOM_A	61	0.56	1620
3SGQ_E:I	E	rigid	2QA9_E	185	2OVO_A	51	0.39	1211
7CEI_A:B	E	rigid	1UNK_D	127	1M08_B	87	0.7	1384
1AY7_A:B	E	rigid	1RGH_B	96	1A19_B	89	0.54	1237
1H9D_A:B	O	rigid	1EAN_A	125	1ILF_A(1)	114	1.32	2121
1HE1_C:A	O	rigid	1MH1_	176	1HE9_A	128	0.93	2113
1JK9_A:B	O	difficult	1QUP_A	219	2JCW_A	153	2.51	2130
1MQ8_A:B	O	medium	1IAM_A	184	1MQ9_A	171	1.76	1253
1RV6_VW:X	O	rigid	1FZV_AB	189	1QSZ_A	92	1.09	1626
1YVB_A:I	E	rigid	2CHU_A	241	1CEW_I	108	0.51	1743
2CFH_A:C	O	medium	1SZ7_A	156	2BJN_A	137	1.55	2384
2OUL_A:B	E	rigid	3BPF_A	236	2NNR_A	107	0.53	1933
2SIC_E:I	E	rigid	1SUP_	275	3SSI_	107	0.36	1617

cat: Complex category labels: E = Enzyme/Inhibitor or Enzyme/Substrate O = Others.

RMSD: RMSD of Ca atoms of interface residues calculated after finding the best superposition of bound and unbound interfaces.

DASA: Change in Accessible Surface Area upon complex formation.

Analysis of Docking Results and Computational Efficiency

The sampled docked complex were analyzed according to ligand RMSD (L_rmsd) and fraction of native contacts (f_{nat}), as defined in CAPRI [49] using the bound complexes as references. To evaluate the capacity of the methods to sample near-native decoys, we calculated the fraction of CAPRI medium (**, $f_{nat} \geq 0.5$ & L_rmsd > 1Å or $0.3 \leq f_{nat} < 0.5$ & L_rmsd ≤ 5) or high (***,$f_{nat} \geq 0.5$ & L_rmsd ≤ 1Å) quality decoys in the collected results [50]. To evaluate the agreement between generated complexes with closest L_rmsd and best f_{nat} compared to the bound complex. To evaluate the scoring energies the interaction score (I_sc) was used which is computed by subtracting the all-atom energy of non-interacting partners from the all-atom energy of the interacting partners in the complex. To compute the energy of non-interacting partners the two binding partners are moved far away from each other while keeping all internal degrees of freedom fixed. To investigate the efficiency of optimizing the scoring energies, we calculated for a given MC step number the average difference of the sampled best score (up to the selected MC step number) and the final most favorable score.

RESULTS AND DISCUSSION

Monte Carlo docking simulations are frequently used to perform protein-protein docking searches or for the refinement of predicted complexes at atomic resolution including limited conformational changes of the partner structures [14–17]. In recent years, enhanced sampling methods to improve the MC search efficiency have been developed. In order to test the performance of such improvements we compare the application of standard MD, parallel tempering REMC, well-tempered replica exchange (WTE-REMC), and well-tempered ensemble combined with 2-dimensional temperature and Hamiltonian replica exchange (WTE-H-REMC) to a set of protein-protein complexes in unbound partner conformation. In each case the MC moves included rigid body translation and rotation as well as side-chain moves (illustrated in Fig 2). For each protocol the same start configuration was used corresponding to a random arrangement of one mobile partner placed approximately 15 Å away from the bound complex geometry. In case of the replica exchange methods only configurations in the reference replica were retained, resulting in approximately 25x2,000 decoys for standard MC protocol (with $2x10^6$ MC steps), 2x2,000 for REMC protocol, 5x2,000 for WTE-REMC protocol and 1x2,000 for WTE-H-REMC protocol, respectively. On a 2.6 GHz AMD Opteron Processor (12 cores), $2x10^6$ MC steps take between 4.5–20 hours. The sampled docking solutions were analyzed in terms of deviation from the known complex geometry (using the root mean square backbone deviation of the mobile ligand partner protein from the bound complex after best superposition of the receptor protein onto the bound complex: L_rmsd) and interaction score (I_sc).

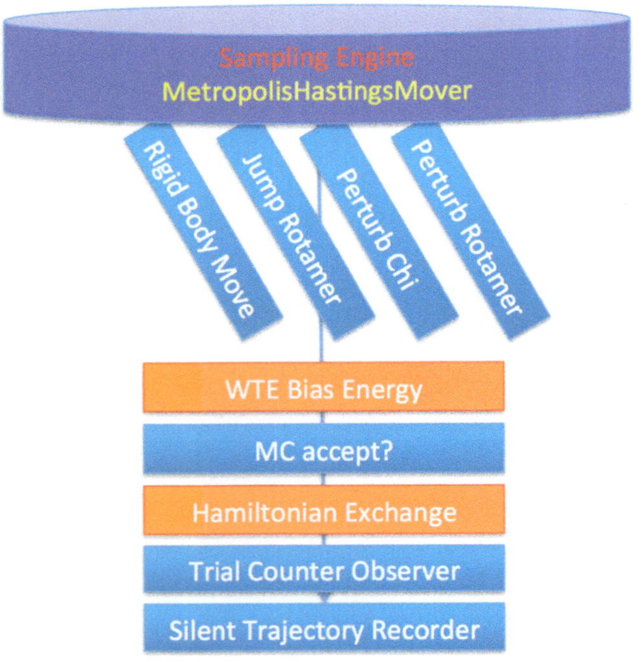

Figure 2. Workflow represented by combination of Rosetta modules and setup of the four docking protocols. The modules in orange, representing the enhanced sampling techniques of replica exchange and well-tempered ensemble, are only applied in the combined protocols.

The protocols were first tested on 10 benchmark targets of the "rigid body" category (with only limited side chain changes upon complex formation). We consider the sampling of medium or high-quality solutions (CAPRI **/*** solutions, defined in Methods section) according to the CAPRI criteria as successful docking refinement. The evolution of the sampling in terms of L_rmsd and I_sc scoring is showcased for two representative targets (pdb1EAW and pdb3SGQ) in Fig 3. For the pdb1EAW-target all methods sample progressively lower (more favorable) force field scores with increasing number of MC steps. However, for the first example only the WTE-H-REMC protocol samples docking solutions with L_rmsd < 5 Å after 2×10^6 MC steps. Only after 10^7 MC steps all techniques except the standard MC technique sample near-native solutions (Fig 3A). For the second example (pdb3SGQ) the MC technique successfully samples solutions with Lrmsd < 5 Å only after 10^7 MC steps whereas all three advanced sampling techniques reach near-native solutions already after 3×10^5 MC steps (Fig 3B). Qualitatively similar trends were observed for all other test cases.

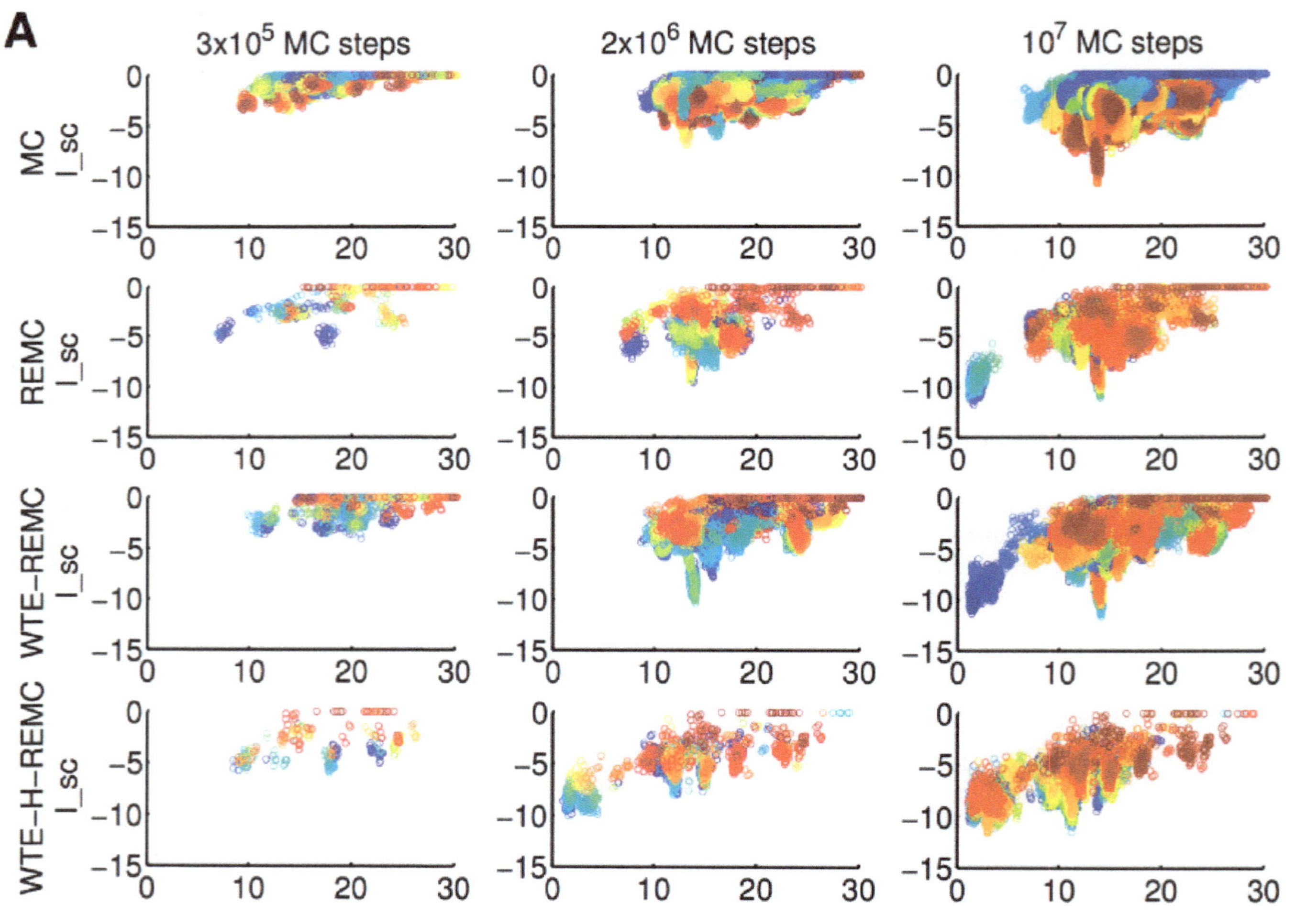

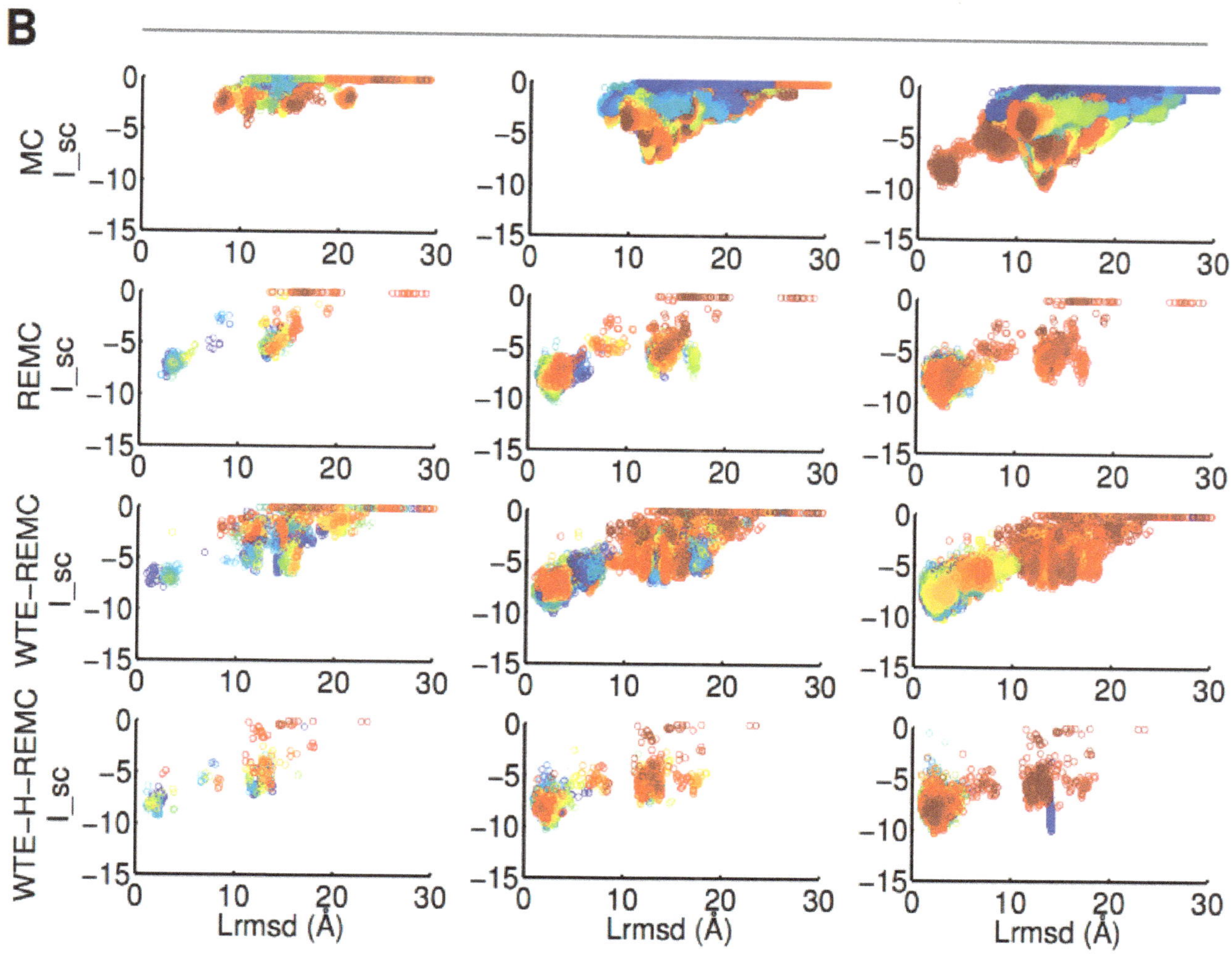

Figure 3. Scatter plot of interaction score I_sc (Rosetta units) vs. L_rmsd **(Å) for the four docking refinement protocols and two representative targets, 1EAW (A) and 3SGQ (B).** The protocol is indicated on the left for each row of plots. The snapshots number is color-coded, that means blue and red dots corresponding to decoys sampled at the beginning and the end of the docking searches in each panel, respectively. The three columns of plots indicate the result after different simulation lengths (indicated on top of each column).

A quantitative comparison of the docking refinement solutions in terms of the fraction of native contacts (f_{nat}) indicates that the WTE-H-REMC method succeeded in all 10 cases in sampling near-native docking solutions with very high quality (Fig 4A). In contrast, the other protocols succeeded only in 3 (MC) or 7 (REMC and WTE-REMC) of the first 10 cases (Fig 4A). Note, that near-native docking solutions are also the best scoring solutions in several but not all docking test cases (e.g. for targets 1KTZ, 3D5S and 7CEI docking solutions with Lrmsd > 10 Å give I_sc scores lower than the solutions closest to the bound docking geometry). The results on the first 10 test cases indicate that the WTE-H-REMC enhanced sampling protocol showed the best performance. For a second test set of 10 targets (including also targets of the "medium" and "difficult" category, see Table 1), only the standard MC and the WTE-H-REMC protocols were compared. Again, the WTE-H-REMC protocol gave better docking results in 5 cases (IJK9, 1MQ8, 2CFH, 2OUL, 2SIC) with lower final I_sc scores and Lrmsd (Fig 4A) compared to the standard MC-method. However, in two cases (1H9D and 1HE1) the standard MC–method reached configurations closer to the bound form compared to the WTE-H-REMC technique. Note, that especially in these cases the score of near-native docking solutions was higher (less favorable) than for alternative docking geometries. Since the search techniques optimize the score (and not deviation from bound structure) it may explain the failure of the WTE-H-REMC technique in these cases.

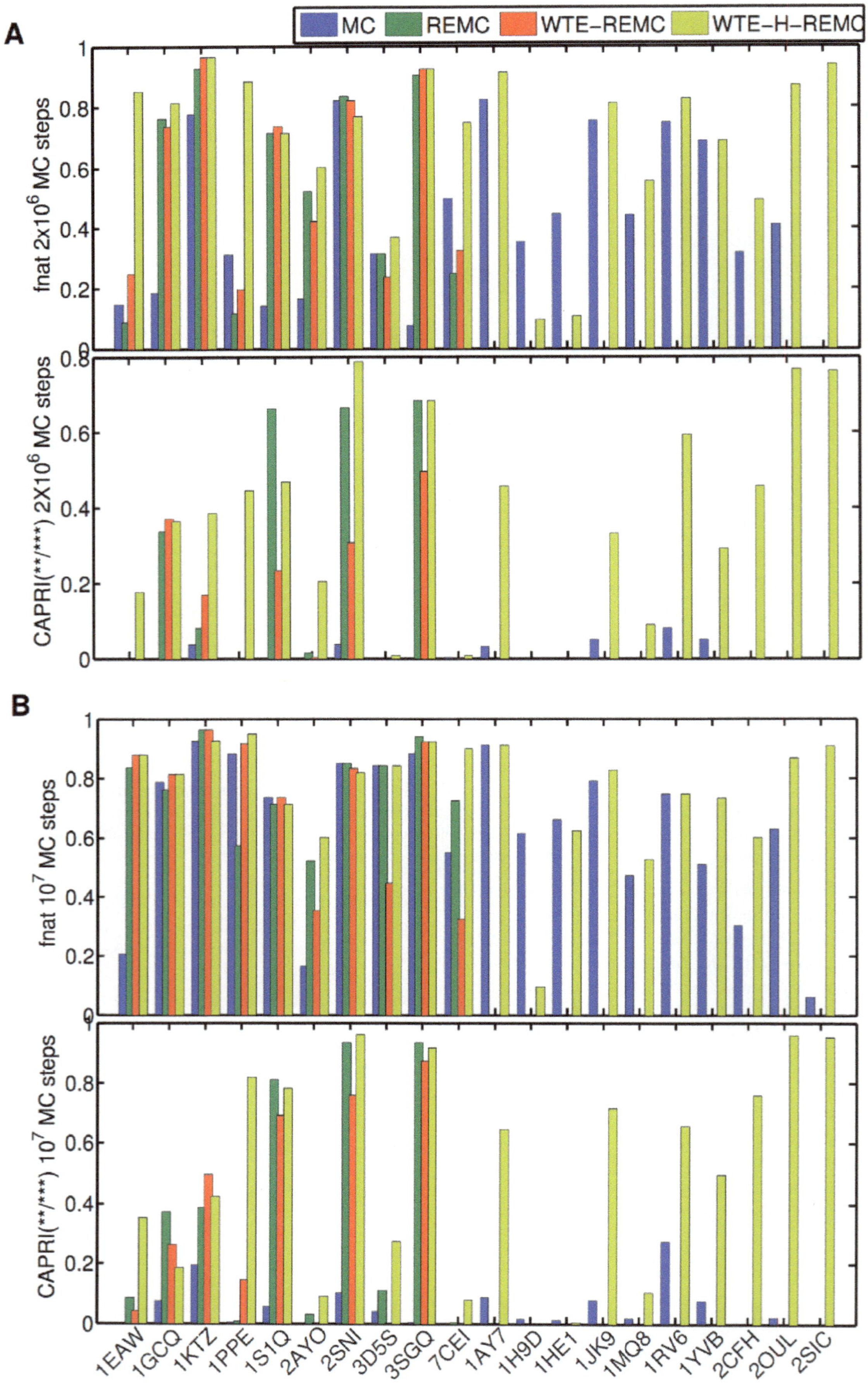

Figure 4. Agreement between sampled docking geometries and the corresponding bound complex. (A, upper panel) Highest fraction of native contacts (f_{nat}) found in the top 10 decoys (according to L_rmsd) sampled in each protocol (2x10[6] MC steps). (A, lower panel) Fraction of CAPRI medium and high quality complexes found for each target and each protocol (the protocols MC, REMC, WTE-REMC and WTE-H-REMC are indicated by different colors). (B) same as in (A) but for the docking refinement runs with 10[7] MC steps.

To check if slower convergence to reach low energy docking solutions was the main reason for the failure of some protocols to reach near-native docking solutions, we increased the trajectory length to 10^7 MC steps. Indeed, the success of MC, REMC, and WTE-REMC protocols to reach near-native docking solutions increased to 8, 10 and 10 out of the first 10 targets, respectively (Fig 4B). The results were further analyzed with respect to fraction of native contacts (f_{nat}) of near –native docking solutions and the maximum quality of predicted docking geometries (Fig 4). Also for these measures and in case of the protocol with $2x10^6$ MC steps the WTE-H-REMC protocol achieves overall the best performance (Fig 4A). For the extended protocol with 10^7 MC steps the quality of solutions in terms of f_{nat} is more similar for all 4 protocols (Fig 4B), indicating that indeed the standard MC technique requires longer searches to achieve convergence compared to the WTE-H-REMC method. The best f_{nat} for protocols MC, REMC and WTE-REMC all increased on average around 17%, getting close to that of the WTE-H-REMC protocol. The best f_{nat} for protocol WTE-H-REMC also increased slightly (~7%, Fig 4B). Fig 3 presents two representative examples on targets 1EAW and 3SGQ of ligand RMSD (L_rmsd) versus interaction score (I_sc). Comparing these data with that from the trajectory of $2x10^6$ MC steps, it shows that the details of the energy landscape sampled by WTE-H-REMC remains similar indicating reasonable convergence within $2x10^6$ MC steps for most cases. Meanwhile, the rigid-body space reached for target 1EAW by protocols MC, REMC and WTE-REMC, and on target 3SGQ by protocol MC, drastically improved (Fig 3).

Since the force field score is the quantity which is optimized during the docking searches (and not the agreement with the bound structure) it is of interest to compare the protocols in terms of the efficiency to optimize the force field score. For each target, the average of the lowest 10 interaction scores (lowest I_sc) sampled up to a given MC step number was considered (using the extended trajectories) and the difference relative to the lowest score found during the search was recorded. The average of this quantity for all 20 cases was calculated and plotted in Fig 5. The enhanced sampling techniques REMC, WTE-REMC and WTE-H-REMC, consistently, reached lower interaction scores than the standard MC method for a given number of MC step (Fig 5A). The WTE-H-REMC technique reached on average lower I_sc than the other three protocols already after ~$3x10^5$ MC steps. Interestingly, the same analysis using the L_rmsd instead of the I_sc yields the same trend, indicating that on average the I_sc score correlates with the L_rmsd (Fig 5B). Low L_rmsd of sampled geometries gives on average (but not for all targets) also a favorable score.

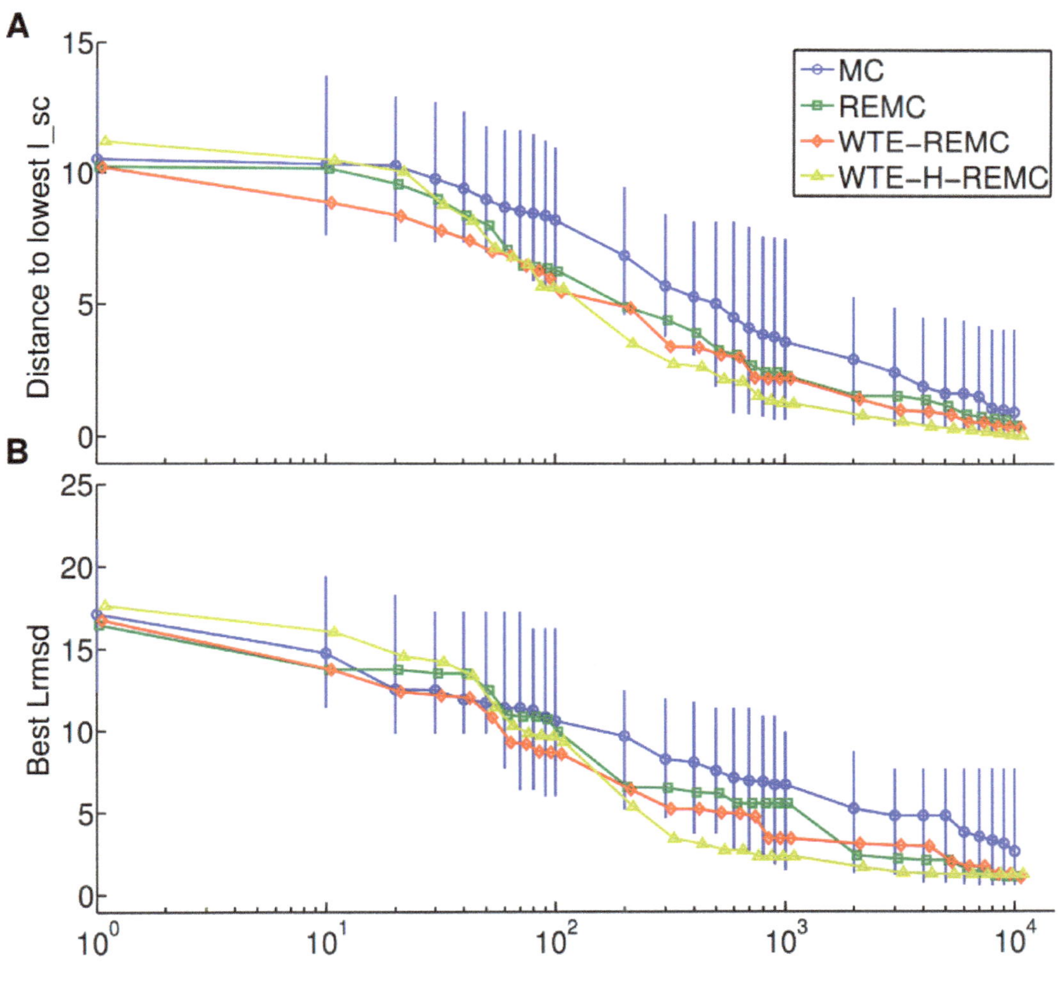

Figure 5. Evolution of I_sc docking interaction score (A) and best sampled Lrmsd (B) vs. MC step number. The MC step number is scaled by x1000. For the interaction score I_sc the smallest difference (sampled up to the selected step number) relative to the lowest scoring complex sampled in the entire docking search is plotted. The variance in sampled scores (up to the considered number of MC steps) is indicated by error bars for the MC protocol. It is of similar magnitude for the other protocols (not shown). For (B) the smallest sampled L_rmsd up to the step number indicated in the x-axis is shown.

Protocol Testing

A subset of three complexes (pdb-entries: 1PPE, 2OUL and 2SIC) was chosen as an independent protocol test set. For two of these three complexes, only the MC and the WTE_H_REMC protocol had been tested in the main work. The four protocols were run using the RosettaScripts interface with parameters as described above, trajectory length of $2x10^6$ MC steps and a newly created starting conformation. In contrast to the previous docking runs, only a single starting conformation was generated for each complex. All the protocols could be successfully executed. A summary of the test results can be found in Table 2. For all three test cases, the enhanced sampling methods yielded structures of lower L_rmsd and higher f_{nat} than standard MC sampling. Enhanced sampling methods generated near-native docking models, whereas standard MC sampling did not yield any structures of CAPRI one star quality or better. The WTE_H_REMC technique was the only method to generate CAPRI three star quality structures for all three cases and thus yielded the best performance on the test set. The results match the previously presented data for these three complexes and thus confirm that the sampling does not depend on the choice of the starting conformation. The randomly generated starting conformations sometimes contained clashes, but the enhanced sampling methods were able to refine them to high-quality solutions. Hence, it might be possible to use enhanced sampling methods also for refinement of docked complexes using other methods than RosettaDock.

Table 2. Results for protocol testing on a subset of three complexes

PDB ID	MC	REMC	WTE-REMC	WTE-H-REMC
1PPE	5.6 Å	1.5 Å	2.2 Å	1.0 Å
	14.5 h	14 h	12.3 h	17.8 h
2OUL	4.2 Å	1.9 Å	2.2 Å	0.7 Å
	21.8 h	18.3 h	18 h	22 h
2SIC	4.9 Å	0.4 Å	2.8 Å	0.5 Å
	25.5 h	22 h	19 h	20 h

For all complexes, the best sampled L_rmsd and the execution time on 27 threads are listed.

CONCLUSIONS

In this work, four different Monte Carlo advanced sampling protocols implemented in RosettaDock to predict the geometry of protein-protein complexes have been compared. For all the protocols (on each target) the same initial protein-protein docking start configurations were used with 15 Å translational displacements and 60° rotation of one partner from the native complex structure. This situation corresponds to the frequent scenario that the interaction region between proteins is approximately known and start configurations are placed close to the approximately known binding region. It is also very useful for directly comparing the docking performance of different approaches at the computational demanding atomistic high resolution level. Note, that the protocol is computationally too demanding for routine applications that require to search over the entire surface of two protein partners. If complete protein surfaces are considered it is also very likely that the scoring function is not accurate enough to pick out near-native solutions as lowest energy complexes. Our results on docking refinement show indeed that the application of advanced sampling schemes improves the docking refinement performance yielding final configurations in better agreement with the bound structure and yield also a much larger fraction of near-native structures compared to regular MC searches. The WTE-H-REMC consistently gave the best performance since it explores the phase space more efficiently due to larger energy fluctuation and due to the added biasing potential that effectively smoothes the landscape and increases the replica exchange rates. An increase of the number of MC steps to 10^7 in the standard MC protocol resulted in improved performance achieving a similar fraction of native contacts of best sampled solutions and similar final docking scores compared to WTE-H-REMC with $2x10^6$ steps. Hence, in most cases a standard MC protocol requires roughly 5 times larger computational demand to achieve the same final docking prediction performance. It should be emphasized that this reflects only a general trend. For some test cases even 10^7 MC steps still gave inferior docking results compared to WTE-H-REMC and still the fraction of the best solutions relative to the total number of sampled geometries is much smaller than for the advanced sampling method. Further improvement might be possible by an adjustment of the bin size in the WTE to collect the history dependent bias energy. However, an even larger gain could be achieved by an improvement of the docking scoring function to increase the gap between ranking near-native and non-native solutions.

ACKNOWLEDGMENTS

The authors thank Drs. Isaure de Beauchene and Sjoerd de Vries for helpful discussions. This work was performed in the framework of SFB 1035. Computer resources for this project have been provided by the Gauss Centre for Supercomputing/ Leibniz Supercomputing Centre under grant pr86pu.

AUTHOR CONTRIBUTIONS

Conceived and designed the experiments: ZZ OL MZ. Performed the experiments: ZZ CS. Analyzed the data: ZZ CS MZ. Contributed reagents/materials/analysis tools: ZZ OL CS MZ. Wrote the paper: ZZ OL CS MZ.

REFERENCES

1. Stumpf MPH, Thorne T, de Silva E, Stewart R, An HJ, Lappe M, et al. Estimating the size of the human interactome. Proc. Natl. Acad Sci USA. 2008; 105: 6959–6964. doi: 10.1073/pnas.0708078105. pmid:18474861

2. Garma L, Mukherjee S, Mitra P, Zhang Y. How Many Protein-Protein Interactions Types Exist in Nature? PLoS ONE. 2012; 7: e38913. doi: 10.1371/journal.pone.0038913. pmid:22719985
3. Baker D. Protein folding, structure prediction and design. Biochem. Soc. Trans. 2014; 42: 225–229. doi: 10.1042/BST20130055. pmid:24646222
4. Parmeggiani F, Huang P- S, Vorobiev S, Xiao R, Park K, Caprari S, et al. A general computational approach for repeat protein design. J Mol Biol. 2015; 427: 563–575. doi: 10.1016/j.jmb.2014.11.005. pmid:25451037
5. London N, Ambroggio X. An accurate binding interaction model in de novo computational protein design of interactions: if you build it, they will bind. J Struct Biol. 2014; 185: 136–146. doi: 10.1016/j.jsb.2013.03.012. pmid:23558036
6. Baker D. Prediction and design of macromolecular structures and interactions. Phil. Trans. R. Soc. Lond. B. 2006; 361: 459–463. pmid:16524834
7. Kortemme T, Joachimiak LA, Bullock AN, Schuler AD, Stoddard BL, Baker D. Computational redesign of protein-protein interaction specificity. Nat Struct Mol Biol. 2004; 11: 371–379. pmid:15034550 doi: 10.1038/nsmb749
8. Huang S-Y. Search strategies and evaluation in protein-protein docking: principles, advances and challenges. Drug Discov Today. 2014; 19: 1081–1096. doi: 10.1016/j.drudis.2014.02.005. pmid:24594385
9. Vajda S, Kozakov D. Convergence and combination of methods in protein–protein docking. Current Opinion in Structural Biology. 2009; 19: 164–170. doi: 10.1016/j.sbi.2009.02.008. pmid:19327983
10. Katchalski-Katzir E, Shariv I, Eisenstein M, Friesem AA, Aflalo C, Vakser IA. Molecular surface recognition: determination of geometric fit between proteins and their ligands by correlation techniques. Proc Natl Acad Sci USA. 1992; 89: 2195–2199. pmid:1549581 doi: 10.1073/pnas.89.6.2195
11. Lawrence MC, Colman PM. Shape complementarity at protein-protein interfaces. J. Mol. Biol. 1994; 234: 946–950. doi: 10.1006/jmbi.1993.1648
12. Zacharias M. Protein-protein docking with a reduced protein model accounting for side-chain flexibility. Protein Sci. 2003; 12: 1271–1282. pmid:12761398 doi: 10.1110/ps.0239303
13. Gray JJ, Moughon S, Wang C, Schueler-Furman O, Kuhlman B, Rohl CA, et al. Protein–Protein Docking with Simultaneous Optimization of Rigid-body Displacement and Side-chain Conformations. J Mol Biol. 2003; 331: 281–299. pmid:12875852 doi: 10.1016/s0022-2836(03)00670-3
14. Zhang Z, Lange OF. Replica Exchange Improves Sampling in Low-Resolution Docking Stage of RosettaDock. PLoS ONE. 2013; 8(8): e72096. doi: 10.1371/journal.pone.0072096. pmid:24009670
15. Lorenzen S, Zhang Y. Monte Carlo refinement of rigid-body protein docking structures with backbone displacement and side-chain optimization. Protein Sci. 2007; 16: 2716–2725. pmid:17965193 doi: 10.1110/ps.072847207
16. Fernández Recio J, Totrov M, Abagyan R. ICM-DISCO docking by global energy optimization with fully flexible side-chains. Proteins: Structure, Function, and Bioinformatics. 2003; 52: 113–117. pmid:12784376 doi: 10.1002/prot.10383
17. Dominguez C, Boelens R, Bonvin AMJJ. HADDOCK: A Protein−Protein Docking Approach Based on Biochemical or Biophysical Information. J Am Chem Soc. 2003; 125: 1731–1737. pmid:12580598 doi: 10.1021/ja026939x
18. Gray JJ. High-resolution protein–protein docking. Current Opinion in Structural Biology. 2006; 16: 183–193. pmid:16546374 doi: 10.1016/j.sbi.2006.03.003
19. London N, Schueler-Furman O. Funnel hunting in a rough terrain: learning and discriminating native energy funnels. Structure. 2008; 16: 269–279. doi: 10.1016/j.str.2007.11.013. pmid:18275818
20. Dill KA, Chan HS. From Levinthal to pathways to funnels. Nat Struct Biol 1997; 4: 10–19. pmid:8989315 doi: 10.1038/nsb0197-10
21. Camacho CJ, Weng Z, Vajda S, DeLisi C. Free energy landscapes of encounter complexes in protein-protein association. Biophys J. 1999; 76: 1166–1178. pmid:10049302 doi: 10.1016/s0006-3495(99)77281-4
22. Swendsen RH, Wang JS. Replica Monte Carlo simulation of spin-glasses. Phys Rev Lett. 1986; 57: 2607–2609. pmid:10033814 doi: 10.1103/physrevlett.57.2607
23. Hansmann UHE. Parallel tempering algorithm for conformational studies of biological molecules. Chem Phys Lett. 1996; 281: 140–150. doi: 10.1016/s0009-2614(97)01198-6
24. Falcioni M, Deem MW. A biased Monte Carlo scheme for zeolite structure solution. J Chem Phys 1999; 110, 1754. doi: 10.1063/1.477812
25. Sugita Y, Okamoto Y. Replica-exchange molecular dynamics method for protein folding. Chem Phys Lett. 1999; 314: 141–151. doi: 10.1016/s0009-2614(99)01123-9
26. Fukunishi H, Watanabe O, Takada S. On the Hamiltonian replica exchange method for efficient sampling of biomolecular systems: Application to protein structure prediction. J Chem Phys. 2002; 116: 9058. doi: 10.1063/1.1472510
27. Hritz J, Oostenbrink C. Hamiltonian replica exchange molecular dynamics using soft-core interactions. J Chem Phys. 2008; 128: 144121. doi: 10.1063/1.2888998. pmid:18412437
28. Luitz MP, Zacharias M. Protein-ligand docking using hamiltonian replica exchange simulations with soft core potentials. J Chem Inf Model. 2014; 54: 1669–1675. doi: 10.1021/ci500296f. pmid:24855894
29. Dantas G, Corrent C, Reichow SL, Havranek JJ, Eletr ZM, Isern NG, et al. High-resolution structural and thermodynamic analysis of extreme stabilization of human procarboxypeptidase by computational

protein design. J Mol Biol. 2007; 366: 1209–1221. pmid:17196978 doi: 10.1016/j.jmb.2006.11.080

30. Yanover C, Schueler-Furman O, Weiss Y. Minimizing and learning energy functions for side-chain prediction. J Comput Biol. 2008; 15: 899–911. doi: 10.1089/cmb.2007.0158. pmid:18707538

31. Yan Q, de Pablo JJ. Hyper-parallel tempering Monte Carlo: Application to the Lennard-Jones fluid and the restricted primitive model. J Chem Phys. 1999; 111: 9509. doi: 10.1063/1.480282

32. Yan Q, de Pablo JJ. Hyperparallel tempering Monte Carlo simulation of polymeric systems. J Chem Phys. 2000; 113: 1276–1282. doi: 10.1063/1.481905

33. Earl DJ, Deem MW. Parallel tempering: Theory, applications, and new perspectives. Phys Chem Chem Phys. 2005; 7: 3910. pmid:19810318 doi: 10.1039/b509983h

34. Rathore N, Chopra M, de Pablo JJ. Optimal allocation of replicas in parallel tempering simulations. J Chem Phys. 2005; 122: 024111. pmid:15638576 doi: 10.1063/1.1831273

35. Katzgraber HG, Trebst S, Huse DA, Troyer M. Feedback-optimized parallel tempering Monte Carlo. J Stat Mech. 2006: P03018–P03018.

36. Ostermeir K, Zacharias M. Advanced replica-exchange sampling to study the flexibility and plasticity of peptides and proteins. Biochim Biophys Acta. 2013; 1834: 847–853. doi: 10.1016/j.bbapap.2012.12.016. pmid:23298543

37. Barducci A, Bonomi M, Parrinello M. Metadynamics. WIREs Comput Mol Sci. 2011; 1: 826–843. doi: 10.1002/wcms.31

38. Laio A, Gervasio FL. Metadynamics: a method to simulate rare events and reconstruct the free energy in biophysics, chemistry and material science. Rep Prog Phys. 2008; 71: 126601. doi: 10.1088/0034-4885/71/12/126601

39. Bonomi M, Parrinello M. Enhanced sampling in the well-tempered ensemble. Phys Rev Lett. 2010; 104: 190601. pmid:20866953 doi: 10.1103/physrevlett.104.190601

40. Deighan M, Bonomi M, Pfaendtner J. Efficient Simulation of Explicitly Solvated Proteins in the Well-Tempered Ensemble. J Chem Theory Comput. 2012; 8: 2189–2192. doi: 10.1021/ct300297t

41. Chaudhury S, Berrondo M, Weitzner BD, Muthu P, Bergman H, Gray JJ. Benchmarking and Analysis of Protein Docking Performance in Rosetta v3.2. PLoS ONE. 2011; 6: e22477. doi: 10.1371/journal.pone.0022477. pmid:21829626

42. Gray JJ, Moughon SE, Kortemme T, Schueler-Furman O, Misura KMS, Morozov AV, et al. Protein-protein docking predictions for the CAPRI experiment. Proteins: Structure, Function, and Bioinformatics. 2003; 52: 118–122. pmid:12784377 doi: 10.1002/prot.10384

43. Rohl CA, Strauss CE, Misura KM, Baker D. Protein structure prediction using Rosetta. Meth Enzymol. 2004; 383: 66–93. pmid:15063647 doi: 10.1016/s0076-6879(04)83004-0

44. Metropolis N, Rosenbluth AW, Rosenbluth MN, Teller AH, Teller E. Equation of State Calculations by Fast Computing Machines. J Chem Phys. 1953; 21: 1087–1092. doi: 10.1063/1.1699114

45. Shapovalov MV, Dunbrack RL Jr. A Smoothed Backbone-Dependent Rotamer Library for Proteins Derived from Adaptive Kernel Density Estimates and Regressions. Structure. 2011; 19: 844–858. doi: 10.1016/j.str.2011.03.019. pmid:21645855

46. Barducci A, Bussi G, Parrinello M. Well-Tempered Metadynamics: A Smoothly Converging and Tunable Free-Energy Method. Phys Rev Lett. 2008; 100: 020603. pmid:18232845 doi: 10.1103/physrevlett.100.020603

47. Hwang H, Vreven T, Janin J, Weng Z. Protein-protein docking benchmark version 4.0. Proteins. 2010; 78: 3111–3114. doi: 10.1002/prot.22830. pmid:20806234

48. Hwang H, Pierce B, Mintseris J, Janin J, Weng Z. Protein-protein docking benchmark version 3.0. Proteins. 2008; 73: 705–709. doi: 10.1002/prot.22106. pmid:18491384

49. Méndez R, Leplae R, De Maria L, Wodak SJ. Assessment of blind predictions of protein–protein interactions: current status of docking methods. Proteins. 2003; 52: 51–67. pmid:12784368 doi: 10.1002/prot.10393

50. Lensink MF, Wodak SJ. Docking and scoring protein interactions: CAPRI 2009. Proteins. 2010; 78: 3073–3084. doi: 10.1002/prot.22818. pmid:20806235

CHAPTER 16

PARALLEL KINETIC MONTE CARLO SIMULATION FRAMEWORK INCORPORATING ACCURATE MODELS OF ADSORBATE LATERAL INTERACTIONS

Jens Nielsen[1], Mayeul d'Avezac[1], James Hetherington[1] and Michail Stamatakis[2]

Jens Nielsen[1], Mayeul d'Avezac[1], James Hetherington[1] and Michail Stamatakis[2]

[1]Research Software Development Team, Research IT Services, University College London, Torrington Place, London WC1E 6BT, United Kingdom

[2]Department of Chemical Engineering, University College London, Torrington Place, London WC1E 7JE, United Kingdom

ABSTRACT

Ab initio kinetic Monte Carlo (KMC) simulations have been successfully applied for over two decades to elucidate the underlying physico-chemical phenomena on the surfaces of heterogeneous catalysts. These simulations necessitate detailed knowledge of the kinetics of elementary reactions constituting the reaction mechanism, and the energetics of the species participating in the chemistry. The information about the energetics is encoded in the formation energies of gas and surface-bound species, and the lateral interactions between adsorbates on the catalytic surface, which can be modeled at different levels of detail. The majority of previous works accounted for only pairwise-additive first nearest-neighbor interactions. More recently, cluster-expansion Hamiltonians incorporating long-range interactions and many-body terms have been used for detailed estimations of catalytic rate [C. Wu, D. J. Schmidt, C. Wolverton, and W. F. Schneider, J. Catal.286, 88 (2012)]. In view of the increasing interest in accurate predictions of catalytic performance, there is a need for general-purpose KMC approaches incorporating detailed cluster expansion models for the adlayer energetics. We have addressed this need by building on the previously introduced graph-theoretical KMC framework, and we have developed Zacros, aFORTRAN2003 KMC package for simulating catalytic chemistries. To tackle the high computational cost in the presence of long-range interactions we introduce parallelization with OpenMP. We further benchmark our framework by simulating a KMC analogue of the NO oxidation system established by Schneider and co-workers [J. Catal.286, 88 (2012)]. We show that taking into account only first nearest-neighbor interactions may lead to large errors in the prediction of the catalytic rate, whereas for accurate estimates thereof, one needs to include long-range terms in the cluster expansion.

INTRODUCTION

The origins of temporal investigations of processes on lattices can be sought in the seminal work by Glauber, who analyzed a Markov process for spin flips in the Ising Model, back in 1963.[1] The subsequent development of kinetic Monte Carlo approaches[2,3] enabled the study of complex time-dependent phenomena in the Ising model such as metastability and dynamic critical phenomena (see for instance Refs. 4–7). Application of kinetic Monte Carlo (KMC) to catalysis on surfaces was pioneered by Ziff, Gulari, and Barshad,[8] who investigated kinetic phase transitions on a model for CO oxidation in 1986. Subsequent studies focused on a variety of temporal phenomena in catalytic systems, such as bistable responses, noise-induced transitions, as well as oscillatory behavior.[9–15]

Citation: Nielsen, J; d'Avezac, M; Hetherington, J; Stamatakis, M; (2013) Parallel kinetic Monte Carlo simulation framework incorporating accurate models of adsorbate lateral interactions. Journal of Chemical Physics, 139 (22) , Article 224706. 10.1063/1.4840395.

Since these early studies, significant developments have taken place. Over the past two decades KMC has evolved to enable the coupling with *ab initio* calculations for a first-principles based simulation of chemistries on catalytic surfaces[16–31] (for reviews, see Refs. 32–34). In these first-principles KMC frameworks the reaction energies and activation barriers are typically obtained from density functional theory (DFT) calculations, and the kinetic parameters are calculated by employing transition state theory (TST) approximations. Thus, within the level of accuracy of DFT and to the extent of validity of TST, the quality of the predictions obtained by KMC depends on "how well" the *ab initio* data have been incorporated into the simulation.

In particular, attractive or repulsive interactions between adsorbates on the catalytic surface have been shown to affect the rates of elementary reactions.[16,35] Early first-principles KMC frameworks used DFT and bond-order conservation (BOC) methods to account for such effects.[16,17]In some instances, such interactions were neglected altogether or modeled by pairwise additive nearest neighbor contributions, due to large computational expense needed for implementing more accurate models.[36] The most general such models consist of cluster-expansion (CE) Hamiltonians, which can accommodate any level of accuracy by taking into account long-range and many-body contributions to the total energy. This approach dates back to the 1980s,[37] and has been applied extensively to study the thermodynamics of adlayer structures.[38–45]Moreover, CEs have been employed to investigate adsorption/desorption dynamics[46–48] and the diffusion of adatoms on metal systems.[49,50]

Only recently however, have cluster expansion Hamiltonians been applied for the estimation of catalytic rates.[51] In an elegant study, Schneider and co-workers investigated the kinetics of NO oxidation on Pt(111) by means of equilibrium Monte Carlo (MC) calculations and appropriate averaging of the microscopic kinetic rates over the lattice. The MC calculations employed a cluster expansion for O on Pt(111)[45]and yielded equilibrium structures of the oxygen adlayer. The dissociative adsorption energy of O_2 was then calculated for each pair of empty sites (using the cluster expansion) and mapped to an activation energy for O_2 dissociation thereon via a Brønsted-Evans-Polanyi (BEP) relationship. Thus, a distribution of activation energies was obtained, from which the average rate of O_2 dissociation was evaluated. To make the connection with NO oxidation, it was assumed that the NO/NO_2 mixture was at equilibrium with adsorbed oxygen, thereby computing the chemical potential of O which is the input to the aforementioned MC calculations.

In view of the increasing interest in accurate predictions of catalytic performance, long-range lateral interactions and possibly many-body contributions are expected to form an integral part of kinetic models of catalysts. We have thus developed a general-purpose KMC framework that incorporates detailed cluster expansion models for the adlayer energetics, building on the previously introduced graph-theoretical KMC approach.[52] Using this framework, one can define a cluster expansion model for the adlayer energetics, along with the forward activation energies of the elementary reactions for the chemistry under investigation. The reverse activation energies are computed through linear BEP relations thereby ensuring that the simulation exhibits microscopic reversibility. To tackle the high cost of computing the energetics in the presence of long-range interactions we introduce parallelization with OpenMP, and benchmark the performance of the framework by simulating a KMC analogue for the NO oxidation model established by Schneider and co-workers.[51]

The rest of the paper is organized as follows. In Sec. II we present a brief overview of the graph-theoretical KMC framework, and present our current development work incorporating cluster expansion Hamiltonians in the framework. Subsequently, we discuss the setting up of a benchmark model for NO oxidation on Pt(111) based on the cluster expansions developed by Schneider and co-workers.[45,51] In Sec. III, we validate our computational framework and investigate the behavior of the model under different operating conditions. We further discuss the performance and scalability of the parallel code. In Sec. IV, we highlight our contributions and discuss the significance of this work.

METHODOLOGY

The components and procedures of the graph-theoretical KMC implementation are discussed in Ref. 52, whereas Refs. 53–56 demonstrate the use of the method in gaining a fundamental understanding of surface processes in a variety of catalytic systems. For a general overview of KMC approaches and applications in heterogeneous catalysis the reader is referred to Ref. 55. Here, we will briefly review the framework and present in detail the calculation of the energetics which is the focus of the present work.

Overview of Graph-Theoretical KMC

The workload of the KMC algorithm consists of simulating a sequence of elementary events, such as adsorption, desorption, diffusion, reaction, which change the configuration of a lattice representing the catalytic surface. In the graph-theoretical KMC, the lattice, adsorbate configurations, elementary reactions, as well as energetic interaction contributions, are all represented by graphs. The input to a simulation consists of parameters controlling the behavior of the program (for instance, sampling times, stopping criteria, etc.), and the components specifying the physics of the system under investigation. For the latter, one has to provide the conditions (T, P, and gas phase composition), a lattice structure, an initial configuration of the lattice, an energetics model containing at minimum the energies of gas species and the binding configurations of adsorbates, and finally, a mechanistic model containing a list of possible elementary events. Note that in our previous work, no energetic model was specified. The code could only handle pairwise additive interactions and microscopic reversibility was manually implemented by correctly listing all interactions between neighboring species and the initial and final state species of each elementary event. In the current work, we introduce cluster-expansion based models for a general and thermodynamically consistent KMC simulation scheme.

As in previous KMC schemes, a simulation is initialized with the given lattice configuration (possibly an empty lattice), and a queue is generated containing all the possible lattice processes for the given configuration. For instance, if the lattice is initialized with one adsorbed molecule, the possible processes could be: adsorption on all empty sites, desorption of the adsorbate, or diffusional hops to each of the neighboring empty sites. The queue just mentioned also stores randomly generated times in which each process can occur. Under constant conditions, each of these random times follows an exponential distribution with a rate parameter equal to the kinetic constant of that process.[57]

After initialization, the algorithm enters a loop at each step of which the most imminent process in the queue (the one with the smallest time of occurrence) is executed. Thus, the reactants are removed from the lattice along with the processes they participate in, and the products are added to the lattice. The latter operation also involves detecting all the possible processes in which the newly added adsorbates can participate and including them in the queue of lattice processes. This detection entails searching the neighborhood of each newly added adsorbate to identify patterns matching an elementary event, e.g., desorption or diffusion. Note that for each of these patterns identified, a rate constant has to be calculated. Thus, one needs to know the activation energy, which in general depends on the local environment due to the lateral interactions. Moreover, the removal of adsorbed reactants from the lattice eliminates energetic interactions, and the addition of adsorbed reaction products introduces new energetic interactions which affect the activation energies and consequently the rates of existing processes. Computing the activation energies can be computationally intensive if the energetics' model contains long-range interactions. Thus, we implemented parallel computing in order to improve the efficiency of these calculations.

The procedures just described occur at every KMC step. By simulating a long sequence of such steps, a KMC trajectory (realization) is generated which can be subsequently post-processed to yield a variety of observables. For instance, counting the number of adsorbates on each site type yields surface coverages from which the most abundant surface intermediate can be found. Counting the number of reactant molecules consumed per site per time allows one to calculate the turnover frequency of the overall reaction. Moreover, the rates of each of the elementary reactions can be calculated by counting the occurrences thereof per unit time. Analysis of this data can reveal the most dominant pathway and sensitivity analysis can elucidate the processes limiting the overall reaction rate, namely the rate determining steps (RDS).[58,59]KMC simulations thus provide a wealth of information that can be used to elucidate the underlying molecular phenomena giving rise to catalysis.

Implementation of Detailed Models of Lateral Interactions

Adsorbate lateral interactions have attracted the interest of experimental and theoretical studies for many years.[60] In catalysis, interactions between reactant species and spectator species in the neighborhood of a microscopic event can give rise to spatial heterogeneity in chemical reactivity.[61] It is thus important to model such effects in order to enable the accurate prediction of catalytic performance.

To this end, it is necessary to formulate a model Hamiltonian, in order to capture the energetics of the adsorbate overlayer. For lattice systems, the most general such Hamiltonian is given by the so-called cluster expansion.[37] Mathematicallly, for a specified lattice and a given function that maps the occupancy of the lattice points to a real number, for instance, the energy for a given configuration of adsorbates, one can in principle construct a cluster expansion that represents this function exactly. Of course, for practical purposes one usually truncates the expansion, which introduces error. The basic idea behind this model is to decompose the energy of a lattice configuration into single-body, two-body, and many-body interaction terms represented by clusters (also known as figures). From a technical standpoint, the state of each site (which in our case represents which adsorbate is bound thereon) is given by a spin variable. One defines a set of basis functions for each site, consisting for instance of Chebyshev polynomials if an orthonormal basis is desired. Then, a set of cluster functions are defined as all the possible products of basis functions for each site in the cluster. To compute the energy of the lattice, one evaluates the linear combination of all cluster functions appropriately weighted with the contribution of each one of them to the overall energy of the system.[37] The method has recently been expanded to accommodate different site types.[62]

To implement a general cluster expansion Hamiltonian in the graph theoretical KMC, we adopt a slightly different formulation which is based on representing each cluster as a graph pattern, and counting the occurrences of that pattern by solving subgraph isomorphism problems. This makes it possible to accommodate patterns in which multidentate species participate. In the graph-theoretical KMC, the state of a single site $\boldsymbol{\sigma}_i$ is given by a 3-element vector,[52]

$$\sigma i \in \{1,2,\ldots,SL\} \times \{0,1,\ldots,NS\} \times \{1,2,\ldots,maxdent\} \forall i \in \{1,\ldots,S_L\} \quad (1)$$

where S_L denotes the number of sites on the lattice, N_S the number of species, maxdent is the maximum number of sites that any species can occupy, for instance if species A binds to one site (monodentate) and species B binds to two (bidentate), maxdent = 2. In vector $\boldsymbol{\sigma}_i$, the first element gives the index of the entity bound to that site; this is done in order to be able to tell the sites occupied by multidentate species. If only monodentate species are present, the maximum number of bound entities is S_L; yet, in cases where multidentate species are bound on the lattice, this first element may range up to values less than S_L. The second element gives the species type, and the third element gives the dentate with which a species is bound to site i. Thus, the state of the overall lattice is given by an $S_L \times 3$ array $\boldsymbol{\sigma}$,[52]

$$\boldsymbol{\sigma} = \left\{ \{\sigma_{i,j}\}_{j=1}^{3} \right\}_{i=1}^{S_L}. \quad (2)$$

In this setting, a graph pattern representing a cluster in the expansion is defined as

$$\mathcal{C}_k = (\Xi_k, \mathcal{E}_k), \quad (3)$$

where Ξ_k represents the set of sites in cluster k (vertexes of the graph pattern), and E_k the neighboring relations between these sites (edges of the graph). Each of the sites just noted can be assigned one out of the possible S_T site types that may exist on the lattice. Thus, if cluster k involves $S_{C,k}$ sites, the type of site k will be

$$\xi_{k,i} \in \{0, 1, 2, \ldots, S_T\}, \qquad \forall i \in \Xi_k = \{1, 2, \ldots, S_{C,k}\}, \quad (4)$$

where the 0 means that a site k may be of any permitted type. In addition, the coverage pattern of cluster k can be defined by the states of all sites in the graph pattern in line with Eqs. (1) and (2),

$$\sigma_{k,i} \in \{\{1, 2, \ldots, S_{C,k}\} \times \{0, 1, \ldots, N_S\} \times \{1, 2, \ldots, \text{maxdent}(\mathbf{d})\}, \&\} \qquad \forall i \in \Xi_k. \quad (5)$$

The above definition allows for the state of site i to be left unspecified, in which case it is denoted by the ampersand symbol (&). This is useful when defining long range interaction patterns in which the intermediate sites can be vacant or occupied by any species. Finally, a set of geometric constraints can be defined by specifying the angles between certain edges in the graph. Thus, for these constraints a set of three sites $\{s_1, s_2, s_3\}$ in the cluster is defined and the angle between the vectors (x_1-x_2, y_1-y_2) and (x_3-x_2, y_3-y_2) is specified to the desired value,

$$\text{Angle}(s_{i,1}, s_{i,2}, s_{i,3}) = \varphi_i \in \{0, 2\pi\} \qquad \forall i \in \left\{1, \ldots, N_k^{AC}\right\}, \quad (6)$$

where the function Angle calculates the angle between the aforementioned vectors, φ_i is the desired angle for geometric constraint i, and N_k^{AC} gives the number of such constraints defined for cluster k.

Similar to the detection of lattice processes in the graph-theoretical KMC,[52] an energetic contribution is a mapping between the vertex sets of C_k (the subgraph) and L=(S,E) (the "large" graph representing the lattice with sites SS and edges E),

$$\mathcal{M} : \Xi_k \rightarrow \mathcal{S}. \quad (7)$$

This mapping has to satisfy the following conditions:

- MM is a subgraph isomorphism, namely, every pair p, q of neighboring sites in pattern C_k is mapped to a pair M(p),M(q) of neighboring sites in the lattice:

$$\forall_{\substack{1 \le p \le S_{C,k} \\ 1 \le q \le S_{C,k}}} \{p, q\} \in \mathcal{C}_k \Rightarrow \{\mathcal{M}(p), \mathcal{M}(q)\} \in \mathcal{E}. \quad (8)$$

- Wherever specified, the angles between the specified edges of pattern C_k are the same as those on the lattice:

$$\text{Angle}(\mathcal{M}(s_{i,1}), \mathcal{M}(s_{i,2}), \mathcal{M}(s_{i,3})) = \varphi_i \quad \forall i \in \{1, \ldots, N_k^{AC}\}. \quad (9)$$

- Wherever specified, the types of sites of the pattern are the same as those on the lattice:

$$s_{\mathcal{M}(p),3} = \xi_{k,p} \qquad \forall 1 \le p \le S_{C,k}. \quad (10)$$

- There is a mapping between the cluster entities and the lattice entities:

$$\mathcal{F} : \{1, 2, \ldots, N_{CE}\} \mapsto \{1, 2, \ldots, N_E\}, \quad (11)$$

such that

$$\boldsymbol{\sigma}_{\mathbf{M(p)}} = (\sigma_{k,p,1}, \sigma_{k,p,2}, \mathcal{F}(\sigma_{k,p,3})) \qquad \forall 1 \le p \le S_{C,k}, \text{ if } \boldsymbol{\sigma}_{\mathbf{M(p)}} \neq \&. \quad (12)$$

Thus, the coverage patterns of the elementary step and the lattice match for sites with specified state.

Note that there are as many such mappings MM for cluster k (Eq. (7)) as the number of occurrences of that cluster on the lattice. This allows for a direct enumeration of the possible energetic cluster contributions.

Let $NCE_k(\boldsymbol{\sigma})$ denote the number of occurrences of cluster C_k for the lattice configuration $\boldsymbol{\sigma}$ and ECI_k the effective cluster interaction parameter, which gives the contribution of this cluster to the total energy. To avoid overestimating the contributions, we define the graph multiplicity of cluster k, GM_k, which can be thought of as a symmetry number for that elementary pattern and gives the number of times the exact same pattern will be counted. For instance, pairwise energetic interactions between adsorbates of the same type occupying the same site types will be counted twice; thus, GM = 2. Then, the total energy of the system will be given as

$$\mathcal{H}(\boldsymbol{\sigma}) = \sum_{k=1}^{N_C} \frac{ECI_k}{GM_k} \cdot NCE_k(\boldsymbol{\sigma}). \quad (13)$$

Equation (13) is the general cluster expansion Hamiltonian giving the energy of the adsorbate overlayer for any configuration.

Computing Reaction Energies and Activation Barriers

In the graph-theoretical KMC, every elementary event (adsorption, desorption, diffusion, reaction) is represented as a graph pattern with specified initial and final coverages. Due to lateral interactions, the reaction energy depends not only on the coverages of the sites involved in that event, but also on the coverages of sites in the neighborhood thereof. The reaction energy is thus given as

$$\Delta E_{rxn}^k(\boldsymbol{\sigma}) = \mathcal{H}(\boldsymbol{\sigma}'(\boldsymbol{\sigma}, k)) - \mathcal{H}(\boldsymbol{\sigma}) + \Delta E_{gas}^k, \quad (14)$$

where $\boldsymbol{\sigma}$ and $\boldsymbol{\sigma}'$ refer to the initial and final coverages of the overall lattice, and ΔE_{gas} is the change in the energy of gas species; for instance, if adsorbed CO and O gave rise to $CO_{2(gas)}$, then ΔE_{gas} would be equal to the energy of CO_2, approximately −3.1 eV referenced to CO and $1/2O_2$ in the gas phase. In practice, one does not need to recalculate from scratch the energy of the overall lattice in the final state H(σ′). Rather, after computing H(σ) the energetic contributions of the reactants are subtracted therefrom, the clusters in which the products can participate are detected and the corresponding contributions are added therein to obtain H(σ′).

Given the reaction energy, the activation energies of the forward and reverse elementary steps must satisfy the following relation to ensure microscopic reversibility (see Figure 1):

$$\Delta E_{rxn}^k(\boldsymbol{\sigma}) = E_{fwd}^{k,\ddagger}(\boldsymbol{\sigma}) - E_{rev}^{k,\ddagger}(\boldsymbol{\sigma}). \quad (15)$$

The forward activation barrier can be parameterized by means of a BEP relation and *ab initio* data.[51] Thus, for a given configuration, the forward activation energy can be expressed by a linear relation that involves the activation energy at the zero coverage limit, and the difference between the reaction energies at the given configuration and the zero coverage limit:

$$E_{fwd}^{k,\ddagger}(\boldsymbol{\sigma}) = \max\left(0, \; \Delta E_{rxn}^k(\boldsymbol{\sigma}), \; E_{fwd,0}^{k,\ddagger} + \omega \cdot \left(\Delta E_{rxn}^k(\boldsymbol{\sigma}) - \Delta E_{rxn,0}^k\right)\right). \quad (16)$$

In the equation above, Ek,‡fwd,0Efwd,0k,‡ and ΔEkrxn,0ΔErxn,0k are the forward activation barrier and reaction energy at the zero coverage limit (see Figure 1), and ω is a parameter referred to as the proximity factor.[63] The max operator ensures that the activation energy will be at least zero, or equal to the reaction energy if the latter is positive (otherwise the transition state energy would lie between the energies of reactants and products). In order to satisfy Eq. (15) the reverse activation energy has to be parameterized as

$$E_{rev}^{k,\ddagger}(\boldsymbol{\sigma}) = \max\left(-\Delta E_{rxn}^{k}(\boldsymbol{\sigma}),\ 0,\ E_{rev,0}^{k,\ddagger} - (1-\omega)\cdot\left(\Delta E_{rxn}^{k}(\boldsymbol{\sigma}) - \Delta E_{rxn,0}^{k}\right)\right), \quad (17)$$

where

$$E_{rev,0}^{k,\ddagger} = E_{fwd,0}^{k,\ddagger} - \Delta E_{rxn,0}^{k}. \quad (18)$$

Equations (16)–(18) allow us to compute the activation barriers in a thermodynamically consistent way, as prescribed by the cluster expansion Hamiltonian(13) and the energetics of the gas species. The rate constant for reaction i can be then computed according to transition state theory as[23,32,33,64,65]

$$k_{TST}^{i} = \frac{k_B \cdot T}{h} \cdot \frac{Q^{\ddagger}}{Q_R} \cdot \exp\left(-\frac{E^{i,\ddagger}(\boldsymbol{\sigma})}{k_B \cdot T}\right), \quad (19)$$

where $Q^{\ddagger}$ and Q_R denote the quasi-partition functions of the transition state and reactants, respectively; h denotes Planck's constant, k_B Boltzmann's constant, and T the temperature. For details on the calculation of the quasi-partition functions and the rate constants for particular elementary events (for instance Eley-Rideal reactions) the reader can consult the supplementary material of Ref. 33.

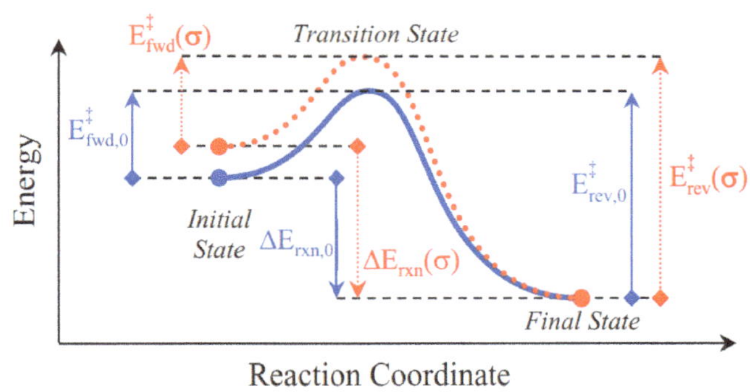

Figure 1. Energy with respect to reaction coordinate showing the energetics- and kinetics-related quantities entering Eqs. (14)–(18). The blue solid path pertains to the reaction energy profile in the limit of zero coverage. For the red dotted path in this example we have assumed that repulsive interactions between the reactants and neighboring adsorbates raise the energy of the initial state.

Note that fitting the *ab initio*-computed activation energies to a BEP scaling relation introduces error. Our KMC approach however does not necessarily require the fitting of the activation energies to the BEP: if the activation energies for specific neighboring configurations are available, one can define several elementary reactions in which the arrangements of the spectator species (within the interaction range) and the corresponding activation energies are explicitly defined. However, this procedure is cumbersome from a practical standpoint, and thus, one resorts to an approximation, such as the one provided by the BEP relation. The latter, while not quantitative, is crucial for the qualitative understanding of the process of interest.

Algorithmic Implementation

As implied by Eq. (13), computing the energy of the lattice necessitates detecting all energetic clusters and summing their contributions to the total energy. For bookkeeping purposes, the graph-theoretical KMC algorithm utilizes a data structure that stores the mapping information for each energetic cluster (namely, the lattice sites involved in the cluster), and also the inverse mapping that gives the indexes of the energetic clusters in which each adsorbed entity participates. This scheme is similar to that used for handling the elementary processes as described in Ref. 52. Thus, at every KMC step one knows the total lattice energy and all the energetic contributions it comprises. This information is used in calculating reaction energies and activation barriers as discussed in the following.

To calculate the activation energy for newly detected lattice processes (for instance an Eley-Rideal reaction step), the algorithm makes a temporary change in the state of the lattice, thereby executing that elementary step. In the final state, it detects the energetic clusters in which the products participate and computes the energy of the lattice from Eq. (13). The difference between final and initial state energies, calculated from Eq. (14), yields the reaction energy for the given configuration $\Delta E_{rxn}^{k}(\boldsymbol{\sigma})$. To find the reaction energy at the zero coverage limit, the algorithm performs a second computation, in which the clusters involving only the reactant entities are taken into account at the initial state, and similarly, the clusters involving only the product entities are considered in the final state, thereby neglecting interactions with neighboring adsorbates. The difference between the two energies thus computed gives $\Delta E_{rxn,0}^{k}$. Subsequently, the activation barriers can be computed via Eq. (16) or (17) depending whether the lattice process in discussion is a forward or a reverse step of an elementary event. Note that the user has to supply as input the forward activation energy at the zero coverage limit $E_{fwd,0}^{k,\ddagger}$, as well as the proximity factor ω. The reverse activation energy at zero coverage $E_{rev,0}^{k,\ddagger}$ is computed from Eq. (18).

The procedures just discussed take place whenever a new lattice process is detected in the course of a KMC simulation. The pseudocode for the whole algorithm has been given in our previous article; here we give a brief outline with specific focus on the calculation of energetics and kinetics.

0. **Start**
1. Define simulation lattice, conditions, participating species, elementary steps, energies of gas species, and cluster expansion Hamiltonian.
2. Initialize the lattice state and set the time clock to t = 0.
3. Detect and store all energetic cluster contributions in the energetics data-structure. Sum them up to obtain the lattice energy.
4. Detect all elementary events that can happen. For each such event, compute the activation energy and kinetic constant, and generate a random time at which it will occur. Put these times in an event-queue.

5. While $t < t_{final}$
   5.a. Find the process μ that will occur next and update the time.
   5.b. Remove the reactants from the lattice. Also remove their energetic contributions from the energetics data-structure and the processes which they participate in from the event-queue.
   5.c. Add the products of process μ in the lattice. Detect their energetic cluster contributions and store them in the energetics data-structure. Update the lattice energy.
   5.d. For every product, detect the elementary events in which it can participate. For each of these events:
   5.d.i. Call subroutine *CalculateKineticConstant* to find the rate for the newly detected event.
   5.d.ii. Given the rate calculated from step 5.d.i, generate a random time for the occurrence of that event and store it in the event-queue.
   5.e. For each event within the neighborhood of the lattice process μ that just took place:
   5.e.i. Call subroutine *CalculateKineticConstant* to find the current rate in the presence of the new environment.
   5.e.ii. Given the rate calculated from step 5.e.i, generate a random time for the occurrence of that event and update the corresponding entry in the event-queue.
6 Repeat
7. **Terminate**

Subroutine *CalculateKineticConstant*

1. Calculate the initial state energy at the given configuration: from the energetics data-structure find the clusters in which reactants and neighboring adsorbates participate and sum up the contributions.
2. Calculate the initial state energy at the zero coverage limit: repeat the above calculation for clusters in which only reactants participate.
3. Make a temporary change in the state of the lattice to represent the final state.
4. Calculate the final state energy at the given configuration: detect the energetic clusters in which products and neighboring adsorbates participate and sum up the contributions.
5. Calculate the final state energy at the zero coverage limit: repeat the above calculation for clusters in which only products participate.
6. Revert the change in the state of the lattice done in step 3 of this subroutine.
7. Calculate the energy of reaction at the given configuration, $\Delta E^{k}_{rxn}(\sigma)$, and the zero coverage limit, $\Delta E^{k}_{rxn,0}$, from Eq. (14).
8. Calculate the activation energy from Eq. (16) if this is a forward step of an elementary event, otherwise use Eqs. (17) and (18).
9. Calculate and return the kinetic constant from Eq. (19).

End Subroutine *CalculateKineticConstant*

E. Computational implementation and parallelization strategy

The above procedures were coded in our FORTRAN2003 software package, Zacros, which is currently available from the UCL online licensing portal, e-Lucid.[66] The package features an easy-to-learn keyword-based language for defining a simulation, and generates detailed output of the lattice state and energetics, reaction occurrence statistics, and numbers of molecules produced or consumed for each of the species participating in the chemistry. In addition, one can run simulations in debugging mode, during which the program provides an account of the changes in the key data structures during the simulation. This information can subsequently be used to efficiently troubleshoot a KMC simulation, if needed.

Using code profiling it was determined that the majority of the computational cost is due to the event time update operations in the loop of step 5.e. Thus, to improve performance, OpenMP parallelization was implemented over that loop. This scheme is applicable to shared memory architectures (for instance a single node in a computational cluster), and is based on the new reaction rate constants being calculated in parallel and stored into thread-private arrays. Subsequently, the random reaction occurrence times are generated serially and inserted into the event-queue, by looping over the threads and the processes assigned to every thread (double loop). This scheme produces simulation results which are identical to those generated by the serial version of the code.

Setting up a Benchmark Model

In a recent article, Schneider and co-workers, presented a set of cluster expansions for atomic oxygen adsorbed on Pt(111).[45] Careful error analysis of these expansions showed that for an accurate representation of the ground states of this system one needs to include up to 8thnearest neighbor pairwise additive interactions and at least two triplets (three-body contributions). Larger expansions were shown to be able to fit the DFT results quantitatively, but this essentially resulted in fitting the DFT error as well (overfitting).

In a subsequent study, Schneider and co-workers used the cluster expansion the strikes the best balance between accuracy and computational efficiency to study NO oxidation on Pt.[51] Thus, they performed equilibrium MC simulations to calculate the distribution of activation energies and the overall rate of O_2 dissociation on the surface, given the chemical potential of O. The latter was calculated from the ratio of NO/NO_2 in the gas phase assuming that this mixture is in quasi-equilibrium with adsorbed O. Thus, the rate of NO oxidation was computed as one-half the rate of O_2 dissociative adsorption. In this work, associative desorption of O adatoms was neglected.

To benchmark our code we set up a KMC model for the NO oxidation using as input the energetic models (cluster expansions) and kinetic parameters by Schneider and co-workers.[45,51,67]

Energetics

For developing the cluster expansion Hamiltonians capturing the adlayer energetics, Schmidt *et al.* used the Alloy Theoretic Automated Toolkit (ATAT) [68] coupled with the Vienna *Ab Initio* Simulation package (VASP).[69–72] Plane-wave, DFT calculations were performed using the PW91 functional and the projector augmented wave (PAW) method.[69,71–74] The slab models contained 4 layers of Pt (with only the bottom one fixed), a layer of oxygen adatoms and four vacuum layers, on $p(1 \times 1)$ up to $p(4 \times 4)$ supercells to investigate a variety of O coverages. This computational scheme was shown to incur a maximum error of 4 meV per adsorbed oxygen. The error due to uncertainty of the O_2 energy calculated from Generalized Gradient Approximation (GGA) was estimated to be 0.2 eV per adsorbed oxygen. For more information about the DFT methodology and its accuracy, the reader is referred to Ref. 45.

Schneider and co-workers[45,51] thus developed four cluster expansion Hamiltonians of increasing accuracy by including 3, 5, 8, and 12 figures (clusters). The parameters reported in that work are for the Ising Hamiltonian formalism, which uses values of ±1 for the state of a single site, denoting spin up or down. For catalytic systems it seems more natural to adopt a lattice-gas Hamiltonian formulation in which the state of a single site takes values of 0 and 1, denoting a vacant or an occupied site accordingly. The mapping from the Ising parameters to the lattice-gas ones is discussed in detail in Sec. I of the supplementary material,[76] along with example calculations, and allows us to calculate the parameters of Table I from those of Table 2 in Ref. 45. These parameters correspond to the energetic contributions of the patterns shown in Figure 2. Note that the triplet 1-2-3b has a top site in the triangle sketched in Figure 2, and therefore, the only acceptable reflections and rotations are the ones drawn therein. Reflections in the horizontal axis would result in an hcp site inside the triangle and the figure would then be the 1-2-3a, which had not been included in this cluster expansion.[45,51]

Table 1. Energetic contribution parameters (in meV) entering the lattice-gas Hamiltonian for O on Pt(111).

Figure	3-Figure CE	5-Figure CE	8-Figure CE	12-Figure CE
H_0/N_L	– 27	45	9	12
Point	– 1200	– 1580	– 1374	– 1368
1NN	300	260	156	140
2NN		84	56	32
3NN		84	12	– 16
4NN			28	28
5NN			28	32
6NN				12
7NN				8
8NN				12
1-1-3			64	56
1-2-3b				16

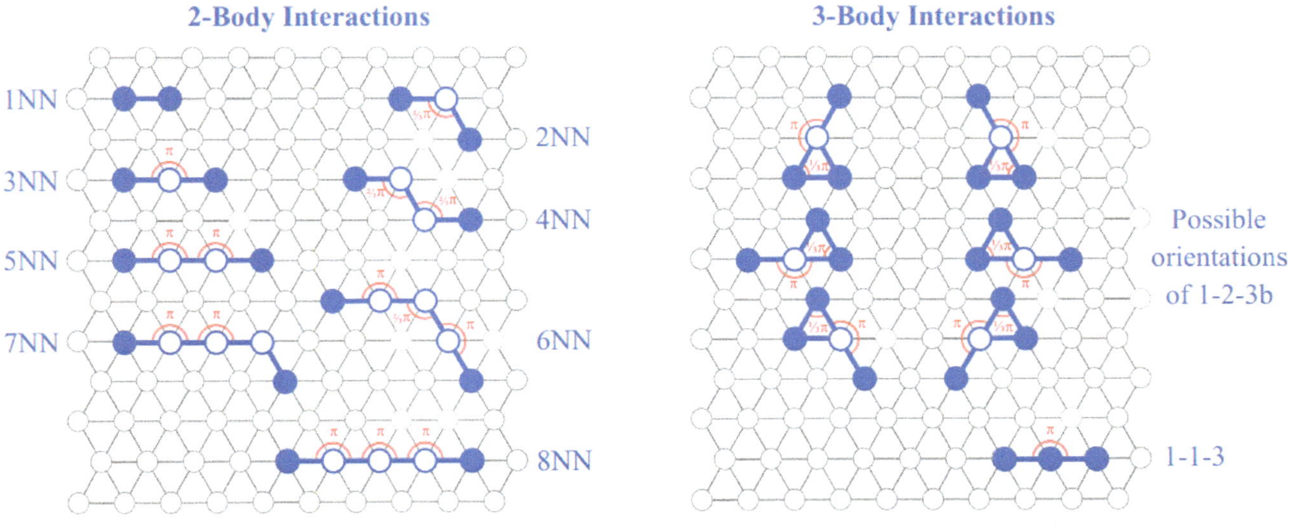

Figure 2. Energetic interaction patterns taken into account in the cluster expansion Hamiltonians. Filled blue circles represent sites occupied by an oxygen adatom. Open blue circles represent sites that may or may not be occupied (their state is represented by the symbol & in Eq. (5)). The graph multiplicity for each of the pairwise interaction patterns is equal to 2. For the triplets, the graph multiplicity of the 1-1-3 pattern is also equal to 2, whereas the 1-2-3b patterns have multiplicities equal to 1.

Kinetics

Oxygen adatoms are bound to fcc sites which is the only site type accounted for, in line with Refs. 45 and 51. In our model, the following reversible elementary events are explicitly considered:

$$NO_{(g)} + O^* \underset{k_{red}}{\overset{k_{oxi}}{\rightleftarrows}} NO_{2(g)} + *, \tag{20}$$

$$O_{2(g)} + * \underset{k_{des}}{\overset{k_{ads}}{\rightleftarrows}} O^* + O^*, \tag{21}$$

$$O^* + * \underset{k_{dif}}{\overset{k_{dif}}{\rightleftarrows}} * + O^*, \tag{22}$$

where the last reaction denotes diffusion of O between neighboring sites. The graph patterns of these reactions as implemented in the graph-theoretical KMC are shown in Figure 3. In the following, we summarize the equations giving the rate constants of the above reactions. For a detailed discussion on how these equations were derived, the reader is referred to Sec. II of the supplementary material.[76]

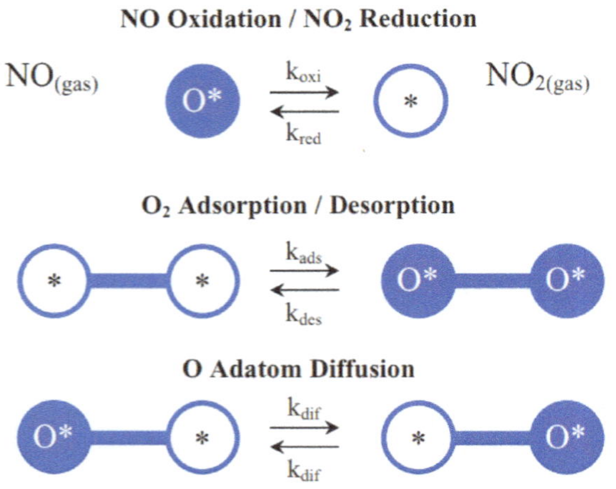

Figure 3. Graph representations of the elementary events taken into account in the NO oxidation model.

Oxidation of NO and Reduction of NO_2

The 1st reaction (NO oxidation and reduction) was considered to be in quasi-equilibrium conditions by Wu *et al.*[51] In the present work, the quasi-equilibrium conditions are ensured by taking the forward and reverse steps of reaction (20) to be fast. Thus, we parameterize the kinetics constants thereof as follows:

$$k_{red} = A' \cdot P_{NO_2},$$

$$k_{oxi} = \left\{ \frac{A' \cdot P_{NO}}{Q^{surf}_{vib,O}} \cdot \exp\left(\frac{ZPE^{surf}_{O}}{k_B \cdot T}\right) \right\} \cdot \exp\left(-\frac{\Delta_f G^o_{NO_2}(T, 1bar) - FE^{surf}_{O} - \Delta_f G^o_{NO}(T, 1bar)}{k_B \cdot T}\right), \tag{23}$$

where A' is a parameter that can be given a high value to ensure quasi-equilibration. The reduction and oxidation rates are proportional to the partial pressures of NO_2 and NO, P_{NO2}, and P_{NO}, because of the participation of these gaseous species in the corresponding reactions. Moreover, $Q^{surf}_{vib,O}$ denotes the vibrational quasi-partition function of O adatoms (frequencies of 429, 380, 377 cm^{-1}, calculated from DFT and used in our calculations),[45] and ZPE^{surf}_{O} the zero point energy thereof. Thus, the terms inside the curly brackets constitute the oxidation pre-exponential entering the KMC input. The activation energy at the zero coverage limit for this reaction is assumed to be negligible, and the reaction energy appearing in the exponent of the last term, is automatically handled by the KMC code. This reaction energy has three contributions, namely, the Gibbs free energies of formation of gas NO_2, $\Delta_f G^o_{NO_2}$ (T,1bar) $\Delta_f G^o_{NO}$T,1bar, calculated from the formulas given in the NIST WebBook,[75] and the energy of formation of the O adlayer with respect to gas phase O_2, FE^{surf}_{O} calculated from the cluster expansions.

Adsorption of O_2

The rate of O_2 adsorption is given as

$$k_{ads} = \frac{\kappa_{ads}}{\sqrt{2 \cdot \pi \cdot m_{O_2} \cdot k_B \cdot T}} \cdot P_{O_2} \cdot \exp\left(-\frac{E^{\ddagger}_{ads}}{k_B \cdot T}\right) = A_{ads} \cdot P_{O_2} \cdot \exp\left(-\frac{E^{\ddagger}_{ads}}{k_B \cdot T}\right). \tag{24}$$

In the above equation the orientational sticking factor κ_{ads} was arbitrarily taken to 1/6 as in Ref. 51. The activation energy for the adsorption step follows the BEP relationship of Ref. 51:

$$E^{\ddagger}_{ads} = \max\left[0, FE^{surf}_{2O} + 2.12\ eV\right], \tag{25}$$

where FE^{surf}_{2O} is the energy of formation of an O adatom pair on neighboring sites on the lattice. In order to implement this BEP relationship within the KMC framework (Eqs. (16) and (17)) we need to find the activation energy at the zero coverage limit and the proximity factor. The former is

$$E^{\ddagger}_{ads,0} = 2 \cdot ECI_{O\_Point} + ECI_{O\_pair\_1NN} + 2.12\ eV, \tag{26}$$

where ECI_{O_Point} is the 1-body term in the cluster expansion Hamiltonian, and $ECI_{O\_pair\_1NN}$ is the first nearest neighbor effective cluster interaction. Note that negative barriers are replaced by a zero value in Eq. (25). Finally, the proximity factor is equal to unity, which is expected as the transition state is a surface bound O_2 precursor.

Desorption of O_2

The rate constant of O_2 desorption is given as

$$k_{des} = \left\{ A_{ads} \cdot \frac{1}{(Q^{surf}_{vib,O})^2} \cdot \exp\left(\frac{2 \cdot ZPE^{surf}_{O}}{k_B \cdot T}\right) \right\} \cdot \exp\left(-\frac{-2 \cdot FE^{surf}_{O}}{k_B \cdot T}\right), \tag{27}$$

where the terms in curly brackets will be the pre-exponential entering the KMC input.

Diffusion of Adsorbed O

An oxygen adatom can diffuse to a neighboring site with a rate that depends on the neighboring environment of the initial and the final site. Since the diffusion process is symmetric, we only consider the forward rate:

$$k_{dif} = \frac{k_B \cdot T}{h} \cdot \frac{Q^{\ddagger}}{Q_R} \cdot \exp\left(-\frac{E^{\ddagger}_{diff} - FE^{surf}_{O}}{k_B \cdot T}\right)$$
$$= A_{dif} \cdot \exp\left(-\frac{E^{\ddagger}_{diff} - FE^{surf}_{O}}{k_B \cdot T}\right). \quad (28)$$

To simulate the fast equilibration of the adsorbate overlayer, we adjusted the pre-exponentials for diffusion, A_{dif}, and NO oxidation/NO_2reduction, A′, such that: (i) the rates of these processes are at least 50 times larger than the O_2 dissociative adsorption (or O_2 associative desorption), and (ii) these processes are in partial equilibrium (forward divided by reverse rate is close to unity).

As an exemplar parameter set, Table II provides the pre-exponentials and activation energies (at the zero coverage limit) for the aforementioned elementary events at 480 K and $\ln(P_{NO2}/P_{NO}) = -4.0$ for the 12-figure cluster expansion.

Table 2. Exemplar kinetic parameters for the NO oxidation KMC model at 480 K and $\ln(P_{NO2}/P_{NO}) = -4.0$ for the 12-figure CE.

Elementary event	Forward prefactor	Reverse prefactor	Proximity factor	Forward barrier (eV) at zero coverage
NO oxidation/NO_2 reduction	5.049×10^{16} bar^{-1} s^{-1}	9.447×10^{10} bar^{-1} s^{-1}	0.00	0.000
O_2 adsorption/desorption	2.444×10^{6} bar^{-1} s^{-1}	6.980×10^{17} s^{-1}	1.00	− 0.476
O diffusion	1.444×10^{4} s^{-1}	1.444×10^{4} s^{-1}	0.50	0.100

RESULTS

Hull of Ground States Computed Through Simulated Annealing

To validate the correct implementation of the cluster expansion Hamiltonian in the graph-theoretical KMC algorithm, we first computed the convex hull of the ground states for O on Pt(111). For each of these calculations, only diffusional hops were simulated at a given coverage. The lattice was initialized with a configuration of oxygen adatoms placed randomly on the lattice, such that the probability of occupancy of any site was equal to the coverage. Such configurations obviously have a high energetic cost, as they are far from the ordered adlayer structures observed at equilibrium. Subsequently, the system was allowed to relax while reducing the temperature linearly in time, thereby performing a simulated annealing calculation. After the system relaxed to a final state at a low temperature, we performed a second simulated annealing step, by restarting the system from that final state and the former initial temperature, and letting it relax again. This simulation setup ensures that the configuration minimizing the energy (ground state) will be found if the cooling rate is low enough.

The results of these calculations are shown in Figure 4. Panel (a) shows the temporal evolution of energy for a simulated annealing calculation on a 42 × 42 lattice (1764 sites) at a coverage equal to 0.35 ML. The energy is observed to drop quickly at initial times, as the adlayer relaxes. Moreover, as the temperature decreases linearly in time, the energy fluctuations are shown to drop, until no more events can happen. Figure 4(b) displays the convex hull of ground states, with some prominent structures noted thereon. These results are in agreement with the previously published ones by Schneider and co-workers,[45] thereby validating our methodology.

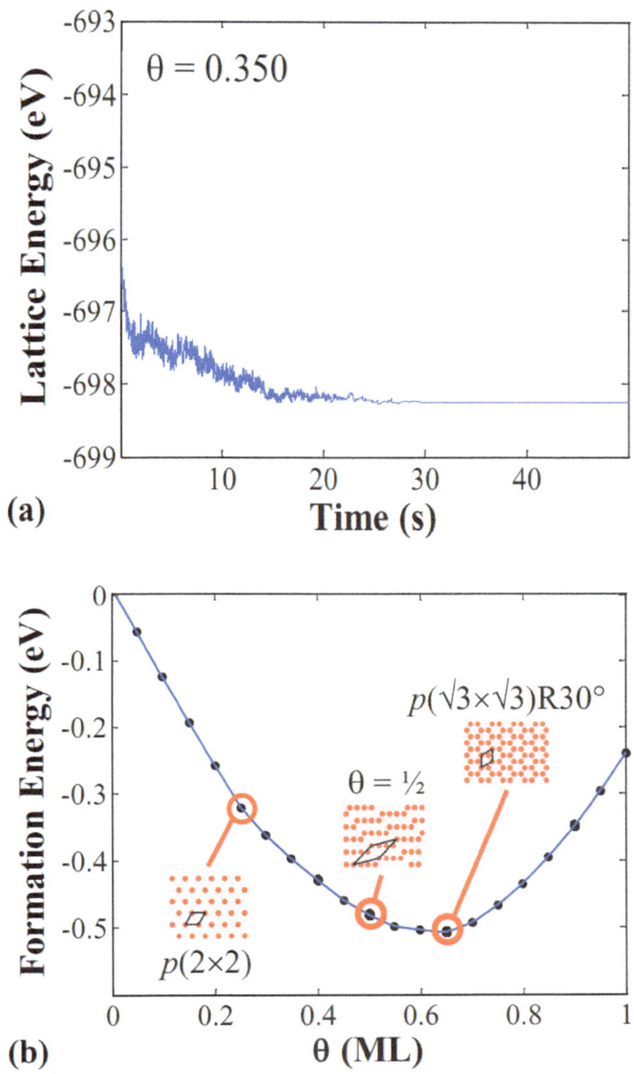

Figure. 4. Results of the simulated annealing calculations. (a) Energy versus time for θ = 0.35, T_{ini} = 100.5 K, T_{ramp} = −2 K/s. The energy fluctuations drop over time and the ground state configuration is reached. (b) Convex hull of ground states. Prominent ground state structures are noted.

Dynamic simulations of NO oxidation and NO_2 reduction

To further test our methodology against the previously published results,[51] we simulated the full dynamic model (reactions (20)–(22), Figure 3) containing, in addition to O diffusion, NO oxidation, NO_2 reduction, and O_2 dissociative adsorption. The reverse process of the latter was neglected for these comparisons, so that the latter are performed on an equal footing. We will come back to this point later.

The results of these simulations are summarized in Figure 5. Panels (a) and (b) show transients of the oxygen coverage and the number of NO_2 molecules produced or consumed per site. For this simulation, the lattice was initialized with a coverage of 0.25, which is lower than the stationary coverage of approximately 0.31. Thus, NO_2 reduction happens rapidly at initial times leaving oxygen adatoms on the surface. After reaction (20) has reached quasi-equilibrium, O_2 dissociation and slow NO oxidation are observed. A statistical analysis of the event occurrence is shown in the bar graph of Figure 5(c), which depicts the forward and reverse rates for each of the elementary steps taken into account in the model. It is evident that NO oxidation and NO_2 reduction are in partial equilibrium. O diffusion is observed to be in equilibrium as well. Since the diffusion step is symmetric with respect to the renumbering of the two sites, the observed partial equilibrium indicates that the energetic interactions of neighboring adsorbates have been correctly implemented in the forward and reverse diffusion rates. Finally, Figure 5(d) shows the turnover frequencies (TOFs) calculated by counting the NO_2 molecules produced per site per time after discarding the initial transient (see Figure 5(a)). Higher NO_2/NO ratios result in higher coverages of oxygen on the surface and thus TOFs. The latter are in excellent agreement with previously published results[51] even though they have been computed with different methodologies.

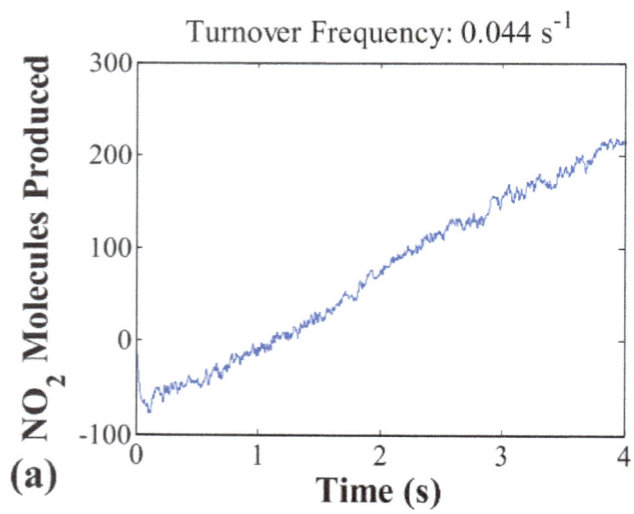

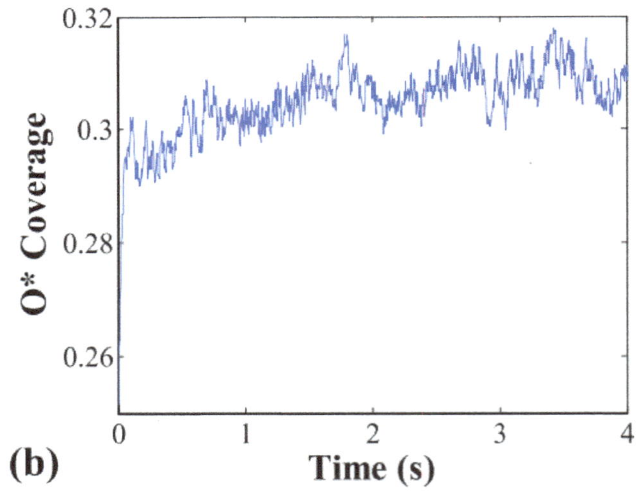

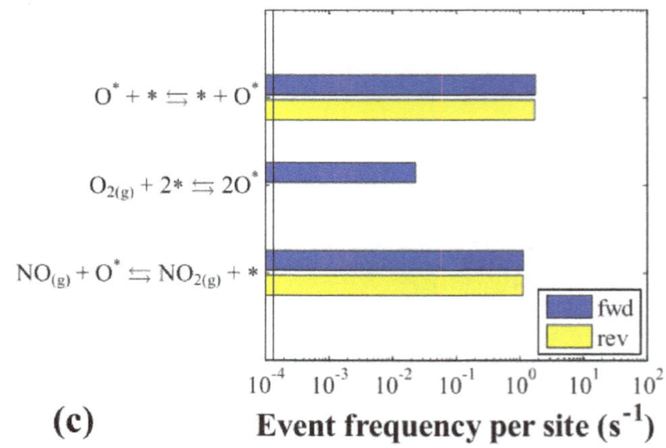

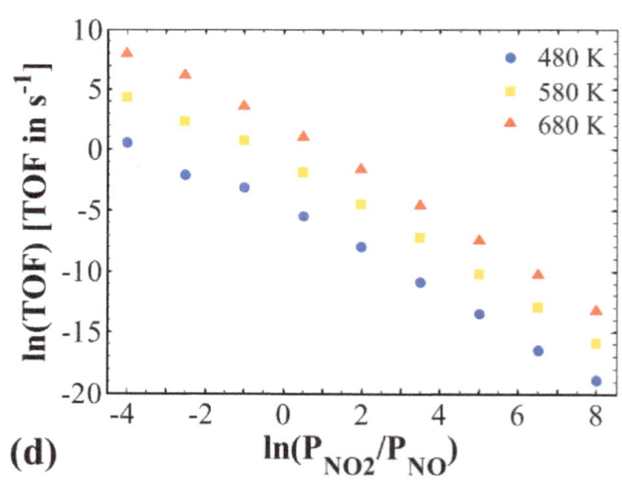

Figure 5. Simulation results for the NO oxidation model with O_2 associative desorption neglected. (a) and (b) Number of NO_2 molecules produced or consumed per site and oxygen coverage, respectively. (c) Statistics of elementary event occurrence, showing the number of events per site per time for the forward and reverse steps. (d) Turnover frequencies (TOFs) for a range of ratios of NO_2/NO partial pressures. Each TOF was calculated from the slope of a plot similar to that of panel (a), after filtering out the initial transient. P = 1 atm and yO2yO2= 0.1 for all simulations.

As previously mentioned, the simulations of Figure 5 neglect O_2 associative desorption for the comparison with published results to be valid. Moreover, from a physics standpoint, practical NO oxidation occurs at conditions at which O_2 desorption is not important. For high NO_2/NO ratios, however, it is expected that the rate of this step will be significant and dominate the mechanism, resulting in NO_2 reduction. One can actually compute the NO_2/NO ratio for which the overall chemistry will be at equilibrium given the molar fraction of O_2:

$$\frac{(y_{NO_2})^2}{(y_{NO})^2 \cdot y_{O_2}} \cdot \frac{1\text{ bar}}{P} = \exp\left(-\frac{\Delta G^o_{\text{overall}}(T, 1\text{ bar})}{k_B \cdot T}\right) \tag{29}$$

and therefore,

$$\log\left(\frac{y_{NO_2}}{y_{NO}}\right) = \frac{1}{2} \cdot \log\left(y_{O_2} \cdot \frac{P}{1\text{bar}}\right) - \frac{\Delta G^o_{\text{overall}}(T, 1\text{bar})}{2 \cdot k_B \cdot T}, \tag{30}$$

(30)

where the change in the Gibbs free energy of the overall reaction $2NO + O_2 \leftrightarrow 2NO_2$ is

$$\Delta G^o_{\text{overall}}(T, 1\text{ bar}) = 2 \cdot G^o_{NO_2}(T, 1\text{ bar}) - 2 \cdot G^o_{NO}(T, 1\text{ bar}) - G^o_{O_2}(T, 1\text{ bar}). \tag{31}$$

Simulations in which the O_2 associative desorption is taken into account are shown in Figure 6. Panel (a) shows the TOFs for a range of conditions, with the critical NO_2 to NO ratio computed from Eq. (30) noted in the plot. We observed that the KMC simulations correctly capture the transition from NO oxidation on the left of the vertical lines, to NO_2 reduction on the right of the lines, where the number of gas NO_2molecules decreases in time. This observation also verifies the correct implementation of the energetics model in the KMC simulation. It is interesting to observe that the net NO_2 reduction rate exhibits a negligible increase for higher NO_2/NO ratios. This can be attributed to the activation barrier for the associative desorption of oxygen being insensitive to the surface coverage thereof. Indeed, the proximity factor for the reversible O_2 adsorption (reaction (21)) is equal to 1 (see Table II), which means that the activation energy for oxygen desorption will be only weakly affected by lateral interactions. In the case of high enough coverages, for which O_2 dissociation is activated, the oxygen desorption barrier will be constant (see Eq. (17)). From a physical standpoint, this insensitivity means that the energetic interactions exerted by neighbors to the two O adatoms about to dissociate are similar to those exerted to the transition state (O_2 precursor). The slight increase in the NO_2reduction rates for NO_2-richer mixtures is due to the higher numbers of neighboring O pairs on the surface, resulting from the higher oxygen coverages in this regime.

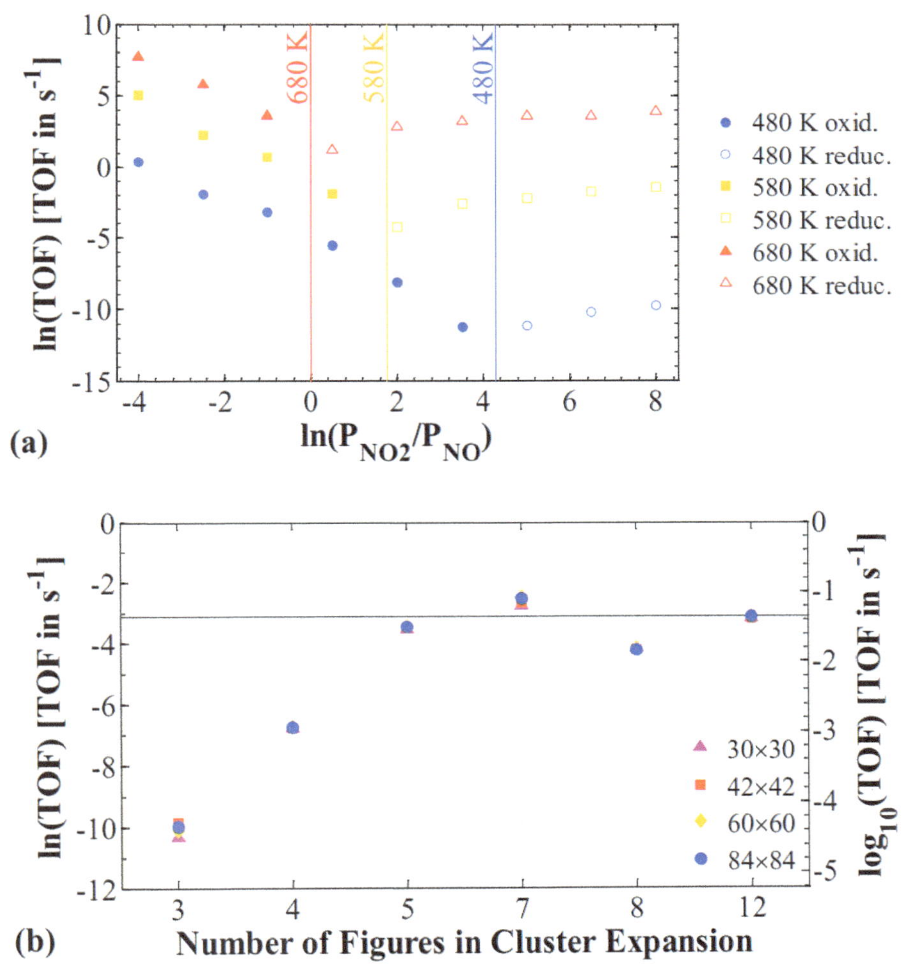

Figure 6. Simulation results for the full NO oxidation model. (a) The vertical dashed lines correspond to the equilibrium NO_2/NO ratios for yO2yO2 = 0.1 and the three different temperatures (480 K, 580 K, and 680 K from right to left). On the left side of each line, NO oxidation is the net reaction, whereas on the right side, NO_2 reduction. (b) Turnover frequency predictions for six cluster expansions and progressively larger lattices. The conditions for these simulations were as follows: T = 480 K, P = 1 bar, log(yNO2yNO2/y_{NO}) = −1.0. The lattice sizes used are noted in the legend, with the largest lattice (84 × 84) having 7056 sites.

To evaluate the performance of the different CEs in predicting the TOFs we performed a set of calculations with progressively more accurate expansions. In addition to the four CEs already presented, we generated two more, capturing up to 2nd and 5th nearest-neighbor interactions (4- and 7-figure CEs, respectively). The latter CEs included at most 2-body terms and were generated by refitting configuration-energy data generated from the 12-figure CE. The configurations were randomly generated for a 10×10 lattice, and coverages uniformly distributed from 0 to 1. The parameters of these CEs are given in Sec. III of the supplementary material.[76]

Figure 6(b) shows the results of simulations of the full NO oxidation NO_2 reduction system at T = 480 K and $\ln(P_{NO2}/P_{NO}) = -1.0$. Simulations were performed for 4 different lattice sizes to exclude the possibility of error originating from small size effects. It is evident that the 3-figure CE, which includes only 1st nearest-neighbor pairwise additive interactions, errs by about 3 orders of magnitude. Inclusion of 2nd nearest-neighbor interactions (4-figure CE), while providing some improvement, still underpredicts the TOF by 36-fold. The remaining cluster expansions (5-figure, 7-figure, 8-figure, and 12-figure CE) give similar results. These observations underscore the importance of including long-range interactions, for our purposes at least up to 3rd nearest-neighbor interactions. Taking into account multi-body terms, while certainly important for the correct prediction of adlayer structures,[43,45] does not seem to be critical for TOF estimations.

Performance Benchmarks for Parallel Algorithm

The calculations involving long-range interactions are extremely computationally intensive: apart from the detection of energetic cluster contributions every time a new event is detected (steps 5.d of the pseudocode in Sec. II D), the algorithm also has to deal with the updates of existing lattice processes after the execution of an elementary event. To improve the performance of the algorithm, these updates were processed in parallel (see Sec. II E). To evaluate the performance of the algorithm we performed two benchmarks: (i) for a given number of threads we investigated how the computational time scales with respect to lattice size; (ii) for a given lattice size, we examined the speedup with respect to the number of threads. All benchmarks were performed on the computational cluster Legion@UCL on type U nodes with 16 cores. Furthermore, we allocated a full node independent of the thread count of the job to prevent any backfilling jobs from interfering with the benchmarking.

The results of the first benchmark are shown in Figure 7, which portrays the computational time required to propagate the system for 1 simulated second with respect to the number of sites in the lattice. We see that the computational time scales almost linearly with respect to the number of sites. This is intuitively expected as the number of lattice processes executed per unit time increases linearly with respect to the size of the system. Moreover, the computational expense of incorporating long range interactions into the KMC simulation becomes evident by the computational time for the 12-figure CE being more than 4 orders of magnitude higher than that of the 3-figure CE. Improving the performance of the algorithm is thus of paramount importance.

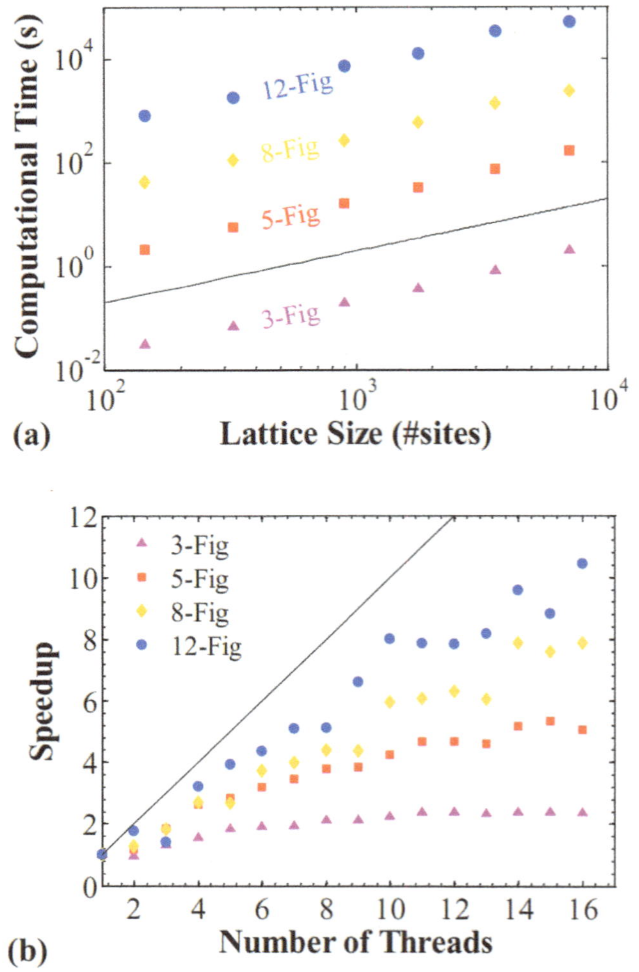

Figure 7. Performance benchmarks: (a) Scaling of the computational time needed for propagating the system 1 simulated second, with respect to the lattice size. The black line is the graph of a linear equation: constant × (number of sites). (b) Speedup with respect to number of threads. The speedup is defined as the ratio of the single-thread computational time with respect to the time of the multi-threaded simulation. The black line is the graph of a line with both slope and intercept equal to unity. Each point gives an average speedup computed from 10 runs.

Thus, Figure 7(b) shows the results of the second benchmark pertinent to the achieved speedup for the parallelized simulations. We observe that the improvement in the performance reaches a plateau as the number of threads increases. To understand why this is the case, we need to recall how our parallelization strategy works: the processes to be updated are first identified serially. Then, for each of these processes, the new activation energy and rate constant are computed; these calculations are the most intense and are performed in parallel. Subsequently, all new rate constants are collected from the different threads; this introduces a synchronization overhead. Finally, the new event occurrence times are computed serially and the queue of lattice processes is updated. If the number of threads is small with respect to the number of processes, the time required for the new rate constant calculations (in parallel) dominates, and the addition of more threads increases the speedup. On the contrary, if the number of threads is large, the synchronization when collecting the new rate constants is the "bottleneck" and the efficiency through parallelization is overshadowed by this synchronization overhead. Increasing the number of threads has a small effect on the speedup, which explains the plateaus observed.

This plateau behavior is reached "more quickly" (i.e., at lower number of threads) for the smaller CEs. This effect can be explained by the fact that the number of lattice processes in need of update after a KMC step increases for longer range interactions. In this case, the workload distribution via parallelization results in high efficiency, as opposed to when the number of updated processes is low, in which case the aforementioned communication overhead is significant.

SUMMARY AND CONCLUSIONS

In view of the increasing interest in accurate predictions of reactions on catalytic surfaces, there is a need for general, flexible, and accurate kinetic modeling approaches. The graph-theoretical KMC method previously developed by Stamatakis and Vlachos[52] allows for the direct incorporation of structurally complex elementary reactions computed by *ab initio* methods into KMC simulation. Thus, steric exclusions due to multidentate binding configurations, as well as intricate neighboring patterns of reactants and products were treated in a natural way within that formalism. The energetic interactions of the adlayer were constrained however to pairwise interactions and detailed balance was cumbersome to implement.

Cluster expansion Hamiltonians can be used to overcome both challenges with the construction of general models of adsorbate energetics, which may include many-body and long-range lateral interactions. Such Hamiltonians have been used extensively in the past to investigate the thermodynamics of adsorbate overlayers but there is only a limited number of studies focusing on the kinetics of reactions occurring in such environments. In this work, we incorporated general cluster expansion Hamiltonians in the graph-theoretical KMC framework, thereby enabling the detail treatment of adlayer energetics in KMC simulation at any desired level of accuracy. Moreover, detailed balance is naturally incorporated into the framework, as the forward and reverse activation energies of a particular elementary step are contained to the thermodynamics of that step for the given neighboring environment. We have implemented this framework in Zacros, a computational KMC package written in FORTRAN2003, available through the UCL online licensing portal e-Lucid.[66] To improve the efficiency of the calculations for energetic models with long range interactions we explored parallelization with OpenMP.

To validate and benchmark our framework we set up a kinetic model for NO oxidation based on previous work by Schneider and co-workers.[51] We performed simulated annealing calculations to verify that KMC correctly predicts the convex hull of ground states, and by means of dynamic simulations at a range of conditions we showed that our calculated TOFs agree with the ones found in the literature. Moreover, we demonstrated the effect of O_2 associative desorption, which was not accounted for in the original model. In this case, the NO_2/NO ratio in which the NO oxidation switches to NO_2 reduction was predicted by the KMC and was found to be in agreement with the ratio predicted purely by the thermodynamics of the gas phase reaction. We finally, showed that for accurate predictions of the TOF one needs to include long range interactions, while the 3-body terms are not critical.

It thus becomes evident that a KMC scheme incorporating detailed models for the adlayer energetics is a powerful tool for investigating surface kinetics. As massively parallel computer architectures become widely available, cluster expansion models with quantitative accuracy with respect to DFT calculations will be easily handled in KMC simulations. This will enable computational studies of the molecular-scale processes on catalytic surfaces at an unprecedented level of detail, unraveling new phenomena for chemistries of theoretical and practical importance.

ACKNOWLEDGMENTS

The research was partially supported by a new faculty start-up grant from the UCL Department of Chemical Engineering. The authors acknowledge the use of the UCL Research Software Development Service (RSD@UCL), as well as the UCL Legion High Performance Computing Facility (Legion@UCL) and associated support services in the completion of this work. M.S. is grateful to Professor William F. Schneider and Mr. Jason Bray for fruitful discussions.

REFERENCES

1. R. J. Glauber, *J. Math. Phys.* **4**(2), 294 (1963).http://dx.doi.org/10.1063/1.1703954
2. H. Müller-Krumbhaar and K. Binder, *J. Stat. Phys.* **8**(1), 1 (1973).http://dx.doi.org/10.1007/BF01008440
3. A. B. Bortz, M. H. Kalos, and J. L. Lebowitz, *J. Comput. Phys.* **17** (1), 10 (1975).http://dx.doi.org/10.1016/0021-9991(75)90060-1
4. P. A. Flinn, *J. Stat. Phys.* **10**(1), 89 (1974).http://dx.doi.org/10.1007/BF01011718
5. E. Stoll, K. Binder, and T. Schneider, *Phys. Rev. B* **8**(7), 3266 (1973).http://dx.doi.org/10.1103/PhysRevB.8.3266
6. E. Stoll and T. Schneider, *Phys. Rev. A* **6**(1), 429 (1972). http://dx.doi.org/10.1103/PhysRevA.6.429
7. J. C. Angles d'Auriac, R. Maynard, and R. Rammal, *J. Stat. Phys.* **28**(2), 307 (1982).http://dx.doi.org/10.1007/BF01012608
8. R. M. Ziff, E. Gulari, and Y. Barshad, *Phys. Rev. Lett.* **56**(24), 2553 (1986). http://dx.doi.org/10.1103/PhysRevLett.56.2553
9. M. Pineda, R. Imbihl, L. Schimansky-Geier, and Ch. Zülicke, *J. Chem. Phys.* **124**(4), 044701 (2006).http://dx.doi.org/10.1063/1.2140709
10. M. Pineda, L. Schimansky-Geier, and R. Imbihl, *Phys. Rev. E* **75**(6), 061107 (2007). http://dx.doi.org/10.1103/PhysRevE.75.061107
11. V. P. Zhdanov and B. Kasemo, *Surf. Sci. Rep.* **20**(3), 113 (1994).http://dx.doi.org/10.1016/0167-5729(94)90009-4
12. V. P. Zhdanov and B. Kasemo, *Surf. Sci.* **405**(1), 27 (1998). http://dx.doi.org/10.1016/S0039-6028(97)01078-9
13. V. P. Zhdanov and B. Kasemo, *Surf. Sci. Rep.* **39**(2–4), 25–104 (2000).http://dx.doi.org/10.1016/S0167-5729(00)00004-2
14. D. J. Liu and J. W. Evans, *J. Chem. Phys.* **117**(15), 7319 (2002).http://dx.doi.org/10.1063/1.1507105

15. D. G. Vlachos, L. D. Schmidt, and R. Aris, *Surf. Sci.* **249**(1–3), 248 (1991).http://dx.doi.org/10.1016/0039-6028(91)90850-R

16. M. Neurock and E. W. Hansen, *Comput. Chem. Eng.* **22**, S1045 (1998).http://dx.doi.org/10.1016/S0098-1354(98)00210-5

17. E. W. Hansen and M. Neurock, *Chem. Eng. Sci.* **54**(15–16), 3411 (1999).http://dx.doi.org/10.1016/S0009-2509(98)00489-8

18. E. W. Hansen and M. Neurock, *J. Catal.* **196**(2), 241 (2000).http://dx.doi.org/10.1006/jcat.2000.3018

19. E. Hansen and M. Neurock, *J. Phys. Chem. B* **105**(38), 9218 (2001).http://dx.doi.org/10.1021/jp0103427

20. K. Reuter and M. Scheffler, *Phys. Rev. Lett.* **90**(4), 046103 (2003). http://dx.doi.org/10.1103/PhysRevLett.90.046103

21. K. Reuter, D. Frenkel, and M. Scheffler, *Phys. Rev. Lett.* **93**(11), 116105 (2004).http://dx.doi.org/10.1103/PhysRevLett.93.116105

22. D. Mei, P. A. Sheth, M. Neurock, and C. M. Smith, *J. Catal.* **242**(1), 1 (2006). http://dx.doi.org/10.1016/j.jcat.2006.05.009

23. K. Reuter and M. Scheffler, *Phys. Rev. B* **73** (4), 045433 (2006). http://dx.doi.org/10.1103/PhysRevB.73.045433

24. J. Rogal, K. Reuter, and M. Scheffler, *Phys. Rev. B* **77**(15), 155410 (2008).http://dx.doi.org/10.1103/PhysRevB.77.155410

25. S. Matera and K. Reuter, *Catal. Lett.* **133**(1–2), 156 (2009).http://dx.doi.org/10.1007/s10562-009-0168-8

26. D. H. Mei, M. Neurock, and C. M. Smith, *J. Catal.* **268**(2), 181 (2009).http://dx.doi.org/10.1016/j.jcat.2009.09.004

27. D. H. Mei, J. Du, and M. Neurock, *Ind. Eng. Chem. Res.* **49**(21), 10364 (2010). http://dx.doi.org/10.1021/ie100999e

28. H. A. Aleksandrov, L. V. Moskaleva, Z. J. Zhao, D. Basaran, Z. X. Chen, D. Mei, and N. Rösch, *J. Catal.* **285**(1), 187 (2012). http://dx.doi.org/10.1016/j.jcat.2011.09.035

29. F. Hess, A. Farkas, A. P. Seitsonen, and H. Over, *J. Comput. Chem.* **33**(7), 757 (2012). http://dx.doi.org/10.1002/jcc.22902

30. L. J. Xu, D. Mei, and G. Henkelman, *J. Chem. Phys.* **131**(24), 244520 (2009). http://dx.doi.org/10.1063/1.3281688

31. P. Kratzer and M. Scheffler, *Phys. Rev. Lett.* **88**(3), 036102 (2002).http://dx.doi.org/10.1103/PhysRevLett.88.036102

32. K. Reuter, "First-principles kinetic Monte Carlo simulations for heterogeneous catalysis: Concepts, status and frontiers," in *Modeling Heterogeneous Catalytic Reactions: From the Molecular Process to the Technical System*, edited by O. Deutschmann (Wiley-VCH, Weinberg, 2010).

33. M. Stamatakis and D. G. Vlachos, *ACS Catal.* **2**(12), 2648 (2012).http://dx.doi.org/10.1021/cs3005709

34. M. K. Sabbe, M.-F. Reyniers, and K. Reuter, *Catal. Sci. Technol.* **2**(10), 2010 (2012). http://dx.doi.org/10.1039/c2cy20261a

35. M. Neurock, R. A. van Santen, W. Biemolt, and A. P. J. Jansen, *J. Am. Chem. Soc.* **116**(15), 6860 (1994). http://dx.doi.org/10.1021/ja00094a046

36. K. Reuter, C. Stampfl, and M. Scheffler, "*Ab initio* thermodynamics and statistical mechanics of surface properties and functions," in *Handbook of Materials Modeling. Part A: Methods* (Springer, Berlin, 2005), p. 149.

37. J. M. Sanchez, F. Ducastelle, and D. Gratias, *Physica A* **128**(1–2), 334 (1984).http://dx.doi.org/10.1016/0378-4371(84)90096-7

38. R. Drautz, R. Singer, and M. Fähnle, *Phys. Rev. B* **67**(3), 035418 (2003). http://dx.doi.org/10.1103/PhysRevB.67.035418

39. H. R. Tang, A. Van der Ven, and B. L. Trout, *Phys. Rev. B* **70**(4), 045420 (2004). http://dx.doi.org/10.1103/PhysRevB.70.045420

40. H. R. Tang, A. Van der Ven, and B. L. Trout, *Mol. Phys.* **102**(3), 273 (2004).http://dx.doi.org/10.1080/0026897042000178088

41. C. Stampfl, *Phase Transitions* **80**(4–5), 311 (2007). http://dx.doi.org/10.1080/01411590701228596

42. C. Lazo and F. J. Keil, *Phys. Rev. B* **79**(24), 245418 (2009).http://dx.doi.org/10.1103/PhysRevB.79.245418

43. S. D. Miller and J. R. Kitchin, *Mol. Simul.* **35**(10–11), 920 (2009).http://dx.doi.org/10.1080/08927020902833137

44. S. Piccinin and C. Stampfl, *Phys. Rev. B* **81**(15), 155427 (2010). http://dx.doi.org/10.1103/PhysRevB.81.155427

45. D. J. Schmidt, W. Chen, C. Wolverton, and W. F. Schneider, *J. Chem. Theory Comput.* **8** (1), 264 (2012). http://dx.doi.org/10.1021/ct200659c

46. J. S. McEwen and A. Eichler, *J. Chem. Phys.* **126**(9), 094701 (2007).http://dx.doi.org/10.1063/1.2464085

47. S. H. Payne, J. S. McEwen, H. J. Kreuzer, and D. Menzel, *Surf. Sci.* **594**(1–3), 240 (2005).http://dx.doi.org/10.1016/j.susc.2005.07.031

48. C. Stampfl, H. J. Kreuzer, S. H. Payne, H. Pfnür, and M. Scheffler, *Phys. Rev. Lett.* **83**(15), 2993 (1999).http://dx.doi.org/10.1103/PhysRevLett.83.2993

49. T. Rehman, M. Jaipal, and A. Chatterjee, *J. Comput. Phys.* **243**, 244 (2013). http://dx.doi.org/10.1016/j.jcp.2013.03.005

50. S. Verma, T. Rehman, and A. Chatterjee, *Surf. Sci.* **613**, 114 (2013). http://dx.doi.org/10.1016/j.susc.2013.03.022

51. C. Wu, D. J. Schmidt, C. Wolverton, and W. F. Schneider, *J. Catal.* **286**, 88 (2012).http://dx.doi.org/10.1016/j.jcat.2011.10.020

52. M. Stamatakis and D. G. Vlachos, *J. Chem. Phys.* **134**(21), 214115 (2011). http://dx.doi.org/10.1063/1.3596751

53. M. D. Marcinkowski, A. D. Jewell, M. Stamatakis, M. B. Boucher, E. A. Lewis, C. J. Murphy, G. Kyriakou, and E. C. H. Sykes, *Nature Mater.* **12**(6), 523 (2013).http://dx.doi.org/10.1038/nmat3620

54. M. Stamatakis, Y. Chen, and D. G. Vlachos, *J. Phys. Chem. C* **115**(50), 24750 (2011). http://dx.doi.org/10.1021/jp2071869

55. M. Stamatakis, M. Christiansen, D. G. Vlachos, and G. Mpourmpakis, *Nano Lett.* **12**(7), 3621 (2012). http://dx.doi.org/10.1021/nl301318b

56. W. Guo, M. Stamatakis, and D. G. Vlachos, *ACS Catal.* **3**(10), 2248 (2013). http://dx.doi.org/10.1021/cs4005166

57. K. A. Fichthorn and W. H. Weinberg, *J. Chem. Phys.* **95** (2), 1090 (1991).http://dx.doi.org/10.1063/1.461138

58. C. T. Campbell, *J. Catal.* **204**(2), 520 (2001).http://dx.doi.org/10.1006/jcat.2001.3396

59. H. Meskine, S. Matera, M. Scheffler, K. Reuter, and H. Metiu, *Surf. Sci.* **603**(10–12), 1724 (2009).http://dx.doi.org/10.1016/j.susc.2008.08.036

60. T. Matsushima, *Surf. Sci.* **558**(1–3), 1 (2004).http://dx.doi.org/10.1016/j.susc.2004.03.060

61. M. Salciccioli, M. Stamatakis, S. Caratzoulas, and D. G. Vlachos, *Chem. Eng. Sci.* **66**(19), 4319 (2011).http://dx.doi.org/10.1016/j.ces.2011.05.050

62. D. Lerch, O. Wieckhorst, L. Hammer, K. Heinz, and S. Müller, *Phys. Rev. B* **78**(12), 121405(R) (2008).http://dx.doi.org/10.1103/PhysRevB.78.121405

63. L. C. Grabow, A. A. Gokhale, S. T. Evans, J. A. Dumesic, and M. Mavrikakis, *J. Phys. Chem. C* **112**(12), 4608 (2008).http://dx.doi.org/10.1021/jp7099702

64. P. Hänggi, P. Talkner, and M. Borkovec, *Rev. Mod. Phys.* **62**, 251 (1990). http://dx.doi.org/10.1103/RevModPhys.62.251

65. P. Pechukas, "Statistical approximations in collision theory," in *Dynamics of Molecular Collisions. Part B*, edited by W. H. Miller (Plenum Press, New York, London, 1976).

66. See http://www.e-lucid.com/i/software/material_modelling/Zacros.html for Zacros: Advanced Lattice-KMC simulation Made Easy.

67. J. M. Bray and W. F. Schneider, "First-principles thermodynamic models in heterogeneous catalysis," in *Computational Catalysis*, edited by A. Asthagiri and M. J. Janik (Royal Society of Chemistry, 2013).

68. A. van de Walle, M. Asta, and G. Ceder, *CALPHAD: Comput. Coupling Phase Diagrams Thermochem.* **26**(4), 539 (2002).http://dx.doi.org/10.1016/S0364-5916(02)80006-2

69. G. Kresse and J. Furthmüller, *Comp. Mater. Sci.* **6**(1), 15 (1996).http://dx.doi.org/10.1016/0927-0256(96)00008-0

70. G. Kresse and J. Furthmüller, *Phys. Rev. B* **54**(16), 11169 (1996). http://dx.doi.org/10.1103/PhysRevB.54.11169

71. G. Kresse and D. Joubert, *Phys. Rev. B* **59**(3), 1758 (1999).http://dx.doi.org/10.1103/PhysRevB.59.1758

72. G. Kresse, M. Marsman, and J. Furthmüller, Vienna *ab initio* simulation package (VASP): The guide, seehttp://cms.mpi.univie.ac.at/vasp/vasp/vasp.html.

73. J. P. Perdew and Y. Wang, *Phys. Rev. B* **45**(23), 13244 (1992).http://dx.doi.org/10.1103/PhysRevB.45.13244

74. P. E. Blöchl, *Phys. Rev. B* **50**(24), 17953 (1994).http://dx.doi.org/10.1103/PhysRevB.50.17953

75. NIST, *National Institute of Standards and Technology—Chemistry WebBook*, see http://webbook.nist.gov/.

76. See supplementary material at http://dx.doi.org/10.1063/1.4840395 for (I) Converting a single-species Ising Hamiltonian to a lattice-gas Hamiltonian, (II) Derivation of kinetic rate constants for NO oxidation model, and (III) Lattice-gas parameters for all cluster expansions.

INDEX

H

I

K

L

M

N

O

P

Q

R

S

T

V

W

X